华章IT | Information Technology

云计算与虚拟化技术丛书

Redefining Practice of Spring Cloud

重新定义
Spring Cloud实战

许进 叶志远 钟尊发 蔡波斯 方志朋 郭芳碧 朱德明 著

机械工业出版社
China Machine Press

图书在版编目（CIP）数据

重新定义 Spring Cloud 实战 / 许进等著 . —北京：机械工业出版社，2018.9（2019.7 重印）
（云计算与虚拟化技术丛书）

ISBN 978-7-111-60939-1

I. 重… II. 许… III. 互联网络 - 网络服务器 IV. TP368.5

中国版本图书馆 CIP 数据核字（2018）第 213178 号

重新定义 Spring Cloud 实战

出版发行：机械工业出版社（北京市西城区百万庄大街 22 号		邮政编码：100037）	
责任编辑：李 艺		责任校对：殷 虹	
印　　刷：北京诚信伟业印刷有限公司		版　　次：2019 年 7 月第 1 版第 5 次印刷	
开　　本：186mm×240mm　1/16		印　　张：41	
书　　号：ISBN 978-7-111-60939-1		定　　价：129.00 元	

凡购本书，如有缺页、倒页、脱页，由本社发行部调换
客服热线：（010）88379426　88361066　　　　　　　投稿热线：（010）88379604
购书热线：（010）68326294　88379649　68995259　　读者信箱：hzit@hzbook.com

版权所有 • 侵权必究
封底无防伪标均为盗版
本书法律顾问：北京大成律师事务所　韩光 / 邹晓东

Foreword 专家推荐

（排名不分先后）

在微服务体系中，Spring Cloud 是目前最热门的构建微服务体系的解决方案，它提供了构建微服务架构的一些基础设施。本书内容上覆盖了 Spring Cloud 的一些主要组件，不仅在如何使用上做了详细的介绍，也从原理上深入浅出地剖析了其中的技术要点，同时也将部分组件与周边的一些开源项目进行了对比，且提供了一些原理分析和相关的示例，是一本不可多得的 Spring Cloud 实战书籍。新手和有微服务实践经验的读者都能从书中得到一些不一样的收获。

张艺辰　腾讯高级研发工程师

本书不仅对 Spring Cloud 各核心组件进行了细致入微的介绍，同时也跳出了框架本身，为微服务的实施和分布式架构所面临的基本问题交出了 Spring Cloud 式答卷，是开发者快速掌握 Spring Cloud 技术栈的神兵利器。不仅如此，本书还凝聚着 Spring Cloud 中国社区的智慧结晶，让我们看到了国人在开源领域的研发力量，可喜可贺。

王鸿飞　百度高级研发工程师

在微服务如火如荼的今天，各种微服务框架层出不穷，而 Spring Cloud 无疑是那颗最闪亮的星。从 Spring Framework 到 Spring Boot，再到如今的 Spring Cloud，Spring 全家桶给众多程序员带来了"真正的春天"。由于分布式和服务化是极具挑战的任务，因此 Spring Cloud 也不可避免的愈加复杂。Spring Cloud 中国社区为 Spring Cloud 的普及做出了巨大贡献，并迅速降低了语言问题所带来的学习门槛。这本书由 Spring Cloud 中国社区倾力打造，书中涵盖了 Spring Cloud 的服务发现、网关、熔断器、配置、全链路监控等最核心组件，并很接地气地详述了 Dubbo 向 Spring Cloud 迁移以及 Spring Cloud 与分布式事务相关内容，值得一看。

张亮　京东金融数据研发负责人/分布式数据库中间件 Sharding-Sphere 负责人

Spring Cloud 提供了完整的微服务技术体系，可以帮助开发者快速实现架构升级。本书完整地介绍了 Spring Cloud 中各个组件的使用方法并深度剖析了其中的原理，深入浅出地帮助开发者快速掌握和理解 Spring Cloud。

李艺恒　腾讯研发工程师

Preface / 序1

随着淘宝业务的高速发展，阿里遇到了性能和效率两大问题，在经过一系列探索之后，从 2007 年开始进行了一系列分布式服务架构改造。其中，内部做服务化的最底层、最核心的两个框架是 Dubbo（2011 年开源）和 HSF。我本人便是从 2010 年开始接触微服务的，由于阿里内部配套比较完善，所以用起来也是得心应手，没有遇到什么问题。直到有一个朋友的公司在用开源版本 Dubbo 遇到问题来咨询我时，我才发现如果单纯用开源的 Dubbo 来搭建一套完整的分布式系统，除了服务治理外还有不少的配套设置需要完善。

2014 年 3 月，Martin Fowler 发表文章《Microservices》，通俗易懂地讲解了什么是微服务架构，自那以后 Microservices（微服务）一词越来越火爆，如今已是最热门的话题之一。Spring Cloud 也是从那个时候开始流行起来，作为新一代的服务框架，它提出了开发"面向云环境的应用程序"的口号，为微服务架构提供了更加全面的技术支持。它的一站式解决方案大大降低了开发成本，让无数开发者欢呼雀跃。本书作者之一许进作为 Spring Cloud 中国社区的创始人之一，一直在中国为 Spring Cloud 布道。在学习 Spring Cloud 时，网上搜索到的很多中文资料都出自这个社区，我与许进也是因此结识。当他把本书书稿发给我时，我本着负责任的态度认真通读了一遍，发现这确实是一本不错的著作，从理论介绍到实战案例，从基础概念到高级特性基本全覆盖了。既适合初学者入门，也适合有一定经验的人查漏补缺。

目前国内还存在大量正在向互联网转型的传统 IT 企业，这些企业转型基本有三种方案：一是采用基于阿里企业级中间件（EDAS）的商业化一站式解决方案；二是基于开源 Dubbo 及生态（阿里已经加大开源投入，诸如像 Nacos 这样的注册、配置中心，像 Sentinel 这样熔断限流组件已经作为 Dubbo 的生态成长起来了）完全自研；第三个就是基于 Spring Cloud 这种"全家桶"的解决方案了。无论采用哪种方案，书中关于微服务的一些思想都是通用的，所以本书同样会让这些转型过程中的架构师有所收获。

最后祝本书大卖，也同样期望作者继续保持开源投入，因为这两样都是在帮助更多的中国开发者成长，也希望从此书受益的开发者能为开源贡献一份自己的力量。

谢吉宝（唐三）
阿里巴巴高级技术专家

序 2 Preface 2

两年多前在唯品会基础架构部的时候，我刚认识许进，记得当时他正在从事高性能网关的研发工作（本书很多篇幅也涉及微服务网关，从中可以看到许进在这块的积累还是比较深入的）。之后我和许进陆陆续续有一些交流，我发现许进是一个偏后端且社区型的工程师，他对基础中间件、微服务架构、开源和技术社区建设抱有浓厚的兴趣和热情，不仅自己参与实际中间件的研发，也和社区中的志愿者一起做开源、写书，同时也经营社区（技术网站，微信群等）。说实话，国内这种社区型工程师还是比较少的，所以我一直比较关注许进，包括他写书的事情，当他提出要我帮忙写一篇序时，我觉得应该支持他一下。

微服务架构正逐渐成为企业应用架构的主流。几年前，Netflix 在大规模生产级微服务落地的基础上，把它的成功经验以开源组件的形式贡献给了社区，Pivotal 则基于 Netflix 的开源组件进行封装简化，推出 Spring Cloud 微服务技术栈，进一步简化了这些组件的使用门槛，大大推进了微服务架构在企业的普及和落地。本书主要介绍 Spring Cloud 微服务栈，各个组件背后的业务需求、技术架构，以及实际的操作案例。除了 Spring Cloud 以外，本书也涉及开源社区内比较成熟的其他微服务组件，还包括分布式系统的架构原理，可以说是深入了解微服务社区生态和理解分布式系统架构原理的一本实践参考。

相信本书对一线架构师拓展技术视野，实践落地微服务架构有所帮助。也感谢许进对 Spring Cloud 和微服务架构在国内的推广所做出的贡献。

杨波

拍拍贷基础架构研发总监，资深架构师，极客时间《微服务架构和实践 160 讲》作者

序 3

现在想来，以前做架构升级的时候，我们往往是被业务的快速发展或者线上爆发的问题逼迫着去改造，然后才会思考为何不能未雨绸缪地来一次战略架构设计，或者是否有哪些好的模式可以帮助我们，而微服务的思想恰好可以解决这个问题。

让我们回顾下曾经的单体应用环境：在业务简单、团队组织成员很少的时候，我们常常把功能都集中于一个应用中，统一部署，统一测试，玩得不亦乐乎。但随着业务迅速发展，组织成员日益增多，我们会将所有的功能集中到一个 Tomcat 中去，每当更新一个功能模块时，势必要更新所有程序，搞不好，还要牵一发动全身，实在难以维护。在单体应用满足不了我们逐渐增长的扩展需求之后，微服务应运而生。它将原来集中于一体的功能拆分出去，比如商品功能、订单功能、用户功能，使其自成体系地发布、运维等，从而解决了单体应用中功能过多、不便维护的弊端。

自从微服务概念诞生以来，各种关于微服务的实践层出不穷，Spring Cloud 作为实现微服务的工具集一直有着举足轻重的作用。这套工具集好比金庸武侠小说中的独孤九剑，招式鲜明而又实用。如果把所有的剑诀都学会，待任督二脉打通之后肯定所向披靡。我们在学习微服务以及 Spring Cloud 实践的过程中倒是不必将每个招式都学会，凭借破剑式、破枪式也能够行走江湖。比如我们可以裁剪里面的工具集，将 Spring Cloud 中的 Zuul、Hystrix、Config 等拿来为我所用，待熟练掌握之后，也一样可以独步微服务应用的"武林"。本书实现了 Spring Cloud 理论和实践的统一，为读者带来详实的指导，便于读者更好地将这套工具集应用到实际业务开发中去。

王新栋
京东商城京麦开放网关技术负责人，资深架构师

前 言 Preface

随着互联网的快速普及，云计算近年来得到蓬勃发展，企业的 IT 环境和架构体系也逐渐发生变革，其中最典型的就是由过去的单体应用架构发展为当今流行的微服务架构。微服务是一种架构风格，其优势是为软件应用开发带来很大的便利，让敏捷开发和复杂的企业应用快速持续交付成为可能。随着微服务架构的流行，很多企业纷纷使用微服务架构来搭建新的系统或者对历史系统进行重构，但是微服务架构的实施和落地会面临很大的挑战。虽然微服务架构的解决方案很多，但是对于如何真正落地微服务架构，目前还没有公认的技术标准和规范。幸运的是，业界已经有一些很有影响力的开源微服务解决方案，比如 2015 年年初，Spring 团队推出的 Spring Cloud，其目标是成为 Java 领域微服务架构落地的标准。Spring Cloud 经过高速迭代和发展，至今已经成为 Java 领域落地微服务架构的推荐解决方案，为企业 IT 架构变革保驾护航。

Spring Cloud 是一个优质的开源项目，它的稳健发展离不开众多开发人员的实践与反馈，开发人员通过一个社区化的平台去交流学习从而使 Spring Cloud 逐渐完善。Spring Cloud 发展到 2016 年，得到国内越来越多的人的关注，但是相应的学习交流平台和材料比较分散，这阻碍了 Spring Cloud 在我国的普及和发展。因此 Spring Cloud 中国社区应运而生。Spring Cloud 中国社区 (http://springcloud.cn) 是国内首个基于 Spring Cloud 微服务体系创建的非盈利技术社区，也是国内首个致力于 Spring Cloud 微服务架构开放交流的社区，是专为 Spring Boot 或 Spring Cloud 技术人员提供分享和交流服务的平台，目的是推动 Spring Cloud 在中国的普及和应用。

为什么写这本书？

Spring Cloud 中国社区自 2016 年创建以来，在北京、上海、深圳、成都等地举办了多次技术沙龙，帮助数万名开发者快速学习 Spring Cloud 并用于生产。为更好地推动 Spring Cloud 在中国的发展，让更多开发者受益，社区针对 Spring Cloud 在国内的使用情况，结合国内上百家企业使用 Spring Cloud 落地微服务架构时遇到的问题和相应的解决方案，特推出本书。

你适合本书吗？

如果你没听说过微服务，也没有听说过 Spring Cloud，或者你正在学习或尝试使用 Spring Cloud 去落地微服务架构，那么这本书会非常适合你。因为本书更加偏实战，书中所讲是一套可落地的解决方案。不管你是初学者、开发人员还是架构师，只要你想使用 Spring Cloud 去落地微服务架构，就可以阅读并学习本书。

本书是如何组织的？

本书共 25 章，按照"核心组件→进阶实战→解决方案"的结构将内容从逻辑上划分为三个部分，具体如下：

第 1～10 章为核心组件部分，主要介绍 Spring Cloud 的核心组件。首先从应用架构的发展历程讲起，介绍了微服务出现的背景，并对微服务架构的落地提出了相应的解决方案。然后分别详细介绍了 Spring Cloud 微服务体系中的核心常用组件，如 Eureka、Feign、Ribbon、Hystrix、Zuul 等。最后通过一个综合案例将前面介绍的组件连接起来，帮助大家融会贯通。

第 11～18 章为进阶实战部分，在核心组件的基础上，对 Config、Consul、认证和鉴权、全链路监控以及 Spring Cloud 生态圈中第二代网关 Spring Cloud Gateway 进行详细阐述，循序渐进、案例驱动，帮助读者加深对组件的理解，更好地掌握相关内容并运用于生产实践中。

第 19～25 章为解决方案部分，主要从解决方案着手，内容包括 Spring Cloud 与 gRPC 的整合方式、版本控制与灰度发布、Spring Cloud 容器化、Dubbo 向 Spring Cloud 的迁移、分布式事务、领域驱动等生产级实用解决方案，为企业 IT 架构微服务化和变革保驾护航。

本书源码及勘误

本书案例源码使用 Java 8 版本编写，并使用 Maven 来构建管理。关于 JDK 的安装、Maven 的安装，以及集成开发环境 IDE 的安装，在这里不再阐述，请读者自行安装。本书的源码托管在 GitHub 上，具体地址为 https://github.com/springcloud/spring-cloud-code.git。可以使用 Git 客户端管理工具，比如通过 SourceTree 直接复制仓库到本地，然后通过 IDE 直接导入。代码模块根据章节内容进行编号，如 ch2-1，其中 ch 是 chapter 的简称，ch2 代表第 2 章，ch2-1 代表第 2 章的第一个聚合案例工程。

由于能力有限，书中难免有不妥之处，如果在学习中遇到问题或者发现 Bug，可以加 QQ（2508203324）或者发邮件到 Software_King@qq.com 与我们交流沟通。

致谢

我们代表 Spring Cloud 中国社区编写本书，我们想把本书送给每一位 Spring Cloud 技术爱好者，帮助大家快速掌握 Spring Cloud 的实战技巧并快速落地实践，为企业创造价值。

感谢与我一起写作的小伙伴们，大家能聚在一起写一本书也是一种缘分，大家的思维碰撞和交流，使本书变得更加完善，更加接地气，更具实战性。

感谢机械工业出版社华章公司的杨福川编辑与他的团队，他们在写作和审校过程中给予的帮助和支持，让我们能顺利地写完本书，跟大家见面。

感谢为本书评审的专家们，他们专业的态度和建议，为本书提供了宝贵的建议，尤其感谢Spring Cloud中国社区核心成员何鹰、刘石明、任浩军、任聪、李云龙等对Spring Cloud中国社区开源项目的贡献和支持。

感谢一直支持Spring Cloud中国社区发展的读者和亲朋好友们，感恩一路有你们，因为有你们的支持和陪伴，社区才能更好地发展，本书才能顺利完成。

最后我想说的是，我们7位作者并不是Spring Cloud微服务落地方面的架构专家，我们只是Spring Cloud微服务架构的实践者。我们只是想重新定义什么是真正意义上的Spring Cloud实战，并把我们自己的实践经验分享给大家，帮助大家解决学习和工作上遇到的问题。三人行，必有我师焉，由于我们学识有限，难免会有不足之处，还请读者不吝赐教，一起交流学习，共同进步。

许进

目 录

专家推荐
序 1
序 2
序 3
前言

第 1 章 微服务与 Spring Cloud ······ 1

1.1 微服务架构概述 ······ 1
    1.1.1 应用架构的发展 ······ 1
    1.1.2 微服务架构 ······ 3
    1.1.3 微服务解决方案 ······ 4
1.2 Spring Cloud 与中间件 ······ 5
    1.2.1 中间件概述 ······ 5
    1.2.2 什么是 Spring Cloud ······ 5
    1.2.3 Spring Cloud 项目模块 ······ 5
    1.2.4 Spring Cloud 与服务治理中间件 ······ 6
    1.2.5 Spring Cloud 与配置中心中间件 ······ 6
    1.2.6 Spring Cloud 与网关中间件 ······ 8
    1.2.7 Spring Cloud 与全链路监控中间件 ······ 9
1.3 Spring Cloud 增强生态 ······ 10
    1.3.1 Spring Cloud 分布式事务 ······ 10
    1.3.2 Spring Cloud 与领域驱动 ······ 10
    1.3.3 Spring Cloud 与 gRPC ······ 11
    1.3.4 Spring Cloud 与 Dubbo 生态融合 ······ 11
1.4 本章小结 ······ 11

第 2 章 Spring Cloud Eureka 上篇 ······ 12

2.1 服务发现概述 ······ 12
    2.1.1 服务发现由来 ······ 12
    2.1.2 Eureka 简介 ······ 14
    2.1.3 服务发现技术选型 ······ 15
2.2 Spring Cloud Eureka 入门案例 ······ 16
2.3 Eureka Server 的 REST API 简介 ······ 20
    2.3.1 REST API 列表 ······ 20
    2.3.2 REST API 实例 ······ 20
2.4 本章小结 ······ 26

第 3 章 Spring Cloud Eureka 下篇 ······ 27

3.1 Eureka 的核心类 ······ 27
    3.1.1 InstanceInfo ······ 27
    3.1.2 LeaseInfo ······ 28
    3.1.3 ServiceInstance ······ 29
    3.1.4 InstanceStatus ······ 29

3.2 服务的核心操作 30
   3.2.1 概述 30
   3.2.2 LeaseManager 30
   3.2.3 LookupService 31
3.3 Eureka 的设计理念 31
   3.3.1 概述 31
   3.3.2 AP 优于 CP 32
   3.3.3 Peer to Peer 架构 33
   3.3.4 Zone 及 Region 设计 34
   3.3.5 SELF PRESERVATION 设计 36
3.4 Eureka 参数调优及监控 36
   3.4.1 核心参数 36
   3.4.2 参数调优 39
   3.4.3 指标监控 41
3.5 Eureka 实战 42
   3.5.1 Eureka Server 在线扩容 42
   3.5.2 构建 Multi Zone Eureka Server 47
   3.5.3 支持 Remote Region 52
   3.5.4 开启 HTTP Basic 认证 58
   3.5.5 启用 https 61
   3.5.6 Eureka Admin 66
   3.5.7 基于 metadata 路由实例 67
3.6 Eureka 故障演练 69
   3.6.1 Eureka Server 全部不可用 69
   3.6.2 Eureka Server 部分不可用 71
   3.6.3 Eureka 高可用原理 73
3.7 本章小结 74

第 4 章 Spring Cloud Feign 的使用扩展 75

4.1 Feign 概述 75
   4.1.1 什么是 Feign 75
   4.1.2 Feign 的入门案例 76
   4.1.3 Feign 的工作原理 78
4.2 Feign 的基础功能 79
   4.2.1 FeignClient 注解剖析 79
   4.2.2 Feign 开启 GZIP 压缩 79
   4.2.3 Feign 支持属性文件配置 80
   4.2.4 Feign Client 开启日志 81
   4.2.5 Feign 的超时设置 82
4.3 Feign 的实战运用 83
   4.3.1 Feign 默认 Client 的替换 83
   4.3.2 Feign 的 Post 和 Get 的多参数传递 86
   4.3.3 Feign 的文件上传 90
   4.3.4 解决 Feign 首次请求失败问题 92
   4.3.5 Feign 返回图片流处理方式 93
   4.3.6 Feign 调用传递 Token 93
4.4 venus-cloud-feign 设计与使用 94
   4.4.1 venus-cloud-feign 的设计 94
   4.4.2 venus-cloud-feign 的使用 96
4.5 本章小结 98

第 5 章 Spring Cloud Ribbon 实战运用 99

5.1 Spring Cloud Ribbon 概述 99
   5.1.1 Ribbon 与负载均衡 99
   5.1.2 入门案例 100
5.2 Spring Cloud Ribbon 实战 105
   5.2.1 Ribbon 负载均衡策略与自定义配置 105
   5.2.2 Ribbon 超时与重试 107
   5.2.3 Ribbon 的饥饿加载 108
   5.2.4 利用配置文件自定义 Ribbon

　　　　客户端 ·· 108
　　5.2.5　Ribbon 脱离 Eureka 的使用 ······ 108
5.3　Spring Cloud Ribbon 进阶 ···················· 109
　　5.3.1　核心工作原理 ····························· 109
　　5.3.2　负载均衡策略源码导读 ··············· 113
5.4　本章小结 ··· 114

第 6 章　Spring Cloud Hystrix 实战运用 ·· 115

6.1　Spring Cloud Hystrix 概述 ····················· 115
　　6.1.1　解决什么问题 ······························· 116
　　6.1.2　设计目标 ······································ 117
6.2　Spring Cloud Hystrix 实战运用 ·············· 118
　　6.2.1　入门示例 ······································ 118
　　6.2.2　Feign 中使用断路器 ····················· 120
　　6.2.3　Hystrix Dashboard ······················· 121
　　6.2.4　Turbine 聚合 Hystrix ···················· 124
　　6.2.5　Hystrix 异常机制和处理 ················ 126
　　6.2.6　Hystrix 配置说明 ·························· 128
　　6.2.7　Hystrix 线程调整和计算 ················ 129
　　6.2.8　Hystrix 请求缓存 ·························· 130
　　6.2.9　Hystrix Request Collapser ············ 134
　　6.2.10　Hystrix 线程传递及并发策略 ······· 137
　　6.2.11　Hystrix 命令注解 ························ 142
6.3　本章小结 ··· 144

第 7 章　Spring Cloud Zuul 基础篇 ··· 145

7.1　Spring Cloud Zuul 概述 ························· 145
7.2　Spring Cloud Zuul 入门案例 ·················· 146
7.3　Spring Cloud Zuul 典型配置 ·················· 149
　　7.3.1　路由配置 ······································ 149
　　7.3.2　功能配置 ······································ 152

7.4　本章小结 ··· 154

第 8 章　Spring Cloud Zuul 中级篇 ··· 155

8.1　Spring Cloud Zuul Filter 链 ··················· 155
　　8.1.1　工作原理 ······································ 155
　　8.1.2　Zuul 原生 Filter ····························· 158
　　8.1.3　多级业务处理 ······························· 160
　　8.1.4　使用 Groovy 编写 Filter ················· 165
8.2　Spring Cloud Zuul 权限集成 ·················· 168
　　8.2.1　应用权限概述 ······························· 168
　　8.2.2　Zuul+OAuth2.0+JWT 实战 ············ 169
8.3　Spring Cloud Zuul 限流 ························· 176
　　8.3.1　限流算法 ······································ 176
　　8.3.2　限流实战 ······································ 177
8.4　Spring Cloud Zuul 动态路由 ·················· 179
　　8.4.1　动态路由概述 ······························· 179
　　8.4.2　动态路由实现原理剖析 ·················· 180
　　8.4.3　基于 DB 的动态路由实战 ··············· 182
8.5　Spring Cloud Zuul 灰度发布 ·················· 185
　　8.5.1　灰度发布概述 ······························· 185
　　8.5.2　灰度发布实战之一 ························ 186
8.6　Spring Cloud Zuul 文件上传 ·················· 189
　　8.6.1　文件上传实战 ······························· 189
　　8.6.2　文件上传乱码解决 ························ 191
8.7　Spring Cloud Zuul 实用小技巧 ··············· 192
　　8.7.1　饥饿加载 ······································ 192
　　8.7.2　请求体修改 ··································· 192
　　8.7.3　使用 okhttp 替换 HttpClient ·········· 193
　　8.7.4　重试机制 ······································ 194
　　8.7.5　Header 传递 ·································· 195
　　8.7.6　整合 Swagger2 调试源服务 ············ 195
8.8　本章小结 ··· 197

第 9 章　Spring Cloud Zuul 高级篇 … 198

9.1　Spring Cloud Zuul 多层负载 … 198
9.1.1　痛点场景 … 198
9.1.2　解决方案 … 198

9.2　Spring Cloud Zuul 应用优化 … 200
9.2.1　概述 … 200
9.2.2　容器优化 … 201
9.2.3　组件优化 … 202
9.2.4　JVM 参数优化 … 203
9.2.5　内部优化 … 204

9.3　Spring Cloud Zuul 原理 & 核心源码解析 … 205
9.3.1　工作原理与生命周期 … 205
9.3.2　Filter 装载与 Filter 链实现 … 208
9.3.3　核心路由实现 … 210

9.4　本章小结 … 213

第 10 章　Spring Cloud 基础综合案例 … 214

10.1　基础框架 … 214
10.1.1　搭建说明 … 214
10.1.2　技术方案 … 214
10.1.3　具体实现 … 215

10.2　实战扩展 … 217
10.2.1　公共包(对象，拦截器，工具类等) … 218
10.2.2　用户上下文对象传递 … 218
10.2.3　Zuul 的 Fallback 机制 … 221

10.3　生产环境各组件参考配置 … 222
10.3.1　Eureka 推荐配置 … 222
10.3.2　Ribbon 推荐配置 … 223
10.3.3　Hystrix 推荐配置 … 223
10.3.4　Zuul 推荐配置 … 223

10.4　本章小结 … 224

第 11 章　Spring Cloud Config 上篇 … 225

11.1　Spring Cloud Config 配置中心概述 … 225
11.1.1　什么是配置中心 … 225
11.1.2　Spring Cloud Config … 227
11.1.3　Spring Cloud Config 入门案例 … 228

11.2　刷新配置中心信息 … 234
11.2.1　手动刷新操作 … 234
11.2.2　结合 Spring Cloud Bus 热刷新 … 237

11.3　本章小结 … 244

第 12 章　Spring Cloud Config 下篇 … 245

12.1　服务端 Git 配置详解与实战 … 245
12.1.1　Git 多种配置详解概述 … 245
12.1.2　Git 中 URI 占位符 … 245
12.1.3　模式匹配和多个存储库 … 250
12.1.4　路径搜索占位符 … 251

12.2　关系型数据库的配置中心的实现 … 251
12.2.1　Spring Cloud Config 基于 MySQL 的配置概述 … 251
12.2.2　Spring Cloud Config 与 MySQL 结合案例 … 252

12.3　非关系型数据库的配置中心的实现 … 255

12.3.1 Spring Cloud Config 基于 MongoDB 的配置概述 ········ 255
12.3.2 Spring Cloud Config MongoDB 案例 ········ 256
12.4 Spring Cloud Config 使用技能 ······· 259
12.5 Spring Cloud Config 功能扩展 ······· 260
12.5.1 客户端自动刷新 ·············· 260
12.5.2 客户端回退功能 ·············· 264
12.5.3 客户端的安全认证机制 JWT ·············· 270
12.6 高可用部分 ························· 285
12.6.1 客户端高可用 ················ 285
12.6.2 服务端高可用 ················ 293
12.7 Spring Cloud 与 Apollo 配置使用 ··· 300
12.7.1 Apollo 简介 ·················· 300
12.7.2 Apollo 具备功能 ············· 300
12.7.3 Apollo 总体架构模块 ········ 300
12.7.4 客户端设计 ··················· 301
12.7.5 Apollo 运行环境方式 ······· 302
12.8 Spring Cloud 与 Apollo 结合使用实战 ··················· 303
12.8.1 Apollo 环境的要求 ·········· 303
12.8.2 Apollo 基础数据导入 ······· 303
12.8.3 创建 config-client-apollo ···· 307
12.8.4 创建 gateway-zuul-apollo ··· 310
12.9 本章总结 ·························· 316

第13章 Spring Cloud Consul 上篇 ··· 317

13.1 Consul 简介 ······················· 317
13.1.1 什么是 Consul ··············· 317
13.1.2 Consul 能做什么 ············ 317
13.1.3 Consul 的安装 ··············· 318
13.1.4 Consul 启动 ················· 318
13.1.5 Consul UI ···················· 319
13.1.6 Consul 实用接口 ············ 319
13.2 Spring Cloud Consul 简介 ········ 319
13.2.1 Spring Cloud Consul 是什么 ··· 319
13.2.2 Spring Cloud Consul 能做什么 ··················· 320
13.2.3 Spring Cloud Consul 入门案例 ······················ 320
13.3 本章小节 ··························· 324

第14章 Spring Cloud Consul 下篇 ··· 325

14.1 Spring Cloud Consul 深入 ········ 325
14.1.1 Spring Cloud Consul 的模块介绍 ························· 325
14.1.2 Spring Cloud Consul Discovery ··· 325
14.1.3 Spring Cloud Consul Config ··· 332
14.2 Spring Cloud Consul 功能重写 ··· 335
14.2.1 重写 ConsulDiscoveryClient ··· 335
14.2.2 重写 ConsulServerList ········ 338
14.3 常见问题排查 ····················· 343
14.3.1 版本兼容的那些坑 ·········· 343
14.3.2 Spring Cloud Consul 的一些问题 ··························· 344
14.4 本章小节 ··························· 346

第15章 Spring Cloud 认证和鉴权 ··· 347

15.1 微服务安全与权限 ··············· 347
15.2 Spring Cloud 认证与鉴权方案 ··· 348
15.2.1 单体应用下的常用方案 ···· 348
15.2.2 微服务下 SSO 单点登录方案 ··· 348

15.2.3　分布式 Session 与网关结合方案⋯⋯349
　　15.2.4　客户端 Token 与网关结合方案⋯⋯349
　　15.2.5　浏览器 Cookie 与网关结合方案⋯⋯350
　　15.2.6　网关与 Token 和服务间鉴权结合⋯⋯350
15.3　Spring Cloud 认证鉴权实战案例⋯⋯351
　　15.3.1　创建 Spring Cloud Gateway 及关联信息⋯⋯351
　　15.3.2　核心的公共工程 core-service⋯⋯353
　　15.3.3　服务提供方工程 provider-service⋯⋯355
　　15.3.4　客户端工程 client-service⋯⋯356
　　15.3.5　运行结果⋯⋯356
15.4　本章小结⋯⋯358

第 16 章　Spring Cloud 全链路监控⋯359

16.1　全链路监控概述⋯⋯359
　　16.1.1　链路监控的原理来源⋯⋯359
　　16.1.2　Sleuth 原理介绍⋯⋯360
　　16.1.3　Brave 和 Zipkin⋯⋯360
16.2　Sleuth 基本用法⋯⋯362
　　16.2.1　Sleuth 对 Feign 的支持⋯⋯365
　　16.2.2　Sleuth 对 RestTemplate 的支持⋯⋯366
　　16.2.3　Sleuth 对多线程的支持⋯⋯367
16.3　Sleuth 深入用法⋯⋯367
　　16.3.1　TraceFilter⋯⋯367

　　16.3.2　Baggage⋯⋯367
　　16.3.3　案例⋯⋯367
16.4　Spring Cloud 与 SkyWalking⋯⋯369
　　16.4.1　Skywalking 概述⋯⋯369
　　16.4.2　SkyWalking 提供主要功能⋯⋯370
　　16.4.3　SkyWalking 主要特性⋯⋯370
　　16.4.4　SkyWalking 整体架构⋯⋯370
16.5　Spring Cloud 与 Skywalking 实战⋯⋯370
　　16.5.1　父工程创建⋯⋯371
　　16.5.2　创建 eureka-server-skywalking 工程⋯⋯372
　　16.5.3　创建 zuul-skywalking⋯⋯373
　　16.5.4　创建 service-a⋯⋯375
　　16.5.5　创建 service-b⋯⋯377
　　16.5.6　SkyWalking Collector 基础环境安装⋯⋯378
　　16.5.7　使用 Agent 启动服务和监控查看⋯⋯382
　　16.5.8　总结⋯⋯385
16.6　Spring Cloud 与 Pinpoint⋯⋯386
　　16.6.1　Pinpoint 概述⋯⋯386
　　16.6.2　Pinpoint 架构模块⋯⋯386
　　16.6.3　Pinpoint 的数据结构⋯⋯386
　　16.6.4　Pinpoint 兼容性⋯⋯387
16.7　Spring Cloud 与 Pinpoint 实战⋯⋯389
　　16.7.1　Pinpoint 基础环境⋯⋯389
　　16.7.2　Collector 和 Web 部署⋯⋯391
　　16.7.3　Agent 启动应用⋯⋯392
　　16.7.4　UI 浏览指标⋯⋯394
　　16.7.5　总结⋯⋯397
16.8　本章总结⋯⋯398

第 17 章　Spring Cloud Gateway 上篇······399

17.1　Spring Cloud Gateway 概述······399
17.1.1　什么是 Spring Cloud Gateway······399
17.1.2　Spring Cloud Gateway 的核心概念······399
17.2　Spring Cloud Gateway 的工作原理······400
17.3　Spring Cloud Gateway 入门案例······401
17.4　Spring Cloud Gateway 的路由断言······404
17.4.1　After 路由断言工厂······404
17.4.2　Before 路由断言工厂······406
17.4.3　Between 路由断言工厂······406
17.4.4　Cookie 路由断言工厂······407
17.4.5　Header 路由断言工厂······408
17.4.6　Host 路由断言工厂······410
17.4.7　Method 路由断言工厂······411
17.4.8　Query 路由断言工厂······411
17.4.9　RemoteAddr 路由断言工厂······412
17.5　Spring Cloud Gateway 的内置 Filter······413
17.5.1　AddRequestHeader 过滤器工厂······413
17.5.2　AddRequestParameter 过滤器······413
17.5.3　RewritePath 过滤器······414
17.5.4　AddResponseHeader 过滤器······415
17.5.5　StripPrefix 过滤器······416
17.5.6　Retry 过滤器······417
17.5.7　Hystrix 过滤器······418
17.6　本章小结······420

第 18 章　Spring Cloud Gateway 下篇······421

18.1　Gateway 基于服务发现的路由规则······421
18.1.1　Gateway 的服务发现路由概述······421
18.1.2　服务发现的路由规则案例······422
18.2　Gateway Filter 和 Global Filter······425
18.2.1　Gateway Filter 和 Global Filter 概述······425
18.2.2　自定义 Gateway Filter 案例······425
18.2.3　自定义 Global Filter 案例······427
18.3　Spring Cloud Gateway 实战······428
18.3.1　Spring Cloud Gateway 权重路由······428
18.3.2　Spring Cloud Gateway 中 Https 的使用技巧······431
18.3.3　Spring Cloud Gateway 集成 Swagger······436
18.3.4　Spring Cloud Gateway 限流······442
18.3.5　Spring Cloud Gateway 的动态路由······450
18.4　Spring Cloud Gateway 源码篇······458
18.4.1　Spring Cloud Gateway 的处理流程······458
18.4.2　Gateway 中 ServerWebExchange 构建分析······459
18.4.3　DispatcherHandler 源码分析······460
18.4.4　RoutePredicateHandlerMapping 源码分析······461
18.4.5　FilteringWebHandler 源码分析······462
18.4.6　执行 Filter 源码分析······463
18.5　本章小结······465

第 19 章　Spring Cloud 与 gRPC 上篇 466

19.1　Spring Cloud 为什么需要 gRPC 466
19.2　gRPC 简介 468
19.3　gRPC 的一些核心概念 469
  19.3.1　服务定义 469
  19.3.2　使用 API 470
  19.3.3　同步 vs 异步 470
19.4　RPC 的生命周期 470
19.5　gRPC 依赖于 Protocol Buffers 472
  19.5.1　Protocol Buffers 的特点 472
  19.5.2　使用 Protocol Buffers 的 Maven 插件 472
  19.5.3　Proto Buffer 语法介绍 475
19.6　gRPC 基于 HTTP2 476
19.7　gRPC 基于 Netty 进行 IO 处理 477
19.8　gRPC 案例实战 478
19.9　本章小结 481

第 20 章　gRPC 在 Spring Cloud 与 gRPC 下篇 482

20.1　gRPC Spring Boot Starter 介绍 482
20.2　gRPC Spring Boot Starter 架构设计 482
20.3　gRPC Spring Boot Starter 源码分析 483
  20.3.1　gRPC Server Spring Boot Starter 源码解析 483
  20.3.2　gRPC Client Spring Boot Starter 源码解析 486
20.4　案例实战 489
  20.4.1　注册中心 489
  20.4.2　链路追踪服务端 490
  20.4.3　gRPC 的 lib 工程 490
  20.4.4　gRPC 服务端 491
  20.4.5　gRPC 客户端 494
20.5　本章小结 497

第 21 章　Spring Cloud 版本控制与灰度发布 498

21.1　背景 498
21.2　常见发布方式 499
  21.2.1　蓝绿发布 499
  21.2.2　滚动发布 500
  21.2.3　灰度发布 500
  21.2.4　对比 501
21.3　版本控制与灰度发布实战 502
  21.3.1　Discovery 项目 503
  21.3.2　实战案例 504
  21.3.3　实战测试 505
21.4　本章小结 509

第 22 章　Spring Cloud 容器化 510

22.1　Java 服务 Docker 化 510
  22.1.1　基础镜像选择 510
  22.1.2　Dockerfile 编写 511
  22.1.3　镜像构建插件 514
  22.1.4　JDK8+ 的 Docker 支持 516
  22.1.5　JDK9+ 镜像优化 517
22.2　Spring Cloud 组件的 Docker 化 519
  22.2.1　Docker 化配置 519
  22.2.2　config-server 的 Docker 化 520
  22.2.3　eureka-server 的 Docker 化 522
  22.2.4　gateway 的 Docker 化 524

	22.2.5	turbine 的 Docker 化	526
	22.2.6	Spring Admin 的 Docker 化	528
	22.2.7	biz-service 的 Docker 化	530
	22.2.8	网卡选择	532
	22.2.9	小结	532
22.3	使用 Kubernetes 管理		532
	22.3.1	概述	532
	22.3.2	本地安装 Kubernetes	533
	22.3.3	部署到 Kubernetes	536
	22.3.4	一键伸缩	544
	22.3.5	滚动升级	547
22.4	本章小结		552

第 23 章　Dubbo 向 Spring Cloud 迁移 ……553

23.1	将 Dubbo 服务纳入 Spring Cloud 体系中		553
	23.1.1	将 Dubbo 项目改造成 Spring Boot 项目	553
	23.1.2	集成 Spring Cloud 组件	554
	23.1.3	将 Dubbo 服务暴露为 RESTful API	555
23.2	将 Spring Cloud 服务 Dubbo 化		556
	23.2.1	服务注册中心	556
	23.2.2	服务提供者	556
	23.2.3	服务消费者	558
	23.2.4	Spring Cloud Dubbo 框架原理	561
23.3	本章小结		562

第 24 章　Spring Cloud 与分布式事务 …563

24.1	概述		563
	24.1.1	ACID	563
	24.1.2	X/Open DTP 模型与 XA 接口	564
	24.1.3	CAP 与 BASE 定理	567
24.2	解决方案		567
	24.2.1	Java 事务编程接口 JTA	567
	24.2.2	分布式事务 TCC 模式	568
	24.2.3	分布式事务 SAGA 模式	570
24.3	实战		572
	24.3.1	Atomikos JTA	572
	24.3.2	TCC for REST	580
	24.3.3	Servicecomb SAGA	594
24.4	本章小结		603

第 25 章　Spring Cloud 与领域驱动实践 ……604

25.1	领域驱动概述		604
	25.1.1	Spring Cloud 与领域驱动	604
	25.1.2	为什么需要领域建模	605
25.2	领域驱动核心概念		606
	25.2.1	实体概述	606
	25.2.2	值对象概述	606
	25.2.3	领域服务	607
	25.2.4	聚合及聚合根	607
	25.2.5	边界上下文	608
	25.2.6	工厂	609
	25.2.7	仓储/资源库	609
	25.2.8	CQRS 架构	610
	25.2.9	领域事件	610
	25.2.10	领域驱动模型的设计步骤	611
25.3	Halo 框架的设计		611
	25.3.1	DDD 应用框架的意义	611
	25.3.2	领域驱动框架现状	612
	25.3.3	Halo 框架概述	612

25.3.4 Halo 框架分层设计 ·············· 613
25.3.5 Halo 框架中的 CQRS 设计 ···· 615
25.3.6 Command 与 Command Bus 设计 ························· 616
25.3.7 Event 与 Event Bus 设计 ········ 619
25.3.8 Extend 扩展点设计 ············ 621
25.3.9 业务身份设计 ················ 623
25.3.10 规范设计 ··················· 624
25.4 Spring Cloud 与 Halo 实战 ·········· 625
    25.4.1 事件风暴寻找模型和聚合 ······ 625
    25.4.2 Spring Cloud 与 Halo 实战案例 ························ 626
    25.4.3 新建二方包工程模块 ·········· 627
    25.4.4 新建 DDD 基础设施层 ········ 629
    25.4.5 新建 DDD 领域层 ············ 630
    25.4.6 新建 DDD 应用层 ············ 632
    25.4.7 启动测试 ···················· 634
25.5 本章小结 ························ 634

第 1 章 Chapter 1

微服务与 Spring Cloud

随着互联网的快速发展，云计算近十年也得到蓬勃发展，企业的 IT 环境和 IT 架构也逐渐在发生变革，从过去的单体应用架构发展为至今广泛流行的微服务架构。微服务是一种架构风格，能给软件应用开发带来很大的便利，但是微服务的实施和落地会面临很大的挑战，因此需要一套完整的微服务解决方案。在 Java 领域，Spring 框架的出现给 Java 企业级软件开发带来了福音，提高了开发效率。在 2014 年底，Spring 团队推出 Spring Cloud，目标使其成为 Java 领域微服务架构落地的标准，发展至今，Spring Cloud 已经成为 Java 领域落地微服务架构的完整解决方案，为企业 IT 架构变革保驾护航。

1.1 微服务架构概述

1.1.1 应用架构的发展

应用是可独立运行的程序代码，提供相对完善的业务功能。目前软件架构有三种架构类型，分别是业务架构、应用架构、技术架构。它们之间的关系是业务架构决定应用架构，技术架构支撑应用架构。架构的发展历程是从单体架构、分布式架构、SOA 架构再到微服务架构，如图 1-1 所示。

图 1-1 架构发展历程

1. 单体应用架构

单体架构在 Java 领域可以理解为一个 Java Web 应用程序，包含表现层、业务层、数据访问层。从 Controller 到 Service 再到 Dao 层，"一杆子捅到底"，没有任何应用拆分，开发完毕之后变成一个超级大型的 War 部署。简单的单体架构水平分层逻辑如图 1-2 所示。

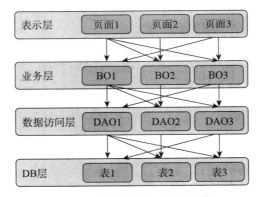

图 1-2　单体架构水平分层逻辑

单体架构的优点：
- 易于开发：开发人员使用当前开发工具在短时间内就可以开发出单体应用。
- 易于测试：因为不需要依赖其他接口，测试可以节约很多时间。
- 易于部署：你只需要将目录部署在运行环境中即可。

单体架构的缺点：
- 灵活度不够：如果程序有任何修改，修改的不只是一个点，而是自上而下地去修改，测试时必须等到整个程序部署完后才能看出效果。在开发过程可能需要等待其他开发人员开发完成后才能完成部署，降低了团队的灵活性。
- 降低系统的性能：原本可以直接访问数据库但是现在多了一层。即使只包含一个功能点，也需要在各个层写上代码。
- 系统启动慢：一个进程包含了所有业务逻辑，涉及的启动模块过多，导致系统的启动时间延长。
- 系统扩展性比较差：增加新东西的时候不能针对单个点增加，要全局性地增加。牵一发而动全身。

2. 分布式架构

什么是传统的分布式架构？简单来说，按照业务垂直切分，每个应用都是单体架构，通过 API 互相调用，如图 1-3 所示。

3. 面向服务的 SOA 架构

面向服务的架构是一种软件体系结构，其应用程序的不同组件通过网络上的通信协议向其他组件提供服务或消费服务，所以也是一种分布式架构。简单来说，SOA 是不同业务建立不同的服务，服务之间的数据交互粗粒度可以通过服务接口分级，这样松散耦合提高服务的可重用性，也让业务逻辑变得可组合，并且每个服务可以根据使用情况做出合理的分布式部署，从而让服务变得规范、高性能、高可用。

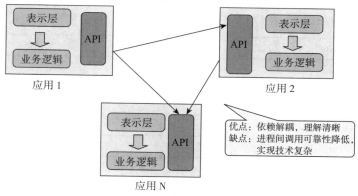

图 1-3　分布式架构

SOA 架构中有两个主要角色：服务提供者（Provider）和服务消费者（Consumer）。阿里开源的 Dubbo 是 SOA 的典型实现。

SOA 架构的优点：

- 把模块拆分，使用接口通信，降低模块之间的耦合度。
- 把项目拆分成若干个子项目，不同的团队负责不同的子项目。
- 增加功能时只需要增加一个子项目，调用其他系统的接口即可。
- 可以灵活地进行分布式部署。

SOA 架构的缺点：系统之间的交互需要使用远程通信，接口开发增加工作量。

1.1.2　微服务架构

微服务架构在某种程度上是 SOA 架构继续发展的下一步。微服务的概念最早源于 Martin Fowler 的一篇文章《Microservices》。总体来说，微服务是一种架构风格，对于一个大型复杂的业务系统，它的业务功能可以拆分为多个相互独立的微服务，各个微服务之间是松耦合的，通过各种远程协议进行同步/异步通信，各微服务均可以被独立部署、扩/缩容以及升/降级。这里对微服务技术选型做了对比，如表 1-1 所示。

表 1-1　微服务架构技术选型对比

	Spring Cloud	Dubbo	Motan	MSEC	其他
功能	微服务完整方案	服务治理框架	服务治理框架	服务开发运营框架	略
通信方式	REST/Http	RPC 协议	RPC/Hessian2	Protocol buffer	grpc、thrift
服务发现/注册	Eureka（AP）	ZK、Nacos	ZK/Conusl	只有服务发现	Etcd
负载均衡	Ribbon	客户端负载	客户端负载	客户端负载	Ngnix+Lua

(续)

	Spring Cloud	Dubbo	Motan	MSEC	其他
容错机制	6种容错策略	6种容错策略	2种容错策略	自动容错	Keepalived、HeartBeat
熔断机制	Hystrix	无	无	提供过载保护	无
配置中心	Spring Cloud Config	Nacos	无	无	Apollo、Nacos
网关	Zuul、Gateway	无	无	无	Kong、自研
服务监控	Hystrix+Turbine	Dubbo+Monitor	无	Monitor	ELK
链路监控	Sleuth+Zipkin	无	无	无	Pinpoint
多语言	Rest 支持多语言	Java	Java	Java、C++、PHP	Java、PHP、Node.js
社区活跃	高（背靠spring）	高（背靠阿里）	一般	未知	略

1.1.3 微服务解决方案

现如今微服务架构十分流行，而采用微服务构建系统也会带来更清晰的业务划分和可扩展性。同时支持微服务的技术栈也是多种多样的。下面介绍两种实现微服务的解决方案。

1. 基于 Spring Cloud 的微服务解决方案

Spring Cloud 的技术选型是中立的，因此可以随需更换搭配使用，基于 Spring Cloud 的微服务落地解决方案可以分为三种，如表 1-2 所示。

表 1-2 基于 Spring Cloud 的三种方案

组件	方案1	方案2	方案3
服务发现	Eureka	Consul	etcd、阿里 Nacos
共用组件	服务间调用组件 Feign、负载均衡组件 Ribbon、熔断器 Hytrix		
网关	性能低：Zuul；性能高：Spring Cloud Gateway		自研网关中间件
配置中心	Spring Cloud Config、携程阿波罗、阿里 Nacos		
全链路监控	zikpin（不推荐）、Pinpoint（不推荐）、Skywalking（推荐）		
搭配使用	参考本书其他章节，比如分布式事务、容器化、Spring Cloud 与 DDD、gRPC		

2. 基于 Dubbo 实现微服务解决方案

2012 年，阿里巴巴在 GitHub 上开源了基于 Java 的分布式服务治理框架 Dubbo，但是 Dubbo 未来的定位并不是要成为一个微服务的全面解决方案，而是专注于 RPC 领域，成为微服务生态体系中的一个重要组件。至于微服务化衍生出的服务治理需求，Dubbo 正在积极适配开源解决方案，并且已经启动独立的开源项目予以支持，比如最近宣布的开源的 Nacos。Nacos

的定位是一个更易于帮助构建云原生应用的动态服务发现、配置和服务管理平台。因此基于 Dubbo 的微服务解决方案是：Dubbo+Nacos+其他。

1.2 Spring Cloud 与中间件

1.2.1 中间件概述

中间件与操作系统、数据库并列为传统基础软件的三驾马车。其中，中间件也是难度极高的软件工程。传统中间件的概念，诞生于上一个"分布式"计算的年代，也就是小规模局域网中的服务器/客户端计算模式，在操作系统之上、应用软件之下的"中间层"软件。

随着互联网的快速发展，以及云计算的出现，企业的 IT 架构正在发生深刻的变革。在这个过程中，软件向大规模互联网云服务演化，无论是操作系统还是数据库都发生了深刻的变化，中间件也在这个过程不断演进和扩大自己的边界。中间件向下屏蔽异构的硬件、软件、网络等计算资源，向上提供应用开发、运行、维护等全生命周期的统一计算环境与管理，属于承上启下的中间连接层，对企业来说有着极其重要的价值。中间件本质上可以归属为技术架构，常见的中间件分别是服务治理中间件（例如：Dubbo 等 RPC 框架）、配置中心、全链路监控、分布式事务、分布式定时任务、消息中间件、API 网关、分布式缓存、数据库中间件等。

1.2.2 什么是 Spring Cloud

在前面的小节介绍了中间件，Spring Cloud 也是一个中间件。它目前由 Spring 官方开发维护，基于 Spring Boot 开发，提供一套完整的微服务解决方案。包括服务注册与发现、配置中心、全链路监控、API 网关、熔断器等选型中立的开源组件，可以随需扩展和替换组装。

Spring Cloud 项目自从推出以来，到目前为止一直在高速迭代。Spring Cloud 技术团队于 2018 年 6 月 19 日发布了 Spring Cloud 的重大里程碑 Finchley 版本。本书也是基于此版本编写，目前 Spring Cloud 在国内外得到广泛使用。Spring Cloud 中国社区发起了 Spring Cloud 在中国的使用情况收集，具体可以参考如下统计链接：https://github.com/SpringCloud/spring-cloud-document/issues/1。

1.2.3 Spring Cloud 项目模块

Spring Cloud 是一个开源项目集合，包括很多子项目。具体的项目地址可以参考 GitHub 组织：https://github.com/spring-cloud。Spring Cloud 的主要组件汇总如表 1-3 所示，因为 Spring Cloud 的子项目居多，每个子项目有自己的版本号，为了对 Spring Cloud 整体进行版本编号，确定一个可用于生产上的版本标识。这些版本采用伦敦地铁站的名字，按名称首字母排序，比如 Dalston 版、Edgware 版、Finchley。但是我们一般都会简称为 D 版、E 版、F 版等。

表 1-3 Spring Cloud 组件列表

组件名称	所属项目	组件分类
Eureka	spring-cloud-netflix	注册中心
Zuul	spring-cloud-netflix	第一代网关
Sidecar	spring-cloud-netflix	多语言
Ribbon	spring-cloud-netflix	负载均衡
Hystrix	spring-cloud-netflix	熔断器
Turbine	spring-cloud-netflix	集群监控
Feign	spring-cloud-openfeign	声明式 HTTP 客户端
Consul	spring-cloud-consul	注册中心
Gateway	spring-cloud-gateway	第二代网关
Sleuth	spring-cloud-seluth	链路追踪
Config	spring-cloud-config	配置中心
Bus	spring-cloud-bus	总线
Pipeline	spring-cloud-pipelines	部署管道
Dataflow	spring-cloud-dataflow	数据处理

1.2.4 Spring Cloud 与服务治理中间件

服务治理中间件包含服务注册与发现、服务路由、负载均衡、自我保护、丰富的治理管理机制等功能。其中服务路由包含服务上下线、在线测试、机房就近选择、A/B 测试、灰度发布等。负载均衡支持根据目标状态和目标权重进行负载均衡。自我保护包括服务降级、优雅降级和流量控制。

Spring Cloud 作为一个服务治理中间件，它的服务治理体系做了高度的抽象，目前支持使用 Eureka、Zookeeper、Consul 作为注册中心，并且预留了扩展接口，而且由于选型是中立的，所以支持无缝替换。在 Spring Cloud 中可以通过 Hystrix 进行熔断自我保护 Fallback，通过 Ribbon 进行负载均衡。本书围绕着 Spring Cloud 的服务治理体系介绍 Eureka、Consul、Ribbon、Hystrix 等实战用法。关于 Spring Cloud 注册中心的对比选型参考，如图 1-4 所示。

1.2.5 Spring Cloud 与配置中心中间件

在单体应用中，我们一般的做法是把属性配置和代码硬编码放在一起，这没有什么问题。但是在分布式系统中，由于存在多个服务实例，需要分别管理每个具体服务工程中的配置，上线需要准备 Check List 并逐个检查每个上线的服务是否正确。在系统上线之后一旦修改某个配置，就需要重启服务。这样开发管理相当麻烦。因此我们需要把分布式系统中的配置信息抽取出来统一管理，这个管理的中间件称为配置中心。配置中心应该具备的功能，分别是支持各种复杂的配置场景，与公司的运维体系和权限管理体系集成，各种配置兼容支持，如图 1-5 所示。

Feature	Consul	Zookeeper	etcd	Euerka
服务健康检查	服务状态，内存，硬盘等	（弱）长连接，keepalive	连接心跳	可配支持
多数据中心	支持	—	—	—
kv存储服务	支持	支持	支持	—
一致性	raft	paxos	raft	—
cap	cp	cp	cp	ap
使用接口（多语言能力）	支持http和dns	客户端	http/grpc	http(sidecar)
watch支持	全量/支持long polling	支持	支持long polling	支持long polling/大部分增量
自身监控	metrics	—	metrics	metrics
安全	acl/https	acl	https支持（弱）	—
Spring Cloud集成	已支持	已支持	已支持	已支持

图 1-4　注册中心选型对比

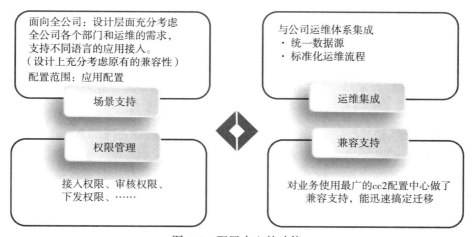

图 1-5　配置中心的功能

Spring Cloud Config 是 Spring Cloud 生态圈中的配置中心中间件，它把应用原本放在本地文件中的配置抽取出来放在中心服务器，从而能够提供更好的管理、发布能力。Spring Cloud Config 基于应用、环境、版本三个维度管理，配置存储支持 Git 和其他扩展存储，且无缝支持 Spring 里 Environment 和 PropertySource 的接口。但是 Spring Cloud Config 的缺点是没有可视化的管控平台，因此会用其他的配置中心中间件取代它管理配置。在本书的第 11 章和第 12 章介绍了 Spring Cloud Config 和携程开源配置中心 Apollo 的实战用法。

1.2.6 Spring Cloud 与网关中间件

1. 网关中间件概述

API Gateway（APIGW/API 网关），顾名思义，是出现在系统边界上一个面向 API 的、串行集中式的强管控服务，这里的边界是企业 IT 系统的边界，可以理解为企业级应用防火墙，主要起到隔离外部访问与内部系统的作用。在微服务概念流行之前，API 网关就已经诞生了，例如银行、证券等领域常见的前置系统，它的设计与出现主要是为了解决访问认证、报文转换、访问统计等问题。网关在微服务架构中所处的位置，如图 1-6 所示。

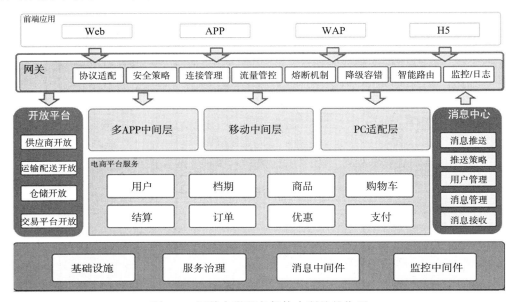

图 1-6 网关在微服务架构中所处的位置

随着微服务架构概念的提出，API 网关成为微服务架构的一个标配组件。作为一个网关中间件，至少具备如下四大功能。

（1）统一接入功能

为各种无线应用提供统一的接入服务，提供一个高性能、高并发、高可靠的网关服务。不仅如此，还要支持负载均衡、容灾切换和异地多活。

（2）协议适配功能

网关对外的请求协议一般是 HTTP 或 HTTP2 协议，而后端提供访问的服务协议要么是 REST 协议要么是 RPC 协议。网关需要根据请求进来的协议进行协议适配，然后协议转发调用不同协议提供的服务。比如网关对外暴露是 HTTP 协议的请求，而网关后端服务可能是 Dubbo 提供的 RPC 服务，可能是 Spring Cloud 提供的 REST 服务，也可能是其他 PHP 编写的服务。当一个 HTTP 请求经过网关时，通过一系列不同功能的 Filter 处理完毕之后，此时就需要进行协议适配，判断应该协议转发调用 RPC 服务，调用 REST 服务，还是调用 PHP 提供的服务。

（3）流量管控功能

网关作为所有请求流量的入口，当请求流量瞬间剧增，比如天猫双 11、双 12 或者其他大促活动，流量会迅速剧增，此时需要进行流量管控、流量调拨。当后端服务出现异常，服务不可用时，需要网关进行熔断和服务降级。在异地多活场景中需要根据请求流量进行分片，路由到不同的机房。

（4）安全防护功能

网关需要对所有请求进行安全防护过滤，保护后端服务。通过与安全风控部门合作，对 IP 黑名单和 URL 黑名单封禁控制，做风控防刷，防恶意攻击等。

2. Spring Cloud 第一代网关 Zuul

Spring Cloud 生态圈的第一代网关是在 Netflix 公司开源的网关组件 Zuul 之上，它基于 Spring Boot 注解，采用 Starter 的方式进行二次封装，可以做到开箱即用。目前 Zuul 融合了 Spring Cloud 提供的服务治理体系，根据配置的路由规则或者默认的路由规则进行路由转发和负载均衡。它可以与 Spring Cloud 生态系统内其他组件集成使用，例如：集成 Hystrix 可以在网关层面实现降级的功能；集成 Ribbon 可以使得整个架构具备弹性伸缩能力；集成 Archaius 可以进行配置管理等。但是 Spring Cloud Zuul 如果需要做一些灰度、降级、标签路由、限流、WAF 封禁，则需要自定义 Filter 做一些定制化实现。

3. Spring Cloud 第二代网关 Gateway

Spring Cloud Zuul 处理每个请求的方式是分别对每个请求分配一个线程来处理。根据参考数据统计，目前 Zuul 最多能达到 1000 至 2000 QPS。在高并发的场景下，不推荐使用 Zuul 作为网关。因此出现了 Spring Cloud 的第二代网关，即 Spring Cloud Gateway。

Spring Cloud Gateway 是 Spring 官方基于 Spring 5.0、Spring Boot 2.0 和 Project Reactor 等技术开发的网关，旨在为微服务架构提供一种简单、有效、统一的 API 路由管理方式。Spring Cloud Gateway 底层基于 Netty 实现（Netty 的线程模型是多线程 reactor 模型，使用 boss 线程和 worker 线程接收并异步处理请求，具有很强大的高并发处理能力），因此 Spring Cloud Gateway 出现的目的就是替代 Netflix Zuul。其不仅提供统一的路由方式，并且基于 Filter 链的方式提供了网关基本的功能，例如，安全、监控/埋点、限流等。在本书的第 17 章和 18 章，详细地介绍了 Spring Cloud Gateway 的使用方法和实战技巧。

1.2.7 Spring Cloud 与全链路监控中间件

众所周知，中大型互联网公司的后台业务系统由众多的分布式应用组成。一个通过浏览器或移动客户端的前端请求到后端服务应用，会经过很多应用系统，并且留下足迹和相关日志信息。但这些分散在每个业务应用主机下的日志信息不利于问题排查和定位问题发生的根本原因。此时就需要利用全链路监控中间件收集、汇总并分析日志信息，进行可视化展示和监控告警。全链路监控中间件应该提供的主要功能包括：

- ❑ 定位慢调用：包括慢 Web 服务（包括 Restful Web 服务）、慢 REST 或 RPC 服务、慢 SQL。
- ❑ 定位各种错误：包括 4××、5××、Service Error。
- ❑ 定位各种异常：包括 Error Exception、Fatal Exception。

- 展现依赖和拓扑：域拓扑、服务拓扑、trace 拓扑。
- Trace 调用链：将端到端的调用，以及附加在这次调用的上下文信息，异常日志信息，每一个调用点的耗时都呈现给用户进行展示。
- 应用告警：根据运维设定的告警规则，扫描指标数据，如违反告警规则，则将告警信息上报到中央告警平台。

全链路监控中间件相关产品发展至今百花齐放，如京东 Hydra、阿里 Eagleye，这两个中间件都吸收了 Dapper/Zipkin 的设计思路，但是目前都未开源。近年，社区又发展出很多调用链监控产品，如国内开源爱好者吴晟（原 OneAPM 工程师，目前在华为）开源并提交到 Apache 孵化器的产品 Skywalking，它同时吸收了 Zipkin/Pinpoint/CAT 的设计思路，支持非侵入式埋点。如果使用 Java 技术栈，希望采用非侵入式的监控，推荐使用 Pinpoint 或者 Skywalking。

本书的相关章节介绍了 Spring Cloud Sleuth 的实战用法，以及如何借助 Pinpoint 或者 Skywalking 实现 Spring Cloud 体系下的全链路监控。

1.3　Spring Cloud 增强生态

1.3.1　Spring Cloud 分布式事务

微服务倡导将复杂的单体应用拆分为若干个功能简单、松耦合的服务，这样可以降低开发难度，增强扩展性，便于敏捷开发。当前微服务被越来越多的开发者推崇，很多互联网行业巨头、开源社区等都开始了微服务的讨论和实践。很多中小型互联网公司，由于经验、技术实力等问题，想要让微服务落地还比较困难。如著名架构师 Chris Richardson 所言，目前存在的主要困难有如下几方面：

1）单体应用拆分为分布式系统后，进程间的通信机制和故障处理措施变得更加复杂。

2）系统微服务化后，一个看似简单的功能，内部可能需要调用多个服务并操作多个数据库实现，服务调用的分布式事务问题变得非常突出。

3）微服务数量众多，其测试、部署、监控等都变得更加困难。

4）随着 RPC 框架的成熟，第一个问题已经逐渐得到解决。例如 HSF、Dubbo 可以支持多种通信协议，Spring Cloud 可以非常好地支持 RESTful 调用。

对于第三个问题，随着 Docker、Devops 技术的发展以及各公有云 PAAS 平台自动化运维工具的推出，微服务的测试、部署与运维变得越来越容易。而对于第二个问题，现在还没有一个通用方案可以很好地解决微服务产生的事务问题。分布式事务问题已经成为微服务落地最大的阻碍，也是最具挑战性的一个技术难题。在本书的第 24 章，会通过案例介绍 Spring Cloud 与分布式事务的各种解决方案。

1.3.2　Spring Cloud 与领域驱动

Spring Cloud 组件很多，更像是一套中间件体系，是实现微服务架构的基础实施。Spring Cloud 作为微服务架构的基础设施，能够快速帮助企业开发者搭建微服务架构。Spring Cloud

解决了框架层面的问题，但是对于业务怎么开发，业务架构怎么治理，架构怎么防腐，怎么解决应用架构的复杂性问题，还需要方法论去指导实践。所以在微服务架构落地的过程中，Spring Cloud 和领域驱动相辅相成显得十分重要。Spring Cloud 解决架构分布式等问题，领域驱动作为业务治理和架构防腐的方法论，两者并驾齐驱，为企业 IT 架构变革与微服务改造保驾护航。在本书的第 25 章将会详细介绍领域驱动 Halo 框架的设计。Halo 框架无缝整合了 Spring Cloud，目前支持 Spring Cloud Finchley 版本，通过案例的方式讲解如何运用 Spring Cloud 和领域驱动来进行业务架构治理和代码防腐。

1.3.3　Spring Cloud 与 gRPC

通过 Spring Cloud 构建微服务应用，大多数开发者使用官方提供的服务调用组件 Feign 来进行内部服务调用通信，这种声明式的 HTTP 客户端使用起来极为简单、优雅、方便。然而 Feign 的底层调用实现走的还是 HTTP 协议，相对于 Dubbo、gRPC 等 RPC 框架走 RPC 协议来说，通过 HTTP 来进行服务之间的调用，性能相对低下。那么我们是否可以通过 Spring Cloud 集成其他 RPC 框架来实现服务之间的高性能调用，答案是肯定的。在本书的第 19 章至 20 章将为大家讲解如何在 Spring Cloud 中集成 Google 开源框架 gRPC。

1.3.4　Spring Cloud 与 Dubbo 生态融合

在微服务架构的实施和落地过程中，我们通常会进行技术选型，做一些对比。很多人都会拿阿里开源的 Dubbo 和 Spring Cloud 进行对比，其本质对比的主要是 REST 和 RPC。其实 Dubbo 和 Spring Cloud 并不在同一个领域，没有可比性。因为 Spring Cloud 是一个完整的微服务解决方案，提供分布式情况下的各种解决方案合集。而 Dubbo 是一款高性能 Java RPC 框架。Spring Cloud 生态与 Dubbo 生态随着发展将会逐渐融合互补。

Spring Cloud 的设计理念是 Integrate Everything，即充分利用现有开源组件，在它们之上设计一套统一规范 / 接口使它们能够接入 Spring Cloud 体系并且能够无缝切换底层实现。最典型的例子就是 DiscoveryClient，只要实现 DiscoveryClient 相关接口，Spring Cloud 的底层注册中心就可以随意更换，Dubbo 的注册中心也有 SPI 规范进行替换。

在 2018 年 6 月 Spring Cloud 中国社区开源了一个名为 spring-cloud-dubbo 项目，该项目的目标是将 Dubbo 融入 Spring Cloud 生态体系中，使微服务之间的调用同时具备 RESTful 和 Dubbo 调用的能力，做到对业务代码无侵入、无感知。若在使用过程中引入 jar 包则在微服务间调用时使用 Dubbo，去掉 jar 包则使用默认的 RESTful。在本书的第 23 章，将会详细讲解 spring-cloud-dubbo 的设计与使用，为 Dubbo 向 Spring Cloud 的无缝迁移提供统一的方法论并指导落地。

1.4　本章小结

本章开始介绍了应用架构发展的历程，讲解了什么是微服务架构，然后针对微服务架构的落地提出了解决方案。接着介绍了什么是中间件，从中间件的角度介绍了 Spring Cloud 的主要组件和类似中间件的区别和联系。为了更好地通过 Spring Cloud 落地微服务架构，介绍了 Spring Cloud 的增强生态。

Chapter 2 第 2 章

Spring Cloud Eureka 上篇

Netflix Eureka（后文简称 Eureka）是由 Netflix 开源的一款基于 REST 的服务发现组件，包括 Eureka Server 及 Eureka Client。从 2012 年 9 月在 GitHub 上发布 1.1.2 版本以来，至今已经发布了 231 次，最新版本为 2018 年 8 月份发布的 1.9.4 版本。期间有进行 2.x 版本的开发，不过由于各种原因内部已经冻结开发，目前还是以 1.x 版本为主。Spring Cloud Netflix Eureka 是 Pivotal 公司为了将 Netflix Eureka 整合于 Spring Cloud 生态系统提供的版本。

本章以 Spring Cloud Finchley 版本展开，对应 Eureka 的 1.9.2 版本，将系统全面地介绍服务发现的由来、Eureka 的核心概念及其设计理念以及在实际企业级开发中的应用技巧。

2.1 服务发现概述

2.1.1 服务发现由来

服务发现及注册中心或是名字服务（后文统一简称服务发现），不是凭空出现的，其演进与软件开发的架构方式的演进有密切关联，大致如下：

1. 单体架构时代

早期的互联网开发，多使用单体架构，服务自成一体，对于依赖的少数外部服务，会采用配置域名的方式访问，比如要使用外部短信供应商的短信发送接口，会使用 appId 和 appKey，调用该供应商的域名接口即可。

2. SOA 架构时代

随着 SOA 架构的流行，公司的内部服务开始从单体架构拆分为粒度较粗的服务化架构，这个时候，依赖的内部服务会比较多，那么内部的服务之间如何相互调用呢？以基于 HTTP 形式暴露服务为例，假设 A 服务部署在 3 台虚拟机上，这 3 个服务实例均有各自的独立内网 ip，此时假设 B 服务要调用 A 服务的接口服务，有几种方式。

方式一：A 服务把这 3 个服务实例的内网 ip 给到消费者 B 服务，这个时候 B 服务在没有 Client 端负载均衡技术的条件下，通常会在 B 服务自己的 Nginx 上配置 A 服务的 upstream，即将 A 服务的 3 个服务实例配置进去，比如：

```
upstream    servicea_api_servers {
    server    192.168.99.100:80 weight=3 max_fails=3 fail_timeout=20s;
    server    192.168.99.101:80 weight=1 max_fails=3 fail_timeout=20s;
    server    192.168.99.102:80 weight=4 max_fails=3 fail_timeout=20s;
}
##......
server {
    listen 80 default_server;
    server_name serviceb.com.cn;
    location /api/servicea/ {
        proxy_pass http://servicea_api_servers/api/ ;
    }
}
```

通过 B 服务自己的 Nginx 来维护 A 服务的具体实例 ip，这种方式缺点比较明显，那就是 B 服务耦合了 A 服务的实现细节，当 A 服务实例扩充或者 ip 地址变化的时候，其下游的消费者都需要去修改这些 ip，非常费劲。

方式二：为了解耦合，采用服务提供方 A 服务自己维护 ip 实例的方式，暴露统一的内网域名给消费者去消费，这样 B 服务只需要配置一个内网域名即可，比如：

```
server {
    listen 80 default_server;
    server_name serviceb.com.cn;
    location /api/servicea/ {
        proxy_pass http://servicea.com.cn/api/ ;
    }
}
```

而 A 服务自己的 Nginx 则自己维护 ip 实例，比如：

```
upstream    servicea_api_servers {
    server    192.168.99.100:80 weight=3 max_fails=3 fail_timeout=20s;
    server    192.168.99.101:80 weight=1 max_fails=3 fail_timeout=20s;
    server    192.168.99.102:80 weight=4 max_fails=3 fail_timeout=20s;
}
##......
server {
    listen 80 default_server;
    server_name servicea.com.cn;
    location /api/ {
        proxy_pass http://servicea_api_servers/api/ ;
    }
}
```

这样即实现了服务提供方与消费者之间的解耦，若 A 服务要变更实例 ip 地址，自己更改自身的 Nginx 配置即可。

3. 微服务时代

在微服务时代，底层运维方式发生了巨大的变化，随着 Docker 的流行，业务服务不再部署在固定的虚拟机上，其 ip 地址也不再固定，这个时候前面的解决方案就显得捉襟见肘了。针对这个问题，不同的思考方式提出了不同的解决方案，这里列举几个。

方案一： 以 Nginx 为例，在没有引入服务注册中心的时候，那就是手工或是通过脚本的方式，在部署的时候去更新 Nginx 的配置文件，然后 reload。抑或是使用 ngx_http_dyups_module 通过 rest api 来在运行时直接更新 upstream 而不需要 reload。

方案二： 将服务注册中心作为一个标配的分布式服务组件，网关等都从服务注册中心获取相关服务的实例信息，实现动态路由。比如 consul-template+Nginx 的方案，通过 consul 监听服务实例变化，然后更新 Nginx 的配置文件，通过 reload 实现服务列表更新。又拍云的 slardar 也是这个思路，不过不是通过 reload 的方式来，而是通过在运行时通过 consul 获取服务列表来实现动态 upstream 的路由。

由此可见，随着服务架构模式以及底层运维方式的变化，服务注册中心逐步在分布式系统架构中占据了一个重要的地位。

2.1.2　Eureka 简介

Eureka 是 Netflix 公司开源的一款服务发现组件，该组件提供的服务发现可以为负载均衡、failover 等提供支持，如图 2-1 所示。Eureka 包括 Eureka Server 及 Eureka Client。Eureka Server 提供 REST 服务，而 Eureka Client 则是使用 Java 编写的客户端，用于简化与 Eureka Server 的交互。

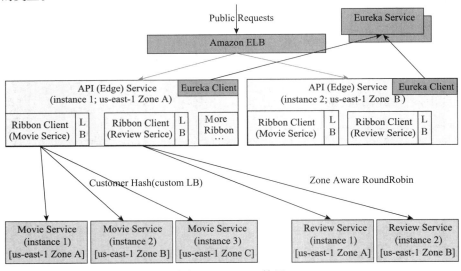

图 2-1　Eureka 简图

Eureka 最初是针对 AWS 不提供中间服务层的负载均衡的限制而设计开发的。AWS Elastic Load Balancer 用来对客户端或终端设备请求进行负载均衡，而 Eureka 则用来对中间层的服务做服务发现，配合其他组件提供负载均衡的能力。

Netflix 为什么要设计 Eureka，而不是直接利用 AWS Elastic Load Balancer 或者 AWS Route 53 呢？其官方文档说明简要如下：

理论上可以使用 AWS Elastic Load Balancer 对内部进行负载均衡，但是这样就会暴露到外网，存在安全性问题，另外 AWS Elastic Load Balancer 是传统的基于代理的负载均衡解决方案，无法直接基于服务元数据信息定制负载均衡算法。因此 Netflix 设计了 Eureka，一方面给内部服务做服务发现，另一方面可以结合 ribbon 组件提供各种个性化的负载均衡算法。

而 AWS Route 53 是一款命名服务，可以给中间层的服务提供服务发现功能，但它是基于 DNS 的服务，传统的基于 DNS 的负载均衡技术存在缓存更新延时问题，另外主要是无法对服务的健康状态进行检查，因此 Netflix 就自己设计了 Eureka。

2.1.3 服务发现技术选型

Jason Wilder 在 2014 年 2 月的时候写了一篇博客《Open-Source Service Discovery》（http://jasonwilder.com/blog/2014/02/04/service-discovery-in-the-cloud/），总结了当时市面上的几类服务发现组件，这里补充上 consul 以及一致性算法，如表 2-1 所示。

表 2-1 服务发现组件对比

名 称	类 型	AP 或 CP	语言	依 赖	集 成	一致性算法
Zookeeper	General	CP	Java	JVM	Client Binding	Paxos
Doozer	General	CP	Go		Client Binding	Paxos
Consul	General	CP	Go		HTTP/DNS Library	Raft
Etcd	General	CP or Mixed (1)	Go		Client Binding/HTTP	Raft
SmartStack	Dedicated	AP	Ruby	haproxy/Zookeeper	Sidekick (nerve/synapse)	
Eureka	Dedicated	AP	Java	JVM	Java Client	
NSQ (lookupd)	Dedicated	AP	Go		Client Binding	
Serf	Dedicated	AP	Go		Local CLI	
Spotify (DNS)	Dedicated	AP	N/A	Bind	DNS Library	
SkyDNS	Dedicated	Mixed (2)	Go		HTTP/DNS Library	

从列表看，有很多服务发现组件可以选择，针对 AP 及 CP，本书主要选取了 Eureka 及 Consul 为代表来阐述。关于 Eureka 及 Consul 的区别，Consul 的官方文档有一个很好的阐述（https://www.consul.io/intro/vs/eureka.html），具体如下：

Eureka Server 端采用的是 P2P 的复制模式，但是它不保证复制操作一定能成功，因此它提供的是一个最终一致性的服务实例视图；Client 端在 Server 端的注册信息有一个带期限的租约，一旦 Server 端在指定期间没有收到 Client 端发送的心跳，则 Server 端会认为 Client 端注册的服务是不健康的，定时任务会将其从注册表中删除。Consul 与 Eureka 不同，Consul 采用 Raft 算法，可以提供强一致性的保证，Consul 的 agent 相当于 Netflix Ribbon + Netflix Eureka

Client，而且对应用来说相对透明，同时相对于 Eureka 这种集中式的心跳检测机制，Consul 的 agent 可以参与到基于 gossip 协议的健康检查，分散了 Server 端的心跳检测压力。除此之外，Consul 为多数据中心提供了开箱即用的原生支持等。

那么基于什么考虑因素可以选择 Eureka 呢，主要有如下几点：
- 选择 AP 而不是 CP，这一点在后面的章节会阐述。
- 如果团队是 Java 语言体系的，则偏好 Java 语言开发的，技术体系上比较统一，出问题也好排查修复，对组件的掌控力较强，方便扩展维护。
- 当然除此之外，更主要的是 Eureka 是 Netflix 开源套件的一部分，跟 zuul，ribbon 等整合的比较好。

2.2 Spring Cloud Eureka 入门案例

下面来让我们体验一下 Eureka 的 Hello World 工程，这里需要用到的组件是 Spring Cloud Netflix Eureka。

1. 创建 Maven 父级 pom 工程

在父工程里面配置好工程需要的父级依赖，目的是为了方便管理与简化配置，如代码清单 2-1 所示。

代码清单2-1　ch2-1\pom.xml

```xml
<dependencyManagement>
    <dependencies>
        <dependency>
            <groupId>org.springframework.cloud</groupId>
            <artifactId>spring-cloud-dependencies</artifactId>
            <version>${spring-cloud.version}</version>
            <type>pom</type>
            <scope>import</scope>
        </dependency>
    </dependencies>
</dependencyManagement>

<dependencies>
    <dependency>
        <groupId>org.springframework.boot</groupId>
        <artifactId>spring-boot-starter-web</artifactId>
    </dependency>

    <dependency>
        <groupId>org.springframework.boot</groupId>
        <artifactId>spring-boot-starter-test</artifactId>
        <scope>test</scope>
    </dependency>
</dependencies>
```

2. 创建 Eureka Server 工程

为避免赘述,在本章中,Eureka 组件配置只展示一次,后续实战演练配置中心皆基于此。

配置 Eureka Server 工程的 pom.xml 文件,只需要添加 spring-cloud-starter-netflix-eureka-server 即可,注意 F 版和之前的版本有些变化,如代码清单 2-2 所示。

代码清单2-2　ch2-1\ch2-1-eureka-server\pom.xml

```xml
<dependencies>
    <dependency>
        <groupId>org.springframework.cloud</groupId>
        <artifactId>spring-cloud-starter-netflix-eureka-server</artifactId>
    </dependency>
</dependencies>
```

对于 Eureka 的启动主类,这里添加相应注解,作为程序的入口,如代码清单 2-3 所示。

代码清单2-3　ch2-1\ch2-1-eureka-server\src\main\java\cn\springcloud\book\Ch21EurekaServerApplication.java

```java
@SpringBootApplication
@EnableEurekaServer
public class Ch21EurekaServerApplication {

    public static void main(String[] args) {
        SpringApplication.run(Ch21EurekaServerApplication.class, args);
    }
}
```

Eureka Server 需要的配置文件,如代码清单 2-4 所示。

代码清单2-4　ch2-1\ch2-1-eureka-server\src\main\resources\application-standalone.yml

```yml
server:
    port: 8761

eureka:
    instance:
        hostname: localhost
    client:
        registerWithEureka: false
        fetchRegistry: false
        serviceUrl:
            defaultZone: http://${eureka.instance.hostname}:${server.port}/eureka/
    server:
        waitTimeInMsWhenSyncEmpty: 0
        enableSelfPreservation: false
```

这里的单机版配置,仅仅是为了演示,切勿用于生产。

3. 创建 Eureka Client 组件工程

配置 Eureka Client 工程的 pom.xml 文件,只需要引入 spring-cloud-starter-netflix-eureka-client 即可,注意也与之前版本不一样,如代码清单 2-5 所示。

代码清单2-5　ch2-1\ch2-1-eureka-client \pom.xml

```xml
<dependencies>
    <dependency>
        <groupId>org.springframework.cloud</groupId>
        <artifactId>spring-cloud-starter-netflix-eureka-client</artifactId>
    </dependency>
</dependencies>
```

添加 Eureka Client 的启动主类，如代码清单 2-6 所示。

代码清单2-6　ch2-1\ch2-1-eureka-client\src\main\java\cn\springcloud\book \Ch21Eureka-ClientApplication.java

```java
@SpringBootApplication
@EnableDiscoveryClient
public class Ch21EurekaClientApplication {

    public static void main(String[] args) {
        SpringApplication.run(Ch21EurekaClientApplication.class, args);
    }
}
```

Eureka Client 的配置文件，如代码清单 2-7 所示。

代码清单2-7　ch2-1\ch2-1-eureka-client\src\main\resources\application-demo.yml

```yaml
server:
    port: 8081

spring:
    application:
        name: demo-client1

eureka:
    client:
        serviceUrl:
            defaultZone: http://localhost:8761/eureka/
```

这里需要说明一下，需要指定 spring.application.name，不然会在 Eureka Server 界面显示为 UNKNOWN。

4. 效果展示

分别启动 eureka-server 及 eureka-client，然后访问 http://localhost:8761，结果如图 2-2 所示。

通过访问 Eureka Server 的 rest api 接口，比如 http://localhost:8761/eureka/apps，返回的结果如下：

```xml
<applications>
    <versions__delta>1</versions__delta>
    <apps__hashcode>UP_1_</apps__hashcode>
    <application>
```

```xml
            <name>DEMO-CLIENT1</name>
            <instance>
                <instanceId>10.2.238.223:demo-client1:8081</instanceId>
                <hostName>10.2.238.223</hostName>
                <app>DEMO-CLIENT1</app>
                <ipAddr>10.2.238.223</ipAddr>
                <status>UP</status>
                <overriddenstatus>UNKNOWN</overriddenstatus>
                <port enabled="true">8081</port>
                <securePort enabled="false">443</securePort>
                <countryId>1</countryId>
                <dataCenterInfo class="com.netflix.appinfo.InstanceInfo$DefaultDataC
                    enterInfo">
                    <name>MyOwn</name>
                </dataCenterInfo>
                <leaseInfo>
                    <renewalIntervalInSecs>30</renewalIntervalInSecs>
                    <durationInSecs>90</durationInSecs>
                    <registrationTimestamp>1529377552795</registrationTimestamp>
                    <lastRenewalTimestamp>1529386376662</lastRenewalTimestamp>
                        <evictionTimestamp>0</evictionTimestamp>
                        <serviceUpTimestamp>1529377552795</serviceUpTimestamp>
                </leaseInfo>
                <metadata>
                    <management.port>8081</management.port>
                </metadata>
                <homePageUrl>http://10.2.238.223:8081/</homePageUrl>
                <statusPageUrl>http://10.2.238.223:8081/actuator/info</statusPageUrl>
                <healthCheckUrl>http://10.2.238.223:8081/actuator/health</
                    healthCheckUrl>
                <vipAddress>demo-client1</vipAddress>
                <secureVipAddress>demo-client1</secureVipAddress>
                <isCoordinatingDiscoveryServer>false</isCoordinatingDiscoveryServer>
                <lastUpdatedTimestamp>1529377552795</lastUpdatedTimestamp>
                <lastDirtyTimestamp>1529377552665</lastDirtyTimestamp>
                <actionType>ADDED</actionType>
            </instance>
        </application>
</applications>
```

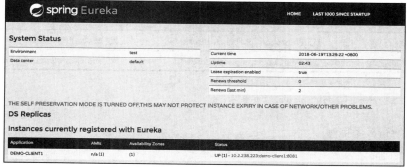

图 2-2　eureka-server 管理界面

2.3 Eureka Server 的 REST API 简介

前面已经介绍了一个最基本的 Eureka Server 及 Eureka Client，它是整个 Spring Cloud 生态里面"服务注册及发现"的一个缩影。这里举的是使用 Java 语言的 Client 端的例子，但是实际 Eureka Server 提供了 REST API，允许非 Java 语言的其他应用服务通过 HTTP REST 的方式接入 Eureka 的服务发现中。本节主要介绍下 Eureka Server 的 REST API 的基本操作。

2.3.1 REST API 列表

Eureka 在 GitHub 的 wiki 上专门写了一篇《Eureka REST operations》来介绍 Eureka Server 的 REST API 接口，Spring Cloud Netflix Eureka 跟 Spring Boot 适配之后，提供的 REST API 与原始的 REST API 有一点点不同，其路径中的 {version} 值固定为 eureka，其他的变化不大，如表 2-2 所示。

表 2-2 Eureka REST API 接口列表

操作	http 动作	描述
注册新的应用实例	POST /eureka/apps/{appId}	可以输入 json 或 xml 格式的 body，成功返回 204
注销应用实例	DELETE /eureka/apps/{appId}/{instanceId}	成功返回 200
应用实例发送心跳	PUT /eureka/apps/{appId}/{instanceId}	成功返回 200，如果 instanceId 不存在返回 404
查询所有实例	GET /eureka/apps	成功返回 200，输出 json 或 xml 格式
查询指定 appId 的实例	GET /eureka/apps/{appId}	成功返回 200，输出 json 或 xml 格式
根据指定 appId 和 instanceId 查询	GET /eureka/apps/{appId}/{instanceId}	成功返回 200，输出 json 或 xml 格式
根据指定 instanceId 查询	GET/eureka/instances/{instanceId}	成功返回 200，输出 json 或 xml 格式
暂停应用实例	PUT/eureka/apps/{appId}/{instanceId}/status?value=OUT_OF_SERVICE	成功返回 200，失败返回 500
恢复应用实例	DELETE /eureka/apps/{appId}/{instanceId}/status?value=UP（value 参数可不传）	成功返回 200，失败返回 500
更新元数据	PUT/eureka/apps/{appId}/{instanceId}/metadata?key=value	成功返回 200，失败返回 500
根据 vip 地址查询	GET/eureka/vips/{vipAddress}	成功返回 200，输出 json 或 xml 格式
根据 svip 地址查询	GET/eureka/svips/{svipAddress}	成功返回 200，输出 json 或 xml 格式

2.3.2 REST API 实例

1. 查询所有应用实例

```
HTTP/1.1 200
Content-Type: application/xml
Transfer-Encoding: chunked
```

```
Date: Tue, 19 Jun 2018 06:22:56 GMT

<applications>
    <versions__delta>1</versions__delta>
    <apps__hashcode>UP_1_</apps__hashcode>
    <application>
        <name>DEMO-CLIENT1</name>
        <instance>
            <instanceId>10.2.238.223:demo-client1:8081</instanceId>
            <hostName>10.2.238.223</hostName>
            <app>DEMO-CLIENT1</app>
            <ipAddr>10.2.238.223</ipAddr>
            <status>UP</status>
            <overriddenstatus>UNKNOWN</overriddenstatus>
            <port enabled="true">8081</port>
            <securePort enabled="false">443</securePort>
            <countryId>1</countryId>
            <dataCenterInfo class="com.netflix.appinfo.InstanceInfo$DefaultDataC
                enterInfo">
                <name>MyOwn</name>
            </dataCenterInfo>
            <leaseInfo>
                <renewalIntervalInSecs>30</renewalIntervalInSecs>
                <durationInSecs>90</durationInSecs>
                <registrationTimestamp>1529377552795</registrationTimestamp>
    <lastRenewalTimestamp>1529389437739</lastRenewalTimestamp>
                <evictionTimestamp>0</evictionTimestamp>
                <serviceUpTimestamp>1529377552795</serviceUpTimestamp>
            </leaseInfo>
            <metadata>
                <management.port>8081</management.port>
            </metadata>
            <homePageUrl>http://10.2.238.223:8081/</homePageUrl>
<statusPageUrl>http://10.2.238.223:8081/actuator/info</statusPageUrl>
<healthCheckUrl>http://10.2.238.223:8081/actuator/health</healthCheckUrl>
            <vipAddress>demo-client1</vipAddress>
            <secureVipAddress>demo-client1</secureVipAddress>
<isCoordinatingDiscoveryServer>false</isCoordinatingDiscoveryServer>
            <lastUpdatedTimestamp>1529377552795</lastUpdatedTimestamp>
            <lastDirtyTimestamp>1529377552665</lastDirtyTimestamp>
            <actionType>ADDED</actionType>
        </instance>
    </application>
</applications>
```

2. 根据 appId 查询

```
curl -i http://localhost:8761/eureka/apps/demo-client1

HTTP/1.1 200
Content-Type: application/xml
```

```
Transfer-Encoding: chunked
Date: Tue, 19 Jun 2018 06:28:34 GMT

<application>
    <name>DEMO-CLIENT1</name>
    <instance>
        <instanceId>10.2.238.223:demo-client1:8081</instanceId>
        <hostName>10.2.238.223</hostName>
        <app>DEMO-CLIENT1</app>
        <ipAddr>10.2.238.223</ipAddr>
        <status>UP</status>
        <overriddenstatus>UNKNOWN</overriddenstatus>
        <port enabled="true">8081</port>
        <securePort enabled="false">443</securePort>
        <countryId>1</countryId>
        <dataCenterInfo class="com.netflix.appinfo.InstanceInfo$DefaultDataCenterInfo">
            <name>MyOwn</name>
        </dataCenterInfo>
        <leaseInfo>
            <renewalIntervalInSecs>30</renewalIntervalInSecs>
            <durationInSecs>90</durationInSecs>
            <registrationTimestamp>1529377552795</registrationTimestamp>
            <lastRenewalTimestamp>1529389797866</lastRenewalTimestamp>
            <evictionTimestamp>0</evictionTimestamp>
            <serviceUpTimestamp>1529377552795</serviceUpTimestamp>
        </leaseInfo>
        <metadata>
            <management.port>8081</management.port>
        </metadata>
        <homePageUrl>http://10.2.238.223:8081/</homePageUrl>
        <statusPageUrl>http://10.2.238.223:8081/actuator/info</statusPageUrl>
        <healthCheckUrl>http://10.2.238.223:8081/actuator/health</healthCheckUrl>
        <vipAddress>demo-client1</vipAddress>
        <secureVipAddress>demo-client1</secureVipAddress>
        <isCoordinatingDiscoveryServer>false</isCoordinatingDiscoveryServer>
        <lastUpdatedTimestamp>1529377552795</lastUpdatedTimestamp>
        <lastDirtyTimestamp>1529377552665</lastDirtyTimestamp>
        <actionType>ADDED</actionType>
    </instance>
</application>
```

查询不到则返回如下结果:

```
curl -i http://localhost:8761/eureka/apps/demo
HTTP/1.1 404
Content-Type: application/xml
Content-Length: 0
Date: Tue, 19 Jun 2018 06:38:14 GMT
```

3. 根据 appId 及 instanceId 查询

查询出数据时跟上面的返回一致,查找不到则返回如下结果:

```
curl -i http://localhost:8761/eureka/apps/demo-client1/notfound
HTTP/1.1 404
```

```
Content-Type: application/xml
Content-Length: 0
Date: Tue, 19 Jun 2018 06:33:42 GMT
```

4. 根据 instanceId 查询

```
curl -i http://localhost:8761/eureka/instances/10.2.238.223:demo-client1:8081
HTTP/1.1 200
Content-Type: application/xml
Transfer-Encoding: chunked
Date: Tue, 19 Jun 2018 06:36:09 GMT

<instance>
    <instanceId>10.2.238.223:demo-client1:8081</instanceId>
    <hostName>10.2.238.223</hostName>
    <app>DEMO-CLIENT1</app>
    <ipAddr>10.2.238.223</ipAddr>
    <status>UP</status>
    <overriddenstatus>UNKNOWN</overriddenstatus>
    <port enabled="true">8081</port>
    <securePort enabled="false">443</securePort>
    <countryId>1</countryId>
    <dataCenterInfo class="com.netflix.appinfo.InstanceInfo$DefaultDataCenterInfo">
        <name>MyOwn</name>
    </dataCenterInfo>
    <leaseInfo>
        <renewalIntervalInSecs>30</renewalIntervalInSecs>
        <durationInSecs>90</durationInSecs>
        <registrationTimestamp>1529377552795</registrationTimestamp>
        <lastRenewalTimestamp>1529390248015</lastRenewalTimestamp>
        <evictionTimestamp>0</evictionTimestamp>
        <serviceUpTimestamp>1529377552795</serviceUpTimestamp>
    </leaseInfo>
    <metadata>
        <management.port>8081</management.port>
    </metadata>
    <homePageUrl>http://10.2.238.223:8081/</homePageUrl>
    <statusPageUrl>http://10.2.238.223:8081/actuator/info</statusPageUrl>
<healthCheckUrl>http://10.2.238.223:8081/actuator/health</healthCheckUrl>
    <vipAddress>demo-client1</vipAddress>
    <secureVipAddress>demo-client1</secureVipAddress>
    <isCoordinatingDiscoveryServer>false</isCoordinatingDiscoveryServer>
    <lastUpdatedTimestamp>1529377552795</lastUpdatedTimestamp>
    <lastDirtyTimestamp>1529377552665</lastDirtyTimestamp>
    <actionType>ADDED</actionType>
</instance>
```

查询不到则返回如下结果：

```
curl -i http://localhost:8761/eureka/instances/demo-instance-id
HTTP/1.1 404
Content-Type: application/xml
Content-Length: 0
Date: Tue, 19 Jun 2018 06:36:51 GMT
```

5. 注册新应用实例

xml 格式提交的实例如下：

```
curl -i -H "Content-Type: application/xml" -H "Content-Length: 773" -H "Accept-
    Encoding: gzip" -X POST -d '<instance>
        <instanceId>client2:8082</instanceId>
        <hostName>127.0.0.1</hostName>
        <app>CLIENT2</app>
        <ipAddr>127.0.0.1</ipAddr>
        <status>UP</status>
        <overriddenstatus>UNKNOWN</overriddenstatus>
        <port enabled="true">8082</port>
        <securePort enabled="false">7002</securePort>
        <countryId>1</countryId>
        <dataCenterInfo class="com.netflix.appinfo.InstanceInfo$DefaultDataCenterInfo">
            <name>MyOwn</name>
        </dataCenterInfo>
        <metadata class="java.util.Collections$EmptyMap"/>
        <vipAddress>client2</vipAddress>
        <secureVipAddress>client2</secureVipAddress>
        <isCoordinatingDiscoveryServer>false</isCoordinatingDiscoveryServer>
        <lastUpdatedTimestamp>1529397327986</lastUpdatedTimestamp>
        <lastDirtyTimestamp>1529397327986</lastDirtyTimestamp>
</instance>' http://localhost:8761/eureka/apps/client2
HTTP/1.1 204
Content-Type: application/xml
Date: Tue, 19 Jun 2018 08:35:59 GMT
```

json 格式提交的实例如下：

```
curl -i -H "Content-Type: application/json" -X POST -d '{
    "instance": {
        "instanceId": "client2:8082",
        "app": "client2",
        "appGroupName": null,
        "ipAddr": "127.0.0.1",
        "sid": "na",
        "homePageUrl": null,
        "statusPageUrl": null,
        "healthCheckUrl": null,
        "secureHealthCheckUrl": null,
        "vipAddress": "client2",
        "secureVipAddress": "client2",
        "countryId": 1,
        "dataCenterInfo": {
            "@class": "com.netflix.appinfo.InstanceInfo$DefaultDataCenterInfo",
            "name": "MyOwn"
        },
        "hostName": "127.0.0.1",
        "status": "UP",
        "leaseInfo": null,
        "isCoordinatingDiscoveryServer": false,
        "lastUpdatedTimestamp": 1529391461000,
        "lastDirtyTimestamp": 1529391461000,
        "actionType": null,
```

```
        "asgName": null,
        "overridden_status": "UNKNOWN",
        "port": {
            "$": 8082,
            "@enabled": "true"
        },
        "securePort": {
            "$": 7002,
            "@enabled": "false"
        },
        "metadata": {
            "@class": "java.util.Collections$EmptyMap"
        }
    }
}' http://localhost:8761/eureka/apps/client2
HTTP/1.1 204
Content-Type: application/xml
Date: Tue, 19 Jun 2018 06:58:18 GMT
```

6. 注销应用实例

```
curl -i -X DELETE http://localhost:8761/eureka/apps/client2/client2:8082
HTTP/1.1 200
Content-Type: application/xml
Content-Length: 0
Date: Tue, 19 Jun 2018 06:59:31 GMT
```

7. 暂停 / 下线应用实例

成功则返回：

```
curl -i -X PUT http://localhost:8761/eureka/apps/demo-client1/10.2.238.223:demo-client1:8081/status\?value\=OUT_OF_SERVICE
HTTP/1.1 200
Content-Type: application/xml
Content-Length: 0
Date: Tue, 19 Jun 2018 07:12:24 GMT
```

此时 Eureka 界面示例如图 2-3 所示。

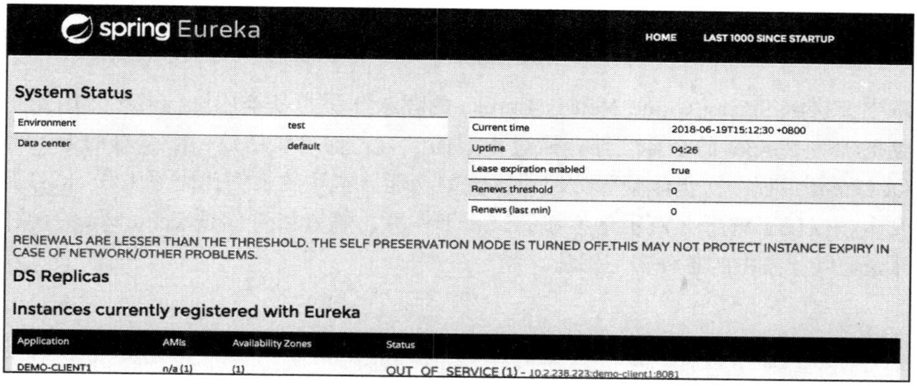

图 2-3　下线服务时 Eureka 界面展示示例

找不到该实例返回：

```
curl -i -X PUT http://localhost:8761/eureka/apps/demo-client1/10.2.238.223:demo-
    client1:8081/status\?value\=OUT_OF_SERVICE
HTTP/1.1 404
Content-Type: application/xml
Content-Length: 0
Date: Tue, 19 Jun 2018 07:11:17 GMT
```

8. 恢复应用实例

```
curl -i -X DELETE http://localhost:8761/eureka/apps/demo-client1/10.2.238.223:demo-
    client1:8081/status
HTTP/1.1 200
Content-Type: application/xml
Content-Length: 0
Date: Tue, 19 Jun 2018 07:14:03 GMT
```

9. 应用实例发送心跳

```
curl -i -X PUT http://localhost:8761/eureka/apps/demo-client1/10.2.238.223:demo-
    client1:8081
HTTP/1.1 200
Content-Type: application/xml
Content-Length: 0
Date: Tue, 19 Jun 2018 07:16:38 GMT
```

10. 修改应用实例元数据

```
curl -i -X PUT http://localhost:8761/eureka/apps/demo-client1/10.2.238.223:demo-
    client1:8081/metadata\?profile\=canary
HTTP/1.1 200
Content-Type: application/xml
Content-Length: 0
Date: Tue, 19 Jun 2018 07:18:18 GMT
```

设置 metadata，其中 key 为 proflie，value 为 canary。

2.4　本章小结

本章主要讲解 Spring Cloud Netflix Eureka 的背景知识与基本用法，对于入门是十分有帮助的，也是学习 Eureka 的必经之路。通过本章的学习，你可以搭建一个基本的 Eureka Server 及 Eureka Client 工程，初步体验服务注册发现的功能。本章的示例代码见工程 ch2-1。在实际业务开发中，仅仅了解怎么使用是不够的，在下一章，笔者将带你领略 Eureka 的设计原理及其在实际企业级开发中的各种实用用法。

第 3 章 Chapter 3

Spring Cloud Eureka 下篇

前面一章介绍了服务发现及 Eureka 的由来，同时展示了最基础的如何搭建 Eureka Server 及 Client 的实例工程，也介绍了 Eureka Server 的基本 REST API 操作。本章将详细介绍 Eureka 更深层次的知识，其中包括 Eureka 的核心设计类、设计思想、参数调优及故障演练。学习完本章后，相信你会在 Eureka 的使用上更加得心应手。

3.1 Eureka 的核心类

3.1.1 InstanceInfo

Eureka 使用 InstanceInfo（com/netflix/appinfo/InstanceInfo.java）来代表注册的服务实例，其主要字段如表 3-1 所示。

表 3-1　InstanceInfo 类字段说明

字　　段	说　　明
instanceId	实例 id
app	应用名
appGroupName	应用所属群组
ipAddr	ip 地址
sid	被废弃的属性，默认 na
port	端口号
securePort	https 的端口号
homePageUrl	应用实例的首页 url

(续)

字段	说明
statusPageUrl	应用实例的状态页 url
healthCheckUrl	应用实例健康检查的 url
secureHealthCheckUrl	应用实例健康检查的 https 的 url
vipAddress	虚拟 ip 地址
secureVipAddress	https 的虚拟 ip 地址
countryId	被废弃的属性,默认为 1,代表 US
dataCenterInfo	dataCenter 信息,Netflix 或者 Amazon 或者 MyOwn
hostName	主机名称
status	实例状态,如 UP、DOWN、STARTING、OUT_OF_SERVICE、UNKNOWN
overriddenstatus	外界需要强制覆盖的状态值,默认为 UNKNOWN
leaseInfo	租约信息
isCoordinatingDiscoveryServer	首先标识是否是 discoveryServer,其实标识该 discoveryServer 是否是响应你请求的实例
metadata	应用实例的元数据信息
lastUpdatedTimestamp	状态信息最后更新时间
lastDirtyTimestamp	实例信息最新的过期时间,在 Client 端用于标识该实例信息是否与 Eureka Server 一致,在 Server 端则用于多个 Eureka Server 之间的信息同步处理
actionType	标识 Eureka Server 对该实例执行的操作,包括 ADDED、MODIFIED、DELETED 这三类
asgName	在 AWS 的 autoscaling group 的名称

可以看到 InstanceInfo 里既有 metadata,也有 dataCenterInfo,还有一个比较重要的 leaseInfo,用来标识该应用实例的租约信息。

3.1.2 LeaseInfo

Eureka 使用 LeaseInfo(com/netflix/appinfo/LeaseInfo.java)来标识应用实例的租约信息,其字段如表 3-2 所示。

表 3-2 LeaseInfo 类字段说明

字段	说明
renewalIntervalInSecs	Client 端续约的间隔周期
durationInSecs	Client 端需要设定的租约的有效时长
registrationTimestamp	Server 端设置的该租约的第一次注册时间
lastRenewalTimestamp	Server 端设置的该租约的最后一次续约时间
evictionTimestamp	Server 端设置的该租约被剔除的时间
serviceUpTimestamp	Server 端设置的该服务实例标记为 UP 的时间

这些参数主要用于标识应用实例的心跳情况，比如约定的心跳周期，租约有效期，最近一次续约的时间等。

3.1.3 ServiceInstance

ServiceInstance（org/springframework/cloud/client/ServiceInstance.java）是 Spring Cloud 对 service discovery 的实例信息的抽象接口，约定了服务发现的实例应用有哪些通用的信息，其方法如表 3-3 所示。

表 3-3　ServiceInstance 接口方法列表

方　　法	说　　明
getServiceId()	服务 id
getHost()	实例的 host
getPort()	实例的端口
isSecure()	实例是否开启 https
getUri()	实例的 uri 地址
getMetadata()	实例的元数据信息
getScheme()	实例的 scheme

由于 Spring Cloud Discovery 适配了 Zookeeper、Consul、Netflix Eureka 等注册中心，因此其 ServiceInstance 定义更为抽象和通用，而且采取的是定义方法的方式。Spring Cloud 对该接口的实现类为 EurekaRegistration（org/springframework/cloud/netflix/eureka/serviceregistry/EurekaRegistration.java），EurekaRegistration 实现了 ServiceInstance 接口，同时还实现了 Closeable 接口，它的作用之一就是在 close 的时候调用 eurekaClient.shutdown() 方法，实现优雅关闭 Eureka Client。

3.1.4 InstanceStatus

InstanceStatus 用于标识服务实例的状态，它是一个枚举，定义如下：

```
public enum InstanceStatus {
    UP, // Ready to receive traffic
    DOWN, // Do not send traffic- healthcheck callback failed
    STARTING, // Just about starting- initializations to be done - do not
    // send traffic
    OUT_OF_SERVICE, // Intentionally shutdown for traffic
    UNKNOWN;

    public static InstanceStatus toEnum(String s) {
        if (s != null) {
            try {
                return InstanceStatus.valueOf(s.toUpperCase());
            } catch (IllegalArgumentException e) {
                // ignore and fall through to unknown
```

```
                logger.debug("illegal argument supplied to InstanceStatus.
                    valueOf: {}, defaulting to {}", s, UNKNOWN);
            }
        }
        return UNKNOWN;
    }
}
```

从定义可以看出，服务实例主要有 UP、DOWN、STARTING、OUT_OF_SERVICE、UNKNOWN 这几个状态。其中 OUT_OF_SERVICE 标识停止服务，即停止接收请求，处于这个状态的服务实例将不会被路由到，经常用于升级部署的场景。

3.2 服务的核心操作

3.2.1 概述

对于服务发现来说，围绕服务实例主要有如下几个重要的操作：
- 服务注册（register）
- 服务下线（cancel）
- 服务租约（renew）
- 服务剔除（evict）

围绕这几个功能，Eureka 设计了几个核心操作类：
- com/netflix/eureka/lease/LeaseManager.java
- com/netflix/discovery/shared/LookupService.java
- com/netflix/eureka/registry/InstanceRegistry.java
- com/netflix/eureka/registry/AbstractInstanceRegistry.java
- com/netflix/eureka/registry/PeerAwareInstanceRegistryImpl.java

Spring Cloud Eureka 在 Netflix Eureka 的基础上，抽象或定义了如下几个核心类：
- org/springframework/cloud/netflix/eureka/server/InstanceRegistry.java
- org/springframework/cloud/client/serviceregistry/ServiceRegistry.java
- org/springframework/cloud/netflix/eureka/serviceregistry/EurekaServiceRegistry.java
- org/springframework/cloud/netflix/eureka/serviceregistry/EurekaRegistration.java
- org/springframework/cloud/netflix/eureka/EurekaClientAutoConfiguration.java
- org/springframework/cloud/netflix/eureka/EurekaClientConfigBean.java
- org/springframework/cloud/netflix/eureka/EurekaInstanceConfigBean.java

其中 LeaseManager 以及 LookupService 是 Eureka 关于服务发现相关操作定义的接口类，前者定义了服务写操作相关的方法，后者定义了查询操作相关的方法。下面我们重点看下这两个类。

3.2.2 LeaseManager

LeaseManager（com/netflix/eureka/lease/LeaseManager.java）接口定义了应用服务实例在服

务中心的几个操作方法：register、cancel、renew、evict。其接口源码如下：

```
public interface LeaseManager<T> {
    void register(T r, int leaseDuration, boolean isReplication);
    boolean cancel(String appName, String id, boolean isReplication);
    boolean renew(String appName, String id, boolean isReplication);
    void evict();
}
```

这里简单介绍下这几个方法：

- Register：用于注册服务实例信息。
- Cancel：用于删除服务实例信息。
- Renew：用于与 Eureka Server 进行心跳操作，维持租约。
- evict 是 Server 端的一个方法，用于剔除租约过期的服务实例信息。

3.2.3 LookupService

LookupService（com/netflix/discovery/shared/LookupService.java）接口定义了 Eureka Client 从服务中心获取服务实例的查询方法。其定义如下：

```
public interface LookupService<T> {
    Application getApplication(String appName);
    Applications getApplications();
    List<InstanceInfo> getInstancesById(String id);
    InstanceInfo getNextServerFromEureka(String virtualHostname, boolean
        secure);
}
```

这个接口主要是给 Client 端用的，其定义了获取所有应用信息、根据应用 id 获取所有服务实例，以及根据 visualHostname 使用 round-robin 方式获取下一个服务实例的方法。

3.3 Eureka 的设计理念

3.3.1 概述

作为一个服务注册及发现中心，主要解决如下几个问题：

1. 服务实例如何注册到服务中心

本质上就是在服务启动的时候，需要调用 Eureka Server 的 REST API 的 register 方法，去注册该应用实例的信息。对于使用 Java 的应用服务，可以使用 Netflix 的 Eureka Client 封装的 API 去调用；对于 Spring Cloud 的应用，可以使用 spring-cloud-starter-netflix-eureka-client，基于 Spring Boot 的自动配置，自动帮你实现服务信息的注册。

2. 服务实例如何从服务中心剔除

正常情况下服务实例在关闭应用的时候，应该通过钩子方法或其他生命周期回调方法去调用 Eureka Server 的 REST API 的 de-register 方法，来删除自身服务实例的信息。另外为了解决服务实例挂掉或其他异常情况没有及时删除自身信息的问题，Eureka Server 要求 Client 端定时进行续约，也就是发送心跳，来证明该服务实例还是存活的、是健康的、是可以调用的。如果租约超过一定时间没有进行续约操作，Eureka Server 端会主动剔除。这一点 Eureka Server 采用的就是分布式应用里头经典的心跳模式。

3. 服务实例信息的一致性问题

由于服务注册及发现中心不可能是单点的，其自身势必有个集群，那么服务实例注册信息如何在这个集群里保持一致呢？这跟 Eureka Server 的架构有关，理解其设计理念有助于后面的实战及调优，下面主要分 AP 优于 CP、Peer to Peer 架构、Zone 及 Region 设计、SELF PRESERVATION 设计四个方面来阐述。

3.3.2 AP 优于 CP

分布式系统领域有个重要的 CAP 理论，该理论由加州大学伯克利分校的 Eric Brewer 教授提出，由麻省理工学院的 Seth Gilbert 和 Nancy Lynch 进行理论证明。该理论提到了分布式系统的 CAP 三个特性：

- ❑ Consistency：数据一致性，即数据在存在多副本的情况下，可能由于网络、机器故障、软件系统等问题导致数据写入部分副本成功，部分副本失败，进而造成副本之间数据不一致，存在冲突。满足一致性则要求对数据的更新操作成功之后，多副本的数据保持一致。
- ❑ Availability：在任何时候客户端对集群进行读写操作时，请求能够正常响应，即在一定的延时内完成。
- ❑ Partition Tolerance：分区容忍性，即发生通信故障的时候，整个集群被分割为多个无法相互通信的分区时，集群仍然可用。

对于分布式系统来说，一般网络条件相对不可控，出现网络分区是不可避免的，因此系统必须具备分区容忍性。在这个前提下分布式系统的设计则在 AP 及 CP 之间进行选择。不过不能理解为 CAP 三者之间必须三选二，它们三者之间不是对等和可以相互替换的。在分布式系统领域，P 是一个客观存在的事实，不可绕过，所以 P 与 AC 之间不是对等关系。

对于 ZooKeeper，它是 "C"P 的，之所以 C 加引号是因为 ZooKeeper 默认并不是严格的强一致，比如客户端 A 提交一个写操作，ZooKeeper 在过半数节点操作成功之后就返回，此时假设客户端 B 的读操作请求到的是 A 写操作尚未同步到的节点，那么读取到的就不是客户端 A 写操作成功之后的数据。如果在使用的时候需要强一致，则需要在读取数据的时候先执行一下 sync 操作，即与 leader 节点先同步下数据，这样才能保证强一致。在极端的情况下发生网络分区的时候，如果 leader 节点不在 non-quorum 分区，那么对这个分区上节点的读写请求将会报错，无法满足 Availability 特性。

Eureka 是在部署在 AWS 的背景下设计的，其设计者认为，在云端，特别是在大规模部署的情况下，失败是不可避免的，可能因为 Eureka 自身部署失败，注册的服务不可用，或者由于网络分区导致服务不可用，因此不能回避这个问题。要拥抱这个问题，就需要 Eureka 在网络分区的时候,还能够正常提供服务注册及发现功能,因此 Eureka 选择满足 Availability 这个特性。Peter Kelley 在《Eureka! Why You Shouldn't Use ZooKeeper for Service Discovery》(https://medium.com/knerd/eureka-why-you-shouldnt-use-zookeeper-for-service-discovery-4932c5c7e764)一文中指出,在实际生产实践中,服务注册及发现中心保留可用及过期的数据总比丢失掉可用的数据好。这样的话,应用实例的注册信息在集群的所有节点间并不是强一致的,这就需要客户端能够支持负载均衡及失败重试。在 Netflix 的生态中,由 ribbon 提供这个功能。

3.3.3 Peer to Peer 架构

一般而言,分布式系统的数据在多个副本之间的复制方式,可分为主从复制和对等复制。

1. 主从复制

主从复制也就是广为人知的 Master-Slave 模式,即有一个主副本,其他副本为从副本。所有对数据的写操作都提交到主副本,最后再由主副本更新到其他从副本。具体更新的方式,还可以细分为同步更新、异步更新、同步及异步混合。

对于主从复制模式来讲,写操作的压力都在主副本上,它是整个系统的瓶颈,但是从副本可以帮主副本分担读请求。

2. 对等复制

即 Peer to Peer 的模式,副本之间不分主从,任何副本都可以接收写操作,然后每个副本之间相互进行数据更新。

对于对等复制模式来讲,由于任何副本都可以接收写操作请求,不存在写操作压力瓶颈。但是由于每个副本都可以进行写操作处理,各个副本之间的数据同步及冲突处理是一个比较棘手的问题。

Eureka Server 采用的就是 Peer to Peer 的复制模式。这里我们分为客户端及服务端两个角度来阐述。

(1) 客户端

Client 端一般通过如下配置 Eureka Server 的 peer 节点:

```
eureka:
    client:
        serviceUrl:
            defaultZone:
http://127.0.0.1:8761/eureka/,http://127.0.0.1:8762/eureka/
```

实际代码里支持 preferSameZoneEureka,即有多个分区的话,优先选择与应用实例所在分区一样的其他服务的实例,如果没找到则默认使用 defaultZone。客户端使用 quarantineSet 维护了一个不可用的 Eureka Server 列表,进行请求的时候,优先从可用的列表中进行选择,如果请求失败则切换到下一个 Eureka Server 进行重试,重试次数默认为 3。

另外为了防止每个 Client 端都按配置文件指定的顺序进行请求造成 Eureka Server 节点请

求分布不均衡的情况，Client 端有个定时任务（默认 5 分钟执行一次）来刷新并随机化 Eureka Server 的列表。

（2）服务端

Eureka Server 本身依赖了 Eureka Client，也就是每个 Eureka Server 是作为其他 Eureka Server 的 Client。在单个 Eureka Server 启动的时候，会有一个 syncUp 的操作，通过 Eureka Client 请求其他 Eureka Server 节点中的一个节点获取注册的应用实例信息，然后复制到其他 peer 节点。

Eureka Server 在执行复制操作的时候，使用 HEADER_REPLICATION 的 http header 来将这个请求操作与普通应用实例的正常请求操作区分开来。通过 HEADER_REPLICATION 来标识是复制请求，这样其他 peer 节点接收到请求的时候，就不会再对它的 peer 节点进行复制操作，从而避免死循环。

Eureka Server 由于采用了 Peer to peer 的复制模式，其重点要解决的另外一个问题就是数据复制的冲突问题。针对这个问题，Eureka 采用如下两个方式来解决：

- lastDirtyTimestamp 标识
- heartbeat

针对数据的不一致，一般是通过版本号机制来解决，最后在不同副本之间只需要判断请求复制数据的版本号与本地数据的版本号高低就可以了。Eureka 没有直接使用版本号的属性，而是采用一个叫作 lastDirtyTimestamp 的字段来对比。

如果开启 SyncWhenTimestampDiffers 配置（默认开启），当 lastDirtyTimestamp 不为空的时候，就会进行相应的处理：

- 如果请求参数的 lastDirtyTimestamp 值大于 Server 本地该实例的 lastDirtyTimestamp 值，则表示 Eureka Server 之间的数据出现冲突，这个时候就返回 404，要求应用实例重新进行 register 操作。
- 如果请求参数的 lastDirtyTimestamp 值小于 Server 本地该实例的 lastDirtyTimestamp 值，如果是 peer 节点的复制请求，则表示数据出现冲突，返回 409 给 peer 节点，要求其同步自己最新的数据信息。

peer 节点之间的相互复制并不能保证所有操作都能够成功，因此 Eureka 还通过应用实例与 Server 之间的 heartbeat 也就是 renewLease 操作来进行数据的最终修复，即如果发现应用实例数据与某个 Server 的数据出现不一致，则 Server 返回 404，应用实例重新进行 register 操作。

3.3.4　Zone 及 Region 设计

由于 Netflix 的服务大部分在 Amazon 上，因此 Eureka 的设计有一部分也是基于 Amazon 的 Zone 及 Region 的基础设施之上。

在 Amazon EC2 托管在全球的各个地方，它用 Region 来代表一个独立的地理区域，比如 Eureka Server 默认设置了 4 个 Region：us-east-1、us-west-1、us-west-2、eu-west-1。Amazon 的部分 Region 代码及名称列表如表 3-4 所示。

表 3-4　Amazon 云 Region 代码及名称列表

Region 代码	名　　称
us-east-1	美国东部（弗吉尼亚北部）
us-east-2	美国东部（俄亥俄州）
us-west-1	美国西部（加利福尼亚北部）
us-west-2	美国西部（俄勒冈）
ca-central-1	加拿大（中部）
eu-central-1	欧洲（法兰克福）
eu-west-1	欧洲（爱尔兰）
eu-west-2	欧洲（伦敦）
eu-west-3	欧洲（巴黎）
ap-northeast-1	亚太区域（东京）
ap-northeast-2	亚太区域（首尔）
ap-northeast-3	亚太区域（大阪当地）
ap-southeast-1	亚太区域（新加坡）
ap-southeast-2	亚太区域（悉尼）
ap-south-1	亚太地区（孟买）
sa-east-1	南美洲（圣保罗）

在每个 Region 下面，还分了多个 AvailabilityZone，一个 Region 对应多个 AvailabilityZone。每个 Region 之间是相互独立及隔离的，默认情况下资源只在单个 Region 之间的 Availability-Zone 进行复制，跨 Region 之间不会进行资源复制。Region 与 AvailabilityZone 之间的关系图 3-1 所示。

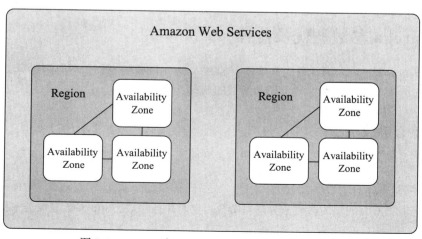

图 3-1　Region 与 AvailabilityZone 之间的关系图

AvailabilityZone 就类似 Region 下面的子 Region，比如 us-east-1 的 Region 可分为 us-east-1a、us-east-1c、us-east-1d、us-east-1e 这几个 AvailabilityZone。AvailabilityZone 可看作 Region 下面的一个个机房，各个机房相对独立，主要是为了 Region 的高可用设计，当同一个 Region 下面的 AvailabilityZone 不可用时，还有其他 AvailabilityZone 可用。

Eureka Server 原生支持了 Region 及 AvailabilityZone，由于资源在 Region 之间默认是不会复制的，因此 Eureka Server 的高可用主要就在于 Region 下面的 AvailabilityZone。

Eureka Client 支持 preferSameZone，也就是获取 Eureka Server 的 serviceUrl 优先拉取跟应用实例同处于一个 AvailabilityZone 的 Eureka Server 地址列表。一个 AvailabilityZone 可以设置多个 Eureka Server 实例，它们之间构成 peer 节点，然后采用 Peer to Peer 的复制模式。

Netflix 的 Ribbon 组件针对多个 AvailabilityZone 提供了 ZoneAffinity 的支持，允许在客户端路由或网关路由时，优先选取与自身实例处于同一个 AvailabilityZone 的服务实例。

3.3.5 SELF PRESERVATION 设计

在分布式系统设计里头，通常需要对应用实例的存活进行健康检查，这里比较关键的问题就是要处理好网络偶尔抖动或短暂不可用时造成的误判。另外 Eureka Server 端与 Client 端之间如果出现网络分区问题，在极端情况下可能会使得 Eureka Server 清空部分服务的实例列表，这个将严重影响到 Eureka Server 的 Availability 属性。因此 Eureka Server 引入了 SELF PRESERVATION 机制。

Eureka Client 端与 Server 端之间有个租约，Client 要定时发送心跳来维持这个租约，表示自己还存活着。Eureka 通过当前注册的实例数，去计算每分钟应该从应用实例接收到的心跳数，如果最近一分钟接收到的续约的次数小于等于指定阈值的话，则关闭租约失效剔除，禁止定时任务剔除失效的实例，从而保护注册信息。

3.4 Eureka 参数调优及监控

若要深入了解一个开源组件，从解读其参数入手，不失为一种好的研读代码的方法，本节首先讲解 Eureka 的核心参数，然后讲解参数的调优及监控。

3.4.1 核心参数

下面主要分为 Client 端及 Server 端两大类来简述一下 Eureka 的几个核心参数。

1. Client 端

这里笔者将 Client 端的参数分为基本参数、定时任务参数、http 参数三大类来梳理。

1）基本参数如表 3-5 所示。

表 3-5　Eureka Client 参数说明

参　　数	默认值	说　　明
eureka.client.availability-zones		告知 Client 有哪些 region 及 availability-zones，支持配置修改运行时生效
eureka.client.filter-only-up-instances	true	是否过滤出 InstanceStatus 为 UP 的实例
eureka.client.region	us-east-1	指定该应用实例所在的 region，AWS datacenters 适用
eureka.client.register-with-eureka	true	是否将该应用实例注册到 Eureka Server
eureka.client.prefer-same-zone-eureka	true	是否优先使用与该应用实例处于相同 zone 的 Eureka Server
eureka.client.on-demand-update-status-change	true	是否将本地实例状态的更新通过 ApplicationInfoManager 实时触发同步（有请求流控限制）到 Eureka Server
eureka.instance.metadata-map		指定应用实例的元数据信息
eureka.instance.prefer-ip-address	false	是否优先使用 ip 地址来替代 host name 作为实例的 hostName 字段值
eureka.instance.lease-expiration-duration-in-seconds	90	指定 Eureka Client 间隔多久需要向 Eureka Server 发送心跳来告知 Eureka Server 该实例还存活

2）定时任务参数如表 3-6 所示。

表 3-6　Eureka Client 定时任务参数

参　　数	默认值	说　　明
eureka.client.cache-refresh-executor-thread-pool-size	2	刷新缓存的 CacheRefreshThread 的线程池大小
eureka.client.cache-refresh-executor-exponential-back-off-bound	10	调度任务执行超时时下次的调度的延时时间
eureka.client.heartbeat-executor-thread-pool-size	2	心跳线程 HeartbeatThread 的线程池大小
eureka.client.heartbeat-executor-exponential-back-off-bound	10	调度任务执行超时时下次的调度的延时时间
eureka.client.registry-fetch-interval-seconds	30	CacheRefreshThread 线程的调度频率
eureka.client.eureka-service-url-poll-interval-seconds	5*60	AsyncResolver.updateTask 刷新 Eureka Server 地址的时间间隔
eureka.client.initial-instance-info-replication-interval-seconds	40	InstanceInfoReplicator 将实例信息变更同步到 Eureka Server 的初始延时时间
eureka.client.instance-info-replication-interval-seconds	30	InstanceInfoReplicator 将实例信息变更同步到 Eureka Server 的时间间隔
eureka.instance.lease-renewal-interval-in-seconds	30	Eureka Client 向 Eureka Server 发送心跳的时间间隔

3）http 参数。Eureka Client 底层 httpClient 与 Eureka Server 通信，提供的相关参数如表 3-7 所示。

表 3-7　Eureka Client http 相关参数

参数	默认值	说明
eureka.client.eureka-server-connect-timeout-seconds	5	连接超时时间
eureka.client.eureka-server-read-timeout-seconds	8	读超时时间
eureka.client.eureka-server-total-connections	200	连接池最大活动连接数（'MaxTotal'）
eureka.client.eureka-server-total-connections-per-host	50	每个 host 能使用的最大连接数（'DefaultMaxPerRoute'）
eureka.client.eureka-connection-idle-timeout-seconds	30	连接池中连接的空闲时间（'connectionIdleTimeout'）

2.Server 端

笔者将 Server 端的参数分为如下几类：基本参数、response cache 参数、peer 相关参数、http 参数。

（1）基本参数

基本参数列表如表 3-8 所示。

表 3-8　eureka server 基本参数

参数	默认值	说明
eureka.server.enable-self-preservation	true	是否开启自我保护模式
eureka.server.renewal-percent-threshold	0.85	指定每分钟需要收到的续约次数的阈值
eureka.instance.registry.expected-number-of-renews-per-min	1	指定每分钟需要收到的续约次数值，实际该值在其中被写死为 count*2，另外也会被更新
eureka.server.renewal-threshold-update-interval-ms	15 分钟	指定 updateRenewalThreshold 定时任务的调度频率，来动态更新 expectedNumber OfRenewsPerMin 及 numberOfRenewsPerMinThreshold 值
eureka.server.eviction-interval-timer-in-ms	60*1000	指定 EvictionTask 定时任务的调度频率，用于剔除过期的实例

（2）response cache 参数

Eureka Server 为了提升自身 REST API 接口的性能，提供了两个缓存：一个是基于 ConcurrentMap 的 readOnlyCacheMap，一个是基于 Guava Cache 的 readWriteCacheMap。其相关参数如表 3-9 所示。

表 3-9　Eureka Server response cache 参数

参数	默认值	说明
eureka.server.use-read-only-response-cache	true	是否使用只读的 response-cache
eureka.server.response-cache-update-interval-ms	30*1000	设置 CacheUpdateTask 的调度时间间隔，用于从 readWriteCacheMap 更新数据到 readOnlyCacheMap。仅仅在 eureka.server.use-read-only-response-cache 为 true 的时候才生效
eureka.server.response-cache-auto-expiration-in-seconds	180	设置 readWriteCacheMap 的 expireAfterWrite 参数，指定写入多长时间后过期

（3）peer 相关参数

peer 相关系数的说明如表 3-10 所示。

表 3-10 Eureka Server peer 相关参数

参 数	默认值	说 明
eureka.server.peer-eureka-nodes-update-interval-ms	10 分钟	指定 peersUpdateTask 调度的时间间隔，用于从配置文件刷新 peerEurekaNodes 节点的配置信息（'eureka.client.serviceUrl 相关 zone 的配置'）
eureka.server.peer-eureka-status-refresh-time-interval-ms	30*1000	指定更新 peer nodes 状态信息的时间间隔（'目前没看到代码中有使用'）

（4）http 参数

Eureka Server 需要与其他 peer 节点进行通信，复制实例信息，其底层使用 httpClient，提供的相关参数如表 3-11 所示。

表 3-11 Eureka Server http 相关参数

参 数	默认值	说 明
eureka.server.peer-node-connect-timeout-ms	200	连接超时时间
eureka.server.peer-node-read-timeout-ms	200	读超时时间
eureka.server.peer-node-total-connections	1000	连接池最大活动连接数（'MaxTotal'）
eureka.server.peer-node-total-connections-per-host	500	每个 host 能使用的最大连接数（'DefaultMaxPerRoute'）
eureka.server.peer-node-connection-idle-timeout-seconds	30	连接池中连接的空闲时间（'connectionIdleTimeout'）

3.4.2 参数调优

1. 常见问题

对于新接触 Eureka 的开发人员来说，一般会有几个困惑：
- 为什么服务下线了，Eureka Server 接口返回的信息还会存在。
- 为什么服务上线了，Eureka Client 不能及时获取到。
- 为什么有时候会出现如下提示：

```
EMERGENCY! EUREKA MAY BE INCORRECTLY CLAIMING INSTANCES ARE UP WHEN THEY'RE
NOT. RENEWALS ARE LESSER THAN THRESHOLD AND HENCE THE INSTANCES ARE NOT
BEING EXPIRED JUST TO BE SAFE
```

2. 解决方法

对于第一个问题，Eureka Server 并不是强一致的，因此 registry 中会存留过期的实例信息，这里头有几个原因：

- 应用实例异常挂掉，没能在挂掉之前告知 Eureka Server 要下线掉该服务实例信息。这个就需要依赖 Eureka Server 的 EvictionTask 去剔除。
- 应用实例下线时有告知 Eureka Server 下线，但是由于 Eureka Server 的 REST API 有 response cache，因此需要等待缓存过期才能更新。
- Eureka Server 由于开启并引入了 SELF PRESERVATION 模式，导致 registry 的信息不会因为过期而被剔除掉，直到退出 SELF PRESERVATION 模式。

针对 Client 下线没有通知 Eureka Server 的问题，可以调整 EvictionTask 的调度频率，比如下面配置将调度间隔从默认的 60 秒，调整为 5 秒：

```
eureka.server.eviction-interval-timer-in-ms=5000
```

针对 response cache 的问题，可以根据情况考虑关闭 readOnlyCacheMap：

```
eureka.server.use-read-only-response-cache=false
```

或者调整 readWriteCacheMap 的过期时间：

```
eureka.server.response-cache-auto-expiration-in-seconds=60
```

针对 SELF PRESERVATION 的问题，在测试环境可以将 enable-self-preservation 设置为 false：

```
eureka.server.enable-self-preservation=false
```

关闭的话，则会提示：

```
THE SELF PRESERVATION MODE IS TURNED OFF.THIS MAY NOT PROTECT INSTANCE EXPIRY
    IN CASE OF NETWORK/OTHER PROBLEMS.
```

或者：

```
RENEWALS ARE LESSER THAN THE THRESHOLD. THE SELF PRESERVATION MODE IS TURNED
    OFF.THIS MAY NOT PROTECT INSTANCE EXPIRY IN CASE OF NETWORK/OTHER PROBLEMS.
```

针对新服务上线，Eureka Client 获取不及时的问题，在测试环境，可以适当提高 Client 端拉取 Server 注册信息的频率，例如下面将默认的 30 秒改为 5 秒：

```
eureka.client.registry-fetch-interval-seconds=5
```

针对 SELF PRESERVATION 问题，下面我们来重点讨论一下。

在实际生产过程中，经常会有网络抖动等问题造成服务实例与 Eureka Server 的心跳未能如期保持，但是服务实例本身是健康的，这个时候如果按照租约剔除机制剔除的话，会造成误判，如果大范围误判的话，可能会导致整个服务注册列表的大部分注册信息被删除，从而没有可用服务。Eureka 为了解决这个问题引入了 SELF PRESERVATION 机制，当最近一分钟接收到的续约的次数小于等于指定阈值的话，则关闭租约失效剔除，禁止定时任务剔除失效的实例，从而保护注册信息。对于开发测试环境，开启这个机制有时候反而会影响系统的持续集成，因此可以通过如下参数关闭该机制：

```
eureka.server.enableSelfPreservation=false
```

在生产环境中，可以把 renewalPercentThreshold 及 leaseRenewalIntervalInSeconds 参数调小一点，进而提高触发 SELF PRESERVATION 机制的门槛，比如：

```
eureka.instance.leaseRenewalIntervalInSeconds=10    ##默认30
eureka.server.renewalPercentThreshold=0.49          ##默认0.85
```

3.4.3 指标监控

Eureka 内置了基于 servo 的指标统计，具体详见 com/netflix/eureka/util/EurekaMonitors.java 类，其统计指标如表 3-12 所示。

表 3-12 Eureka 监控指标

指标名称	说 明
renewCounter	自启动以来收到的总续约次数
cancelCounter	自启动以来收到的总取消租约次数
getAllCacheMissCounter	自启动以来查询 registry 的总次数 ('AbstractInstanceRegistry.getApplications')
getAllCacheMissDeltaCounter	自启动以来 delta 查询 registry 的总次数 ('AbstractInstanceRegistry.getApplicationDeltas')
getAllWithRemoteRegionCacheMissCounter	自启动以来使用 remote region 查询 registry 的总次数 ('AbstractInstanceRegistry.getApplicationsFromMultipleRegions')
getAllWithRemoteRegionCacheMissDeltaCounter	自启动以来使用 remote region 及 delta 方式查询 registry 的总次数 ('AbstractInstanceRegistry.getApplicationDeltasFromMultipleRegions')
getAllDeltaCounter	自启动以来查询 delta 的总次数 ('/{version}/apps/delta')
getAllDeltaWithRemoteRegionCounter	自启动以来传递 regions 查询 delta 的总次数 ('/{version}/apps/delta')
getAllCounter	自启动以来查询 ('/{version}/apps') 的总次数
getAllWithRemoteRegionCounter	自启动以来传递 regions 参数查询 ('/{version}/apps') 的总次数
getApplicationCounter	自启动以来请求 /{version}/apps/{appId} 的总次数
registerCounter	自启动以来 register 的总次数
expiredCounter	自启动以来剔除过期实例的总次数
statusUpdateCounter	自启动以来 statusUpdate 的总次数
statusOverrideDeleteCounter	自启动以来 deleteStatusOverride 总次数
cancelNotFoundCounter	自启动以来收到 cancel 请求时对应实例找不到的次数
renewNotFoundexpiredCounter	自启动以来收到 renew 请求时对应实例找不到的次数
numOfRejectedReplications	--
numOfFailedReplications	--
numOfRateLimitedRequests	由于 rate limiter 被丢弃的请求数量
numOfRateLimitedRequestCandidates	如果开启 rate limiter 的话，将被丢弃的请求数
numOfRateLimitedFullFetchRequests	开启 rate limiter 时请求全量 registry 被丢弃的请求数
numOfRateLimitedFullFetchRequestCandidates	如果开启 rate limiter 时请求全量 registry 将被丢弃的请求数

Spring Boot 2.x 版本改为使用 Micrometer，不再支持 Netflix Servo，转而支持了 Netflix

Servo 的替代者 Netflix Spectator。不过对于 Servo，可以通过 DefaultMonitorRegistry.getInstance().getRegisteredMonitors() 来获取所有注册了的 Monitor，进而获取其指标值。

3.5 Eureka 实战

3.5.1 Eureka Server 在线扩容

（1）准备工作

由于我们需要动态去修改配置，因此这里引入 config-server 工程，如代码清单 3-1 所示。

代码清单3-1　ch3-1\ch3-1-config-server\pom.xml

```xml
<dependencies>
    <dependency>
        <groupId>org.springframework.cloud</groupId>
        <artifactId>spring-cloud-config-server</artifactId>
    </dependency>
</dependencies>
```

启动类如代码清单 3-2 所示：

代码清单3-2　ch3-1\ch3-1-config-server\src\main\java\cn\springcloud\book\Ch31ConfigServerApplication.java

```java
@SpringBootApplication
@EnableConfigServer
public class Ch31ConfigServerApplication {
    public static void main(String[] args) {
        SpringApplication.run(Ch31ConfigServerApplication.class, args);
    }
}
```

配置文件如代码清单 3-3 所示：

代码清单3-3　ch3-1\ch3-1-config-server\src\main\resources\bootstrap.yml

```yaml
spring:
  application:
    name: config-server
  profiles:
    active: native
server:
  port: 8888
```

这里为了简单演示，我们使用 native 的 proflie，即使用文件来存储配置，默认放在 resources\config 目录下。

另外由于要演示 Eureka Server 的动态扩容，这里还建立了一个 eureka-server 工程及 eureka-client 工程，分别见 ch3-1\ch3-1-eureka-server、ch3-1\ch3-1-eureka-client。与第 2 章的工程的区别在于这里引入了 spring-cloud-starter-config，另外两个工程都添加了一个 QueryController 用于实验，如代码清单 3-4 所示。

代码清单3-4　ch3-1\ch3-1-eureka-server\src\main\cn\springcloud\book\controller\QueryController.java

```java
@RestController
@RequestMapping("/query")
public class QueryController {
    @Autowired
    EurekaClientConfigBean eurekaClientConfigBean;

    @GetMapping("/eureka-server")
    public Object getEurekaServerUrl(){
        return eurekaClientConfigBean.getServiceUrl();
    }
}
```

(2) 1个 Eureka Server

eureka-client 的配置文件如代码清单 3-5 所示：

代码清单3-5　ch3-1\ch3-1-config-server\src\main\resources\config\eureka-client.yml

```
server:
    port: 8081
spring:
    application:
        name: eureka-client1
eureka:
    client:
        serviceUrl:
            defaultZone: http://localhost:8761/eureka/ # one eureka server
```

eureka-server 的配置文件如代码清单 3-6 所示：

代码清单3-6　ch3-1\ch3-1-config-server\src\main\resources\config\eureka-server-peer1.yml

```
server:
    port: 8761
eureka:
    instance:
        hostname: localhost
        preferIpAddress: true
    client:
        registerWithEureka: true
        fetchRegistry: true
        serviceUrl:
            defaultZone: http://localhost:8761/eureka/ # one eureka server
    server:
        waitTimeInMsWhenSyncEmpty: 0
        enableSelfPreservation: false
```

然后分别启动 config-server、eureka-server（使用 peer1 的 profile）、eureka-client，打开 localhost:8761，可以观察注册的实例。

(3) 2个 Eureka Server

现在我们开始把 Eureka Server 的实例扩充一下，在 ch3-1-eureka-server 工程目录下，使用

peer2 的 profile 启动第二个 Eureka Server，命令行如下：

```
mvn spring-boot:run -Dspring.profiles.active=peer2
```

其配置文件如代码清单 3-7 所示：

代码清单3-7 ch3-1\ch3-1-config-server\src\main\resources\config\eureka-server-peer2.yml

```yaml
server:
    port: 8762

eureka:
    instance:
        hostname: localhost
        preferIpAddress: true
    client:
        registerWithEureka: true
        fetchRegistry: true
        serviceUrl:
            defaultZone: http://localhost:8761/eureka/ # two eureka server
    server:
        waitTimeInMsWhenSyncEmpty: 0
        enableSelfPreservation: false
```

第二个 Eureka Server 启动起来之后，要修改 eureka-client.yml、eureka-server-peer1.yml 的配置文件：

其中 eureka-client.yml 变为：

```yaml
server:
    port: 8081

spring:
    application:
        name: eureka-client1

eureka:
    client:
        serviceUrl:
            defaultZone: http://localhost:8761/eureka/,http://localhost:8762/eureka/ # two eureka server
```

修改 eureka.client.serviceUrl，新增第二个 Eureka Server 的地址。

eureka-peer1.yml 变为：

```yaml
server:
    port: 8761

spring:
    application:
        name: eureka-server
eureka:
    instance:
        hostname: localhost
        preferIpAddress: true
    client:
        registerWithEureka: true
        fetchRegistry: true
```

```yaml
    serviceUrl:
      defaultZone: http://localhost:8762/eureka/ # two eureka server
server:
      waitTimeInMsWhenSyncEmpty: 0
      enableSelfPreservation: false
```

修改 eureka.client.serviceUrl,指向第二个 Eureka Server。

之后就是重启 config-server,使配置生效。然后使用如下命令分别刷新 eureka-client 以及 eureka-server-peer1,加载新配置:

```
~ curl -i -X POST localhost:8761/actuator/refresh
HTTP/1.1 200
Content-Type: application/vnd.spring-boot.actuator.v2+json;charset=UTF-8
Transfer-Encoding: chunked
Date: Wed, 20 Jun 2018 11:10:10 GMT

["eureka.client.serviceUrl.defaultZone"]%
~ curl -i -X POST localhost:8081/actuator/refresh
HTTP/1.1 200
Content-Type: application/vnd.spring-boot.actuator.v2+json;charset=UTF-8
Transfer-Encoding: chunked
Date: Wed, 20 Jun 2018 11:10:17 GMT

["eureka.client.serviceUrl.defaultZone"]%
```

之后分别调用 queryController 的方法,如下:

```
~ curl -i http://localhost:8761/query/eureka-server
HTTP/1.1 200
Content-Type: application/json;charset=UTF-8
Transfer-Encoding: chunked
Date: Wed, 20 Jun 2018 11:12:07 GMT

{"defaultZone":"http://localhost:8762/eureka/"}%
→ ~ curl -i http://localhost:8081/query/eureka-server
HTTP/1.1 200
Content-Type: application/json;charset=UTF-8
Transfer-Encoding: chunked
Date: Wed, 20 Jun 2018 11:12:15 GMT

{"defaultZone":"http://localhost:8761/eureka/,http://localhost:8762/eureka/"}%
```

可以看到 Eureka Client 端已经成功识别到两个 Eureka Server,而原来 peer1 的 Eureka Server 请求的 eureka.client.serviceUrl 也指向了 peer2,Eureka Server 扩容成功。

(4) 3 个 Eureka Server

将 Eureka Server 扩容到 3 个实例的话,这里使用 peer3 配置,如代码清单 3-8 所示。

代码清单3-8　ch3-1\ch3-1-config-server\src\main\resources\config\eureka-server-peer3.yml

```yaml
server:
  port: 8763

eureka:
  instance:
```

```yaml
        hostname: localhost
        preferIpAddress: true
    client:
        registerWithEureka: true
        fetchRegistry: true
        serviceUrl:
            defaultZone: http://localhost:8761/eureka/,http://localhost:8762/eureka/ # three eureka server
    server:
            waitTimeInMsWhenSyncEmpty: 0
            enableSelfPreservation: false
```

启动命令如下：

```
mvn spring-boot:run -Dspring.profiles.active=peer3
```

接下来修改 eureka-client.yml 配置，如下所示：

```yaml
server:
    port: 8081

spring:
    application:
        name: eureka-client1

eureka:
    client:
        serviceUrl:
            defaultZone: http://localhost:8761/eureka/,http://localhost:8762/eureka/,http://localhost:8763/eureka/ # three eureka server
```

这里新增了 peer3 的 Eureka Server 地址。

修改 eureka-server-peer1.yml，如下所示：

```yaml
server:
    port: 8761

spring:
    application:
        name: eureka-server
eureka:
    instance:
        hostname: localhost
        preferIpAddress: true
    client:
        registerWithEureka: true
        fetchRegistry: true
        serviceUrl:
            defaultZone: http://localhost:8762/eureka/,http://localhost:8763/eureka/ # three eureka server
    server:
            waitTimeInMsWhenSyncEmpty: 0
            enableSelfPreservation: false
```

其 eureka.client.serviceUrl.defaultZone 指向了 peer2 和 peer3。

修改 eureka-server-peer2.yml 的配置如下：

```yaml
server:
    port: 8762

eureka:
    instance:
        hostname: localhost
        preferIpAddress: true
    client:
        registerWithEureka: true
        fetchRegistry: true
        serviceUrl:
            defaultZone: http://localhost:8761/eureka/,http://localhost:8763/
                eureka/ # three eureka server
    server:
        waitTimeInMsWhenSyncEmpty: 0
        enableSelfPreservation: false
```

其 eureka.client.serviceUrl.defaultZone 指向了 peer1 和 peer3。

接下来重启 config-server，然后分别刷新 Eureka Client、Eureka Server 的 peer1 和 peer2 的配置，如下所示：

```
~ curl -i -X POST http://localhost:8081/actuator/refresh
HTTP/1.1 200
Content-Type: application/vnd.spring-boot.actuator.v2+json;charset=UTF-8
Transfer-Encoding: chunked
Date: Wed, 20 Jun 2018 11:24:57 GMT

["eureka.client.serviceUrl.defaultZone"]%
→ ~ curl -i -X POST http://localhost:8761/actuator/refresh
HTTP/1.1 200
Content-Type: application/vnd.spring-boot.actuator.v2+json;charset=UTF-8
Transfer-Encoding: chunked
Date: Wed, 20 Jun 2018 11:25:17 GMT

["eureka.client.serviceUrl.defaultZone"]%
→ ~ curl -i -X POST http://localhost:8762/actuator/refresh
HTTP/1.1 200
Content-Type: application/vnd.spring-boot.actuator.v2+json;charset=UTF-8
Transfer-Encoding: chunked
Date: Wed, 20 Jun 2018 11:25:36 GMT

["eureka.client.serviceUrl.defaultZone"]%
```

之后分别访问各个实例的 /query/eureka-server，可以看到 Eureka Server 的列表已经成功变为 3 个，扩容成功。

本小节举例的 Eureka Server 在线扩容，需要依赖配置中心的动态刷新功能，具体的就是 /actuator/refresh 这个 endpoint。这里为了方便，使用的 config-server 是 native 的 profile，因此修改后重启才生效，如果是使用 git 仓库，则无须重启 config-server。

3.5.2 构建 Multi Zone Eureka Server

前面的小节简单介绍了 Eureka 的 Zone 及 Region 设计，这里我们来演示下如何构建 Multi Zone 的 Eureka Server，同时演示下默认的 ZoneAffinity 特性。

1. Eureka Server 实例

这里我们启动四个 Eureka Server 实例，配置两个 zone：zone1 及 zone2，每个 zone 都有两个 Eureka Server 实例，这两个 zone 配置在同一个 region：region-east 上。

它们的配置文件分别如代码清单 3-9、代码清单 3-10、代码清单 3-11、代码清单 3-12 所示：

代码清单3-9　ch3-2\ch3-2-eureka-server\src\main\resources\application-zone1a.yml

```yaml
server:
    port: 8761
spring:
    application:
        name: eureka-server
eureka:
    instance:
        hostname: localhost
        preferIpAddress: true
        metadataMap.zone: zone1
    client:
        register-with-eureka: true
        fetch-registry: true
        region: region-east
        service-url:
            zone1: http://localhost:8761/eureka/,http://localhost:8762/eureka/
            zone2: http://localhost:8763/eureka/,http://localhost:8764/eureka/
        availability-zones:
            region-east: zone1,zone2
    server:
        waitTimeInMsWhenSyncEmpty: 0
        enableSelfPreservation: false
```

代码清单3-10　ch3-2\ch3-2-eureka-server\src\main\resources\application-zone1b.yml

```yaml
server:
    port: 8762
spring:
    application:
        name: eureka-server
eureka:
    instance:
        hostname: localhost
        preferIpAddress: true
        metadataMap.zone: zone1
    client:
        register-with-eureka: true
        fetch-registry: true
        region: region-east
        service-url:
            zone1: http://localhost:8761/eureka/,http://localhost:8762/eureka/
            zone2: http://localhost:8763/eureka/,http://localhost:8764/eureka/
        availability-zones:
            region-east: zone1,zone2
    server:
        waitTimeInMsWhenSyncEmpty: 0
        enableSelfPreservation: false
```

代码清单3-11 ch3-2\ch3-2-eureka-server\src\main\resources\application-zone2a.yml

```yaml
server:
    port: 8763
spring:
    application:
        name: eureka-server
eureka:
    instance:
        hostname: localhost
        preferIpAddress: true
        metadataMap.zone: zone2
    client:
        register-with-eureka: true
        fetch-registry: true
        region: region-east
        service-url:
            zone1: http://localhost:8761/eureka/,http://localhost:8762/eureka/
            zone2: http://localhost:8763/eureka/,http://localhost:8764/eureka/
        availability-zones:
            region-east: zone1,zone2
    server:
        waitTimeInMsWhenSyncEmpty: 0
        enableSelfPreservation: false
```

代码清单3-12 ch3-2\ch3-2-eureka-server\src\main\resources\application-zone2b.yml

```yaml
server:
    port: 8764
spring:
    application:
        name: eureka-server
eureka:
    instance:
        hostname: localhost
        preferIpAddress: true
        metadataMap.zone: zone2
    client:
        register-with-eureka: true
        fetch-registry: true
        region: region-east
        service-url:
            zone1: http://localhost:8761/eureka/,http://localhost:8762/eureka/
            zone2: http://localhost:8763/eureka/,http://localhost:8764/eureka/
        availability-zones:
            region-east: zone1,zone2
    server:
        waitTimeInMsWhenSyncEmpty: 0
        enableSelfPreservation: false
```

上面我们配置了四个 Eureka Server 的配置文件，可以看到我们通过 eureka.instance.metadataMap.zone 设置了每个实例所属的 zone。接下来分别使用这四个 proflie 启动这四个 Eureka Server，如下：

```
mvn spring-boot:run -Dspring.profiles.active=zone1a
mvn spring-boot:run -Dspring.profiles.active=zone1b
mvn spring-boot:run -Dspring.profiles.active=zone2a
mvn spring-boot:run -Dspring.profiles.active=zone2b
```

2. Eureka Client 实例

这里我们配置两个 Eureka Client，分别属于 zone1 及 zone2，其配置文件如代码清单 3-13、代码清单 3-14、代码清单 3-15 所示：

代码清单3-13　ch3-2\ch3-2-eureka-client\src\main\resources\application.yml

```yaml
management:
  endpoints:
    web:
      exposure:
        include: '*'
```

这里我们暴露所有的 endpoints，方便后面验证。

代码清单3-14　ch3-2\ch3-2-eureka-client\src\main\resources\application-zone1.yml

```yaml
server:
  port: 8081
spring:
  application:
    name: client
eureka:
  instance:
    metadataMap.zone: zone1
  client:
    register-with-eureka: true
    fetch-registry: true
    region: region-east
    service-url:
      zone1: http://localhost:8761/eureka/,http://localhost:8762/eureka/
      zone2: http://localhost:8763/eureka/,http://localhost:8764/eureka/
    availability-zones:
      region-east: zone1,zone2
```

代码清单3-15　ch3-2\ch3-2-eureka-client\src\main\resources\application-zone2.yml

```yaml
server:
  port: 8082
spring:
  application:
    name: client
eureka:
  instance:
    metadataMap.zone: zone2
  client:
    register-with-eureka: true
    fetch-registry: true
    region: region-east
    service-url:
```

```
            zone1: http://localhost:8761/eureka/,http://localhost:8762/eureka/
            zone2: http://localhost:8763/eureka/,http://localhost:8764/eureka/
        availability-zones:
            region-east: zone1,zone2
```

接着使用如下命令分别启动这两个 client：

```
mvn spring-boot:run -Dspring.profiles.active=zone1
mvn spring-boot:run -Dspring.profiles.active=zone2
```

3. Zuul Gateway 实例

这里我们新建一个 zuul 工程来演示 Eureka 使用 metadataMap 的 zone 属性时的 ZoneAffinity 特性。

配置文件如代码清单 3-16、3-17 所示：

代码清单3-16 ch3-2\ch3-2-zuul-gateway\src\main\resources\application-zone1.yml

```
server:
    port: 10001
eureka:
    instance:
        metadataMap.zone: zone1
    client:
        register-with-eureka: true
        fetch-registry: true
        region: region-east
        service-url:
            zone1: http://localhost:8761/eureka/,http://localhost:8762/eureka/
            zone2: http://localhost:8763/eureka/,http://localhost:8764/eureka/
        availability-zones:
            region-east: zone1,zone2
```

代码清单3-17 ch3-2\ch3-2-zuul-gateway\src\main\resources\application-zone2.yml

```
server:
    port: 10002
eureka:
    instance:
        metadataMap.zone: zone2
    client:
        register-with-eureka: true
        fetch-registry: true
        region: region-east
        service-url:
            zone1: http://localhost:8761/eureka/,http://localhost:8762/eureka/
            zone2: http://localhost:8763/eureka/,http://localhost:8764/eureka/
        availability-zones:
            region-east: zone1,zone2
```

接下来使用这两个 profile 分别启动 gateway 如下：

```
mvn spring-boot:run -Dspring.profiles.active=zone1
mvn spring-boot:run -Dspring.profiles.active=zone2
```

4. 验证 ZoneAffinity

访问 http://localhost:10001/client/actuator/env，部分结果如下：

```
{
    "activeProfiles": [
        "zone1"
    ] //......
}
```

访问 http://localhost:10002/client/actuator/env，部分结果如下：

```
{
    "activeProfiles": [
        "zone2"
    ] //......
}
```

可以看到，通过请求 gateway 的 /client/actuator/env，访问的是 Eureka Client 实例的 /actuator/env 接口，处于 zone1 的 gateway 返回的 activeProfiles 为 zone1，处于 zone2 的 gateway 返回的 activeProfiles 为 zone2。从这个表象看 gateway 路由时对 client 的实例是 ZoneAffinity 的。有兴趣的读者可以去研读源码，看看 gateway 是如何实现 Eureka 的 ZoneAffinity 的。

3.5.3 支持 Remote Region

1. Eureka Server 实例

这里我们配置 4 个 Eureka Server，分 4 个 zone，然后属于 region-east、region-west 两个 region，其 region-east 的配置文件如代码清单 3-18、3-19 所示：

代码清单3-18　ch3-3\ch3-3-eureka-server\src\main\resources\application-zone1.yml

```yaml
server:
    port: 8761
spring:
    application:
        name: eureka-server
eureka:
    server:
        waitTimeInMsWhenSyncEmpty: 0
        enableSelfPreservation: false
        remoteRegionUrlsWithName:
            region-west: http://localhost:8763/eureka/
    client:
        register-with-eureka: true
        fetch-registry: true
        region: region-east
        service-url:
            zone1: http://localhost:8761/eureka/
            zone2: http://localhost:8762/eureka/
        availability-zones:
            region-east: zone1,zone2
    instance:
        hostname: localhost
        metadataMap.zone: zone1
```

代码清单3-19　ch3-3\ch3-3-eureka-server\src\main\resources\application-zone2.yml

```yaml
server:
    port: 8762
spring:
    application:
        name: eureka-server
eureka:
    server:
        waitTimeInMsWhenSyncEmpty: 0
        enableSelfPreservation: false
        remoteRegionUrlsWithName:
            region-west: http://localhost:8763/eureka/
    client:
        register-with-eureka: true
        fetch-registry: true
        region: region-east
        service-url:
            zone1: http://localhost:8761/eureka/
            zone2: http://localhost:8762/eureka/
        availability-zones:
            region-east: zone1,zone2
    instance:
        hostname: localhost
        metadataMap.zone: zone2
```

这里 zone1 及 zone2 属于 region-east，配置其 remote-region 为 region-west，具体如代码清单 3-20、代码清单 3-21 所示。

代码清单3-20　ch3-3\ch3-3-eureka-server\src\main\resources\application-zone3-region-west.yml

```yaml
server:
    port: 8763
spring:
    application:
        name: eureka-server
eureka:
    server:
        waitTimeInMsWhenSyncEmpty: 0
        enableSelfPreservation: false
        remoteRegionUrlsWithName:
            region-east: http://localhost:8761/eureka/
    client:
        register-with-eureka: true
        fetch-registry: true
        region: region-west
        service-url:
            zone3: http://localhost:8763/eureka/
            zone4: http://localhost:8764/eureka/
        availability-zones:
            region-west: zone3,zone4
    instance:
        hostname: localhost
        metadataMap.zone: zone3
```

代码清单3-21 ch3-3\ch3-3-eureka-server\src\main\resources\application-zone4-region-west.yml

```yaml
server:
    port: 8764
spring:
    application:
        name: eureka-server
eureka:
    server:
        waitTimeInMsWhenSyncEmpty: 0
        enableSelfPreservation: false
        remoteRegionUrlsWithName:
            region-east: http://localhost:8761/eureka/
    client:
        register-with-eureka: true
        fetch-registry: true
        region: region-west
        service-url:
            zone3: http://localhost:8763/eureka/
            zone4: http://localhost:8764/eureka/
        availability-zones:
            region-west: zone3,zone4
    instance:
        hostname: localhost
        metadataMap.zone: zone4
```

这里 zone3 及 zone4 属于 region-west，配置其 remote-region 为 region-east。

由于源码里 EurekaServerConfigBean 的 remoteRegionAppWhitelist 默认为 null，而 getRemoteRegionAppWhitelist（String regionName）方法会直接调用，如果不设置，则会报空指针异常：

```
2018-06-25 16:09:43.414 ERROR 13032 --- [nio-8764-exec-2] o.a.c.c.C.[.[./].
    [dispatcherServlet]      : Servlet.service() for servlet [dispatcherServlet]
    in context with path [] threw exception

java.lang.NullPointerException: null
    at org.springframework.cloud.netflix.eureka.server.EurekaServerConfigBean.
        getRemoteRegionAppWhitelist(EurekaServerConfigBean.java:226) ~[spring-
        cloud-netflix-eureka-server-2.0.0.RELEASE.jar:2.0.0.RELEASE]
    at com.netflix.eureka.registry.AbstractInstanceRegistry.shouldFetchFro
        mRemoteRegistry(AbstractInstanceRegistry.java:795) ~[eureka-core-
        1.9.2.jar:1.9.2]
    at com.netflix.eureka.registry.AbstractInstanceRegistry.getApplicationsFr
        omMultipleRegions(AbstractInstanceRegistry.java:767) ~[eureka-core-
        1.9.2.jar:1.9.2]
    at com.netflix.eureka.registry.AbstractInstanceRegistry.getApplicationsFr
        omAllRemoteRegions(AbstractInstanceRegistry.java:702) ~[eureka-core-
        1.9.2.jar:1.9.2]
    at com.netflix.eureka.registry.AbstractInstanceRegistry.getApplications(Abst
        ractInstanceRegistry.java:693) ~[eureka-core-1.9.2.jar:1.9.2]
    at com.netflix.eureka.registry.PeerAwareInstanceRegistryImpl.getSortedA
        pplications(PeerAwareInstanceRegistryImpl.java:555) ~[eureka-core-
        1.9.2.jar:1.9.2]
```

为了避免空指针，需要初始下该对象，如代码清单3-22所示：

代码清单3-22 ch3-3\ch3-3-eureka-server\src\main\java\cn\springcloud\book\config\RegionConfig.java

```java
@Configuration
@AutoConfigureBefore(EurekaServerAutoConfiguration.class)
public class RegionConfig {

    @Bean
    @ConditionalOnMissingBean
    public EurekaServerConfig eurekaServerConfig(EurekaClientConfig clientConfig) {
        EurekaServerConfigBean server = new EurekaServerConfigBean();
        if (clientConfig.shouldRegisterWithEureka()) {
            // Set a sensible default if we are supposed to replicate
            server.setRegistrySyncRetries(5);
        }
        server.setRemoteRegionAppWhitelist(new HashMap<>());
        return server;
    }
}
```

然后使用如下命令启动 4 个 Eureka Server：

```
mvn spring-boot:run -Dspring.profiles.active=zone1
mvn spring-boot:run -Dspring.profiles.active=zone2
mvn spring-boot:run -Dspring.profiles.active=zone3-region-west
mvn spring-boot:run -Dspring.profiles.active=zone4-region-west
```

2. Eureka Client 实例

这里我们配置 4 个 Eureka Client，也是分了 4 个 zone，属于 region-east、region-west 两个 region。其 region-east 的配置文件如代码清单 3-23、3-24 所示：

代码清单3-23 ch3-3\ch3-3-eureka-client\src\main\resources\application-zone1.yml

```yaml
server:
    port: 8071
spring:
    application.name: demo-client
eureka:
    client:
        prefer-same-zone-eureka: true
        region: region-east
        service-url:
            zone1: http://localhost:8761/eureka/
            zone2: http://localhost:8762/eureka/
        availability-zones:
            region-east: zone1,zone2
    instance:
        metadataMap.zone: zone1
```

代码清单3-24 ch3-3\ch3-3-eureka-client\src\main\resources\application-zone2.yml

```yaml
server:
    port: 8072
spring:
    application.name: demo-client
```

```yaml
eureka:
    client:
        prefer-same-zone-eureka: true
        region: region-east
        service-url:
            zone1: http://localhost:8761/eureka/
            zone2: http://localhost:8762/eureka/
        availability-zones:
            region-east: zone1,zone2
    instance:
        metadataMap.zone: zone2
```

zone1 及 zone2 属于 region-east，zone3 及 zone4 属于 region-west，具体如代码清单 3-25、3-26 所示。

代码清单3-25　ch3-3\ch3-3-eureka-client\src\main\resources\application-zone3.yml

```yaml
server:
    port: 8073
spring:
    application.name: demo-client
eureka:
    client:
        prefer-same-zone-eureka: true
        region: region-west
        service-url:
            zone3: http://localhost:8763/eureka/
            zone4: http://localhost:8764/eureka/
        availability-zones:
            region-west: zone3,zone4
    instance:
        metadataMap.zone: zone3
```

代码清单3-26　ch3-3\ch3-3-eureka-client\src\main\resources\application-zone4.yml

```yaml
server:
    port: 8074
spring:
    application.name: demo-client
eureka:
    client:
        prefer-same-zone-eureka: true
        region: region-west
        service-url:
            zone3: http://localhost:8763/eureka/
            zone4: http://localhost:8764/eureka/
        availability-zones:
            region-west: zone3,zone4
    instance:
        metadataMap.zone: zone4
```

然后使用如下命令启动 4 个 Client：

```
mvn spring-boot:run -Dspring.profiles-active=zone1
mvn spring-boot:run -Dspring.profiles-active=zone2
```

```
mvn spring-boot:run -Dspring.profiles-active=zone3
mvn spring-boot:run -Dspring.profiles-active=zone4
```

3. Zuul Gateway 实例

这里我们使用 2 个 zuul gateway 实例来演示 fallback 到 remote region 的应用实例的功能。这两个 gateway，一个属于 region-east，一个属于 region-west。其配置分别如代码清单 3-27、3-28 所示：

代码清单3-27 ch3-3\ch3-3-3-zuul-gateway \src\main\resources\application-zone1.yml

```
server:
    port: 10001
eureka:
    instance:
        metadataMap.zone: zone1
    client:
        register-with-eureka: true
        fetch-registry: true
        region: region-east
        service-url:
            zone1: http://localhost:8761/eureka/
            zone2: http://localhost:8762/eureka/
        availability-zones:
            region-east: zone1,zone2
```

代码清单3-28 ch3-3\ch3-3-zuul-gateway \src\main\resources\application-zone.yml

```
server:
    port: 10002
eureka:
    instance:
        metadataMap.zone: zone3
    client:
        register-with-eureka: true
        fetch-registry: true
        region: region-west
        service-url:
            zone3: http://localhost:8763/eureka/
            zone4: http://localhost:8764/eureka/
        availability-zones:
            region-west: zone3,zone4
```

然后使用如下命令启动 gateway：

```
mvn spring-boot:run -Dspring.profiles.active=zone1
mvn spring-boot:run -Dspring.profiles.active=zone3-region-west
```

启动之后访问：

```
curl -i http://localhost:10001/demo-client/actuator/env
curl -i http://localhost:10002/demo-client/actuator/env
```

可以看到如 3.5.2 节讲的 zoneAffinity 特性，zone1 的 gateway 访问的是 zone1 的 demo-client，zone3 的 gateway 访问的是 zone3 的 demo-client。

接下来关闭到 zone1 及 zone2 的 Eureka Client，再继续访问 curl –i http://localhost:10001/demo-client/actuator/env，可以看到经过几个报错之后，自动 fallback 到了 remote-region 的 zone3 或者 zone4 的实例，实现了类似异地多活自动转移请求的效果。有兴趣的读者可以阅读相关源码，看下 Eureka Server 到底是怎么实现跨 region 的 fallback 的。

3.5.4 开启 HTTP Basic 认证

在实际生产部署的过程中，往往需要考虑一个安全问题，比如 Eureka Server 自己有暴露 REST API，如果没有安全认证，别人就可以通过 REST API 随意修改信息，造成服务异常。这一小节，我们来看一看 Eureka Server 是如何启用 HTTP Basic 校验的，以及 Eureka Client 是如何配置相应鉴权信息的。

1. Eureka Server 配置

要启动 Eureka Server 的 HTTP Basic 认证，则需要引入 spring-boot-starter-security，如代码清单 3-29 所示。

代码清单3-29　ch3-4\ch3-4-eureka-server\pom.xml

```xml
<dependency>
    <groupId>org.springframework.boot</groupId>
    <artifactId>spring-boot-starter-security</artifactId>
</dependency>
```

另外需要在配置文件中指定账户密码，这块可以跟 config-server 的加密功能结合，如代码清单 3-30 所示。

代码清单3-30　ch3-4\ch3-4-eureka-server\src\main\resources\application-security.yml

```yaml
server:
    port: 8761

spring:
    security:
        basic:
            enabled: true
        user:
            name: admin
            password: Xk38CNHigBP5jK75
eureka:
    instance:
        hostname: localhost
    client:
        registerWithEureka: false
        fetchRegistry: false
        serviceUrl:
            defaultZone: http://${eureka.instance.hostname}:${server.port}/eureka/
    server:
        waitTimeInMsWhenSyncEmpty: 0
        enableSelfPreservation: false
```

另外，由于 spring-boot-starter-security 默认开启了 csrf 校验，对于 Client 端这类非界面应

用来说不合适,但是又没有配置文件的方式可以禁用,需要自己通过 Java 的配置文件禁用下,如代码清单 3-31 所示。

代码清单3-31 ch3-4\ch3-4-eureka-server\src\main\java\cn\springcloud\book\config\SecurityConfig.java

```java
@EnableWebSecurity
public class SecurityConfig extends WebSecurityConfigurerAdapter {

    @Override
    protected void configure(HttpSecurity http) throws Exception {
        super.configure(http);
        http.csrf().disable();
    }
}
```

然后使用 security 的 profile 启动 Eureka Server:

```
mvn spring-boot:run -Dspring.profiles.active=security
```

然后如下所示访问:

```
curl -i http://localhost:8761/eureka/apps
HTTP/1.1 401
Set-Cookie: JSESSIONID=D7D019318B2E5D011C3000759659FE1C; Path=/; HttpOnly
WWW-Authenticate: Basic realm="Realm"
X-Content-Type-Options: nosniff
X-XSS-Protection: 1; mode=block
Cache-Control: no-cache, no-store, max-age=0, must-revalidate
Pragma: no-cache
Expires: 0
X-Frame-Options: DENY
Content-Type: application/json;charset=UTF-8
Transfer-Encoding: chunked
Date: Mon, 25 Jun 2018 09:11:07 GMT

{"timestamp":"2018-06-25T09:11:07.832+0000","status":401,"error":"Unauthorized","message":"Unauthorized","path":"/eureka/apps"}
```

可以看到,没有传递 Authorization 的 header,返回 401。

接下来使用 HTTP Basic 的账号密码传递 Authorization 的 header,如下:

```
curl -i --basic -u admin:Xk38CNHigBP5jK75 http://localhost:8761/eureka/apps
HTTP/1.1 200
Set-Cookie: JSESSIONID=8B745BEA3606E8F4856A6197407D8433; Path=/; HttpOnly
X-Content-Type-Options: nosniff
X-XSS-Protection: 1; mode=block
Cache-Control: no-cache, no-store, max-age=0, must-revalidate
Pragma: no-cache
Expires: 0
X-Frame-Options: DENY
Content-Type: application/xml
Transfer-Encoding: chunked
Date: Mon, 25 Jun 2018 09:28:29 GMT
```

```xml
<applications>
    <versions__delta>1</versions__delta>
    <apps__hashcode></apps__hashcode>
</applications>
```

可以看到请求成功返回。

2. Eureka Client 配置

由于 Eureka Server 开启了 HTTP Basic 认证，Eureka Client 也需要配置相应的账号信息来传递，这里我们通过配置文件来指定，相关的密码也结合 config-server 的加密功能来加密，如代码清单 3-32 所示。

代码清单3-32　ch3-4\ch3-4-eureka-client\src\main\resources\application-security.yml

```yaml
server:
    port: 8081

spring:
    application:
        name: client1

eureka:
    client:
        security:
            basic:
                user: admin
                password: Xk38CNHigBP5jK75
        serviceUrl:
            defaultZone: http://${eureka.client.security.basic.user}:${eureka.client.security.basic.password}@localhost:8761/eureka/
```

然后使用如下命令启动：

```
mvn spring-boot:run -Dspring.profiles.active=security
```

之后执行如下命令查看：

```
curl -i --basic -u admin:Xk38CNHigBP5jK75 http://localhost:8761/eureka/apps
HTTP/1.1 200
Set-Cookie: JSESSIONID=0CCE2E3E092CF499CF509F1B799ED837; Path=/; HttpOnly
X-Content-Type-Options: nosniff
X-XSS-Protection: 1; mode=block
Cache-Control: no-cache, no-store, max-age=0, must-revalidate
Pragma: no-cache
Expires: 0
X-Frame-Options: DENY
Content-Type: application/xml
Transfer-Encoding: chunked
Date: Mon, 25 Jun 2018 09:22:49 GMT

<applications>
    <versions__delta>1</versions__delta>
    <apps__hashcode>UP_1_</apps__hashcode>
    <application>
```

```xml
        <name>CLIENT1</name>
        <instance>
            <instanceId>10.2.238.79:client1:8081</instanceId>
            <hostName>10.2.238.79</hostName>
            <app>CLIENT1</app>
            <ipAddr>10.2.238.79</ipAddr>
            <status>UP</status>
            <overriddenstatus>UNKNOWN</overriddenstatus>
            <port enabled="true">8081</port>
            <securePort enabled="false">443</securePort>
            <countryId>1</countryId>
            <dataCenterInfo class="com.netflix.appinfo.InstanceInfo$DefaultDataC
                enterInfo">
                <name>MyOwn</name>
            </dataCenterInfo>
            <leaseInfo>
                <renewalIntervalInSecs>30</renewalIntervalInSecs>
                <durationInSecs>90</durationInSecs>
                <registrationTimestamp>1529917089388</registrationTimestamp>
                <lastRenewalTimestamp>1529918469122</lastRenewalTimestamp>
                <evictionTimestamp>0</evictionTimestamp>
                <serviceUpTimestamp>1529917089389</serviceUpTimestamp>
            </leaseInfo>
            <metadata>
                <management.port>8081</management.port>
                <jmx.port>49950</jmx.port>
            </metadata>
            <homePageUrl>http://10.2.238.79:8081/</homePageUrl>
            <statusPageUrl>http://10.2.238.79:8081/actuator/info</statusPageUrl>
            <healthCheckUrl>http://10.2.238.79:8081/actuator/health</healthCheckUrl>
            <vipAddress>client1</vipAddress>
            <secureVipAddress>client1</secureVipAddress>
            <isCoordinatingDiscoveryServer>false</isCoordinatingDiscoveryServer>
            <lastUpdatedTimestamp>1529917089389</lastUpdatedTimestamp>
            <lastDirtyTimestamp>1529917088523</lastDirtyTimestamp>
            <actionType>ADDED</actionType>
        </instance>
    </application>
</applications>
```

可以看到 Client 已经注册成功。

3.5.5 启用 https

对于上面开启 HTTP Basic 认证来说，从安全角度讲，基于 base64 编码很容易被抓包然后破解，如果暴露在公网会非常不安全，这里就讲述一下如何在 Eureka Server 及 Client 开启 https，来达到这个目的。

1. 证书生成

```
keytool -genkeypair -alias server -storetype PKCS12 -keyalg RSA -keysize 2048
    -keystore server.p12 -validity 3650
输入密钥库口令:
再次输入新口令:
```

```
您的名字与姓氏是什么?
  [Unknown]:
您的组织单位名称是什么?
  [Unknown]:
您的组织名称是什么?
  [Unknown]:
您所在的城市或区域名称是什么?
  [Unknown]:
您所在的省/市/自治区名称是什么?
  [Unknown]:
该单位的双字母国家/地区代码是什么?
  [Unknown]:
CN=Unknown, OU=Unknown, O=Unknown, L=Unknown, ST=Unknown, C=Unknown是否正确?
  [否]:   Y
```

这里使用的密码是 Spring Cloud,然后会在当前目录下生成一个名为 server.p12 的文件。

下面生成 Client 端使用的证书:

```
keytool -genkeypair -alias client -storetype PKCS12 -keyalg RSA -keysize 2048
    -keystore client.p12 -validity 3650
输入密钥库口令:
再次输入新口令:
您的名字与姓氏是什么?
  [Unknown]:
您的组织单位名称是什么?
  [Unknown]:
您的组织名称是什么?
  [Unknown]:
您所在的城市或区域名称是什么?
  [Unknown]:
您所在的省/市/自治区名称是什么?
  [Unknown]:
该单位的双字母国家/地区代码是什么?
  [Unknown]:
CN=Unknown, OU=Unknown, O=Unknown, L=Unknown, ST=Unknown, C=Unknown是否正确?
  [否]:   Y
```

这里使用的密码是 Client,然后会在当前目录下生成一个名为 client.p12 的文件。

下面分别导出两个 p12 的证书,如下:

```
keytool -export -alias server -file server.crt --keystore server.p12
输入密钥库口令:
存储在文件 <server.crt> 中的证书

keytool -export -alias client -file client.crt --keystore client.p12
输入密钥库口令:
存储在文件 <client.crt> 中的证书
```

接下来将 server.crt 文件导入 client.p12 中,使 Client 端信任 Server 的证书:

```
keytool -import -alias server -file server.crt -keystore client.p12
输入密钥库口令:
所有者:CN=Unknown, OU=Unknown, O=Unknown, L=Unknown, ST=Unknown, C=Unknown
发布者: CN=Unknown, OU=Unknown, O=Unknown, L=Unknown, ST=Unknown, C=Unknown
序列号: 5249cc11
```

```
有效期为Mon Jun 25 18:53:20 CST 2018 至Thu Jun 22 18:53:20 CST 2028
证书指纹：
    MD5:    4D:27:25:0E:A2:6A:7A:0C:81:D2:89:35:12:61:3E:16
    SHA1:   A3:5E:8E:09:F8:B1:44:9C:B5:AC:AB:2E:F2:7A:58:95:7F:02:69:C4
    SHA256: 93:3B:9F:CA:74:D3:88:19:69:7F:65:E0:4F:DF:E0:71:C6:3E:5F:BC:FF:7F:4
            F:0F:39:43:D7:22:A6:87:96:8C
签名算法名称: SHA256withRSA
主体公共密钥算法: 2048 位RSA密钥
版本: 3

扩展:

#1: ObjectId: 2.5.29.14 Criticality=false
SubjectKeyIdentifier [
KeyIdentifier [
0000: BE A8 E5 3D D6 8E 58 47    CB C4 17 2A 8D 4F 50 1D    ...=..XG...*.OP.
0010: 83 B8 3E 24                                           ..>$
]
]

是否信任此证书？[否]：  Y
证书已添加到密钥库中
```

这里需要输入的是 client.p12 的密钥。

然后将 client.crt 导入 server.p12 中，使得 Server 信任 Client 的证书：

```
keytool -import -alias client -file client.crt -keystore server.p12
输入密钥库口令：
所有者：CN=Unknown, OU=Unknown, O=Unknown, L=Unknown, ST=Unknown, C=Unknown
发布者：CN=Unknown, OU=Unknown, O=Unknown, L=Unknown, ST=Unknown, C=Unknown
序列号：5c68914d
有效期为Tue Jun 26 09:30:47 CST 2018 至Fri Jun 23 09:30:47 CST 2028
证书指纹：
    MD5:    4C:0F:35:63:CE:47:A9:C5:90:7C:B2:7D:07:CE:67:DC
    SHA1:   E2:E0:DD:E3:3F:84:DF:21:F5:FB:CA:F5:A9:FB:3C:CD:08:AE:3E:C3
    SHA256: 65:B4:49:C0:1D:C3:7B:0C:1B:4D:13:67:91:1F:5E:18:6F:F7:0E:AD:64:D4:D
            9:11:97:DB:55:BB:D4:E3:3F:D2
签名算法名称: SHA256withRSA
主体公共密钥算法: 2048 位RSA密钥
版本: 3

扩展:

#1: ObjectId: 2.5.29.14 Criticality=false
SubjectKeyIdentifier [
KeyIdentifier [
0000: 1F 49 7D 24 2F E0 7B 2E    F2 F7 19 A2 48 23 4D 73    .I.$/.......H#Ms
0010: 1D DC 99 0B                                           ....
]
]

是否信任此证书？[否]：  Y
证书已添加到密钥库中
```

这里需要输入的是 server.p12 的密钥。

2. Eureka Server 配置

把生成的 server.p12 放到 Maven 工程的 resources 目录下，然后指定相关配置如代码清单 3-33 所示。

代码清单3-33 ch3-5\ch3-5-eureka-server\src\main\resources\application-https.yml

```yaml
server:
    port: 8761
    ssl:
        enabled: true
        key-store: classpath:server.p12
        key-store-password: springcloud
        key-store-type: PKCS12
        key-alias: server
eureka:
    instance:
        hostname: localhost
        securePort: ${server.port}
        securePortEnabled: true
        nonSecurePortEnabled: false
        homePageUrl: https://${eureka.instance.hostname}:${server.port}/
        statusPageUrl: https://${eureka.instance.hostname}:${server.port}/
    client:
        registerWithEureka: false
        fetchRegistry: false
        serviceUrl:
            defaultZone: https://${eureka.instance.hostname}:${server.port}/eureka/
    server:
        waitTimeInMsWhenSyncEmpty: 0
        enableSelfPreservation: false
```

这里主要是指定 server.ssl 配置，以及 eureka.instance 的 securePortEnabled 及 eureka.instance.securePort 配置。

使用 https 的 profile 启动如下：

```
mvn spring-boot:run -Dspring.profiles.active=https
```

之后访问 https://localhost:8761/，可以看到 https 已经启用。

3. Eureka Client 配置

把生成的 client.p12 放到 Maven 工程的 resources 目录下，然后指定相关配置如代码清单 3-34 所示。

代码清单3-34 ch3-5\ch3-5-eureka-client\src\main\resources\application-https.yml

```yaml
server:
    port: 8081

spring:
    application:
        name: client1

eureka:
```

```yaml
client:
    securePortEnabled: true
    ssl:
        key-store: client.p12
        key-store-password: client
    serviceUrl:
        defaultZone: https://localhost:8761/eureka/
```

这里我们没有指定整个应用实例启用 https，仅仅是开启访问 Eureka Server 的 https 配置。通过自定义 eureka.client.ssl.key-store 以及 eureka.client.ssl.key-store-password 两个属性，指定 Eureka Client 访问 Eureka Server 的 sslContext 配置，这里需要在代码里指定 DiscoveryClient.DiscoveryClientOptionalArgs，配置如代码清单 3-35 所示。

代码清单3-35 ch3-5\ch3-5-eureka-client\src\main\java\cn\springcloud\book\config\EurekaHttpsClientConfig.java

```java
@Configuration
public class EurekaHttpsClientConfig {

    @Value("${eureka.client.ssl.key-store}")
    String keyStoreFileName;

    @Value("${eureka.client.ssl.key-store-password}")
    String keyStorePassword;

    @Bean
    public DiscoveryClient.DiscoveryClientOptionalArgs discoveryClientOptionalArgs()
        throws CertificateException, NoSuchAlgorithmException, KeyStoreException,
        IOException, KeyManagementException {
        EurekaJerseyClientImpl.EurekaJerseyClientBuilder builder = new
            EurekaJerseyClientImpl.EurekaJerseyClientBuilder();
        builder.withClientName("eureka-https-client");
        SSLContext sslContext = new SSLContextBuilder()
            .loadTrustMaterial(
                this.getClass().getClassLoader().getResource(keyStoreFileName),
                keyStorePassword.toCharArray()
            )
            .build();
        builder.withCustomSSL(sslContext);

        builder.withMaxTotalConnections(10);
        builder.withMaxConnectionsPerHost(10);

        DiscoveryClient.DiscoveryClientOptionalArgs args = new DiscoveryClient.
            DiscoveryClientOptionalArgs();
        args.setEurekaJerseyClient(builder.build());
        return args;
    }
}
```

然后使用如下命令启动：

```
mvn spring-boot:run -Dspring.profiles.active=https
```

查询 Eureka Server 可以看到已经成功注册上：

```
curl --insecure https://localhost:8761/eureka/apps
<applications>
    <versions__delta>1</versions__delta>
    <apps__hashcode>UP_1_</apps__hashcode>
    <application>
        <name>CLIENT1</name>
        <instance>
            <instanceId>10.2.238.208:client1:8081</instanceId>
            <hostName>10.2.238.208</hostName>
            <app>CLIENT1</app>
            <ipAddr>10.2.238.208</ipAddr>
            <status>UP</status>
            <overriddenstatus>UNKNOWN</overriddenstatus>
            <port enabled="true">8081</port>
            <securePort enabled="false">443</securePort>
            <countryId>1</countryId>
            <dataCenterInfo class="com.netflix.appinfo.InstanceInfo$DefaultDataC
                enterInfo">
                <name>MyOwn</name>
            </dataCenterInfo>
            <leaseInfo>
                <renewalIntervalInSecs>30</renewalIntervalInSecs>
                <durationInSecs>90</durationInSecs>
                <registrationTimestamp>1529977873082</registrationTimestamp>
                <lastRenewalTimestamp>1529978503987</lastRenewalTimestamp>
                <evictionTimestamp>0</evictionTimestamp>
                <serviceUpTimestamp>1529977873082</serviceUpTimestamp>
            </leaseInfo>
            <metadata>
                <management.port>8081</management.port>
            </metadata>
            <homePageUrl>http://10.2.238.208:8081/</homePageUrl>
            <statusPageUrl>http://10.2.238.208:8081/actuator/info</statusPageUrl>
            <healthCheckUrl>http://10.2.238.208:8081/actuator/health</
                healthCheckUrl>
            <vipAddress>client1</vipAddress>
            <secureVipAddress>client1</secureVipAddress>
            <isCoordinatingDiscoveryServer>false</isCoordinatingDiscoveryServer>
            <lastUpdatedTimestamp>1529977873082</lastUpdatedTimestamp>
            <lastDirtyTimestamp>1529977872969</lastDirtyTimestamp>
            <actionType>ADDED</actionType>
        </instance>
    </application>
</applications>
```

3.5.6 Eureka Admin

由于在 Spring Cloud 的技术栈中，Eureka Server 是微服务架构中必不可少的基础组件，所以 Spring Cloud 中国社区开放了 Eureka Server 在线服务供大家学习或测试，访问地址为 http://eureka.springcloud.cn/。

除此之外，Spring Cloud 中国社区为 Eureka 注册中心开源了一个节点监控、服务动态启停的管控平台：Eureka Admin，其在 GitHub 的地址为 https://github.com/SpringCloud/eureka-admin/。

该项目的页面预览图如图 3-2 所示。

图 3-2　eureka admin 首页

通过管控平台的界面可以对注册的服务实例进行上线、下线、停止等操作，省得自己再去调用 Eureka Server 的 REST API，非常方便。

3.5.7　基于 metadata 路由实例

对于 Eureka 来说，最常见的就是通过 metadata 属性，进行灰度控制或者是不宕机升级。这里结合 Netflix Ribbon 的例子，介绍一下这类应用场景的实现。

1. ILoadBalancer 接口

Netflix Ribbon 的 ILoadBalancer 接口定义了 loadBalancer 的几个基本方法，如下：

```
public interface ILoadBalancer
    public void addServers(List<Server> newServers);
    public Server chooseServer(Object key);
    public void markServerDown(Server server);
    @Deprecated
    public List<Server> getServerList(boolean availableOnly);
    public List<Server> getReachableServers();
    public List<Server> getAllServers();
}
```

可以看到这里有个 chooseServer 方法，用于从一堆服务实例列表中进行过滤，选取一个 Server 出来，给客户端请求用。

在 Ribbon 中，ILoadBalancer 选取 Server 的逻辑主要由一系列 IRule 来实现。

2. IRule 接口

```
public interface IRule{
    public Server choose(Object key);
    public void setLoadBalancer(ILoadBalancer lb);
    public ILoadBalancer getLoadBalancer();
}
```

最常见的 IRule 接口有 RoundRobinRule，采用轮询调度算法规则来选取 Server，其主要代码如下：

```
public Server choose(ILoadBalancer lb, Object key) {
    if (lb == null) {
        log.warn("no load balancer");
        return null;
    }
```

```
Server server = null;
int count = 0;
while (server == null && count++ < 10) {
    List<Server> reachableServers = lb.getReachableServers();
    List<Server> allServers = lb.getAllServers();
    int upCount = reachableServers.size();
    int serverCount = allServers.size();

    if ((upCount == 0) || (serverCount == 0)) {
        log.warn("No up servers available from load balancer: " + lb);
        return null;
    }

    int nextServerIndex = incrementAndGetModulo(serverCount);
    server = allServers.get(nextServerIndex);

    if (server == null) {
        /* Transient. */
        Thread.yield();
        continue;
    }

    if (server.isAlive() && (server.isReadyToServe())) {
        return (server);
    }

    // Next.
    server = null;
}

if (count >= 10) {
    log.warn("No available alive servers after 10 tries from load balancer: "
            + lb);
}
return server;
}
```

3. MetadataAwarePredicate

这里，由于我们需要根据实例的 metadata 进行过滤，因此，可以自定义实现自己的 rule。Netflix 提供了 PredicateBasedRule，可以基于 Guava 的 Predicate 进行过滤。jmnarloch 在《Spring Cloud: Ribbon dynamic routing》(https://jmnarloch.wordpress.com/2015/11/25/spring-cloud-ribbon-dynamic-routing/ ⊖) 中给出了针对 metadata 过滤的 rule，如下：

```
public class MetadataAwarePredicate extends DiscoveryEnabledPredicate {

    @Override
    protected boolean apply(DiscoveryEnabledServer server) {
        final RibbonFilterContext context = RibbonFilterContextHolder.
           getCurrentContext();
        final Set<Map.Entry<String, String>> attributes = Collections.
```

⊖ 注：若打不开，请使用网络代理。

```
            unmodifiableSet(context.getAttributes().entrySet());
        final Map<String, String> metadata = server.getInstanceInfo().
            getMetadata();
        return metadata.entrySet().containsAll(attributes);
    }
}
```

这个 Predicate 将 Server 的 metadata 跟上下文传递的 attributes 信息进行匹配，全部匹配上才返回 true。比如 attributes 的 map 有个 entry，key 是 env，value 是 canary，表示该实例是 canary 实例，如果请求上下文要求路由到 canary 实例，可以从 request url 参数或者 header 中标识这个路由请求，然后携带到上下文中，最后由 Predicate 进行判断，完成整个 ILoadBalancer 的 choose。

3.6　Eureka 故障演练

Chaos Engineering，中文翻译为混沌工程，它通过一系列可控的实验模拟真实世界可能出现的故障，来暴露系统的缺陷，从而驱动工程师去构建更具弹性的服务。常见的手段有：随机关闭依赖服务、模拟增加依赖服务的延时、模拟机器故障、网络中断等。对这方面有兴趣的读者可以查看《Chaos Engineering：Building Confidence in System Behavior through Experiments》（https://www.oreilly.com/webops-perf/free/chaos-engineering.csp）这本书。

这里我们手工模拟如下几个故障场景，通过这些故障来阐述 Eureka 的高可用。

3.6.1　Eureka Server 全部不可用

假设 Eureka Server 全部挂掉，这里我们假设是同时都挂掉，然后分为下面两个场景来讨论。

1. 应用服务启动前不可用

如果 Eureka Server 在应用服务启动之前挂掉或者没有启动的话，那么应用可以正常启动，但是会有报错信息，如下：

```
2018-05-28 22:16:16.603 ERROR 2048 --- [nfoReplicator-0] c.n.d.s.t.d.Redirectin
    gEurekaHttpClient        : Request execution error

com.sun.jersey.api.client.ClientHandlerException: java.net.ConnectException:
    Connection refused
    at com.sun.jersey.client.apache4.ApacheHttpClient4Handler.
        handle(ApacheHttpClient4Handler.java:187) ~[jersey-apache-client4-
        1.19.1.jar:1.19.1]
    at com.sun.jersey.api.client.filter.GZIPContentEncodingFilter.handle(GZIPCon
        tentEncodingFilter.java:123) ~[jersey-client-1.19.1.jar:1.19.1]
    at com.netflix.discovery.EurekaIdentityHeaderFilter.handle(EurekaIdentityHea
        derFilter.java:27) ~[eureka-client-1.8.8.jar:1.8.8]
    at com.sun.jersey.api.client.Client.handle(Client.java:652) ~[jersey-client-
        1.19.1.jar:1.19.1]
    at com.sun.jersey.api.client.WebResource.handle(WebResource.java:682)
        ~[jersey-client-1.19.1.jar:1.19.1]
    at com.sun.jersey.api.client.WebResource.access$200(WebResource.java:74)
```

```
            ~[jersey-client-1.19.1.jar:1.19.1]
        at com.sun.jersey.api.client.WebResource$Builder.post(WebResource.java:570)
            ~[jersey-client-1.19.1.jar:1.19.1]
        at com.netflix.discovery.shared.transport.jersey.AbstractJerseyEurekaHttpClient.
            register(AbstractJerseyEurekaHttpClient.java:56) ~[eureka-client-
            1.8.8.jar:1.8.8]
        at com.netflix.discovery.shared.transport.decorator.EurekaHttpClientDecorator$1.
            execute(EurekaHttpClientDecorator.java:59) [eureka-client-1.8.8.jar:1.8.8]
        at com.netflix.discovery.shared.transport.decorator.MetricsCollectingEureka
            HttpClient.execute(MetricsCollectingEurekaHttpClient.java:73) ~[eureka-
            client-1.8.8.jar:1.8.8]
```

由于连不上 Eureka Server，自然访问不了 service registry 的服务注册信息，不能与其他服务交互。针对这种情况，Eureka Server 设计了一个 eureka.client.backup-registry-impl 属性，可以配置在启动时 Eureka Server 访问不到的情况下，从这个 back registry 读取服务注册信息，作为 fallback。该 backup-registry-impl 比较适合服务端提供负载均衡或者服务 ip 地址相对固定的场景。实例如下：

```java
public class StaticBackupServiceRegistry implements BackupRegistry {

    private Applications localRegionApps = new Applications();

    public StaticBackupServiceRegistry() {
        Application orgApplication = new Application("org");
        InstanceInfo orgInstance1 = InstanceInfo.Builder.newBuilder()
            .setAppName("org-service")
            .setVIPAddress("org-service")
            .setSecureVIPAddress("org-service")
            .setInstanceId("org-instance-1")
            .setHostName("192.168.99.100")
            .setIPAddr("192.168.99.100")
            .setPort(9090)
            .setDataCenterInfo(new MyDataCenterInfo(DataCenterInfo.Name.MyOwn))
            .setStatus(InstanceInfo.InstanceStatus.UP)
            .build();
        InstanceInfo orgInstance2 = InstanceInfo.Builder.newBuilder()
            .setAppName("org-service")
            .setVIPAddress("org-service")
            .setSecureVIPAddress("org-service")
            .setInstanceId("org-instance-1")
            .setHostName("192.168.99.100")
            .setIPAddr("192.168.99.100")
            .setPort(9091)
            .setDataCenterInfo(new MyDataCenterInfo(DataCenterInfo.Name.MyOwn))
            .setStatus(InstanceInfo.InstanceStatus.UP)
            .build();
        orgApplication.addInstance(orgInstance1);
        orgApplication.addInstance(orgInstance2);
        localRegionApps.addApplication(orgApplication);
    }

    @Override
    public Applications fetchRegistry() {
        return localRegionApps;
```

```
    }

    @Override
    public Applications fetchRegistry(String[] includeRemoteRegions) {
        //ignore remote regions
        return localRegionApps;
    }
}
```

2. 应用服务运行时不可用

Eureka Client 在本地内存中有个 AtomicReference<Applications> 类型的 localRegionApps 变量，来维护从 Eureka Server 拉取回来的注册信息。Client 端有个定时任务 CacheRefreshThread，会定时从 Server 端拉取注册信息更新到本地，如果 Eureka Server 在应用服务运行时挂掉的话，本地的 CacheRefreshThread 会抛出异常，本地的 localRegionApps 变量不会得到更新。

异常输出实例如下：

```
2018-05-29 23:12:59.888 ERROR 1908 --- [freshExecutor-0] c.n.d.s.t.d.Redirectin
    gEurekaHttpClient        : Request execution error

com.sun.jersey.api.client.ClientHandlerException: java.net.ConnectException:
    Connection refused (Connection refused)
    at com.sun.jersey.client.apache4.ApacheHttpClient4Handler.
        handle(ApacheHttpClient4Handler.java:187) ~[jersey-apache-client4-
        1.19.1.jar:1.19.1]
    at com.sun.jersey.api.client.filter.GZIPContentEncodingFilter.handle(GZIPCon
        tentEncodingFilter.java:123) ~[jersey-client-1.19.1.jar:1.19.1]
    at com.netflix.discovery.EurekaIdentityHeaderFilter.handle(EurekaIdentityHea
        derFilter.java:27) ~[eureka-client-1.8.8.jar:1.8.8]
    at com.sun.jersey.api.client.Client.handle(Client.java:652) ~[jersey-client-
        1.19.1.jar:1.19.1]
    at com.sun.jersey.api.client.WebResource.handle(WebResource.java:682)
        ~[jersey-client-1.19.1.jar:1.19.1]
    at com.sun.jersey.api.client.WebResource.access$200(WebResource.java:74)
        ~[jersey-client-1.19.1.jar:1.19.1]
    at com.sun.jersey.api.client.WebResource$Builder.get(WebResource.java:509)
        ~[jersey-client-1.19.1.jar:1.19.1]
```

可以看到名为 freshExecutor-0 的线程请求 Eureka Server 时抛出了异常。

3.6.2 Eureka Server 部分不可用

这里我们分 Client 端及 Server 端两个方面来阐述。

1. Client 端

Client 端有个定时任务（'AsyncResolver.updateTask'）去拉取 serviceUrl 的变更，如果配置文件有改动，运行时可以动态变更。拉取完之后，Client 端会随机化 Server 的 list。例如：

```
eureka:
    client:
        serviceUrl:
```

```
defaultZone: http://host1:8761/eureka/,http://host2:8762/eureka/,http://
host3:8763/eureka/
```

第一次拉取的时候可能是按配置的顺序，如 host1、host2、host3 这样，之后定时任务更新会随机化一次，变为 host2、host1、host3 这样。

而 Client 端在请求 Server 的时候，维护了一个不可用的 Eureka Server 列表（'quarantineSet，在 Connection error 或者 5xx 的情况下会被列入该列表，当该列表的大小超过指定阈值则会重新清空'）；对可用的 Server 列表（'一般为拉取回来的 Server 列表剔除不可用的列表，如果剔除之后为空，则不会做剔除处理'），采用 RetryableEurekaHttpClient 进行请求，numberOfRetries 为 3。也就是说，如果 Eureka Server 有一台挂掉，则会被纳入不可用列表。那么这个时候获取的服务注册信息是来自健康的 Eureka Server。

2. Server 端

Eureka Server 之间相互成为 peer node，如果 Eureka Server 有一台挂了，则 Eureka Server 之间的 replication 会受影响。

PeerEurekaNodes 有个定时任务（peersUpdateTask），会去从配置文件拉取 availabilityZones 及 serviceUrl 信息，然后在运行时更新 peerEurekaNodes 信息。如果在一台 Eureka Server 挂掉的时候，人工介入更改 Eureka Server 的 serviceUrl 信息，则可以主动剔除挂掉的 peerNode。

如果未能及时剔除，则会报错：

```
2018-06-03 23:00:59.069 ERROR 4562 --- [get_localhost-3] c.n.e.cluster.
    ReplicationTaskProcessor    : Network level connection to peer localhost;
    retrying after delay

com.sun.jersey.api.client.ClientHandlerException: java.net.ConnectException:
    Connection refused
    at com.sun.jersey.client.apache4.ApacheHttpClient4Handler.
        handle(ApacheHttpClient4Handler.java:187) ~[jersey-apache-client4-
        1.19.1.jar:1.19.1]
    at com.netflix.eureka.cluster.DynamicGZIPContentEncodingFilter.handle(Dynami
        cGZIPContentEncodingFilter.java:48) ~[eureka-core-1.8.8.jar:1.8.8]
    at com.netflix.discovery.EurekaIdentityHeaderFilter.handle(EurekaIdentityHea
        derFilter.java:27) ~[eureka-client-1.8.8.jar:1.8.8]
    at com.sun.jersey.api.client.Client.handle(Client.java:652) ~[jersey-client-
        1.19.1.jar:1.19.1]
    at com.sun.jersey.api.client.WebResource.handle(WebResource.java:682)
        ~[jersey-client-1.19.1.jar:1.19.1]
    at com.sun.jersey.api.client.WebResource.access$200(WebResource.java:74)
        ~[jersey-client-1.19.1.jar:1.19.1]
    at com.sun.jersey.api.client.WebResource$Builder.post(WebResource.java:570)
        ~[jersey-client-1.19.1.jar:1.19.1]
    at com.netflix.eureka.transport.JerseyReplicationClient.submitBatchUpdates(J
        erseyReplicationClient.java:116) ~[eureka-core-1.8.8.jar:1.8.8]
    at com.netflix.eureka.cluster.ReplicationTaskProcessor.process(ReplicationTa
        skProcessor.java:80) ~[eureka-core-1.8.8.jar:1.8.8]
    at com.netflix.eureka.util.batcher.TaskExecutors$BatchWorkerRunnable.
        run(TaskExecutors.java:187) [eureka-core-1.8.8.jar:1.8.8]
    at java.lang.Thread.run(Thread.java:745) [na:1.8.0_71]
```

Eureka Server 目前没有对 peerEurekaNodes 进行健康检查，源码中可以看到 eureka.server.peer-eureka-status-refresh-time-interval-ms 配置，不过没看到使用的地方，推测这个配置是给后续对 peer node 进行健康检查用。

3.6.3　Eureka 高可用原理

由于 Eureka 是基于部署在 Amazon 的背景下设计的，因此其原生支持了 Amazon 的 Region 及 AvailabilityZone。

这里分为如下几点来阐述一下：

1）Region，默认情况下资源在 Region 之间是不会复制的。不过 Eureka Client 提供了 fetch-remote-regions-registry 配置，这个配置在 dataCenterInfo 是 Amazon 时才生效，其作用是拉取远程 Region 的注册信息到本地。

2）AvailabilityZone，默认 Eureka Client 的 eureka.client.prefer-same-zone-eureka 配置为 true，也就是在拉取 serviceUrl 的时候，优先选取与应用实例处于同一个 zone 的 Eureka Server 列表。

```
eureka:
    client:
        register-with-eureka: true
        fetch-registry: true
        region: region-east
        service-url:
            zone1: http://localhost:8761/eureka/
            zone2: http:// localhost:8762/eureka/
        availability-zones:
            region-east: zone1,zone2
    instance:
        metadataMap.zone: zone1
```

比如上述配置，client 端配置了一个 region，该 region 下面有两个 zone，分别是 zone1 和 zone2，而该应用实例的 metadataMap 信息中 zone 为 zone1，即该应用实例会选取 service-url.zone1 这个 server 列表来进行注册及查询等操作。

如果 eureka.client.prefer-same-zone-eureka 配置为 false，则默认返回的 ZoneOffset 为 0，即取 availZones 的第一个 zone。

1. Client 端高可用

从上面的故障演练，我们可以看到 Client 端的高可用可以分为如下几点。

1）在 Client 启动之前，如果没有 Eureka Server，则可以通过配置 eureka.client.backup-registry-impl 从备份 registry 读取关键服务的信息。

2）在 client 启动之后，若运行时出现 Eureka Server 全部挂掉的情况：本地内存有 localRegion 之前获取的数据，在 Eureka Server 都不可用的情况下，从 Server 端定时拉取注册信息回来更新的线程 CacheRefreshThread 会执行失败，本地 localRegion 信息不会被更新。

3）在 client 启动之后，若运行时出现 Eureka Server 出现部分挂掉的情况：这种情况，如果预计恢复时间比较长，可以人工介入，可以通过配置文件剔除挂掉的 Eureka Server 地址，Client 端会定时刷新 serviceUrl，不过一般 Client 太多，这类操作不太方便，所以一般也不需要这样做，因为 Client 端维护了一个 Eureka Server 的不可用列表，一旦请求发生 Connection error 或者 5xx 的情况则会被列入该列表，当该列表的大小超过指定阈值则会重新清空。在重新清空的情况下，Client 默认是采用 RetryableEurekaHttpClient 进行请求，numberOfRetries 为 3，因此也能够在一定程度保障 Client 的高可用。

2. Server 端高可用

由于 Eureka Server 采用的是 peer to peer 的架构模式，因而也就无所谓 Server 端的高可用，主要的高可用都在 Client 端进行处理了。不过 Server 端也支持了跨 region 基本的高可用，可以通过配置 remoteRegionUrlsWithName 来支持拉取远程 region 的实例信息，如果当期 region 要访问的服务实例都挂了，那么 Server 端就会 fallback 到远程 region 的该服务实例。具体在这段代码中：eureka-core-1.9.2-sources.jar!/com/netflix/eureka/registry/AbstractInstanceRegistry.java，disableTransparentFallback 默认为 false，有兴趣的读者可以深入了解下。

```
public Applications getApplications() {
    boolean disableTransparentFallback = serverConfig.disableTransparentFallback
        ToOtherRegion();
    if (disableTransparentFallback) {
        return getApplicationsFromLocalRegionOnly();
    } else {
        return getApplicationsFromAllRemoteRegions();    // Behavior of falling
            back to remote region can be disabled.
    }
}
```

另外一点就是 Eureka Server 有 SELF PRESERVATION 机制，在现实环境中，网络抖动、闪断等因素可能会造成 Client 端未能及时向 Server 端续约，按照租约的相关约定，Server 端就需要剔除该实例信息，如果大面积出现这类情况，那么可能是误判，因此 SELF PRESERVATION 机制引入一个阈值，如果最近一分钟接收到的续约的次数小于指定阈值的话，则关闭租约失效剔除，禁止定时任务剔除失效的实例，从而保护注册信息。

3.7　本章小结

本章讲解了 Eureka 的核心类及核心操作方法，以及它的设计理念。之后通过解读 Eureka 的相关参数，针对一些常见问题，给出了调优的方案，同时针对指标的监控也进行相应解读。后面的实战部分，分别列举了 Eureka Server 在线扩容、构建 Multi Zone Eureka Server、支持 Remote Region、开启 HTTP Basic 认证、启用 https、基于 Eureka metadata 来实现灰度控制及平滑升级等案例；最后通过 Eureka 的故障演练来讲解 Eureka 的高可用架构，相信读者会对 Eureka 有更深刻的理解。

第 4 章

Spring Cloud Feign 的使用扩展

在使用 Spring Cloud 开发微服务应用时，各个服务提供者都是以 HTTP 接口的形式对外提供服务，因此在服务消费者调用服务提供者时，底层通过 HTTP Client 的方式访问。当然我们可以使用 JDK 原生的 URLConnection、Apache 的 HTTP Client、Netty 的异步 HTTP Client、Spring 的 RestTemplate 去实现服务间的调用。但是最方便、最优雅的方式是通过 Spring Cloud Open Feign 进行服务间的调用。Spring Cloud 对 Feign 进行了增强，使 Feign 支持 Spring MVC 的注解，并整合了 Ribbon 等，从而让 Feign 的使用更加方便。本章将从原理、实战分别对 Feign 扩展增强的角度进行剖析，帮助开发者快速掌握生产级别的 Feign 实战技巧。

4.1 Feign 概述

4.1.1 什么是 Feign

Feign 是一个声明式的 Web Service 客户端。它的出现使开发 Web Service 客户端变得很简单。使用 Feign 只需要创建一个接口加上对应的注解，比如：FeignClient 注解。Feign 有可插拔的注解，包括 Feign 注解和 JAX-RS 注解。Feign 也支持编码器和解码器，Spring Cloud Open Feign 对 Feign 进行增强支持 Spring MVC 注解，可以像 Spring Web 一样使用 HttpMessageConverters 等。

Feign 是一种声明式、模板化的 HTTP 客户端。在 Spring Cloud 中使用 Feign，可以做到使用 HTTP 请求访问远程服务，就像调用本地方法一样的，开发者完全感知不到这是在调用远程方法，更感知不到在访问 HTTP 请求。接下来介绍一下 Feign 的特性，具体如下：

❏ 可插拔的注解支持，包括 Feign 注解和 JAX-RS 注解。
❏ 支持可插拔的 HTTP 编码器和解码器。

- 支持 Hystrix 和它的 Fallback。
- 支持 Ribbon 的负载均衡。
- 支持 HTTP 请求和响应的压缩。Feign 是一个声明式的 Web Service 客户端，它的目的就是让 Web Service 调用更加简单。它整合了 Ribbon 和 Hystrix，从而不需要开发者针对 Feign 对其进行整合。Feign 还提供了 HTTP 请求的模板，通过编写简单的接口和注解，就可以定义好 HTTP 请求的参数、格式、地址等信息。Feign 会完全代理 HTTP 的请求，在使用过程中我们只需要依赖注入 Bean，然后调用对应的方法传递参数即可。

Open Feign 地址：https://github.com/OpenFeign/feign

Spring Cloud Open Feign 地址：https://github.com/spring-cloud/spring-cloud-openfeign

4.1.2 Feign 的入门案例

在前面一节介绍了什么是 Feign 以及 Feign 具有的特性，接下来将会通过一个入门案例，直观地感受一下如何使用 Feign 快速开发微服务应用。本案例通过编写 FeignClient 指定调用 URL 的方式，根据查询的参数搜索 GitHub 上面的仓库信息。

1）创建 Maven 工程如 ch4-1/ch4-1-hello 所示，配置主要 Maven 依赖如代码清单 4-1 所示：

代码清单4-1　ch4-1/ch4-1-hello/pom.xml

```xml
<dependencies>
    <dependency>
        <groupId>org.springframework.boot</groupId>
        <artifactId>spring-boot-starter-web</artifactId>
    </dependency>
    <!-- Spring Cloud Open Feign的Starter的依赖 -->
    <dependency>
        <groupId>org.springframework.cloud</groupId>
        <artifactId>spring-cloud-starter-openfeign</artifactId>
    </dependency>
</dependencies>
```

2）创建主入口程序 SpringCloudFeignApplication 并添加 @EnableFeignClients 注解，如代码清单 4-2 所示：

代码清单4-2　ch4-1/ch4-1-hello/src/main/java/cn/springcloud/book/feign/SpringCloudFeignApplication.java

```java
@SpringBootApplication
@EnableFeignClients
public class SpringCloudFeignApplication {

    public static void main(String[] args) {
        SpringApplication.run(SpringCloudFeignApplication.class, args);
    }
}
```

其中的 @EnableFeignClients 注解表示当程序启动时，会进行包扫描，扫描所有带 @FeignCleint 的注解的类并进行处理。

3）编写名为 HelloFeignService 的 FeignClient，如代码清单 4-3 所示：

代码清单4-3　ch4-1/ch4-1-hello/src/main/java/cn/springcloud/book/feign/service/HelloFeignService.java

```java
@FeignClient(name = "github-client", url = "https://api.github.com", configuration
    = HelloFeignServiceConfig.class)
public interface HelloFeignService {

    /**
     * content: {"message":"Validation Failed","errors":[{"resource":"Search",
     *     "field":"q","code":"missing"}],
     * "documentation_url":"https://developer.github.com/v3/search"}
     * @param queryStr
     * @return
     */
    @RequestMapping(value = "/search/repositories", method = RequestMethod.GET)
    String searchRepo(@RequestParam("q") String queryStr);
}
```

在 HelloFeignService 中通过 @FeignClient 注解手动指定 URL=https://api.github.com，该调用地址用于根据传入字符串搜索 GitHub 上相关的仓库信息。HelloFeignService 最终会根据指定的 URL 和 @RequestMapping 对应的方法，转换成最终的请求地址 https://api.github.com/search/repositories?q=spring-cloud-dubbo，请求该 URL 可以看到返回 GitHub 上对应的仓库信息，如图 4-1 所示。

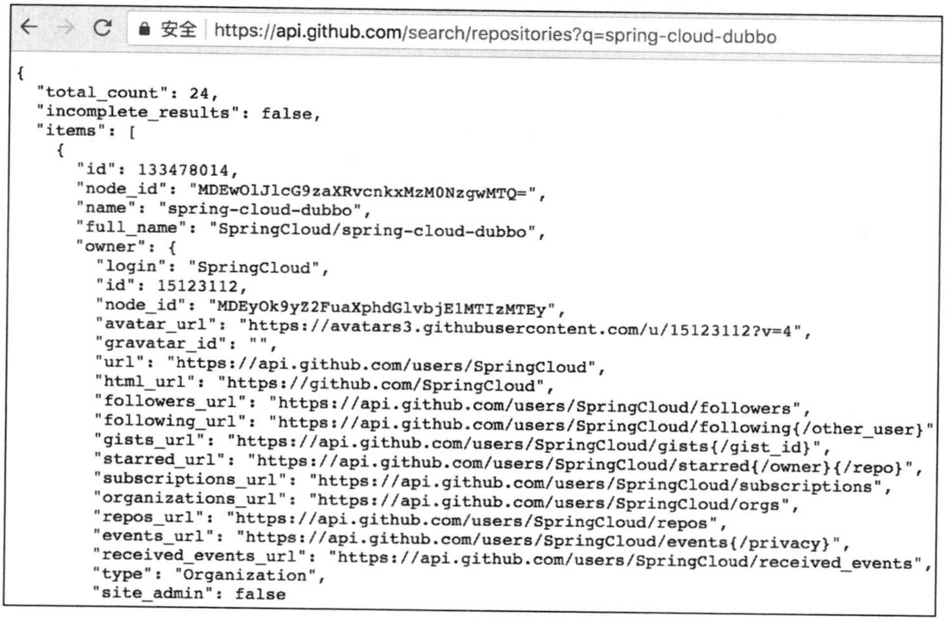

```
{
  "total_count": 24,
  "incomplete_results": false,
  "items": [
    {
      "id": 133478014,
      "node_id": "MDEwOlJlcG9zaXRvcnkxMzM0NzgwMTQ=",
      "name": "spring-cloud-dubbo",
      "full_name": "SpringCloud/spring-cloud-dubbo",
      "owner": {
        "login": "SpringCloud",
        "id": 15123112,
        "node_id": "MDEyOk9yZ2FuaXphdGlvbjE1MTIzMTEy",
        "avatar_url": "https://avatars3.githubusercontent.com/u/15123112?v=4",
        "gravatar_id": "",
        "url": "https://api.github.com/users/SpringCloud",
        "html_url": "https://github.com/SpringCloud",
        "followers_url": "https://api.github.com/users/SpringCloud/followers",
        "following_url": "https://api.github.com/users/SpringCloud/following{/other_user}",
        "gists_url": "https://api.github.com/users/SpringCloud/gists{/gist_id}",
        "starred_url": "https://api.github.com/users/SpringCloud/starred{/owner}{/repo}",
        "subscriptions_url": "https://api.github.com/users/SpringCloud/subscriptions",
        "organizations_url": "https://api.github.com/users/SpringCloud/orgs",
        "repos_url": "https://api.github.com/users/SpringCloud/repos",
        "events_url": "https://api.github.com/users/SpringCloud/events{/privacy}",
        "received_events_url": "https://api.github.com/users/SpringCloud/received_events",
        "type": "Organization",
        "site_admin": false
```

图 4-1　查询 GitHub 返回信息

4）编写 HelloFeignController 并依赖注入 FeignClient。

FeignClient 编写完毕之后，可以通过 @Autowired 注解把编写的 FeignClient 依赖注入对应的调用位置，如代码清单 4-4 所示：

代码清单4-4 ch4-1/ch4-1-hello/src/main/java/cn/springcloud/book/feign/controller/HelloFeignController.java

```java
@RestController
public class HelloFeignController {

    @Autowired
    private HelloFeignService helloFeignService;

    // 服务消费者对位提供的服务
    @GetMapping(value = "/search/github")
    public String searchGithubRepoByStr(@RequestParam("str") String queryStr) {
        return helloFeignService.searchRepo(queryStr);
    }
}
```

5）启动主应用程序 SpringCloudFeignApplication.java，打开浏览器访问如下 URL：http://localhost：8010/search/github?str=spring-cloud-dubbo，返回结果信息如图 4-2 所示。

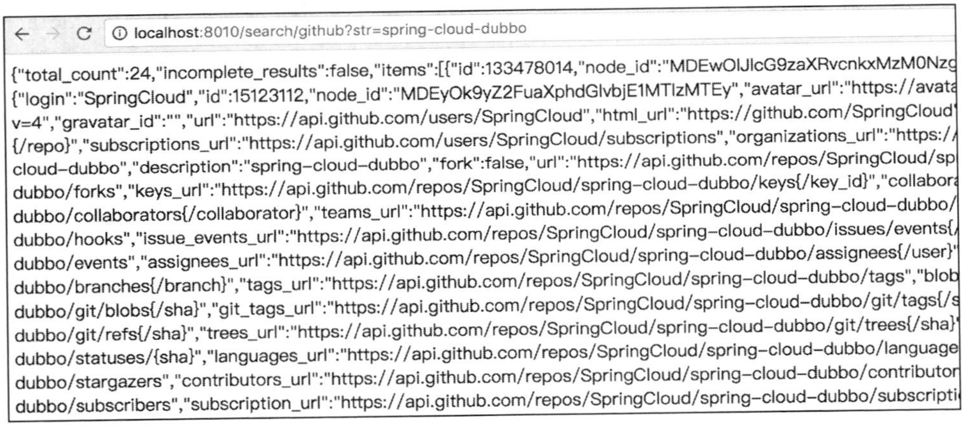

图 4-2　Feign 方式查询 GitHub 返回信息

4.1.3　Feign 的工作原理

通过上面的入门案例可以快速掌握 Feign 的基本使用，下面介绍一下 Feign 的工作原理：

- 在开发微服务应用时，我们会在主程序入口添加 @EnableFeignClients 注解开启对 Feign Client 扫描加载处理。根据 Feign Client 的开发规范，定义接口并加 @FeignClient 注解。
- 当程序启动时，会进行包扫描，扫描所有 @FeignClient 的注解的类，并将这些信息注入 Spring IOC 容器中。当定义的 Feign 接口中的方法被调用时，通过 JDK 的代理的方式，来生成具体的 RequestTemplate。当生成代理时，Feign 会为每个接口方法创建一个 RequetTemplate 对象，该对象封装了 HTTP 请求需要的全部信息，如请求参数名、请求方法等信息都是在这个过程中确定的。
- 然后由 RequestTemplate 生成 Request，然后把 Request 交给 Client 去处理，这里指的 Client 可以是 JDK 原生的 URLConnection、Apache 的 Http Client，也可以是 Okhttp。最后 Client 被封装到 LoadBalanceClient 类，这个类结合 Ribbon 负载均衡发起服务之间的调用。

4.2 Feign 的基础功能

4.2.1 FeignClient 注解剖析

FeignClient 注解被 @Target（ElementType.TYPE）修饰，表示 FeignClient 注解的作用目标在接口上。当打开 org.springframework.cloud.openfeign.FeignClient 这个注解定义类的时候，可以看到 FeignClient 注解对应的属性。

FeignClient 注解的常用属性归纳如下：
- name：指定 FeignClient 的名称，如果项目使用了 Ribbon，name 属性会作为微服务的名称，用于服务发现。
- url：url 一般用于调试，可以手动指定 @FeignClient 调用的地址。
- decode404：当发生 404 错误时，如果该字段为 true，会调用 decoder 进行解码，否则抛出 FeignException。
- configuration：Feign 配置类，可以自定义 Feign 的 Encoder、Decoder、LogLevel、Contract。
- fallback：定义容错的处理类，当调用远程接口失败或超时时，会调用对应接口的容错逻辑，fallback 指定的类必须实现 @FeignClient 标记的接口。
- fallbackFactory：工厂类，用于生成 fallback 类示例，通过这个属性我们可以实现每个接口通用的容错逻辑，减少重复的代码。
- path：定义当前 FeignClient 的统一前缀。

4.2.2 Feign 开启 GZIP 压缩

Spring Cloud Feign 支持对请求和响应进行 GZIP 压缩，以提高通信效率，这里介绍两种不同的配置方式，即 application.yml 和 application.properties 的配置方式。具体的案例代码如 ch4-1/ch4-1-gzip 所示，案例代码比较简单可以直接运行学习，下面将对 Feign 开启 GZIP 压缩的配置进行说明。

1）具体的 application.yml 配置信息如代码清单 4-5 所示。

代码清单4-5　/ch4-1/ch4-1-gzip/src/main/resources/application.yml

```
feign:
  compression:
    request:
      enabled: true
```

```
            mime-types: text/xml,application/xml,application/json # 配置压缩支持的
               MIME TYPE
            min-request-size: 2048    # 配置压缩数据大小的下限
      response:
            enabled: true # 配置响应GZIP压缩
```

等价的 application.properties 配置信息如下：

```
# 配置请求GZIP压缩
feign.compression.request.enabled=true
# 配置响应GZIP压缩
feign.compression.response.enabled=true
# 配置压缩支持的MIME TYPE
feign.compression.request.mime-types=text/xml,application/xml,application/json
# 配置压缩数据大小的下限
feign.compression.request.min-request-size=2048
```

提示 在 Spring Cloud Finchley.M9 版本中开启 Feign GZIP 压缩时，应用会启动报错。目前官方已修复 bug，具体扩展阅读可以参考下面的 issue 地址：https://github.com/spring-cloud/spring-cloud-openfeign/issues/14。

2）由于开启 GZIP 压缩之后，Feign 之间的调用通过二进制协议进行传输，返回值需要修改为 ResponseEntity<byte[]> 才可以正常显示，否则会导致服务之间的调用结果乱码，对应的 Feign 的 Client 处理如代码清单 4-6 所示：

代码清单4-6　ch4-1/ch4-1-gzip/src/main/java/cn/springcloud/book/feign/service/HelloFeignService.java

```
@RequestMapping(value = "/search/repositories", method = RequestMethod.GET)
ResponseEntity<byte[]> searchRepo(@RequestParam("q") String queryStr);
```

提示 Spring Cloud 的 Finchley.RC2 版 Feign 开启 GZIP 压缩时会出现乱码，但官方已经给出解决方案，扩展阅读可以参考：https://github.com/spring-cloud/spring-cloud-openfeign/issues/33。

3）启动主应用程序 SpringCloudFeignApplication.java，打开浏览器访问如下 URL：http://localhost：8011/search/github?str=spring-cloud-dubbo，返回结果如图 4-3 所示。

图 4-3　Feign 开启压缩测试结果

4.2.3　Feign 支持属性文件配置

1. 对单个指定特定名称的 Feign 进行配置

@FeignClient 的配置信息可以通过 application.properties 或 application.yml 来配置，application.yml 示例配置信息如下：

```yaml
feign:
  client:
    config:
      feignName:  #需要配置的FeignName
        connectTimeout: 5000      #连接超时时间
        readTimeout: 5000         #读超时时间设置
        loggerLevel: full         #配置Feign的日志级别
        errorDecoder: com.example.SimpleErrorDecoder #Feign的错误解码器
        retryer: com.example.SimpleRetryer #配置重试
        requestInterceptors:      #配置拦截器
          - com.example.FooRequestInterceptor
          - com.example.BarRequestInterceptor
        decode404: false
        encoder: com.example.SimpleEncoder    #Feign的编码器
        decoder: com.example.SimpleDecoder    #Feign的解码器
        contract: com.example.SimpleContract  #Feign的Contract配置
```

2. 作用于所有 Feign 的配置方式

@EnableFeignClients 注解上有个 defaultConfiguration 属性，我们可以将默认配置写成一个类，比如这个配置类叫 DefaultFeignConfiguration.java，在主程序的启动入口用 defaultConfiguration 来引用配置，示例配置如下所示：

```java
@SpringBootApplication
@EnableFeignClients(defaultConfiguration =DefaultFeignConfiguration.class )
public class ConsumerApplication {
    public static void main(String[] args) {
        SpringApplication.run(ConsumerApplication.class, args);
    }
}
```

如果想使用 application.yml 或 application.properties 来作用于所有 Feign 也是可以的，示例在 application.yml 中的配置如下所示：

```yaml
feign:
  client:
    config:
      default:
        connectTimeout: 5000
        readTimeout: 5000
        loggerLevel: basic
```

注意，如果通过 Java 代码的方式配置过 Feign，然后又通过属性文件的方式配置 Feign，属性文件中 Feign 的配置会覆盖 Java 代码的配置。但是可以配置 feign.client.default-to-properties=false 来改变 Feign 配置生效的优先级。

4.2.4　Feign Client 开启日志

Feign 为每一个 FeignClient 都提供了一个 feign.Logger 实例，可以在配置中开启日志，开启方式比较简单，分为两步。

第一步：在 application.yml 或者 application.properties 中配置日志输出。

在 application.yml 中设置日志输出级别，如代码清单 4-6 所示：

代码清单4-7　ch4-1/ch4-1-hello/src/main/resources/application.yml

```
logging:
    level:
        cn.springcloud.book.feign.service.HelloFeignService: debug
```

第二步：通过 Java 代码的方式在主程序入口类中配置日志 Bean，代码如下所示：

```
@Bean
Logger.Level feignLoggerLevel() {
    return Logger.Level.FULL;
}
```

也可以通过创建带有 @Configuration 注解的类，去配置日志 bean，如代码清单 4-8 所示：

代码清单4-8　ch4-1/ch4-1-hello/src/main/java/cn/springcloud/book/feign/config/HelloFeignServiceConfig.java

```
@Configuration
public class HelloFeignServiceConfig {

    /**
     *
     * Logger.Level的具体级别如下：
     *     NONE：不记录任何信息
     *     BASIC：仅记录请求方法、URL以及响应状态码和执行时间
     *     HEADERS：除了记录BASIC级别的信息外，还会记录请求和响应的头信息
     *     FULL：记录所有请求与响应的明细，包括头信息、请求体、元数据
     * @return
     */
    @Bean
    Logger.Level feignLoggerLevel() {
        return Logger.Level.FULL;
    }
}
```

4.2.5　Feign 的超时设置

Feign 的调用分两层，即 Ribbon 的调用和 Hystrix 的调用，高版本的 Hystrix 默认是关闭的。

```
feign.RetryableException: Read timed out executing POST http://******
    at feign.FeignException.errorExecuting(FeignException.java:67)
    at feign.SynchronousMethodHandler.executeAndDecode(SynchronousMethodHandler.java:104)
    at feign.SynchronousMethodHandler.invoke(SynchronousMethodHandler.java:76)
    at feign.ReflectiveFeign$FeignInvocationHandler.invoke(ReflectiveFeign.java:103)
    at com.sun.proxy.$Proxy113.getBaseRow(Unknown Source)
Caused by: java.net.SocketTimeoutException: Read timed out
```

1）如果出现上面的报错信息，说明 Ribbon 处理超时，此时设置 Ribbon 的配置信息如下即可。

```
#请求处理的超时时间
ribbon.ReadTimeout: 120000
#请求连接的超时时间
ribbon.ConnectTimeout: 30000
```

2)如果开启 Hystrix,Hystrix 的超时报错信息如下所示:

```
com.netflix.hystrix.exception.HystrixRuntimeException: FeignDemo#demo() timed-
    out and no fallback available.
    at com.netflix.hystrix.AbstractCommand$22.call(AbstractCommand.java:819)
    at com.netflix.hystrix.AbstractCommand$22.call(AbstractCommand.java:804)
    at rx.internal.operators.OperatorOnErrorResumeNextViaFunction$4.onError(Ope
        ratorOnErrorResumeNextViaFunction.java:140)
    at rx.internal.operators.OnSubscribeDoOnEach$DoOnEachSubscriber.
        onError(OnSubscribeDoOnEach.java:87)
```

看到上面的报错信息,说明 Hystrix 超时报错,此时设置 Hystrix 的配置信息如下所示:

```
feign.hystrix.enabled: true
hystrix熔断机制
hystrix:
    shareSecurityContext: true
    command:
        default:
            circuitBreaker:
                sleepWindowInMilliseconds: 100000
                forceClosed: true
            execution:
                isolation:
                    thread:
                        timeoutInMilliseconds: 600000
```

4.3 Feign 的实战运用

4.3.1 Feign 默认 Client 的替换

Feign 在默认情况下使用的是 JDK 原生的 URLConnection 发送 HTTP 请求,没有连接池,但是对每个地址会保持一个长连接,即利用 HTTP 的 persistence connection。我们可以用 Apache 的 HTTP Client 替换 Feign 原始的 HTTP Client,通过设置连接池、超时时间等对服务之间的调用调优。Spring Cloud 从 Brixtion.SR5 版本开始支持这种替换,接下来介绍一下如何用 HTTP Client 和 okhttp 去替换 Feign 默认的 Client。

1. 使用 HTTP Client 替换 Feign 默认 Client

使用 HTTP Client 替换 Feign 的默认 Client 非常简单,具体可以参考案例 ch4-3/ch4-3-httpclient。下面介绍一下主要的替换步骤。

1)创建 ch4-3-httpclient 工程,引入主要的 Maven 依赖,如代码清单 4-9 所示:

代码清单4-9　ch4-3/ch4-3-httpclient/pom.xml

```xml
<dependencies>
    <dependency>
        <groupId>org.springframework.boot</groupId>
        <artifactId>spring-boot-starter-web</artifactId>
    </dependency>
    <!-- Spring Cloud OpenFeign的Starter的依赖 -->
    <dependency>
        <groupId>org.springframework.cloud</groupId>
        <artifactId>spring-cloud-starter-openfeign</artifactId>
    </dependency>

    <!-- 使用Apache HttpClient替换Feign原生httpclient -->
    <dependency>
        <groupId>org.apache.httpcomponents</groupId>
        <artifactId>httpclient</artifactId>
    </dependency>

    <dependency>
        <groupId>com.netflix.feign</groupId>
        <artifactId>feign-httpclient</artifactId>
        <version>8.17.0</version>
    </dependency>
</dependencies>
```

2）在 application.yml 中配置让 Feign 启动时加载 HTTP Client 替换默认的 Client，如代码清单 4-10 所示：

代码清单4-10　ch4-3/ch4-3-httpclient/src/main/resources/application.yml

```yaml
server:
    port: 8010
spring:
    application:
        name: ch4-3-httpclient
feign:
    httpclient:
        enabled: true
```

3）启动 ch4-3/ch4-3-httpclient/ 下的 SpringCloudFeignApplication.java 主程序访问 http：//localhost：8010/search/github?str=spring-cloud-openfeign，结果如图 4-4 所示。

图 4-4　使用 HTTP Client 替换测试结果

2. 使用 okhttp 替换 Feign 默认的 Client

HTTP 是目前比较通用的网络请求方式，用来访问请求交换数据，有效地使用 HTTP 可以使应用访问速度变得更快，更节省带宽。okhttp 是一个很棒的 HTTP 客户端，具有以下功能和特性。

- 支持 SPDY，可以合并多个到同一个主机的请求。
- 使用连接池技术减少请求的延迟 (如果 SPDY 是可用的话)。
- 使用 GZIP 压缩减少传输的数据量。
- 缓存响应避免重复的网络请求。

接下来，介绍一下如何使用 okhttp 替换 Feign 默认的 Client，具体的案例代码如 ch4-3/ch4-3-okhttp 所示，下面介绍最主要的实现步骤。

1）创建名为 ch4-3-okhttp 的 Maven 工程，然后引入 okhttp 的 Maven 依赖，如代码清单 4-11 所示：

代码清单4-11　ch4-3/ch4-3-okhttp/pom.xml

```xml
<dependencies>
    <dependency>
        <groupId>org.springframework.boot</groupId>
        <artifactId>spring-boot-starter-web</artifactId>
    </dependency>
    <!-- Spring Cloud OpenFeign的Starter的依赖 -->
    <dependency>
        <groupId>org.springframework.cloud</groupId>
        <artifactId>spring-cloud-starter-openfeign</artifactId>
    </dependency>
    <dependency>
        <groupId>io.github.openfeign</groupId>
        <artifactId>feign-okhttp</artifactId>
    </dependency>
</dependencies>
```

2）开启 okhttp 为 Feign 默认的 Client。

在 application.yml 配置开启替换配置，如代码清单 4-12 所示：

代码清单4-12　ch4-3/ch4-3-okhttp/src/main/resources/application.yml

```yaml
server:
    port: 8011
spring:
    application:
        name: ch4-3-okhttp

feign:
    httpclient:
        enabled: false
    okhttp:
        enabled: true
```

3）okHttpClient 是 okhttp 的核心功能的执行者，可以通过 OkHttpClient client=new OkHttp

Client();来创建默认的 OkHttpClient 对象，也可以使用如下代码来构建自定义的 OkHttpClient 对象，上面配置信息只给出了常用的设置项，其他设置项比较复杂，有兴趣的读者可以自己扩展阅读学习，在这里就不再进行展开，具体代码如代码清单 4-13 所示：

代码清单4-13 ch4-3/ch4-3-okhttp/src/main/java/cn/springcloud/book/feign/config/
FeignOkHttpConfig.java

```java
@Configuration
@ConditionalOnClass(Feign.class)
@AutoConfigureBefore(FeignAutoConfiguration.class)
public class FeignOkHttpConfig {
    @Bean
    public okhttp3.OkHttpClient okHttpClient(){
        return new okhttp3.OkHttpClient.Builder()
                //设置连接超时
                .connectTimeout(60, TimeUnit.SECONDS)
                //设置读超时
                .readTimeout(60, TimeUnit.SECONDS)
                //设置写超时
                .writeTimeout(60,TimeUnit.SECONDS)
                //是否自动重连
                .retryOnConnectionFailure(true)
                .connectionPool(new ConnectionPool())
                //构建OkHttpClient对象
                .build();
    }
}
```

4）启动 ch4-3/ch4-3-okhttp 下的 SpringCloudFeignApplicatio.java 主程序，访问 http://localhost:8011/search/github?str=spring-cloud-openfeign，返回结果如图 4-5 所示。

图 4-5　使用 OkHttpClient 替换结果

4.3.2　Feign 的 Post 和 Get 的多参数传递

在实际项目开发过程中，我们使用 Feign 实现服务与服务之间的调用。但是在很多情况下，多参数传递是无法避免的。下面我们分两种情况，讨论如何在 GET 或 POST 情况下使用 Feign 进行多参数传递。Feign 中的 Post 和 Get 的多参数传递的聚合工程清单如表 4-1 所示。

表 4-1　多参数传递工程合集

工程名	端口	描述
ch4-5-eureka-server	8761	eureka-server 注册中心
ch4-5-consumer	8011	GET 或 POST 请求多参数传递的服务消费者，集成了 Swagger 用于测试
ch4-5-provider	8012	GET 或 POST 请求多参数传递的服务提供者

由于 ch4-5-eureka-server 比较简单，是一个简单的 Eureka Server 注册中心，这里不再进行过多阐述。本案例中主要介绍在服务提供者和服务消费者之间怎样通过 Feign 进行 Get 请求和 Post 请求的多参数传递。

众所周知，在 Web 开发中 Spring MVC 是支持 GET 方法直接绑定 POJO 的，但是 Feign 的实现并未覆盖所有 Spring MVC 的功能，目前解决方式有很多，最常见的解决方式如下：

- 把 POJO 拆散成一个一个单独的属性放在方法参数里。
- 把方法参数变成 Map 传递。
- 使用 GET 传递 @RequestBody，但此方式违反 Restful 规范。

本案中我们介绍一种最佳的实践方式，即通过 Feign 拦截器的方式处理。

1）通过实现 Feign 的 RequestInterceptor 中的 apply 方法来进行统一拦截转换处理 Feign 中的 GET 方法多参数传递的问题。而 Feign 进行 Post 多参数传递相比进行 Get 多参数传递来说比较简单。Feign 拦截器具体的代码如代码清单 4-14 所示：

代码清单 4-14　ch4-5/ch4-5-consumer/src/main/java/cn/springcloud/book/feign/Interceptor/FeignRequestInterceptor.java

```java
@Component
public class FeignRequestInterceptor implements RequestInterceptor {
    @Autowired
    private ObjectMapper objectMapper;
    @Override
    public void apply(RequestTemplate template) {
        // feign不支持GET方法传POJO, json body转query
        if (template.method().equals("GET") && template.body() != null) {
            try {
                JsonNode jsonNode = objectMapper.readTree(template.body());
                template.body(null);

                Map<String, Collection<String>> queries = new HashMap<>();
                buildQuery(jsonNode, "", queries);
                template.queries(queries);
            } catch (IOException e) {
                //提示:根据实践项目情况处理此处异常，这里不做扩展。
                e.printStackTrace();
            }
        }
    }
}
```

2）集成 Swagger 编写服务消费者用于调用 Feign 进行 Get 或 Post 多参数传递，如代码清单 4-15 所示：

代码清单4-15　ch4-5/ch4-5-consumer/src/main/java/cn/springcloud/book/feign/controller/UserController.java

```java
@RestController
@RequestMapping("/user")
public class UserController {
    @Autowired
    private UserFeignService userFeignService;
    /**
     * 用于演示Feign的Get请求多参数传递
     * @param user
     * @return
     */
    @RequestMapping(value = "/add", method = RequestMethod.POST)
    public String addUser( @RequestBody @ApiParam(name="用户",value="传入json格式
        ",required=true) User user){
        return userFeignService.addUser(user);
    }
    /**
     * 用于演示Feign的Post请求多参数传递
     * @param user
     * @return
     */
    @RequestMapping(value = "/update", method = RequestMethod.POST)
    public String updateUser( @RequestBody @ApiParam(name="用户",value="传入json
        格式",required=true) User user){
        return userFeignService.updateUser(user);
    }
}
```

3）编写 Feign Client 用于 GET 的多参数传递，如代码清单 4-16 所示：

代码清单4-16　ch4-5/ch4-5-consumer/src/main/java/cn/springcloud/book/feign/service/UserFeignService.java

```java
@FeignClient(name = "ch4-5-provider")
public interface UserFeignService {
    @RequestMapping(value = "/user/add", method = RequestMethod.GET)
    public String addUser(User user);
    @RequestMapping(value = "/user/update", method = RequestMethod.POST)
    public String updateUser(@RequestBody User user);
}
```

从上面代码中看到可以通过 UserFeignService 中的 addUser() 方法多参数传递 User 对象进行增加用户操作，通过 updateUser() 方法多参数传递 User 进行更新操作。

4）编写服务提供者接收 Feign 的 GET 请求传过来的 User 对象，如代码清单 4-17 所示：

代码清单4-17　ch4-5/ch4-5-provider/src/main/java/cn/springcloud/book/feign/controller/UserController.java

```java
@RestController
@RequestMapping("/user")
public class UserController {
    @RequestMapping(value = "/add", method = RequestMethod.GET)
```

```
    public String addUser(User user){
        return "hello,"+user.getName();
    }
    @RequestMapping(value = "/update", method = RequestMethod.POST)
    public String updateUser( @RequestBody User user){
        return "hello,"+user.getName();
    }
}
```

5）分别依次按顺序启动 ch4-5-eureka-server、ch4-5-provider、ch4-5-consumer 应用。访问 http://localhost:8011/swagger-ui.html 可以看到 Swagger 的访问测试页面，如图 4-6 所示。

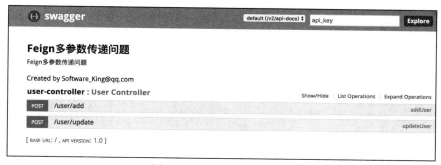

图 4-6　Swagger 测试页面

6）测试 Feign 的 GET 多参数传递，JSON 请求数据示例如图 4-7 所示，通过 Swagger 的操作界面单击 try it out 可以看到返回调用结果 hello，zhangsan。

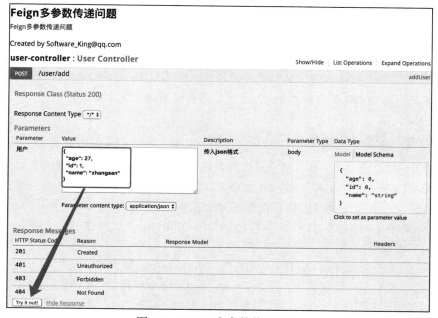

图 4-7　POST 多参数传递配置

可以在 FeignRequestInterceptor#apply 设置断点进行 debug 调试，学习怎么处理 Feign 的 Get 多参数传递。

7）测试 Feign 的 Post 多参数提交，JSON 数据示例如图 4-8 所示。

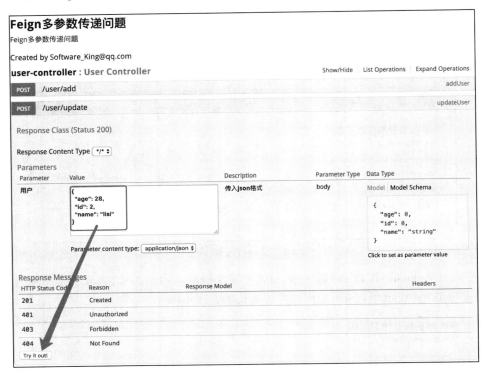

图 4-8　Swagger 测试页面

通过 Swagger 的操作界面单击 try it out 可以看到返回调用结果 hello，lisi。

4.3.3　Feign 的文件上传

Feign 早先不支持文件上传，后来虽支持但仍有缺陷，需要一次性完整地读到内存在编码发送。在早期的 Spring Cloud 中，Feign 本身是没有上传文件功能的，要想实现文件上传功能，需要编写 Encoder 去实现文件上传。现在 Feign 官方提供了子项目 feign-form（https://github.com/OpenFeign/feign-form），其中实现了上传所需的 Encoder，有兴趣的读者可以自行阅读源码，学习其实现原理。

下面将会使用 feign-form 结合 Feign 实现文件的上传。案例将会使用服务提供者 feign-file-server 去模拟一个文件服务器，通过 feign-upload-client 应用结合 Swagger 自动生成的表单上传文件，然后把上传的文件发送到文件服务器。示例工程如表 4-2 所示。

表 4-2 文件上传工程列表

工　程　名	端口	描　　述
ch4-4-eureka-server	8761	eureka-server 注册中心
ch4-4-feign-file-server	8012	模拟文件服务器，服务提供者
ch4-4-feign-upload-client	8011	模拟文件表单上传，通过 Feign Client 发送文件到文件服务器

由于 ch4-4-eureka-server 是一个简单的 Eureka Server，可用于模拟注册中心，由于前面的章节已经做了详细的介绍，在这里将不再阐述。

如 ch4-4-feign-upload-client 工程所示，在这个工程中集成了 Swagger 模拟表单上传文件到 Controller 层，然后通过依赖注入的方式调用 Feign 编写的 FileUploadFeignService 的 Feign Client，然后通过服务注册与发现从 Eureka Server 上找到文件上传服务提供者，将上传的文件发送到文件服务器。

下面来介绍使用 Feign Client 快速实现文件上传的几个主要核心步骤。

1. 编写 Feign 文件上传的客户端

创建 FileUploadFeignService 接口，如代码清单 4-18 所示：

代码清单4-18 ch4-4/ch4-4-feign-upload-client/src/main/java/cn/springcloud/book/feign/service/FileUploadFeignService.java

```java
@FeignClient(value = "feign-file-server", configuration = FeignMultipartSupportConfig.class)
public interface FileUploadFeignService {
    /***
     * 1.produces,consumes必填
     * 2.注意区分@RequestPart和RequestParam, 不要将
     * @RequestPart(value = "file") 写成@RequestParam(value = "file")
     * @param file
     * @return
     */
    @RequestMapping(method = RequestMethod.POST, value = "/uploadFile/server",
        produces = {MediaType.APPLICATION_JSON_UTF8_VALUE},
        consumes = MediaType.MULTIPART_FORM_DATA_VALUE)
    public  String fileUpload(@RequestPart(value = "file") MultipartFile file);
}
```

如上述代码所示，使用 Feign 开发了一个文件上传的客户端，需要注意以下几点：

1）produces、consumes 必填。

2）注意区分 @RequestPart 和 RequestParam，不要将 @RequestPart（value = "file"）写成 @RequestParam（value = "file"）。

2. 编写 Feign 文件上传的服务端

如 ch4-4-feign-file-server 工程所示，该工程是用于模拟一个文件服务器提供者，用于接收从 Feign 客户端发送过来的文件的，FeignUploadController 的代码如代码清单 4-19 所示：

代码清单4-19　ch4-4/ch4-4-feign-file-server/src/main/java/cn/springcloud/book/feign/controller/FeignUploadController.java

```
@RestController
public class FeignUploadController {
    @PostMapping(value = "/uploadFile/server", consumes = MediaType.MULTIPART_
        FORM_DATA_VALUE)
    public String fileUploadServer(MultipartFile file ) throws Exception{
        return file.getOriginalFilename();
    }
}
```

3. 上传文件

依次按顺序启动 ch4-4-eureka-server、ch4-4-feign-file-server、ch4-4-feign-upload-client 三个应用，打开浏览器访问 http://localhost:8011/swagger-ui.html，如图 4-9 所示，当单击 Try it out 按钮的时候可以看到文件已经发送到后台文件服务器，并返回文件名的结果。

图 4-9　Feign 文件上传测试

4.3.4　解决 Feign 首次请求失败问题

当 Feign 和 Ribbon 整合了 Hystrix 之后，可能会出现首次调用失败的问题，造成该问题出现的原因分析如下：

Hystrix 默认的超时时间是 1 秒，如果超过这个时间尚未做出响应，将会进入 fallback 代码。由于 Bean 的装配以及懒加载机制等，Feign 首次请求都会比较慢。如果这个响应时间大于 1 秒，就会出现请求失败的问题。下面以 feign 为例，介绍三种方法处理 Feign 首次请求失败的问题。

方法一 将 Hystrix 的超时时间改为 5 秒，配置如下所示：

```
hystrix.command.default.execution.isolation.thread.timeoutInMilliseconds: 5000
```

方法二 禁用 Hystrix 的超时时间，配置如下所示：

```
hystrix.command.default.execution.timeout.enabled: false
```

方法三 使用 Feign 的时候直接关闭 Hystrix，该方式不推荐使用。

```
feign.hystrix.enabled: false
```

说明　针对 Feign 首次请求失败的问题，可参考如下链接：https://github.com/spring-cloud/spring-cloud-netflix/issues/768

4.3.5　Feign 返回图片流处理方式

通过 Feign 返回图片一般为字节数组，如下列代码所示：

```java
/**
 * 生成图片验证码
 * @param imagekey
 * @return
 */
@RequestMapping(value = "createImageCode")
public byte[] createImageCode(@RequestParam("imagekey") String imagekey);
```

在使用 Feign 的过程中可以将流转成字节数组传递，但是因为 Controller 层的返回不能直接返回 byte，因此需要将 Feign 的返回值修改为 response，示例代码如下：

```java
@RequestMapping(value = "createImageCode")
public Response createImageCode(@RequestParam("imagekey") String imagekey);
```

4.3.6　Feign 调用传递 Token

在进行认证鉴权的时候，不管是 jwt，还是 security，当使用 Feign 时就会发现外部请求到 A 服务的时候，A 服务是可以拿到 Token 的，然而当服务使用 Feign 调用 B 服务时，Token 就会丢失，从而认证失败。解决方法相对比较简单，需要做的就是在 Feign 调用的时候，向请求头里面添加需要传递的 Token。

我们只需要实现 Feign 提供的一个接口 RequestInterceptor，假设我们在验证权限的时候放在请求头里面的 key 为 oauthToken，先获取当前请求中的 key 为 oauthToken 的 Token，然后放到 Feign 的请求 Header 上，示例代码如代码清单 4-20 所示（像这种通用的代码确定传递的 key 之后，建议统一到通用的二方库里使用）。

代码清单4-20　ch4-5/ch4-5-consumer/src/main/java/cn/springcloud/book/feign/Interceptor/FeignTokenInterceptor.java

```java
/**
 * Feign统一Token拦截器
 */
```

```java
@Component
public class FeignTokenInterceptor implements RequestInterceptor {

    @Override
    public void apply(RequestTemplate requestTemplate) {
        if(null==getHttpServletRequest()){
            return;
        }
        //将获取Token对应的值往下面传
        requestTemplate.header("oauthToken", getHeaders(getHttpServletRequest()
            ).get("oauthToken"));
    }
    private HttpServletRequest getHttpServletRequest() {
        try {
            return ((ServletRequestAttributes) RequestContextHolder.getRequest-
                Attributes()).getRequest();
        } catch (Exception e) {
            return null;
        }
    }
}
```

4.4　venus-cloud-feign 设计与使用

4.4.1　venus-cloud-feign 的设计

为了方便，API 中使用 Feign 替代 RestTemplate 手动调用，在写 Feign 接口的时候，想用 Spring MVC 注解只在 Feign 接口写一遍，然后实现类实现此接口即可。但是 Spring MVC 不支持实现接口中方法参数上的注解（支持继承类、方法上的注解）。在上面的小节中通过 Feign 拦截器的方式解决了 Get 请求多参数传递的问题。为了解决上述两个问题，Spring Cloud 中国社区对 spring-cloud-openfeign 进行增强，命名为 venus-cloud-feign。项目开源地址为：https://github.com/springcloud/venus-cloud-feign，目前已安装到 Maven 中央仓库，支持 Spring Cloud 的版本为 Finchley.RELEASE。下面介绍一下 venus-cloud-feign 的设计和使用。整个开源项目模块设计如表 4-3 所示。

表 4-3　venus-cloud-feign 项目模块

模　　块	描　　述
venus-cloud-feign	工程父模块
venus-cloud-feign-core	核心类，包含自动配置模块，Feign 拦截器，基于 Feign 扩展的 Spring MVC 的 ArgumentResolvers
venus-cloud-feign-dependencies	第三方 Maven 依赖
venus-cloud-starter-feign	starter 模块
venus-cloud-feign-sample	工程示例
docs	项目文档

1. Spring MVC Controller 中的方法不支持继承实现 Feign 接口中方法参数上的注解问题

为了解决 Spring MVC 支持继承实现接口中方法参数上的注解，此时的解决办法是使用 @Configuration 配置类添加如下代码，扩展 spring 默认的 rgumentResolvers，代码可以参考 venus-cloud-feign-core 模块下面的 VenusSpringMvcContract.java 和 VenusFeignAutoConfig.java 文件，核心代码如下所示：

```java
@PostConstruct
public void modifyArgumentResolvers() {
    List<HandlerMethodArgumentResolver> list = new ArrayList<>(adapter.
        getArgumentResolvers());
    
    // PathVariable支持接口注解
    list.add(0, new PathVariableMethodArgumentResolver() {
        @Override
        public boolean supportsParameter(MethodParameter parameter) {
            return super.supportsParameter(interfaceMethodParameter(parameter,
                PathVariable.class));
        }
        
        @Override
        protected NamedValueInfo createNamedValueInfo(MethodParameter parameter)
{
            return super.createNamedValueInfo(interfaceMethodParameter(paramet
                er, PathVariable.class));
        }
    });
    
    // RequestHeader支持接口注解
    list.add(0, new RequestHeaderMethodArgumentResolver(beanFactory) {
        @Override
        public boolean supportsParameter(MethodParameter parameter) {
            return super.supportsParameter(interfaceMethodParameter(parameter,
                RequestHeader.class));
        }
        
        @Override
        protected NamedValueInfo createNamedValueInfo(MethodParameter parameter) {
            return super.createNamedValueInfo(interfaceMethodParameter(paramet
                er, RequestHeader.class));
        }
    });
    
    //其他代码省略
}
```

2. Feign 不支持 GET 方法传递 POJO 的问题

由于 Spring MVC 是支持 GET 方法直接绑定 POJO 的，只是 Feign 实现并未覆盖所有 Spring MVC 的功能，其他的解决方案要么是把 POJO 拆散成一个一个单独的属性放在方法参数里传输，要么就是把方法参数变成 Map 传输，要么是要违反 Restful 规范 GET 传递 @RequestBody 等，下面通过实现 Feign 拦截器提供一种最优雅的处理方式，代码可以查看：cn.springcloud.feign 包下的 VenusRequestInterceptor.java 文件，读者可以自行调试源码学习。

4.4.2 venus-cloud-feign 的使用

上面小节介绍了 venus-cloud-feign 的主要设计细节，为了方便使用开源项目已经将其封装成自动配置和 starter，详细代码可以查看 venus-cloud-starter-feign 模块。使用 venus-cloud-feign 只需要引入 venus-cloud-starter-feign 对应的 maven 依赖即可。下面将会通过一个案例 ch4-6 介绍如何使用 venus-cloud-feign，也可以直接运行 ch4-6 源码工程进行学习。

1. 通过 Feign 编写服务提供者，并将服务注册到 Eureka Server 上

1）创建二方包工程如 ch4-6-provider-client 所示，在 pom.xml 中配置 venus-cloud-feign 的核心依赖如代码清单 4-21 所示：

代码清单4-21　ch4-6/ch4-6-provider-client/pom.xml

```xml
<dependency>
    <groupId>cn.springcloud.feign</groupId>
    <artifactId>venus-cloud-starter-feign</artifactId>
    <version>1.0.0</version>
</dependency>
<!-- Spring Cloud OpenFeign的Starter的依赖 -->
<dependency>
    <groupId>org.springframework.cloud</groupId>
    <artifactId>spring-cloud-starter-openfeign</artifactId>
</dependency>
```

2）在 ch4-6-provider-client 中定义服务之间调用的 Feign 接口，如代码清单 4-22 所示：

代码清单4-22　ch4-6/ch4-6-provider-client/src/main/java/cn/springcloud/book/feign/service/UserService.java

```java
@FeignClient(name = "ch4-6-provider")
public interface UserService {
    @RequestMapping(value = "/user/add", method = RequestMethod.GET)
    public String addUser(User user);

    @RequestMapping(value = "/user/update", method = RequestMethod.POST)
    public String updateUser(@RequestBody User user);
}
```

3）创建 ch4-6-provider 服务提供者的实现工程，配置二方包 ch4-6-provider-client 依赖到 pom.xml 中，如代码清单 4-23 所示：

代码清单4-23　ch4-6/ch4-6-provider/pom.xml

```xml
<dependency>
    <groupId>cn.springcloud.book</groupId>
    <artifactId>ch4-6-provider-client</artifactId>
    <version>1.0-SNAPSHOT</version>
</dependency>
<dependency>
    <groupId>org.springframework.cloud</groupId>
    <artifactId>spring-cloud-starter-netflix-eureka-client</artifactId>
</dependency>
```

4）编写 Feign 接口对应的服务提供者实现，如代码清单 4-24 所示：

代码清单4-24　ch4-6/ch4-6-provider/src/main/java/cn/springcloud/book/feign/controller/UserController.java

```
@RestController
public class UserController implements UserService {

    @Override
    public String addUser(User user){
        return "hello,"+user.getName();
    }

    @Override
    public String updateUser(User user){
        return "hello,"+user.getName();
    }

}
```

从上面代码可以看出 UserController 实现了 Feign 定义的接口 UserService，覆盖实现了 Feign 定义的接口，但是没有在 Controller 的方法上加 Spring MVC 相关的注解，这就是 venus-cloud-feign 对 spring cloud feign 功能扩展的最好体现之一。

2. 创建服务消费者工程

1）创建服务消费者工程并引入二方包 ch4-6-provider-client 到 ch4-6-consumer 工程的 pom 中，核心依赖如代码清单 4-25 所示：

代码清单4-25　ch4-6/ch4-6-consumer/pom.xml

```xml
<dependency>
    <groupId>cn.springcloud.book</groupId>
    <artifactId>ch4-6-provider-client</artifactId>
    <version>1.0-SNAPSHOT</version>
</dependency>
<dependency>
    <groupId>cn.springcloud.feign</groupId>
    <artifactId>venus-cloud-starter-feign</artifactId>
    <version>1.0.0</version>
</dependency>
<!-- Spring Cloud OpenFeign的Starter的依赖 -->
<dependency>
    <groupId>org.springframework.cloud</groupId>
    <artifactId>spring-cloud-starter-openfeign</artifactId>
</dependency>
```

2）依赖注入服务提供者提供的接口，进行服务消费调用，示例代码如代码清单 4-26 所示：

代码清单4-26　ch4-6/ch4-6-consumer/src/main/java/cn/springcloud/book/feign/controller/UserController.java

```
@RestController
@RequestMapping("/user")
```

```
public class UserController {
    @Autowired
    private UserService userService;
    @RequestMapping(value = "/add", method = RequestMethod.POST)
    public String addUser( @RequestBody @ApiParam(name="用户",value="传入json格式
        ",required=true) User user){
        return userService.addUser(user);
    }
    @RequestMapping(value = "/update", method = RequestMethod.POST)
    public String updateUser( @RequestBody @ApiParam(name="用户",value="传入json
        格式",required=true) User user){
        return userService.updateUser(user);
    }
}
```

3. 测试

分别按顺序启动 ch4-6-eureka-server，ch4-6-provider，ch4-6-consumer 进行测试。打开浏览器访问 http://localhost:8011/swagger-ui.html，可以看到 Swagger 测试页面如图 4-10 所示，前面小节介绍过测试方法，这里不再阐述。

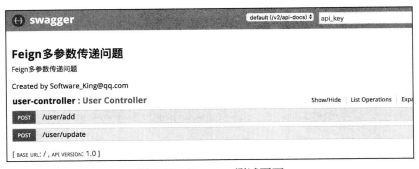

图 4-10　Swagger 测试页面

4.5　本章小结

本章通过案例的方式简单介绍了 Feign 的基础功能、实现原理和基础用法，然后介绍了 Feign 的实战运用，比如 Feign 默认 Client 的替换，最佳方式实现 Feign 的 Post 和 Get 多参数传递，Feign 的文件上传，以及统一异常处理等，最后通过设计 venus-cloud-feign 组件将实战项目中通用的 Feign 增强和扩展功能封装成 starter，做到开箱即用，更大程度地实现项目组件的复用，提高软件开发效率。

第 5 章 Spring Cloud Ribbon 实战运用

Ribbon 是 Netflix 公司开发的一个负载均衡组件，它在云服务体系中起着至关重要的作用，诞生于 2013 年 1 月，一直是 Netflix 活跃度较高的项目。Pivotal 公司将其整合进入 Spring Cloud 生态，正式命名为 Spring Cloud Ribbon（后文简称 Ribbon），它是 Spring Cloud 微服务体系弹性扩展的基础组件，与其他组件结合可以发挥出强大的作用。此外，丰富的负载均衡策略、重试机制、支持多协议的异步与响应式模型、容错、缓存与批处理等功能可以让你在构建自己的微服务架构时显得游刃有余。本章以 Spring Cloud Finchley 版本展开，对应 Ribbon 的 2.2.5 版本，将详细阐述 Ribbon 的应用方式与核心原理。

5.1 Spring Cloud Ribbon 概述

5.1.1 Ribbon 与负载均衡

负载均衡（Load Balance），即利用特定方式将流量分摊到多个操作单元上的一种手段，它对系统吞吐量与系统处理能力有着质的提升，毫不夸张地说，当今极少有企业没有用到负载均衡器或是负载均衡策略的。我们提到负载均衡，往往第一个想到的就是 Nginx，或许还会想到 LVS，这些大家都不陌生，且不管它们的使用方式，工作在什么层次，本质还是对于流量的疏导。

业界对于负载均衡有不少分类，最常见的有软负载与硬负载，代表产品是 Nginx 与 F5；还有一组分类方式笔者认为最能体现出 Ribbon 与传统负载均衡的差别，那就是集中式负载均衡与进程内负载均衡。集中式负载均衡指位于因特网与服务提供者之间，并负责把网络请求转发到各个提供单位，这时候 Nginx 与 F5 就可以划为一类了，也可以称为服务端负载均衡，其原理如图 5-1 所示；进程内负载均衡是指从一个实例库选取一个实例进行流量导入，在微服务的范畴内,实例库一般是存储在 Eureka、Consul、Zookeeper、etcd 这样的注册中心,而此时的负载均衡器就是类似 Ribbon 的 IPC(Inter-Process Communication,进程间通信)组件,因此,进程内负载均衡也叫作客户端负载均衡,其原理如图 5-2 所示。

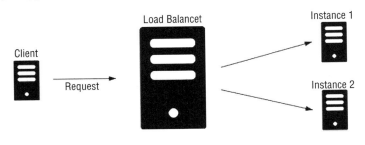

图 5-1　服务端负载均衡

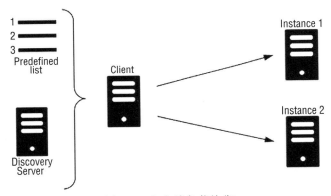

图 5-2　客户端负载均衡

Spring Cloud 官方是这样描述 Ribbon 的:

```
Ribbon is a client-side load balancer that gives you a lot of control over the
    behavior of HTTP and TCP clients.
```

Ribbon 是一个客户端负载均衡器,它赋予了应用一些支配 HTTP 与 TCP 行为的能力,可以得知,这里的客户端负载均衡(许多人称之为后端负载均衡)也是进程内负载均衡的一种。它在 Spring Cloud 生态内是一个不可缺少的组件,少了它,服务便不能横向扩展,这显然是有违云原生 12 要素的。此外,Feign 与 Zuul 中已经默认集成了 Ribbon,在我们的服务之间凡是涉及调用的,都可以集成它并应用,从而使我们的调用链具备良好的伸缩性。

5.1.2　入门案例

在对 Ribbon 有了初步了解之后,我们来看一看它的 Hello World 案例,由于客户端负载均衡需要从注册中心获取服务列表,所以需要集成 Eureka,在本案例中,使用默认配置呈现负载均衡效果。

1. 创建 Maven 父级 pom 工程

在父工程里面配置好工程需要的父级依赖,方便管理与简化配置,让子工程的配置文件变得简洁,这里使用的 Spring Cloud 版本是 Finchley,对应的 Spring Boot 版本是 2.0.3。详细 pom 文件请查阅本书配套源码 ch5-1\pom.xml,由于篇幅原因,这里就不列出了。

2. 创建 Eureka 组件工程

Eureka 工程的 pom.xml 文件,需要添加 spring-cloud-starter-netflix-eureka-server 依赖,注意 F 版和之前的版本有些变化,如代码清单 5-1 所示。

代码清单5-1　ch5-1\ch5-1-eureka-server\pom.xml

```
<dependency>
    <groupId>org.springframework.cloud</groupId>
    <artifactId>spring-cloud-starter-netflix-eureka-server</artifactId>
</dependency>
```

下面是 Eureka 的启动主类,添加注解 @SpringBootApplication,声明 Spring Boot 程序的启动入口,添加注解 @EnableEurekaServer,声明一个 Eureka Server,如代码清单 5-2 所示。

代码清单5-2　ch5-1\ch5-1-eureka-server\src\main\java\cn\springcloud\book\EurekaServerApplication.java

```
@SpringBootApplication
@EnableEurekaServer
public class EurekaServerApplication {

    public static void main(String[] args) {
        SpringApplication.run(EurekaServerApplication.class, args);
    }
}
```

最后是 Eureka Server 需要的配置文件,如代码清单 5-3 所示。

代码清单5-3　ch5-1\ch5-1-eureka-server\src\main\resources\bootstrap.yml

```
server:
    port: 8888
eureka:
    instance:
        hostname: localhost
    client:
        registerWithEureka: false
        fetchRegistry: false
        serviceUrl:
            defaultZone: http://${eureka.instance.hostname}:${server.port}/eureka/
```

3. 创建源服务工程 client-a

为了测试 Ribbon 的负载均衡功能,必须要有一个源服务,并且它需要开启多个实例,在每个实例中需要有一个标识来识别每次的调用是到了不同的服务实例上,我们可以使用一份代码,采取改变端口号的方式启动多次,就能开启多个相同服务实例了。由于需要注册到 Eureka

Server，pom.xml 需要引入 Eureka 客户端依赖，如代码清单 5-4 所示。

代码清单5-4　　ch5-1\ch5-1-client-a\pom.xml

```xml
<dependency>
    <groupId>org.springframework.cloud</groupId>
    <artifactId>spring-cloud-starter-netflix-eureka-client</artifactId>
</dependency>
```

启动主类声明该服务被注册中心发现即可，如代码清单 5-5 所示。

代码清单5-5　　ch5-1\ch5-1-client-a\src\main\java\cn\springcloud\book\ClientAApplication.java

```java
@SpringBootApplication
@EnableDiscoveryClient
public class ClientAApplication {

    public static void main(String[] args) {
        SpringApplication.run(ClientAApplication.class, args);
    }
}
```

在配置文件中标注服务名为 client-a，指定注册中心地址与初始的端口号，后面开启多实例只需要修改端口号即可，如代码清单 5-6 所示。

代码清单5-6　　ch5-1\ch5-1-client-a\src\main\resources\bootstrap.yml

```yaml
server:
    port: 7070
spring:
    application:
        name: client-a
eureka:
    client:
        serviceUrl:
            defaultZone: http://${eureka.host:127.0.0.1}:${eureka.port:8888}/eureka/
    instance:
        prefer-ip-address: true
```

最后是一个用于测试的 Controller，在该 API 中除了计算结果之外，还携带当前实例的端口号，如代码清单 5-7 所示。

代码清单5-7　　ch5-1\ch5-1-client-a\src\main\java\cn\springcloud\book\controller\TestController.java

```java
@RestController
public class TestController {

    @GetMapping("/add")
    public String add(Integer a, Integer b, HttpServletRequest request){
        return " From Port: "+ request.getServerPort() + ", Result: " + (a + b);
    }
}
```

4. 创建 Ribbon 客户端工程

要使用 Ribbon，需要在 pom.xml 文件中引入依赖 spring-cloud-starter-netflix-ribbon，另外，作为 Eureka 客户端，需要引入 spring-cloud-starter-netflix-eureka-client，如代码清单 5-8 所示。

代码清单5-8　ch5-1\ch5-1-ribbon-loadbalancer\pom.xml

```xml
<dependency>
    <groupId>org.springframework.cloud</groupId>
    <artifactId>spring-cloud-starter-netflix-ribbon</artifactId>
</dependency>
<dependency>
    <groupId>org.springframework.cloud</groupId>
    <artifactId>spring-cloud-starter-netflix-eureka-client</artifactId>
</dependency>
```

启动主类中除了声明服务实例被 Eureka 注册中心发现之外，还需要注入 RestTemplate 的 Bean，并且添加注解 @LoadBalanced，声明该 RestTemplate 用于负载均衡，如代码清单 5-9 所示。

代码清单5-9　ch5-1\ch5-1-ribbon-loadbalancer\src\main\java\cn\springcloud\book\RibbonLoadbalancerApplication.java

```java
@SpringBootApplication
@EnableDiscoveryClient
public class RibbonLoadbalancerApplication {

    public static void main(String[] args) {
        SpringApplication.run(RibbonLoadbalancerApplication.class, args);
    }

    @Bean
    @LoadBalanced
    public RestTemplate restTemplate() {
        return new RestTemplate();
    }

}
```

在配置文件中配置端口号，注册中心地址即可，如代码清单 5-10 所示。

代码清单5-10　ch5-1\ch5-1-ribbon-loadbalancer\src\main\resources\bootstrap.yml

```yml
spring:
  application:
    name: ribbon-loadbalancer
server:
  port: 7777
eureka:
  client:
    serviceUrl:
      defaultZone: http://${eureka.host:127.0.0.1}:${eureka.port:8888}/eureka/
  instance:
    prefer-ip-address: true
```

Ribbon 客户端需要创建一个 API 来调用 client-a 源服务的那个自定义 API，这里就需要用到 RestTemplate 来调用。RestTemplate 的一种使用方式如代码清单 5-11 所示，http 后面的 ip:port 变成了具体的服务名。

代码清单5-11　ch5-1\ch5-1-ribbon-loadbalancer\src\main\java\cn\springcloud\book\controller\TestController.java

```java
@RestController
public class TestController {

    @Autowired
    private RestTemplate restTemplate;

    @GetMapping("/add")
    public String add(Integer a, Integer b) {
        String result = restTemplate.getForObject("http://CLIENT-A/add?a=" + a
            + "&b=" + b, String.class);
        System.out.println(result);
        return result;
    }
}
```

5. 测试

1）启动 Eureka Server 之后，修改 client-a 服务的端口号为 7070 与 7075 分别启动，访问 http://localhost:8888，如图 5-3 所示，可以清晰看到已经启动了两个名为 CLIENT-A 的服务实例。

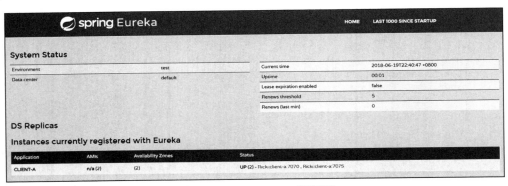

图 5-3　Eureka Server 运行图

2）访问 http://localhost:7777/add?a=100&b=300，会有两种结果，如图 5-4、5-5 所示。

图 5-4　postman 调用结果一

图 5-5 postman 调用结果二

由于我们在程序代码中也实现打印了调用结果，控制台打印信息如图 5-6 所示。

从图 5-6 中可以明显看出，Ribbon 默认使用轮询的方式访问源服务，此外 Ribbon 对服务实例节点的增减也能动态感知。

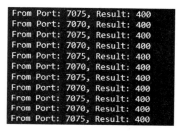

图 5-6 IDE console 打印结果

5.2　Spring Cloud Ribbon 实战

5.2.1　Ribbon 负载均衡策略与自定义配置

在 5.1 节我们已经基本了解了 Ribbon 在 Spring Cloud 微服务体系中的角色以及最简默认工作方式，相信读者在阅读完 5.1 节之后，免不了有许多猜测，迫切地想要知道 Ribbon 的负载均衡策略都有哪些，以及要怎么样使用它们呢？本节就大家最关心的"负载均衡"与策略的自定义配置展开探讨。

谈到 Ribbon，就会谈到负载均衡，而谈到负载均衡，就会衍生出多种负载均衡策略。我们熟知的 Nginx 有一些非常实用的负载均衡策略，比如：轮询（Round Robin）、权重（Weight）、ip_hash 等，丰富的策略能够让我们在构建应用时有充分的余地，可以根据业务场景选择最合适的策略。在 Ribbon 中，也有丰富的负载均衡策略可供选择，一共有 7 种，如表 5-1 所示。

表 5-1　Ribbon 负载均衡策略

策 略 类	命 名	描　　述
RandomRule	随机策略	随机选择 server
RoundRobinRule	轮询策略	按顺序循环选择 server
RetryRule	重试策略	在一个配置时间段内当选择 server 不成功，则一直尝试选择一个可用的 server
BestAvailableRule	最低并发策略	逐个考察 server，如果 server 断路器打开，则忽略，再选择其中并发连接最低的 server
AvailabilityFilteringRule	可用过滤策略	过滤掉一直连接失败的被标记为 circuit tripped 的 server，过滤掉那些高并发连接的 server（active connections 超过配置的阈值）

（续）

策　略　类	命　名	描　述
ResponseTimeWeightedRule	响应时间加权策略	根据 server 的响应时间分配权重。响应时间越长，权重越低，被选择到的概率就越低；响应时间越短，权重越高，被选择到的概率就越高。这个策略很贴切，综合了各种因素，如：网络、磁盘、IO 等，这些因素直接影响着响应时间
ZoneAvoidanceRule	区域权衡策略	综合判断 server 所在区域的性能和 server 的可用性轮询选择 server，并且判定一个 AWS Zone 的运行性能是否可用，剔除不可用的 Zone 中的所有 server

上一节的例子中我们看到了使用 Ribbon 做负载均衡的效果，默认是轮询策略（很多博客讲的是 ZoneAvoidanceRule 策略，其实 RibbonClientConfiguration 类已经被废弃掉了，BaseLoadBalancer 中设置的是 RoundRobinRule），如果我们想要使用其他策略该怎样做呢？下面我们来看一看。

1. 全局策略设置

使用 Ribbon 的时候想要全局更改负载均衡策略，需要加一个配置类，如代码清单 5-12 所示。

代码清单5-12　ch5-2\ch5-2-ribbon-loadbalancer\src\main\java\cn\springcloud\book\config\TestConfiguration.java

```java
@Configuration
public class TestConfiguration {

    @Bean
    public IRule ribbonRule() {
        return new RandomRule();
    }
}
```

就这样一个简单的配置，加上之后凡是通过 Ribbon 的请求都会按照配置的规则来进行，需要注意，代码在后续内容会有改动，读者如想测试可以还原。7 种自带负载均衡策略，使用只需在这里新建出来即可，后续章节可以这样配置，也可以使用自实现策略。

2. 基于注解的策略设置

如果我们想要针对某一个源服务设置其特有的策略，可以通过使用 @RibbonClient 注解，在使用它之前，我们需要对全局设置的代码进行改变，如代码清单 5-13 所示。

代码清单5-13　ch5-2\ch5-2-ribbon-loadbalancer\src\main\java\cn\springcloud\book\config\TestConfiguration.java

```java
@Configuration
@AvoidScan
public class TestConfiguration {

    @Autowired
    IClientConfig config;
```

```
    @Bean
    public IRule ribbonRule(IClientConfig config) {
        return new RandomRule();
    }

}
```

这里的 @AviodScan 注解是一个空的声明，稍后说明。注入的 IClientConfig 是针对客户端的配置管理器，使用 @RibbonClient 注解时尤其需要注意它的作用。最后我们在启动类加上 @RibbonClient 注解，来对源服务进行负载约束，如代码清单 5-14 所示。

代码清单5-14 ch5-2\ch5-2-ribbon-loadbalancer\src\main\java\cn\springcloud\book\ RibbonLoadbalancerApplication.java

```
@RibbonClient(name = "client-a", configuration = TestConfiguration.class)
@ComponentScan(excludeFilters = {@ComponentScan.Filter(type = FilterType.
    ANNOTATION, value = {AvoidScan.class})})
```

@RibbonClient 的使用方式比较直白，配置的意思是对 client-a 服务使用的策略是经过 TestConfiguration 所配置的。此外这里使用 @ComponentScan 注解的意思是让 Spring 不去扫描被 @AvoidScan 注解标记的配置类，因为我们的配置是对单个源服务生效的，所以不能应用于全局，如果不排除，启动就会报错。

此外，也可以使用 @RibbonClients 注解来对多个源服务进行策略指定。其使用方式如下：

```
@RibbonClients(value = {
    @RibbonClient(name = "client-a", configuration = TestConfiguration.class),
    @RibbonClient(name = "client-b", configuration = TestConfiguration.class)
})
```

3. 基于配置文件的策略设置

如果不喜欢用注解，也可以使用配置文件来对源服务负载策略进行配置，其基本语法是 <client name>.ribbon.*，使用它几乎可以不用注解形式的任何配置代码，推荐使用。如：我们可以使用下述配置对 client-a 服务使用随机策略。

```
client-a:
  ribbon:
    NFLoadBalancerRuleClassName: com.netflix.loadbalancer.RandomRule
```

5.2.2 Ribbon 超时与重试

使用 HTTP 发起请求，免不了要经历极端环境，此时对调用进行时限控制以及时限之后的重试尤为重要。注意，F 版中 Ribbon 的重试机制是默认开启的，需要添加对于超时时间与重试策略的配置，如代码清单 5-15 所示。

代码清单5-15 ch5-2\ch5-2-ribbon-loadbalancer\src\main\resources\bootstrap.yml

```
client-a:
  ribbon:
```

```
ConnectTimeout: 30000
ReadTimeout: 30000
MaxAutoRetries: 1 #对第一次请求的服务的重试次数
MaxAutoRetriesNextServer: 1 #要重试的下一个服务的最大数量（不包括第一个服务）
OkToRetryOnAllOperations: true
```

5.2.3 Ribbon 的饥饿加载

Ribbon 在进行客户端负载均衡的时候并不是在启动时就加载上下文，而是在实际请求的时候才去创建，因此这个特性往往会让我们的第一次调用显得颇为疲软乏力，严重的时候会引起调用超时。所以我们可以通过指定 Ribbon 具体的客户端的名称来开启饥饿加载，即在启动的时候便加载所有配置项的应用程序上下文，如代码清单 5-16 所示。

代码清单5-16 ch5-2\ch5-2-ribbon-loadbalancer\src\main\resources\bootstrap.yml

```
ribbon:
    eager-load:
        enabled: true
        clients: client-a, client-b, client-c
```

5.2.4 利用配置文件自定义 Ribbon 客户端

Ribbon 在 1.2.0 版本之后，就可以使用配置文件来定制 Ribbon 客户端了，其实质也就是使用配置文件来指定一些默认加载类，从而更改 Ribbon 客户端的行为方式，并且使用这种方式优先级最高，优先级高于使用注解 @RibbonClient 指定的配置和源码中加载的相关 Bean。我们可以通过表 5-2 中给出的配置来设置。

表 5-2 通过配置文件自定义 Ribbon 客户端的语法说明

配 置 项	说 明
<clientName>.ribbon.NFLoadBalancerClassName	指定 ILoadBalancer 的实现类
<clientName>.ribbon.NFLoadBalancerRuleClassName	指定 IRule 的实现类
<clientName>.ribbon.NFLoadBalancerPingClassName	指定 IPing 的实现类
<clientName>.ribbon.NIWSServerListClassName	指定 ServerList 的实现类
<clientName>.ribbon.NIWSServerListFilterClassName	指定 ServerListFilter 的实现类

可以使用 Ribbon 自带实现类，也可以自实现。下面是一个对 client 源服务的相关定制示例：

```
client:
    ribbon:
        NIWSServerListClassName: com.netflix.loadbalancer.ConfigurationBasedServerList
        NFLoadBalancerRuleClassName: com.netflix.loadbalancer.WeightedResponseTimeRule
```

5.2.5 Ribbon 脱离 Eureka 的使用

在默认情况下，Ribbon 客户端会从 Eureka 注册中心读取服务注册信息列表，来达到一种动态负载均衡的功能。但是有一种情况下则不会推荐这种方式,即如果 Eureka 是一个供很多人使用的公共注册中心(比如社区公益 Eureka,地址:http://eureka.springcloud.cn),此时极易产生服务侵入性问题,所以就不要从 Eureka 中读取服务列表了,而应该在 Ribbon 客户端自行指定源服务地址,让 Ribbon 脱离 Eureka 来使用。

为了达到这一目的,首先需要在 Ribbon 中禁用 Eureka 的功能:

```
ribbon:
    eureka:
        enabled: false
```

然后对源服务设定地址列表:

```
client:
    ribbon:
        listOfServers: http://localhost:7070, http://localhost:7071
```

5.3 Spring Cloud Ribbon 进阶

5.3.1 核心工作原理

我们已经了解了 Ribbon 的一些使用方式,现在让我们来看一看它的核心功能是如何被实现的,关于这部分内容,笔者着重讲解官方文档中提到的那些上层接口,以及 Ribbon 从启动到负载均衡器选择服务实例的过程。

官方文档中提到了 Ribbon 中的核心接口,它们共同定义了 Ribbon 的行为特性,如表 5-3 所示。

表 5-3 Ribbon 核心接口

接口	描述	默认实现类
IClientConfig	定义 Ribbon 中管理配置的接口	DefaultClientConfigImpl
IRule	定义 Ribbon 中负载均衡策略的接口	ZoneAvoidanceRule
IPing	定义定期 ping 服务检查可用性的接口	DummyPing
ServerList<Server>	定义获取服务列表方法的接口	ConfigurationBasedServerList
ServerListFilter<Server>	定义特定期望获取服务列表方法的接口	ZonePreferenceServerListFilter
ILoadBalancer	定义负载均衡选择服务的核心方法的接口	ZoneAwareLoadBalancer
ServerListUpdater	为 DynamicServerListLoadBalancer 定义动态更新服务列表的接口	PollingServerListUpdater

可以说,Ribbon 完全是基于这些接口建立起来的,它们就是 Ribbon 的骨架,对它们追根溯源,可以掌握 Ribbon 的底层实现,关于它们的解读,这里就不详细展开了,知道每一个接口的作用之后,依靠强大的 IDE,相信读者掌握它们并不是难事。其实很多读者对于 Ribbon 最关心的是它是如何做到使用 RestTemplate 达到负载均衡的,下面我们对这部分重要原理进行讲解。

在上一节中，我们知道了如何使用 Ribbon 来对客户端实例进行负载均衡，其基本的使用方式都需要注入一个 RestTemplate 的 Bean，并且使用 @LoadBalanced 注解才能使其具备负载均衡的能力，RestTemplate 在 Spring 中由来已久，为什么一个简单的注解就可以让它具备如此强悍的功能呢？打开其源码如下：

```
/**
 * Annotation to mark a RestTemplate bean to be configured to use a LoadBalancerClient
 * @author Spencer Gibb
 */
@Target({ ElementType.FIELD, ElementType.PARAMETER, ElementType.METHOD })
@Retention(RetentionPolicy.RUNTIME)
@Documented
@Inherited
@Qualifier
public @interface LoadBalanced {
}
```

这里注释写得比较清楚了，这个注解是标记一个 RestTemplate 来使用 LoadBalancerClient，LoadBalancerClient 是什么？搜索发现：

```
public interface LoadBalancerClient extends ServiceInstanceChooser {

    <T> T execute(String serviceId, LoadBalancerRequest<T> request) throws
        IOException;

    <T> T execute(String serviceId, ServiceInstance serviceInstance,
        LoadBalancerRequest<T> request) throws IOException;

    URI reconstructURI(ServiceInstance instance, URI original);
}
```

它是扩展自 ServiceInstanceChooser 接口的，ServiceInstanceChooser 内容如下：

```
public interface ServiceInstanceChooser {

    ServiceInstance choose(String serviceId);
}
```

这两个接口源码中的注释写得都相当详细，这里考虑到篇幅原因就不贴出了，笔者这里对每个方法结合注释进行讲解。

- ServiceInstance choose（String serviceId）：根据 serviceId，结合负载均衡器选择一个服务实例。
- <T> T execute（String serviceId, LoadBalancerRequest<T> request）：使用来自 LoadBalancer 的 ServiceInstance 为指定的服务执行请求。
- <T> T execute（String serviceId, ServiceInstance serviceInstance, LoadBalancerRequest<T> request）：使用来自 LoadBalancer 的 ServiceInstance 为指定的服务执行请求，是上一个方法的重载，在实现类中可以看到它们的关系，其实就是前一个方法的细节实现。
- URI reconstructURI（ServiceInstance instance，URI original）：使用主机 ip 和 port 构建特定的 URI 以供 Ribbon 内部使用。Ribbon 使用具有逻辑服务名称的 URI 作为 host,

例如 http://myservice/path/to/service。

由这些方法可以看出，这两个接口非同一般。有了这个线索，我们继续查找它在哪里被初始化，结果在同一个包下面发现了 LoadBalancerAutoConfiguration，阅读发现，它正是 Ribbon 功能的核心配置类，其部分重要源码如下：

```
@Configuration
@ConditionalOnClass(RestTemplate.class)
@ConditionalOnBean(LoadBalancerClient.class)
@EnableConfigurationProperties(LoadBalancerRetryProperties.class)
public class LoadBalancerAutoConfiguration {

    @Bean
    @ConditionalOnMissingBean
    public LoadBalancerRequestFactory loadBalancerRequestFactory(LoadBalancerCl
        ient loadBalancerClient) {
        return new LoadBalancerRequestFactory(loadBalancerClient, transformers);
    }

    @Configuration
    @ConditionalOnMissingClass("org.springframework.retry.support.RetryTemplate")
    static class LoadBalancerInterceptorConfig {
        @Bean
        public LoadBalancerInterceptor ribbonInterceptor(
            LoadBalancerClient loadBalancerClient,
            LoadBalancerRequestFactory requestFactory) {
            return new LoadBalancerInterceptor(loadBalancerClient, requestFactory);
        }

        @Bean
        @ConditionalOnMissingBean
        public RestTemplateCustomizer restTemplateCustomizer(final LoadBalancer-
            Interceptor loadBalancerInterceptor) {
            return restTemplate -> {
                List<ClientHttpRequestInterceptor> list = new ArrayList<>(
                    restTemplate.getInterceptors());
                list.add(loadBalancerInterceptor);
                restTemplate.setInterceptors(list);
            };
        }
    }
}
```

从类的注解就可以看出，它配置加载的时机一是当前工程环境必须有 RestTemplate 的实例，二是在工程环境中必须初始化了 LoadBalancerClient 的实现类。整个配置类只截取了重要部分，其中，LoadBalancerRequestFactory 用于创建 LoadBalancerRequest 供 LoadBalancer-Interceptor 使用，它在低版本中是没有的；LoadBalancerInterceptorConfig 中则维护了 Load-BalancerInterceptor 与 RestTemplateCustomizer 的实例，它们的作用如下：

❑ LoadBalancerInterceptor：拦截每一次 HTTP 请求，将请求绑定进 Ribbon 负载均衡的生命周期。

❑ RestTemplateCustomizer：为每个 RestTemplate 绑定 LoadBalancerInterceptor 拦截器。
LoadBalancerInterceptor 看来已经很接近我们要找的答案了，我们截取其源码：

```java
public class LoadBalancerInterceptor implements ClientHttpRequestInterceptor {
    private LoadBalancerClient loadBalancer;
    private LoadBalancerRequestFactory requestFactory;

    @Override
    public ClientHttpResponse intercept(final HttpRequest request, final byte[]
        body,final ClientHttpRequestExecution execution) throws IOException {
        final URI originalUri = request.getURI();
        String serviceName = originalUri.getHost();
        return this.loadBalancer.execute(serviceName, requestFactory.
            createRequest(request, body, execution));
    }
}
```

原来它是利用 ClientHttpRequestInterceptor 来对每次 HTTP 请求进行拦截的，此类是 Spring 中维护的请求拦截器，实现它的 intercept 方法就可以使请求进入方法体，从而做一些处理。可以看出这里把请求拦截下来之后使用了 LoadBalancerClient 的 execute 方法来处理请求，顺便提一下，由于我们在 RestTemplate 中使用的 URI 是形如 http://myservice/path/to/service 的，所以这里的 getHost() 方法实际取到的就是服务名 myservice。LoadBalancerClient 接口只有一个实现类，即 RibbonLoadBalancerClient，这里的 execute 方法体如下：

```java
@Override
public <T> T execute(String serviceId, LoadBalancerRequest<T> request) throws
    IOException {
    ILoadBalancer loadBalancer = getLoadBalancer(serviceId);
    Server server = getServer(loadBalancer);
    if (server == null) {
        throw new IllegalStateException("No instances available for " + serviceId);
    }
    RibbonServer ribbonServer = new RibbonServer(serviceId, server, isSecure(server,
        serviceId), serverIntrospector(serviceId).getMetadata(server));

    return execute(serviceId, ribbonServer, request);
}
```

首先要得到一个 ILoadBalancer，再使用它去得到一个 Server，顾名思义，这个 Server 就是具体服务实例的封装了，实际上查看源码也验证了笔者的猜测，所以，getServer (loadBalancer) 就是发生负载均衡过程的地方！我们再看一看它的实现：

```java
protected Server getServer(ILoadBalancer loadBalancer) {
    if (loadBalancer == null) {
        return null;
    }
    return loadBalancer.chooseServer("default"); // TODO: better handling of key
}
```

运用 ILoadBalancer 的 chooseServer 方法，查看 ILoadBalancer 的实现类的该方法实现：

```java
public Server chooseServer(Object key) {
    if (counter == null) {
        counter = createCounter();
    }
    counter.increment();
```

```
        if (rule == null) {
            return null;
        } else {
            try {
                return rule.choose(key);
            } catch (Exception e) {
                logger.warn("LoadBalancer [{}]: Error choosing server for key {}",
                    name, key, e);
                return null;
            }
        }
    }
```

我们发现，rule.choose(key) 中的 rule 其实就是 IRule，至此，拦截的 HTTP 请求与负载均衡策略得以关联起来。

5.3.2　负载均衡策略源码导读

Ribbon 作为一个 Spring Cloud 微服务体系中的负载均衡器，理解其实现原理就显得格外重要，本章之中已经有部分内容对它有所涉及，由于策略较多，详细解读会占大量篇幅，本节就只引导读者去读懂源码。IRule 是定义 Ribbon 负载均衡策略的父接口，所有策略都是基于它实现的，笔者整理了 Ribbon 的负载均衡家族体系，如图 5-7 所示。

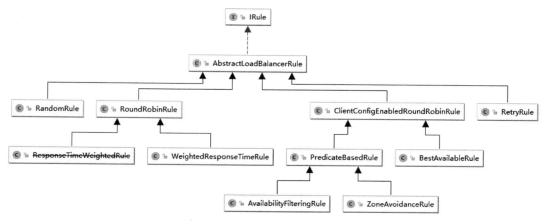

图 5-7　Ribbon 负载均衡家族体系

Ribbon 内部一共提供了 7 种可选的负载均衡策略，可以说应对微服务分布式的复杂环境是十分实用的，从简单的随机与轮询，到具有区域权衡能力的策略，为不同的场景提供合适的策略。要了解这个体系，首先且看 IRule 的源码：

```
public interface IRule{

    public Server choose(Object key);

    public void setLoadBalancer(ILoadBalancer lb);
```

```
    public ILoadBalancer getLoadBalancer();
}
```

IRule 接口一共定义了三个方法，实现类实现 choose 方法会加入具体的负载均衡策略逻辑。另外两个方法与 ILoadBalancer 关联起来，从 5.3.1 节我们知道，在调用过程中，Ribbon 是通过 ILoadBalancer 来关联 IRule 的，ILoadBalancer 的 chooseServer 方法会转换为调用 IRule 的 choose 方法，抽象类 AbstractLoadBalancerRule 实现了这两个方法，从而将 ILoadBalancer 与 IRule 关联起来。再往下走就是具体的实现类了，每种实现逻辑就需要读者自行去阅读了。

5.4 本章小结

本章讲解了负载均衡的一些背景分类，以及 Ribbon 客户端的基本配置与使用。阅读本章之后，你可以了解 Ribbon 在 Spring Cloud 微服务体系中的地位，可以说，"横向扩展"能力就全仰仗它了。Ribbon 的可配置性比较强大，若内部配置不能满足需求，读者可自行根据相关规则增加适用的扩展。

第 6 章 Chapter 6

Spring Cloud Hystrix 实战运用

本章将系统全面地介绍 Spring Cloud Hystrix 的相关知识点，包括 Hystrix、设计目标，以及 Spring Cloud Hystrix 的基础用法和实战技巧。

6.1 Spring Cloud Hystrix 概述

Hystrix 是由 Netflix 开源的一个针对分布式系统容错处理的开源组件，如图 6-1 所示。2011–2012 年相继诞生和成熟，Netflix 公司很多项目都使用了它，Hystrix 单词意为"豪猪"，浑身有刺来保护自己，Hystrix 库就是这样一个用来捍卫应用程序健康的利器。

图 6-1 Hystrix 官方简图

Hystrix 官方代码托管在 https://github.com/Netflix/Hystrix，大家可以登录查看相关的信息，官方首页上写着这么一段话：

Hystrix is a latency and fault tolerance library designed to isolate points of access to remote systems, services and 3rd party libraries, stop cascading failure and enable resilience in complex distributed systems where failure is inevitable.

大意是：Hystrix 是一个延迟和容错库，旨在隔离远程系统、服务和第三方库，阻止级联故障，在复杂的分布式系统中实现恢复能力。

6.1.1 解决什么问题

为什么要使用 Hystrix，它可以解决什么样的问题，这里引用官方 Hystrix 的图来解释一下，在应用拆分成多个服务的情况下：

1）当用户需要请求 A、P、H、I 四个服务获取数据时，在平时正常流量情况下，系统稳定运行，如图 6-2 所示。

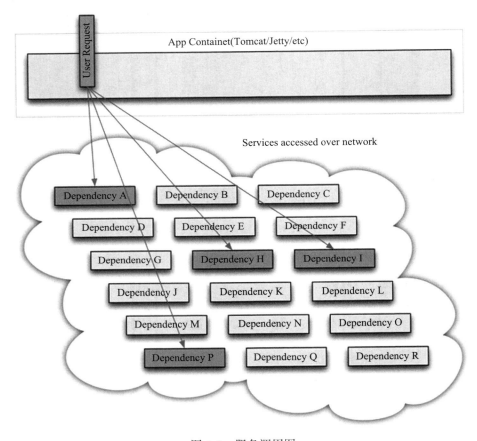

图 6-2　服务调用图

2）某一天公司进行促销，效果明显，使得超越平时几倍数量的用户涌入进来，对其中服务 I 并发超过 50+，这时对服务 I 出现一定程度的影响，逐渐导致 CPU、内存占用过高等问题，结果导致服务 I 延迟，响应过慢，如图 6-3 所示。

3）随着压力持续增加，服务 I 承受不住压力或发生其他内部错误导致机器内部资源耗尽，请求堆积等情况使服务 I 彻底宕机不可用，更糟糕的情况是其他服务对 I 有依赖，导致其他服务也同样出现请求堆积、资源占用等问题，这时会导致整个系统出现大面积的延迟或瘫痪，直到整个系统不可用，如图 6-4 所示。

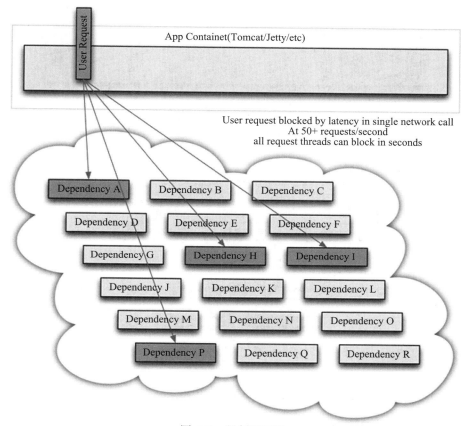

图 6-3　服务调用图

总结：在这种多系统和微服务的情况下，需要一种机制来处理延迟和故障，并保护整个系统处于可用稳定的状态，此时就是 Hystrix 大显身手的时候了。

详细图和官方解释可见官方 Wiki：https://github.com/Netflix/Hystrix/wiki。

6.1.2　设计目标

关于设计目标，这里引用官方给出的定义：

❑ Give protection from and control over latency and failure from dependencies accessed (typically over the network) via third-party client libraries.
❑ Stop cascading failures in a complex distributed system.
❑ Fail fast and rapidly recover.
❑ Fallback and gracefully degrade when possible.
❑ Enable near real-time monitoring，alerting，and operational control.

翻译如下：

1）通过客户端库对延迟和故障进行保护和控制。

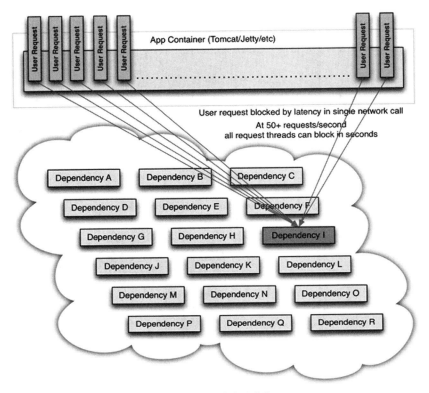

图 6-4　服务调用图

2）在一个复杂的分布式系统中停止级联故障。
3）快速失败和迅速恢复。
4）在合理的情况下回退和优雅地降级。
5）开启近实时监控、告警和操作控制。

Hystrix 底层大量使用了 Rxjava，本书暂不解析源代码，如果想了解源代码可以先了解下 Rxjava，对学习 Hystrix 底层源码会有很大的帮助。

6.2　Spring Cloud Hystrix 实战运用

下面用一个最简单的例子演示下 Hystrix 的简单入门。

6.2.1　入门示例

1. 创建 client-service 工程，引入相关依赖

引入 hystrix 的包，如代码清单 6-1 所示：

代码清单6-1　ch6-1\ch6-1-client-service\pom.xml

```xml
<dependency>
    <groupId>org.springframework.cloud</groupId>
    <artifactId>spring-cloud-starter-netflix-hystrix</artifactId>
</dependency>
```

2. 启用断路器模式

启用断路器模式如代码清单 6-2 所示，开启断路器注解，在启动类添加 @EnableHystrix 注解。

代码清单6-2　ch6-1\ch6-1-client-service\src\main\java\cn\springcloud\book\ClientApplication.java

```java
@SpringBootApplication
@EnableHystrix
@EnableDiscoveryClient
public class ClientApplication {
    public static void main(String[] args) {
        SpringApplication.run(ClientApplication.class, args);
    }
}
```

3. 增加 @HystrixCommand 和降级 fallback

HystrixCommand 使用方法如代码清单 6-3 所示：

代码清单6-3　ch6-1\ch6-1-client-service\src\main\java\cn\springcloud\book\service\impl\UserService.java

```java
@HystrixCommand(fallbackMethod="defaultUser")
public String getUser(String username) throws Exception {
    if(username.equals("spring")) {
        return "this is real user";
    }else {
        throw new Exception();
    }
}

/**
 * 出错则调用该方法返回预设友好错误
 * @param username
 * @return
 */
public String defaultUser(String username) {
    return "The user does not exist in this system";
}
```

4. 效果展示

调用接口，当用户名等于 spring 时返回正确的信息，当不是 spring 时则抛出异常，降级处理返回友好的提示。

浏览器执行：http://localhost:8888/getUser?username=spring

返回结果：This is real user

执行：http://localhost:8888/getUser?username=hello

返回结果：The user does not exist in this system

6.2.2 Feign 中使用断路器

在 Feign 中，默认是自带 Hystrix 的功能的，在很老的版本中默认是打开的，从最近的几个版本开始默认被关闭了，所以需要自己通过配置文件打开它。

1. 创建 consumer 工程，定义接口

使用 @FeignClient 定义接口，配置降级回退类 UserServiceFallback，接口如代码清单 6-4 所示。

代码清单6-4 ch6-2-consumer-service\src\main\java\cn\springcloud\book\service\IUserService.java

```java
@FeignClient(name="sc-provider-service",fallback= UserServiceFallback.class)
public interface IUserService {
    @RequestMapping(value = "/getUser",method = RequestMethod.GET)
    public String getUser(@RequestParam("username") String username);
}
```

2. 创建降级 fallback 类

Fallback 类如代码清单 6-5 所示：

代码清单6-5 ch6-2-consumer-service\src\main\java\cn\springcloud\book\service\impl\UserServiceFallback.java

```java
@Component
public class UserServiceFallback implements IUserService{
    /**
     * 出错则调用该方法返回友好错误
     */
    public String getUser(String username){
        return "The user does not exist in this system, please confirm username";
    }
}
```

3. 结果展示

当不启动 sc-provider-service 工程，并 feign.hystrix.enabled=false 时，访问 http://localhost:8888/getUser?username=hello，结果报 500：

```
There was an unexpected error (type=Internal Server Error, status=500).
```

打开 feign.hystrix 的标志位为 true，再重启访问，结果显示定义的友好提示：

```
The user does not exist in this system, please confirm username
```

这时发现 Hystrix 已经产生了作用。

6.2.3 Hystrix Dashboard

Hystrix Dashboard 仪表盘是根据系统一段时间内发生的请求情况来展示的可视化面板，这些信息是每个 HystrixCommand 执行过程中的信息，这些信息是一个指标集合和具体的系统运行情况，图 6-5 所示为 Hystrix Dashboard 的入口界面图。

图 6-5　Hystrix Dashboard 界面图

在 Spring Cloud 里搭建出这样一个功能也是非常方便的，按照这几个步骤即可。

1. 搭建相关工程

创建 eureka、hello-service、provider-service 三个工程，其中 provider-service 提供了一个接口返回信息。因为 hystrix 的指标是需要端口进行支撑的，所以要增加 actuator 依赖，并公开 hystrix.stream 端点以便能被顺利地访问到，如代码清单 6-6 所示：

代码清单6-6　ch6-3\pom.xml

```
1.依赖
<dependency>
    <groupId>org.springframework.boot</groupId>
    <artifactId>spring-boot-starter-actuator</artifactId>
</dependency>
2.配置
management:
  security:
    enabled: false
  endpoints:
    web:
      exposure:
        include: hystrix.stream
feign:
  hystrix:
    enabled: true
```

因为 feignClient 绑定了 hystrix，这里在 hello-service 中定义了一个 ProviderService 的 FeignClient 负责调用接口，如代码清单 6-7 所示：

代码清单6-7　ch6-3\ch6-3-hello-service\src\main\java\cn\springcloud\book\hello\service\dataservice\ProviderService.java

```
@FeignClient(name = "sc-provider-service")
public interface ProviderService {
    @RequestMapping(value = "/getDashboard", method = RequestMethod.GET)
        public List<String> getProviderData();
}
```

2. 搭建 Hystrix Dashboard 工程

新建 Hystrix Dashboard 工程，增加 Hystrix Dashboard 依赖，如代码清单 6-8 所示：

代码清单6-8　ch6-3\ch6-3-hystrix-dashboard\pom.xml

```xml
<dependency>
    <groupId>org.springframework.cloud</groupId>
    <artifactId>spring-cloud-starter-netflix-hystrix-dashboard</artifactId>
</dependency>
```

增加 @EnableHystrixDashboard 注解，如代码清单 6-9 所示：

代码清单6-9　ch6-3\ch6-3-hystrix-dashboard\src\main\java\cn\springcloud\book\Hystrix DashboardApplication.java

```java
@EnableHystrixDashboard
public class HystrixDashboardApplication {
    public static void main(String[] args) {
        SpringApplication.run(HystrixDashboardApplication.class, args);
    }
}
```

3. Hystrix Dashboard 运行结果和解释

启动 Hystrix Dashboard 和 hello-service 这两个工程，浏览器访问：http：//localhost：9000/hystrix，即可看到图 6-6 所示的 Hystrix Dashboard 的首页。

图 6-6　Hystrix Dashboard 界面图

界面上写的有三个地址，分别是三种监控方式：

- 默认集群监控，通过 http://turbine-hostname:port/turbine.stream 访问。
- 指定的集群：http://turbine-hostname:port/turbine.stream?cluster=[clusterName]。
- 单个应用：http://hystrix-app:port/hystrix.stream。

由于是在本地启动的应用，这时对上面的 9091 的 hello-service 进行监控，访问 http://localhost:9091/actuator/hystrix.stream，需要注意的是，这里的地址是 actuator，因为目前 F 版的 SpringBoot 版本是 2.0，所有端点的前面都加了 actuator，所以在输入地址时，应该加上 actuator，不然会访问不到，进去后会出现 loading 的字样，这时访问接口 http://localhost:9091/getProviderData，会看到如图 6-7 所示的失败消息。

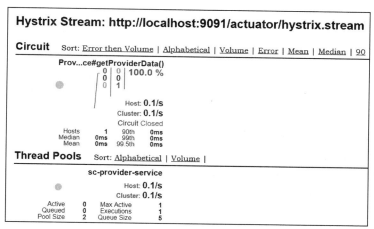

图 6-7　Hystrix Dashboard 仪表盘

这是因为没有启动 provider-service 服务，此时启动 provider-service 工程，再进行访问，发现请求成功，会显示如图 6-8 所示的成功消息。

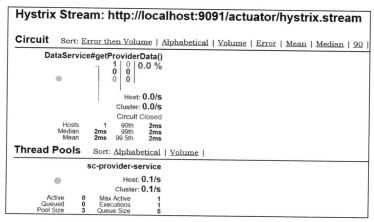

图 6-8　Hystrix Dashboard 仪表盘

这时再进行测试，关掉 provider-service 服务，快速并连续地发送失败请求，会发现断路器被打开，界面会看到 Circuit 的值为 open，并且报错信息也会提示 ProviderService#getProviderData() short-circuited，如图 6-9 所示。

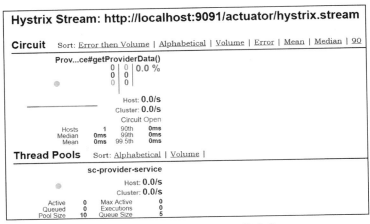

图 6-9 Hystrix Dashboard 仪表盘

关于 Hystrix Dashboard 的监控图，这里来解释下具体的意思。

圆圈：它是代表流量的大小和流量的健康，有绿色、黄色、橙色、红色这几个颜色，通过这些颜色的标识，可以快速发现故障、具体的实例、请求压力等。

曲线：它代表 2 分钟内流量的变化，可以根据它发现流程的浮动趋势。

圆圈右边的数字，见图 6-10 所示。

Hystrix Dashboard 页面左边第 1 列数字代表了请求的成功、熔断数、错误的请求、超时的请求、线程池拒绝数、失败的请求和最近 10 秒内错误的比率，见图 6-11 所示。

图 6-10 Hystrix Dashboard 仪表盘指标（一）

图 6-11 Hystrix Dashboard 仪表盘指标（二）

Host&Cluster：代表机器和集群的请求频率。
Circuit：断路器状态，open/closed。
Hosts&Median&Mean&：集群下的报告，百分位延迟数。
Thread Pools：线程池的指标，核心线程池指标，队列大小等。

6.2.4　Turbine 聚合 Hystrix

在上节讲了单个实例的 Hystrix Dashboard，但在整个系统和集群的情况下不是特别有用，所以需要一种方式来聚合整个集群下的监控状况，Turbine 就是来聚合所有相关的 hystrix.stream 流的方案，然后在 Hystrix Dashboard 中显示。

1. 搭建相关工程和接口

沿用上一节的三个工程，eureka、hello-service、provider-service，并在 provider-service 里增加接口，如代码清单 6-10 所示：

代码清单6-10 ch6-3\ch6-3-provider-service\src\main\java\cn\springcloud\book\provider\controller\ConsumerService.java

```java
@FeignClient(name = "sc-hello-service")
public interface ConsumerService {
    @RequestMapping(value = "/helloService", method = RequestMethod.GET)
    public String getHelloServiceData();
}
```

2. 搭建 Turbine

新建 turbine 工程，在 pom 里增加依赖，如代码清单 6-11 所示：

代码清单6-11 ch6-3\ch6-3-turbine\pom.xml

```xml
<dependency>
    <groupId>org.springframework.cloud</groupId>
    <artifactId>spring-cloud-starter-netflix-turbine</artifactId>
</dependency>
```

增加 @EnableTurbine 注解，如代码清单 6-12 所示：

代码清单6-12 ch6-3\ch6-3-turbine\src\main\java\cn\springcloud\book\ TurbineApplication.java

```java
@SpringBootApplication
@EnableDiscoveryClient
@EnableTurbine
@EnableHystrixDashboard
public class TurbineApplication {
    public static void main(String[] args) {
        SpringApplication.run(TurbineApplication.class, args);
    }
}
```

新增配置属性，turbine.app-config 设置需要收集监控信息的服务名，turbine.cluster-name-expression 设置集群名称，如代码清单 6-13 所示：

代码清单6-13 ch6-3\ch6-3-turbine\src\main\resources\application.yml

```
turbine:
    appConfig: sc-hello-service,sc-provider-service
    clusterNameExpression: "'default'"
```

增加完成后启动 turbine 服务，这时访问 http://localhost:9088/hystrix，选择进行集群监控，输入之前介绍的第一种模式 cluster，输入 http://localhost:9088/turbine.stream。

访问 hello-service 和 provider-service 的接口：

- http://localhost:9091/getProviderData
- http://localhost:8099/getHelloService

访问成功后可以看到如图 6-12 所示的界面，这时就是多个服务的监控状态。

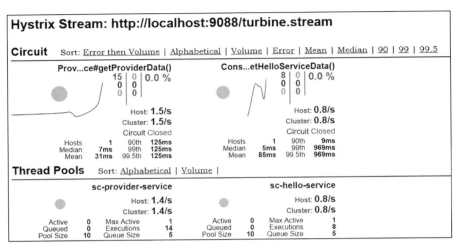

图 6-12　turbine 仪表盘

6.2.5　Hystrix 异常机制和处理

Hystrix 的异常处理中，有 5 种出错的情况下会被 fallback 所截获，从而触发 fallback，这些情况是：

- FAILURE：执行失败，抛出异常。
- TIMEOUT：执行超时。
- SHORT_CIRCUITED：断路器打开。
- THREAD_POOL_REJECTED：线程池拒绝。
- SEMAPHORE_REJECTED：信号量拒绝。

有一种类型的异常是不会触发 fallback 且不会被计数进入熔断的，它是 BAD_REQUEST，会抛出 HystrixBadRequestException，这种异常一般对应的是由非法参数或者一些非系统异常引起的，对于这类异常可以根据响应创建对应的异常进行异常封装或者直接处理，这里通过例子测试一下。

用 PSFallbackBadRequestExpcetion 类抛出 HystrixBadRequestException 不会被 fallback，如代码清单 6-14 所示：

代码清单6-14　ch6-3\ch6-3-hystrix-exception-service\src\main\java\cn\springcloud\book\ex\service\dataservice\PSFallbackBadRequestExpcetion.java

```
@Override
    protected String run() throws Exception {
        throw new HystrixBadRequestException("HystrixBadRequestException error");
}

@Override
protected String getFallback() {
```

```
        return "invoke HystrixBadRequestException fallback method:  ";
    }
```

使用另外的 PSFallbackOtherExpcetion 类抛出 Exception，会被 fallback 触发，如代码清单 6-15 所示：

代码清单6-15　ch6-3\ch6-3-hystrix-exception-service\src\main\java\cn\springcloud\book\ex\service\dataservice\PSFallbackOtherExpcetion.java

```
@Override
protected String run() throws Exception {
    throw new Exception("this command will trigger fallback");
}

@Override
protected String getFallback() {
    return "invoke PSFallbackOtherExpcetion fallback method";
}
```

在 @HystrixCommand 里获取异常信息也很容易，参数里面带 Throwable 即可获得具体的信息，如代码清单 6-16 所示：

代码清单6-16　ch6-3\ch6-3-hystrix-exception-service\src\main\java\cn\springcloud\book\ex\controlle\ExceptionController.java

```
@GetMapping("/getFallbackMethodTest")
@HystrixCommand(fallbackMethod = "fallback")
public String getFallbackMethodTest(String id){
    throw new RuntimeException("getFallbackMethodTest failed");
}

public String fallback(String id, Throwable throwable) {
    log.error(throwable.getMessage());
    return "this is fallback message";
}
```

继承 @HystrixCommand 的命令可以通过方法来获取异常，如代码清单 6-17 所示：

代码清单6-17　ch6-3\ch6-3-hystrix-exception-service\src\main\java\cn\springcloud\book\ex\service\dataservice\PSFallbackBadRequestExpcetion.java

```
@Override
protected String getFallback() {
    System.out.println(getFailedExecutionException().getMessage());
    return "invoke HystrixBadRequestException fallback method:  ";
}
```

在 Feign Client 中可以用 ErrorDecoder 实现对这类异常的包装，在实际的使用中，很多时候调用接口会抛出这些 400-500 之间的错误，此时可以通过它进行封装。

6.2.6 Hystrix 配置说明

Hystrix 的配置比较多,具体可以参考官方的地址(https://github.com/Netflix/Hystrix/wiki/Configuration)来了解很多配置的作用。这里列一些常用的和可能需要改动的配置。

隔离策略,HystrixCommandKey,如果不配置,则默认为方法名:

默认值:`THREAD`
默认属性:`hystrix.command.default.execution.isolation.strategy`
实例属性:`hystrix.command.HystrixCommandKey.execution.isolation.strategy`

配置 hystrixCommand 命令执行超时时间,以毫秒为单位:

默认值:`1000`
默认属性:
`hystrix.command.default.execution.isolation.thread.timeoutInMilliseconds`
实例属性:
`hystrix.command.HystrixCommandKey.execution.isolation.thread.timeoutInMilliseconds`

hystrixCommand 命令执行是否开启超时:

默认值:`true`
默认属性:`hystrix.command.default.execution.timeout.enabled`
实例属性:`hystrix.command.HystrixCommandKey.execution.timeout.enabled`

超时时是否应中断执行操作:

默认值:`true`
默认属性:
`hystrix.command.default.execution.isolation.thread.interruptOnTimeout`
实例属性:
`hystrix.command.HystrixCommandKey.execution.isolation.thread.interruptOnTimeout`

信号量请求数,当设置为信号量隔离策略时,设置最大允许的请求数:

默认值:`10`
默认属性:
`hystrix.command.default.execution.isolation.semaphore.maxConcurrentRequests`
实例属性:
`hystrix.command.HystrixCommandKey.execution.isolation.semaphore.maxConcurrentRequests`

Circuit Breaker 设置打开 fallback 并启动 fallback 逻辑的错误比率。

默认值:`50`
默认属性:`hystrix.command.default.circuitBreaker.errorThresholdPercentage`
实例属性:
`hystrix.command.HystrixCommandKey.circuitBreaker.errorThresholdPercentage`

强制打开断路器,拒绝所有请求:

默认值:`false`
默认属性:`hystrix.command.default.circuitBreaker.forceOpen`
实例属性:`hystrix.command.HystrixCommandKey.circuitBreaker.forceOpen`

当为线程隔离时,线程池核心大小:

默认值:`10`
默认属性:`hystrix.threadpool.default.coreSize`

实例属性：hystrix.threadpool.HystrixThreadPoolKey.coreSize

当 Hystrix 隔离策略为线程池隔离模式时，最大线程池大小的配置如下，在 1.5.9 版本中还需要配置 allowMaximumSizeToDivergeFromCoreSize 为 true。

默认值：10
默认属性：hystrix.threadpool.default.maximumSize
实例属性：hystrix.threadpool.HystrixThreadPoolKey.maximumSize

allowMaximumSizeToDivergeFromCoreSize，此属性允许配置 maximumSize 生效：

默认值：false
默认属性：hystrix.threadpool.default.allowMaximumSizeToDivergeFromCoreSize
实例属性：
hystrix.threadpool.HystrixThreadPoolKey.allowMaximumSizeToDivergeFromCoreSize

说明 在真实的应用过程中，一般会对超时时间、线程池大小、信号量等进行修改，具体要结合业务进行分析，默认 Hystrix 的超时时间为 1 秒，但在实际的运用过程中。发现 1 秒有些过短，通常会设置 5~10 秒左右，对于一些需要同步文件上传等业务则会更长，如果配置了 Ribbon 的时间，其超过时间也需要和 Ribbon 的时间配合实用，一般情况下 Ribbon 的时间应短于 Hystrix 超时时间。

6.2.7　Hystrix 线程调整和计算

在实际使用过程中会涉及很多服务，可能有些服务使用的线程池大，有些服务使用的小，有些服务的超时时间长，有些又短，所以 Hystrix 官方也提供了一些方法供大家来计算和调整这些配置，总的宗旨是，通过自我预判的配置先发布到生产或测试，然后查看它具体的运行情况，再调整为更符合业务的配置，通常的做法是：

1）超时时间默认为 1000ms，如果业务明显超过 1000ms，则根据自己的业务进行修改。
2）线程池默认为 10，如果你知道确实要使用更多时可以调整。
3）金丝雀发布，如果成功则保持。
4）在生产环境中运行超过 24 小时。
5）如果系统有警告和监控，那么可以依靠它们捕捉问题。
6）运行 24 小时之后，通过延迟百分位和流量来计算有意义的最低满足值。
7）在生产或者测试环境中实时修改值，然后用仪表盘监控。
8）如果断路器产生变化和影响，则需再次确认这个配置。

这里列的 8 点是一些比较实用的方法论，可以用来探测符合系统的配置，官方用了一个图标识了一个典型的例子，如图 6-13 所示：

这里 Threadpool 的大小是 10，它有一个计算方法：

每秒请求的峰值×99%的延迟百分比（请求的响应时间）＋预留缓冲的值

所以这个例子为：30×0.2s＝6＋预留缓冲的值＝10，这里预留了 4 个线程数。
Thread Timeout：预留了一个足够的时间，250ms，然后加上重试一次的中位数值。

Connect Timeout & Read Timeout：100ms 和 250ms，这两个值的设置方法远高于中位数的值，以适应大多数请求。

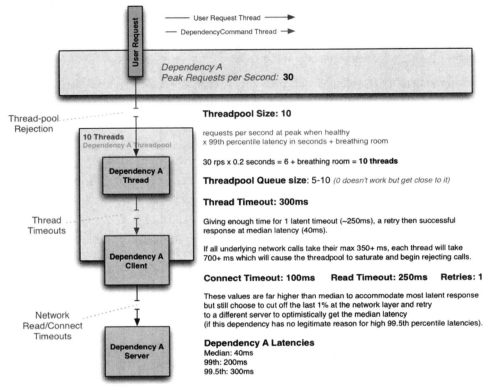

图 6-13　请求配置解析图

在实际的生产测试过程中，大家在配置每个服务时可以根据官方推荐的这些方法来测试自己的业务需要的数值，这样产生最适合的配置。

6.2.8　Hystrix 请求缓存

Hystrix 请求缓存是 Hystrix 在同一个上下文请求中缓存请求结果，它与传统理解的缓存有一定区别，Hystrix 的请求缓存是在同一个请求中进行，在进行第一次调用结束后对结果缓存，然后接下来同参数的请求将会使用第一次的结果，缓存的生命周期只是在这一次请求中有效。

使用 HystrixCommand 有两种方式，第一种是继承，第二种是直接注解。缓存也同时支持这两种，具体做法如下。

1. 初始化请求上下文

前面章节提到，Hystrix 的缓存在一次请求内有效，这要求请求要在一个 Hystrix 上下文里，不然在使用缓存的时候 Hystrix 会报一个没有初始化上下文的错误，可以使用 filter 过滤器和 Interceptor 拦截器进行初始化，在这里使用一个拦截器来举例，实现 HandlerInterceptor。分别在 preHandle 和 afterCompletion 执行这两段代码，另外还需要在启动上增加开启 Hystrix 的注解 @EnableHystrix：

初始化上下文，如代码清单 6-18 所示：

代码清单6-18 ch6-4\ch6-4-hystrix-cache\src\main\java\cn\springcloud\book\hystrix\config\ CacheContextInterceptor.java

```
@Override
    public boolean preHandle(HttpServletRequest request, HttpServletResponse
        respone, Object arg2) throws Exception {
        context = HystrixRequestContext.initializeContext();
        return true;
    }
```

关闭上下文，如代码清单 6-19 所示：

代码清单6-19 ch6-4\ch6-4-hystrix-cache\src\main\java\cn\springcloud\book\hystrix\config\ CacheContextInterceptor.java

```
@Override
public void afterCompletion(HttpServletRequest request, HttpServletResponse
respone, Object arg2, Exception arg3)
    throws Exception {
        context.shutdown();
    }
```

2. 使用类来开启缓存

使用类的方式很简单，只要继承 HystrixCommand，然后重写它的 getCacheKey 方法即可，以保证对于同一个请求返回同样的键值，对于清除缓存，则调用 HystrixRequestCache 类的 clean 方法即可，关键代码如代码清单 6-20 所示：

代码清单6-20 ch6-4\ch6-4-hystrix-cache\src\main\java\cn\springcloud\book\hystrix\service\HelloCommand.java

```
@Override
protected String run() throws Exception {
String json= restTemplate.getForObject("http://sc-provider-service/getUser/{1}",
    String.class, id);
    System.out.println(json);
    return json;
}

    @Override
    protected String getCacheKey() {
        return String.valueOf(id);
    }

    public static void cleanCache(Long id){
        HystrixRequestCache.getInstance(HystrixCommandKey.Factory.asKey
            ("springCloudCacheGroup"), HystrixConcurrencyStrategyDefault.
            getInstance()).clear(String.valueOf(id));
    }
```

运行代码和结果，如代码清单 6-21 所示：

代码清单6-21 ch6-4\ch6-4-hystrix-cache\src\main\java\cn\springcloud\book\hystrix\
controller\CacheController.java

```java
@RequestMapping(value = "/getUserIdByExtendCommand/{id}", method = RequestMethod.
    GET)
    public String getUserIdByExtendCommand(@PathVariable("id") Integer id) {
        HelloCommand one = new HelloCommand(restTemplate,id);
        one.execute();
        logger.info("from cache:   " + one.isResponseFromCache());
        HelloCommand two = new HelloCommand(restTemplate,id);
        two.execute();
        logger.info("from cache:   " + two.isResponseFromCache());
        return "getUserIdByExtendCommand success";
    }
```

访问 http://localhost:5566/getUserIdByExtendCommand/1，调用两次 execute，结果发现控制台只打印一次，并且可以用 Hystrix 的默认方法 isResponseFromCache 来看是否来自缓存，从打印结果来看，第二次的请求确实来自于缓存数据，缓存已经成功：

```
{"username":"Toy","password":"123456","age":10}
CacheController          : from cache:    false
CacheController          : from cache:    true
```

3. 使用注解来开启缓存

Hystrix 也提供了注解来使用缓存机制，且更为方便和快捷。使用 @CacheResult 和 @CacheRemove 即可缓存和清除缓存，缓存接口代码如代码清单 6-22 所示：

代码清单6-22 ch6-4\ch6-4-hystrix-cache\src\main\java\cn\springcloud\book\hystrix\
service\ HelloService.java

```java
@CacheResult
@HystrixCommand
public String hello(Integer id) {
    String json = restTemplate.getForObject("http://sc-provider-service/getUser/
        {1}", String.class, id);
    System.out.println(json);
    return json;
}
```

运行代码和结果，如代码清单 6-23 所示：

代码清单6-23 ch6-4\ch6-4-hystrix-cache\src\main\java\cn\springcloud\book\hystrix\
controller\CacheController.java

```java
@RequestMapping(value = "/getUser/{id}", method = RequestMethod.GET)
    public String getUserId(@PathVariable("id") Integer id) {
        helloService.hello(id);
        helloService.hello(id);
        return "getUser success";
    }
```

访问 http://localhost:5566/getUser/1，调用两次 hello 结果，发现只打印出一条数据，第二次的请求从缓存中读取了，缓存效果成功：

```
{"username":"Toy","password":"123456","age":10}
```

4. 使用注解来清除缓存

可以使用 @CacheRemove 注解清除缓存：

在使用缓存结果时使用 commandKey 参数来指定 hystrixCommand 的 key，在清除缓存时，可以直接附加这个值来清除指定参数，如代码清单 6-24 所示：

代码清单6-24　ch6-4\ch6-4-hystrix-cache\src\main\java\cn\springcloud\book\hystrix\service\HelloCommand.java

```java
@CacheResult
@HystrixCommand(commandKey = "getUser")
public String getUserToCommandKey(@CacheKey Integer id) {
String json=restTemplate.getForObject("http://sc-provider-service/getUser/{1}",
String.class, id);
    System.out.println(json);
    return json;
}

@CacheRemove(commandKey="getUser")
@HystrixCommand
public String updateUser(@CacheKey Integer id) {
    System.out.println("删除getUser缓存");
    return "update success";
}
```

运行代码和结果，如代码清单 6-25 所示：

代码清单6-25　ch6-4\ch6-4-hystrix-cache\src\main\java\cn\springcloud\book\hystrix\controller\CacheController.java

```java
@RequestMapping(value = "/getAndUpdateUser/{id}", method = RequestMethod.GET)
    public String getAndUpdateUser(@PathVariable("id") Integer id) {
        //调用接口并缓存数据
        helloService.getUserToCommandKey(id);
        helloService.getUserToCommandKey(id);
        //清除缓存
        helloService.updateUser(id);
        //再调用接口
        helloService.getUserToCommandKey(id);
        helloService.getUserToCommandKey(id);
        return " getAndUpdateUser success";
}
```

运行结果，在没有缓存的情况下，打印了一次，第二次取得缓存数据，然后清除缓存后又打印了一次，最后一次又有了缓存。

```
{"username":"Toy","password":"123456","age":10}
```

删除 getUser 缓存：

```
{"username":"Toy","password":"123456","age":10}
```

5. 小结

1）这里先解释下常用的三个注解：
- @CacheResult：使用该注解后结果会被缓存，同时它要和 @HystrixCommand 注解一起使用，注解参数为 cacheKeyMethod。
- @CacheRemove：清除缓存，需要指定 commandKey，注解参数为 commandKey、cacheKeyMethod。
- @CacheKey：指定请求命令参数，默认使用方法所有参数作为 key，注解属性为 value。

一般在查询接口上使用 CacheResult，在更新接口中加上 @CacheRemove 删除缓存。

2）注意事项：

在一些请求量大或者重复调用接口的情况下，可以利用缓存机制有效地减轻请求的压力，但在利用 Hystrix 缓存时有几个方面需要注意：
- 需要开启 @EnableHystrix。
- 需要初始化 HystrixRequestContext。
- 在指定了 HystrixCommand 的 commandKey 后，在 @CacheRemove 也要指定 commandKey。

6.2.9 Hystrix Request Collapser

Request Collapser 是 Hystrix 推出的针对多个请求调用单个后端依赖做的一种优化和节约网络开销的方法。

引用官方的这张图来看一下，如图 6-14 所示，当发起 5 个请求时，在请求没有聚合和合并的情况下，是每个请求单独开启一个线程，并开启一个网络链接进行调用，这都会加重应用程序的负担和网络开销，并占用 hystrix 的线程连接池，当使用 Collapser 把请求都合并起来时，则只需要一个线程和一个连接的开销，这大大减少了并发和请求执行所需要的线程数和网络连接数，尤其在一个时间段内有非常多的请求情况下能极大地提高资源利用率。

使用注解进行请求合并

使用 Request Collapser 也可以通过继承类和注解的形式来实现，这里主要讲一下使用起来非常方便的注解形式：

1）和缓存类似，用拦截器实现 Hystrix 上下文的初始化和关闭，如代码清单 6-26 所示。

代码清单6-26 ch6-4\ch6-4-collapsing\src\main\java\cn\springcloud\book\hystrix\config\HystrixContextInterceptor.java

```java
@Override
public boolean preHandle(HttpServletRequest request, HttpServletResponse
    respone, Object arg2) throws Exception {
    context = HystrixRequestContext.initializeContext();
    return true;
}
```

```
@Override
public void afterCompletion(HttpServletRequest request, HttpServletResponse
    respone, Object arg2, Exception arg3) throws Exception {
        context.shutdown();
}
```

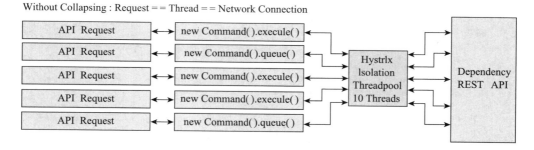

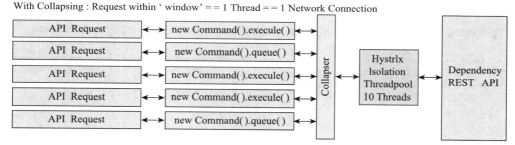

图 6-14　请求对比图

2）实现一个 Future 异步返回值的方法，在这个方法上配置请求合并的注解，之后外部通过调用这方法来实现请求的合并，注意，这个方法必须是 Future 异步返回值的，否则无法合并请求，如代码清单 6-27 所示：

代码清单6-27　ch6-4\ch6-4-collapsing\src\main\java\cn\springcloud\book\hystrix\service\CollapsingService.java

```
@HystrixCollapser(batchMethod = "collapsingList", collapserProperties = {
    @HystrixProperty(name="timerDelayInMilliseconds", value = "1000")})
public Future<Animal> collapsing(Integer id) {
    return null;
}
```

@HystrixCollapser 注解代表开启请求合并，调用该方法时，实际上运行的是 collapsingList 方法，且利用 HystrixProperty 指定 timerDelayInMilliseconds，这属性代表合并多少 ms 内的请求，暂时配置为 1000，代表 1000ms，也就是 1s 内的请求，如果不配置的话，默认是 10ms，collapsingList 方法如下，返回一个批量的 Animal 对象，这里需要注意的是，Feign 调用的话还不支持 Collapser，如代码清单 6-28 所示：

代码清单6-28　ch6-4\ch6-4-collapsing\src\main\java\cn\springcloud\book\hystrix\service\CollapsingService.java

```java
@HystrixCommand
public List<Animal> collapsingList(List<Integer> animalParam) {
    System.out.println("collapsingList当前线程" + Thread.currentThread().
        getName());
    System.out.println("当前请求参数个数:" + animalParam.size());
    List<Animal> animalList = new ArrayList<Animal>();
    for (Integer animalNumber : animalParam) {
        Animal animal = new Animal();
        animal.setName("Cat - " + animalNumber);
        animal.setSex("male");
        animal.setAge(animalNumber);
        animalList.add(animal);
    }
    return animalList;
}
```

3）写一个接口进行测试，连续调用两次 collapsing 方法，如代码清单 6-29 所示：

代码清单6-29　ch6-4\ch6-4-collapsing\src\main\java\cn\springcloud\book\hystrix\CollapsingController.java

```java
@RequestMapping("/getAnimal")
public String getAnimal() throws Exception {
    Future<Animal> user = collapsingService.collapsing(1);
    Future<Animal> user2 = collapsingService.collapsing(2);
    System.out.println(user.get().getName());
    System.out.println(user2.get().getName());
    return "Success";
}
```

启动工程，访问 http://localhost:5555/getAnimal。可以看到实际调用了 collapsingList，输出了当前线程名、请求的参数和结果，一共合并了两个请求，达到了预期目的。

```
collapsingList当前线程hystrix-CollapsingService-1
当前请求参数个数:2
Cat - 1
Cat - 2
```

上面演示了多个请求是如何合并的，但是这都是在同一次请求中，如果请求两次接口都是不同线程运行的，那么如何合并整个应用中的请求呢，即如何对所有线程请求中的多次服务进行合并呢？这时只要修改一个属性配置，@HystrixCollapser 注解的 Scope 属性，配置成 Global 即可实现。

Scope 属性有两个值：一个是 Request，也是默认的，一个是 Global。

增加一个 Scope 等于 Global 的方法，如代码清单 6-30 所示：

代码清单6-30　ch6-4\ch6-4-collapsing\src\main\java\cn\springcloud\book\hystrix\service\CollapsingService.java

```java
@HystrixCollapser(batchMethod = "collapsingListGlobal",scope = Scope.GLOBAL,
    collapserProperties = {
```

```
            @HystrixProperty(name="timerDelayInMilliseconds", value = "10000")})
    public Future<Animal> collapsingGlobal(Integer id) {
        return null;
    }
    @HystrixCommand
    public List<Animal> collapsingListGlobal(List<Integer> animalParam) {
        System.out.println("collapsingListGlobal当前线程" + Thread.currentThread().
            getName());
        System.out.println("当前请求参数个数:" + animalParam.size());
        List<Animal> animalList = new ArrayList<Animal>();
        for (Integer animalNumber : animalParam) {
            Animal animal = new Animal();
            animal.setName("Dog - " + animalNumber);
            animal.setSex("male");
            animal.setAge(animalNumber);
            animalList.add(animal);
        }
        return animalList;
    }
```

增加一个调用接口调用此方法，如代码清单 6-31 所示：

代码清单6-31 ch6-4\ch6-4-collapsing\src\main\java\cn\springcloud\book\hystrix\ CollapsingController.java

```
@RequestMapping("/getAnimalGolbal")
    public String getAnimalGolbal() throws Exception {
        Future<Animal> user = collapsingService.collapsingGlobal(1);
        Future<Animal> user2 = collapsingService.collapsingGlobal(2);
        System.out.println(user.get().getName());
        System.out.println(user2.get().getName());
        return "Success";
    }
```

打开浏览器连续调用两次 http://localhost:5555/getAnimalGolbal，运行结果就出来了，会把所有请求都合并在一次线程中，但 Request 作用域会运行两次线程来分别运行两次请求。

```
collapsingListGlobal当前线程hystrix-CollapsingService-2
当前请求参数个数:4
Dog - 1
Dog - 2
Dog - 1
Dog - 2
```

总结：Hystrix Request Collapser 主要用于请求合并的场景，在一个简单的系统中，这种场景我们可能会很少碰到，所以对于请求合并，我们一般使用的场景是：当在某个时间内有大量或并发的相同请求时，则适合用请求合并；而如果在某个时间内只有很少的请求，且延迟也不高，则使用请求合并反而会增加复杂度和延迟，因为对于 Collapser 本身 Hystrix 也是需要时间进行批处理的。

6.2.10 Hystrix 线程传递及并发策略

Hystrix 会对请求进行封装，然后管理请求的调用，从而实现断路器等多种功能。Hystrix

提供了两种隔离模式来进行请求的操作，一种是信号量，一种是线程隔离。如果是信号量，则 Hystrix 在请求的时候会获取到一个信号量，如果成功拿到，则继续进行请求，请求在一个线程中执行完毕。如果是线程隔离，Hystrix 会把请求放入线程池中执行，这时就有可能产生线程的变化，从而导致线程 1 的上下文数据在线程 2 里不能正常拿到。

这里通过一个例子说明。

1. 新建请求接口和本地线程持有对象

建立一个 ThreadLocal 来保存用户的信息，通常在微服务里，会把当前请求的上下文数据放入本地线程变量，便于方便使用及销毁，如代码清单 6-32 所示：

代码清单6-32　ch6-4\ch6-4-hystrix-thread-context\src\main\java\cn\springcloud\book\hystrix\config\HystrixThreadLocal.java

```java
public class HystrixThreadLocal {
    public static ThreadLocal<String> threadLocal = new ThreadLocal<>();
}
```

这里定义一个接口，做两件事：

1）请求入口打印当前线程 ID，并利用上面的 ThreadLocal 放入用户信息。

2）为了兼容其他情况，在使用 Feign 调用的时候，通常会使用 RequestContextHolder 拿到上下文属性，在此也进行测试一下，如代码清单 6-33 所示：

代码清单6-33　ch6-4\ch6-4-hystrix-thread-context\src\main\java\cn\springcloud\book\hystrix\controller\ThreadContextController.java

```java
@RequestMapping(value = "/getUser/{id}", method = RequestMethod.GET)
public String getUser(@PathVariable("id") Integer id) {
    //第一种测试，放入上下文对象
    HystrixThreadLocal.threadLocal.set("userId : "+ id);
    //第二种测试，利用RequestContextHolder放入对象测试
    RequestContextHolder.currentRequestAttributes().setAttribute("userId",
        "userId : "+ id, RequestAttributes.SCOPE_REQUEST);
    log.info("Current thread: " + Thread.currentThread().getId());
    log.info("Thread local: " + HystrixThreadLocal.threadLocal.get());
    log.info("RequestContextHolder: " + RequestContextHolder.currentRequest-
        Attributes().getAttribute("userId", RequestAttributes.SCOPE_REQUEST));
    //调用
    String user = threadContextService.getUser(id);
    return user;
}
```

2. 测试没有线程池隔离模式下，获取用户信息

定义一个后台服务获取之前放入的用户 id，然后进行请求，如代码清单 6-34 所示：

代码清单6-34　ch6-4\ch6-4-hystrix-thread-context\src\main\java\cn\springcloud\book\hystrix\service\ThreadContextService.java

```java
public String getUser(Integer id) {
```

```
        log.info("ThreadContextService, Current thread : " + Thread.currentThread().
            getId());
        log.info("ThreadContextService, ThreadContext object : " + HystrixThreadLocal.
            threadLocal.get());
        System.out.println(RequestContextHolder.currentRequestAttributes().
            getAttribute("context", RequestAttributes.SCOPE_REQUEST).toString());
        String json = restTemplate.getForObject("http://sc-provider-service/getUser/
            {1}", String.class, id);
        return json;
    }
```

启动程序，运行 http://localhost:3333/getUser/5555，可以看到线程的 id 都是一样的，线程变量也是传入 5555，请求上下文的持有对象也可以顺利拿到。

```
ThreadContextController, Current thread: 29
ThreadContextController, Thread local: userId : 5555
ThreadContextController, RequestContextHolder: userId : 5555

ThreadContextService, Current thread : 29
ThreadContextService, ThreadContext object : userId : 5555
ThreadContextService, RequestContextHolder : userId : 5555
```

3. 测试有线程池隔离模式下，获取用户信息

在上面的 getUser 接口加上 @HystrixCommand，利用 Hystrix 接管。

启动程序，运行 http://localhost:3333/getUser/5555，会发现，进入的线程 id 是 32，当到达后台服务的时候，线程 id 变成了 50，说明线程池的隔离已经生效了，是重新启动的线程进行请求的，然后线程的变量也都丢了，RequestContextHolder 中也报错了，大概的意思是没有线程变量绑定，成功地重现了父子线程数据传递的问题。

```
ThreadContextController, Current thread: 32
ThreadContextController, Thread local: userId : 5555
ThreadContextController, RequestContextHolder: userId : 5555

ThreadContextService, Current thread : 50
ThreadContextService, ThreadContext object : null
java.lang.IllegalStateException: No thread-bound request found:
```

4. 解决方案

根据上面测试出来的问题，解决此问题有两种办法：

1）修改 Hystrix 隔离策略，使用信号量，直接修改配置文件即可，但 Hystrix 默认是线程池隔离，加上从真实的项目情况看，大部分都是使用线程池隔离，所以这种方案不太推荐。属性：hystrix.command.default.execution.isolation.strategy。

2）Hystrix 官方推荐了一种方式，就是使用 HystrixConcurrencyStrategy。

摘取部分官方描述，使用 HystrixConcurrencyStrategy 实现 wrapCallable 方法，对于依赖 ThreadLocal 状态以实现应用程序功能的系统至关重要，也就是说使用 HystrixConcurrencyStrategy 覆盖 wrapCallable 方法即可：

```
Concurrency Strategy:
...
You can implement the HystrixConcurrencyStrategy class with the following:
The wrapCallable() method allows you to decorate every Callable executed by
    Hystrix. This can be essential to systems that rely upon ThreadLocal state
    for application functionality. The wrapping Callable can capture and copy
    state from parent to child thread as needed.
...
```

官方地址为 https://github.com/Netflix/Hystrix/wiki/Plugins#concurrency-strategy。

再看一下 wrapCallable 的代码注释，大致意思就是，提供了一个在请求执行前包装的机会，可以注入自己定义的动作，比如复制线程状态。

```
/**
 * Provides an opportunity to wrap/decorate a {@code Callable<T>} before
     execution.
 * <p>
 * This can be used to inject additional behavior such as copying of thread
     state (such as {@link ThreadLocal}).
 * <p>
 * <b>Default Implementation</b>
 * <p>
 * Pass-thru that does no wrapping.
 * @param callable
 *     {@code Callable<T>} to be executed via a {@link ThreadPoolExecutor}
 * @return {@code Callable<T>} either as a pass-thru or wrapping the one given
 */
```

至此，可以通过重写这个方法来实现想要的封装线程参数的办法。这里来实现这种方式，但由于不太清楚它的用法，可以参考一下已经实现过的这样方式的接口，从接口的继承关系来看，有 HystrixConcurrencyStrategyDefault、SecurityContextConcurrencyStrategy、SleuthHystrixConcurrencyStrategy，默认的实现比较简单，可以参考后两个来写自己的策略，如代码清单 6-35 所示：

代码清单6-35 ch6-4\ch6-4-hystrix-thread-context\src\main\java\cn\springcloud\book\hystrix\config\SpringCloudHystrixConcurrencyStrategy.java

```
@Override
public <T> Callable<T> wrapCallable(Callable<T> callable) {
    return new HystrixThreadCallable<>(callable, RequestContextHolder.getRe
        questAttributes(),HystrixThreadLocal.threadLocal.get());
}
```

这里包装了一个 HystrixThreadCallable 类，在执行请求前包装 HystrixThreadCallable 对象，该对象的构造函数是希望传递的 RequestContextHolder 和自定义的 HystrixThreadLocal，然后重写 wrapCallable 方法，在调用前把需要的对象信息设置进去，这样在下一个新线程中就可以拿到了，如代码清单 6-36 所示：

代码清单6-36 ch6-4\ch6-4-hystrix-thread-context\src\main\java\cn\springcloud\book\hystrix\config\HystrixThreadCallable.java

```java
@Override
    public S call() throws Exception {
        try {
            RequestContextHolder.setRequestAttributes(requestAttributes);
            HystrixThreadLocal.threadLocal.set(params);
            return delegate.call();
        } finally {
            RequestContextHolder.resetRequestAttributes();
            HystrixThreadLocal.threadLocal.remove();
        }
    }
}
```

启动程序，查看运行结果，访问 http://localhost:3333/getUser/5555，可以看见如下信息，不同的线程也能顺利地拿到上个线程传递的信息：

```
ThreadContextController, Current thread: 35
ThreadContextController, Thread local: userId : 5555
ThreadContextController, RequestContextHolder: userId : 5555

ThreadContextService, Current thread : 61
ThreadContextService, ThreadContext object : userId : 5555
ThreadContextService, RequestContextHolder : userId : 5555
```

此时发现使用了 Hystrix 的线程池隔离，然后也能顺利地实现既定目标。

5. 并发策略

上面虽然看起来解决了问题，但仔细查看 HystrixPlugins 的 registerConcurrencyStrategy 方法，发现这个方法只能被调用一次，不然就会报错，这就导致无法与其他的并发策略一起使用：

```java
/**
 * Register a {@link HystrixConcurrencyStrategy} implementation as a global
 *            override of any injected or default implementations.
 * 
 * @param impl
 *            {@link HystrixConcurrencyStrategy} implementation
 * @throws IllegalStateException
 *            if called more than once or after the default was initialized (if
 *                usage occurs before trying to register)
 */
public void registerConcurrencyStrategy(HystrixConcurrencyStrategy impl) {
    if (!concurrencyStrategy.compareAndSet(null, impl)) {
        throw new IllegalStateException("Another strategy was already
            registered.");
    }
}
```

此时需要修改代码，把其他并发策略注入进去，达到并存的目的，如 sleuth 的并发策略也做了同样的事情，具体的做法就是在构造此并发策略时，找到之前已经存在的并发策略，并保留在类的属性中，在调用过程中，返回之前并发策略的相关信息，如请求变量、连接池、阻塞

队列等请求进来时,既不会影响之前的并发策略,也可以包装需要的请求信息,如代码清单 6-37 所示:

代码清单6-37 ch6-4\ch6-4-hystrix-thread-context\src\main\java\cn\springcloud\book\hystrix\config\SpringCloudHystrixConcurrencyStrategy.java

```
    private HystrixConcurrencyStrategy delegateHystrixConcurrencyStrategy;
    @Override
    public <T> Callable<T> wrapCallable(Callable<T> callable) {
        return new HystrixThreadCallable<>(callable, RequestContextHolder.getRe
            questAttributes(),HystrixThreadLocal.threadLocal.get());
    }
    @Override
    public ThreadPoolExecutor getThreadPool(HystrixThreadPoolKey threadPoolKey,
            HystrixProperty<Integer> corePoolSize,
            HystrixProperty<Integer> maximumPoolSize,
            HystrixProperty<Integer> keepAliveTime, TimeUnit unit,
            BlockingQueue<Runnable> workQueue) {
        return this.delegateHystrixConcurrencyStrategy.getThreadPool(threadPoolKey,
            corePoolSize, maximumPoolSize,keepAliveTime, unit, workQueue);
    }
    @Override
    public ThreadPoolExecutor getThreadPool(HystrixThreadPoolKey threadPoolKey,
            HystrixThreadPoolProperties threadPoolProperties) {
        return this.delegateHystrixConcurrencyStrategy.getThreadPool(threadPoolKey,
            threadPoolProperties);
    }
    @Override
    public BlockingQueue<Runnable> getBlockingQueue(int maxQueueSize) {
        return this.delegateHystrixConcurrencyStrategy.getBlockingQueue(maxQueueSize);
    }
    @Override
    public <T> HystrixRequestVariable<T> getRequestVariable(
            HystrixRequestVariableLifecycle<T> rv) {
        return this.delegateHystrixConcurrencyStrategy.getRequestVariable(rv);
    }
```

6.2.11　Hystrix 命令注解

这里主要介绍 HystrixCommand 和 HystrixObservableCommand 注解。

1. HystrixCommand

主要的意思就是封装执行的代码,然后具有故障延迟容错、断路器和统计等功能,但它是阻塞命令,另外也是可以和 Observable 共用的。

代码:com.netflix.hystrix. HystrixCommand

```
Used to wrap code that will execute potentially risky functionality (typically
    meaning a service call over the network)
with fault and latency tolerance, statistics and performance metrics capture,
    circuit breaker and bulkhead functionality.
This command is essentially a blocking command but provides an Observable facade
    if used with observe()
```

2. HystrixObservableCommand

代码：com.netflix.hystrix.HystrixObservableCommand
主要的意思和上面差不多，但主要区别是，它是一个非阻塞的调用模式。

```
Used to wrap code that will execute potentially risky functionality (typically
    meaning a service call over the network)
with fault and latency tolerance, statistics and performance metrics capture,
    circuit breaker and bulkhead functionality.
This command should be used for a purely non-blocking call pattern. The caller of
    this command will be subscribed to the Observable<R> returned by the run()
    method.
```

Hystrix 在使用过程中除了 HystrixCommand，还有 HystrixObservableCommand，这两个命令有很多共同点，如都支持故障和延迟容错、断路器、指标统计。

但两者也有很多区别：

1）HystrixCommand 默认是阻塞式的，可以提供同步和异步两种方式，但 HystrixObservableCommand 是非阻塞的，默认只能是异步的。

2）HystrixCommand 的方法是 run，HystrixObservableCommand 执行的是 construct。

3）HystrixCommand 一个实例一次只能发一条数据出去，HystrixObservableCommand 可以发送多条数据。

通过之前很多例子你肯定对 HystrixCommand 有了一定理解，这里介绍它的一些属性，如图 6-15 所示。

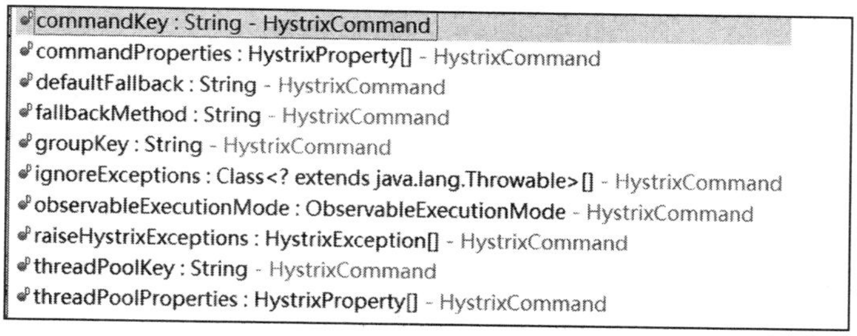

```
commandKey : String - HystrixCommand
commandProperties : HystrixProperty[] - HystrixCommand
defaultFallback : String - HystrixCommand
fallbackMethod : String - HystrixCommand
groupKey : String - HystrixCommand
ignoreExceptions : Class<? extends java.lang.Throwable>[] - HystrixCommand
observableExecutionMode : ObservableExecutionMode - HystrixCommand
raiseHystrixExceptions : HystrixException[] - HystrixCommand
threadPoolKey : String - HystrixCommand
threadPoolProperties : HystrixProperty[] - HystrixCommand
```

图 6-15　HystrixCommand 属性图

commandKey：全局唯一的标识符，如果不配则默认是方法名。

defaultFallback：默认的 fallback 方法，该函数不能有入参，返回值和方法保持一致，但 fallbackMethod 优先级更高。

fallbackMethod：指定的处理回退逻辑的方法，必须和 HystrixCommand 在同一个类里，方法的参数要保持一致。

ignoreExceptions：HystrixBadRequestException 不会触发 fallback，这里定义的就是你不希望哪些异常被 fallback 而是直接抛出。

commandProperties：配置一些命名的属性，如执行的隔离策略等。

threadPoolProperties：用来配置线程池相关的属性。

groupKey：全局唯一标识服务分组的名称，内部会根据这个兼职来展示统计数、仪表盘等信息，默认的线程划分是根据这命令组的名称来进行的，一般会在创建 HystrixCommond 时指定命令组来实现默认的线程池划分。

threadPoolKey：对服务的线程池信息进行设置，用于 HystrixThreadPool 监控、metrics、缓存等用途。

6.3　本章小结

本章主要介绍了 Hystrix 的一些常用的使用方法和实战技巧，工程使用过程中主要还是一些配置、方法、属性注解等方面的注意点比较多，希望能给大家带来帮助。

第 7 章 Chapter 7

Spring Cloud Zuul 基础篇

Zuul 是由 Netflix 孵化的一个致力于"网关"解决方案的开源组件。从 2012 年 3 月以来，其陆续发布了 Zuul1.0 与 Zuul2.0 版本，后经 Pivotal 公司发现并整合于 Spring Cloud 生态系统，即现在的 Spring Cloud Zuul（后文简称 Zuul），在动态路由、监控、弹性、服务治理以及安全方面起着举足轻重的作用。本章以 Spring Cloud Finchley 版本展开，对应 Zuul 版本则为 1.3.1，系统全面地介绍 Zuul 的基本构建方式以及在实际企业级开发中的应用技巧，为使读者能够更加深入地了解，后续还有工作原理和核心源码解读章节。

7.1 Spring Cloud Zuul 概述

1. 诞生背景

区别于传统巨石应用（Monolith Application），微服务架构将后端拆解成许多个单独的应用，这种方式在近年已经风靡整个软件世界，各大公司争相重构应用，拆解服务。在享受到面向服务的开发便捷性之后，也出现了一个新的问题：看似清晰的服务拆分，实则杂乱无章，有时候完成一个业务逻辑，需要到不同主机和不同端口上面调取接口，这是一件很痛苦的事情。于是一个面向服务治理、服务编排的组件出现了——微服务网关。由此，作为 Netflix 的早期服务化后端应用程序前门项目，Zuul 应运而生，如图 7-1 所示。

2. Zuul 简介

关于 Zuul 的定义，Netflix 官方解释得很好：

Zuul is the front door for all requests from devices and web sites to the backend of the Netflix streaming application.

Zuul 是从设备和网站到后端应用程序所有请求的前门，为内部服务提供可配置的对外 URL 到服务的映射关系，基于 JVM 的后端路由器。其具备以下功能：

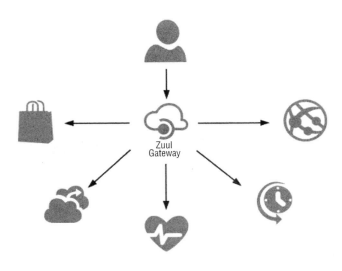

图 7-1 Zuul 简图

- 认证与鉴权
- 压力控制
- 金丝雀测试
- 动态路由
- 负载削减
- 静态响应处理
- 主动流量管理

其底层基于 Servlet，本质组件是一系列 Filter 所构成的责任链，后续章节会有相关原理剖析。并且 Zuul 的逻辑引擎与 Filter 可用其他基于 JVM 的语言编写，比如 Groovy。

细心的读者可能会发现，Spring Cloud Finchley 继续沿用 Netflix Zuul 1.x 版本，不是已经出现了 2.x 版本了吗？为什么不用？笔者这里简单总结下，由于 Zuul 2.x 版本改动相较 1.x 还是较大，考虑到整个生态的稳定性，以及使用者升级版本会遇到的种种问题，虽然 2.x 底层使用 Netty 性能更好，Finchley 版还是继续使用 1.x 版本。另外，由于 Spring Cloud Gateway 已经孵化成功，相较 Zuul 在功能以及性能上都有明显提升，大胆推测一下，Pivotal 公司正在走一条"去 Netflix 化"的路线，关于 Spring Cloud Gateway，在本书第 18、19 章有详细介绍。

7.2 Spring Cloud Zuul 入门案例

下面来让我们体验一下 Zuul 的 Hello World，这里需要用到的 Spring Cloud 组件是 Eureka 与 Zuul，另外再使用一个普通服务作为 Zuul 路由的下级服务，来模拟真实开发中的一次路由过程。

1. 创建 Maven 父级 pom 工程

在父工程里面配置好工程需要的父级依赖，目的是为了方便管理与简化配置，详细 pom

文件请查阅本书配套源码 ch7-2\pom.xml。

2. 创建 Eureka 组件工程
Eureka 组件工程参见第 5 章，为避免赘述，第 7 章和第 8 章 Eureka 组件配置皆基于此。

3. 创建 Zuul Server 组件工程
Zuul Server 工程的 pom.xml 文件，只需要引入 spring-cloud-starter-netflix-zuul，注意也与之前版本不一样，如代码清单 7-1 所示。

代码清单7-1　ch7-2\ch7-2-zuul-server\pom.xml

```xml
<dependencies>
    <dependency>
        <groupId>org.springframework.cloud</groupId>
        <artifactId>spring-cloud-starter-netflix-zuul</artifactId>
    </dependency>
    <dependency>
        <groupId>org.springframework.cloud</groupId>
        <artifactId>spring-cloud-starter-netflix-eureka-client</artifactId>
    </dependency>
</dependencies>
```

添加 Zuul Server 的启动主类，如代码清单 7-2 所示。

代码清单7-2　ch7-2\ch7-2-zuul-server\src\main\java\cn\springcloud\book\eureka\ZuulServerApplication.java

```java
@SpringBootApplication
@EnableDiscoveryClient
@EnableZuulProxy
public class ZuulServerApplication {

    public static void main(String[] args) {
        SpringApplication.run(ZuulServerApplication.class, args);
    }
}
```

Zuul Server 的配置文件，如代码清单 7-3 所示。

代码清单7-3　ch7-2\ch7-2-zuul-server\src\main\resources\bootstrap.yml

```yaml
spring:
  application:
    name: zuul-server
server:
  port: 5555
eureka:
  client:
    serviceUrl:
      defaultZone: http://${eureka.host:127.0.0.1}:${eureka.port:8888}/eureka/
  instance:
    prefer-ip-address: true
zuul:
  routes:
    client-a:
      path: /client/**
      serviceId: client-a
```

这里需要说明一下，设置 Zuul 组件的端口为 5555，指定注册中心，最后 5 行是本例的核心，它的意思是，将所有 /client 开头的 URL 映射到 client-a 这个服务中去。我们在请求的时候就可以不用请求实际的服务，转而请求这个 5555 端口的 Zuul 服务组件，/client 即一次服务路由的规则。

4. 创建普通下游服务 ch7-2-client-a

创建一个下游普通服务，来测试 Zuul Server 路由的路由功能，如代码清单 7-4 所示。

代码清单7-4　ch7-2\ch7-2-client-a\pom.xml

```xml
<dependencies>
    <dependency>
        <groupId>org.springframework.cloud</groupId>
        <artifactId>spring-cloud-starter-netflix-eureka-client</artifactId>
    </dependency>
</dependencies>
```

启动主类，添加服务发现的注解，如代码清单 7-5 所示。

代码清单7-5　ch7-2\ch7-2-client-a\src\main\java\cn\springcloud\book\ClientAApplication.java

```java
@SpringBootApplication
@EnableDiscoveryClient
public class ClientAApplication {

    public static void main(String[] args) {
        SpringApplication.run(ClientAApplication.class, args);
    }
}
```

下游服务的配置文件，如代码清单 7-6 所示。

代码清单7-6　ch7-2\ch7-2-client-a\src\main\resources\bootstrap.yml

```yml
server:
    port: 7070
spring:
    application:
        name: client-a
eureka:
    client:
        serviceUrl:
            defaultZone: http://${eureka.host:127.0.0.1}:${eureka.port:8888}/eureka/
    instance:
        prefer-ip-address: true
```

创建一个测试接口，如代码清单 7-7 所示。

代码清单7-7　ch7-2\ch7-2-client-a\src\main\java\cn\springcloud\book\controller\TestController.java

```java
@RestController
public class TestController {

    @GetMapping("/add")
    public Integer add(Integer a, Integer b){
```

```
        return a + b;
    }
}
```

在普通服务里面做了一个很简单的 get 接口，接收两个参数 a 与 b，将它们相加后返回。下面我们来看看实际运行效果怎么样。

5. 效果展示

依次启动 Eureka、Zuul、普通服务 client-a，使用 postman 分两次调取接口。

直接调取接口，如图 7-2 所示：

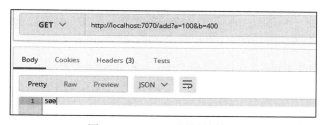

图 7-2　postman 调用结果一

经过 Zuul 网关调取接口，如图 7-3 所示。

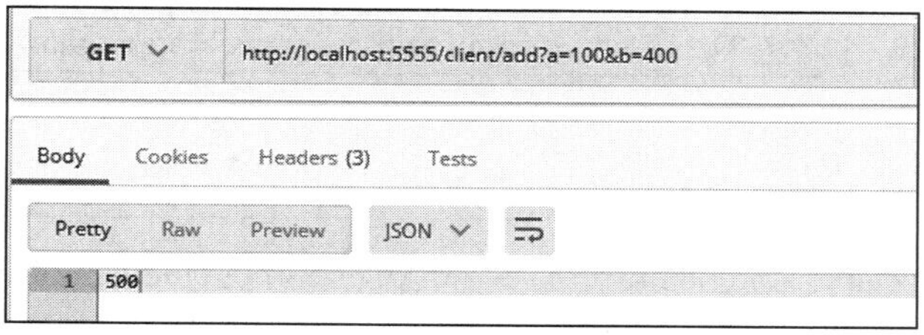

图 7-3　postman 调用结果二

由结果可知，我们在调取 http://localhost:5555/client/add?a=100&b=400 接口的时候，实际上是调取 http://localhost:7070/add?a=100&b=400 接口。这是因为我们在 Zuul 配置文件指定了路由规则，当向 Zuul Server 发起请求的时候，它会去 Eureka 注册中心拉取服务列表，如果发现有指定的路由映射规则，就会按照规则路由到相应的服务接口上去。

7.3　Spring Cloud Zuul 典型配置

前面已经介绍了一个最基本的网关示例，它是整个 Spring Cloud 生态里面"路由 – 服务"的一个缩影，后续的所有功能无非是在源头或者过程锦上添花，本节主要讲述 Zuul 的一些基本典型配置，包括路由与一些增强配置。

7.3.1　路由配置

既然 Zuul 作为一个微服务"路由器"，那么它的路由功能一定是丰富多彩的，这里列举几种常用的功能，示例代码详见本书配套代码 ch7-3。

1. 路由配置简化与规则

（1）单实例 serviceId 映射

前例中：

```
zuul:
    routes:
        client-a:
            path: /client/**
            serviceId: client-a
```

是一个从 /client/** 到 client-a 服务的一个映射规则，其实我们可以把它简化成一个比较简单的配置，如代码清单 7-8 所示：

代码清单7-8   ch7-3\ch7-3-zuul-server\src\main\resources\application-example3.yml

```
zuul:
    routes:
        client-a: /client/**
```

另外，还有一种更加简单的映射规则，映射规则与 serviceId 都不用写，如代码清单 7-9 所示：

代码清单7-9   ch7-3\ch7-3-zuul-server\src\main\resources\application-example2.yml

```
zuul:
    routes:
        client-a:
```

在这种情况下，Zuul 会为 client-a 添加一个默认的映射规则 /client-a/**，相当于：

```
zuul:
    routes:
        client-a:
            path: /client-a/**
            serviceId: client-a
```

（2）单实例 url 映射

除了路由到服务外，还能路由到物理地址，将 serviceId 替换为 url 即可，如代码清单 7-10 所示：

代码清单7-10   ch7-3\ch7-3-zuul-server\src\main\resources\application-example7.yml

```
zuul:
    routes:
        client-a:
            path: /client/**
            url: http://localhost:7070 #client-a的地址
```

（3）多实例路由

在默认情况下，Zuul 会使用 Eureka 中集成的基本负载均衡功能，如果想要使用 Ribbon 的负载均衡功能，就需要指定一个 serviceId，此操作需要禁止 Ribbon 使用 Eureka，在 E 版之后，新增了负载均衡策略的配置，如代码清单 7-11 所示：

代码清单7-11 ch7-3\ch7-3-zuul-server\src\main\resources\application-example8.yml

```yaml
zuul:
  routes:
    ribbon-route:
      path: /ribbon/**
      serviceId: ribbon-route
ribbon:
  eureka:
    enabled: false     #禁止Ribbon使用Eureka
ribbon-route:
  ribbon:
    NIWSServerListClassName: com.netflix.loadbalancer.ConfigurationBasedServerList
    NFLoadBalancerRuleClassName: com.netflix.loadbalancer.RandomRule         #Ribbon LB Strategy
    listOfServers: localhost:7070,localhost:7071           #client services for Ribbon LB
```

（4）forward 本地跳转

有时候我们在 Zuul 中会做一些逻辑处理，在网关中写好一个接口，如代码清单 7-12 所示：

代码清单7-12 ch7-3\ch7-3-zuul-server\src\main\java\cn\springcloud\book\controller\TestController.java

```java
@RestController
public class TestController {

    @GetMapping("/client")
    public String add(Integer a, Integer b){
        return "本地跳转: " + (a + b);
    }
}
```

如果我们希望在访问 /client 接口的时候跳转到这个方法上来处理，就需要用到 Zuul 的本地跳转，配置如代码清单 7-13 所示：

代码清单7-13 ch7-3\ch7-3-zuul-server\src\main\resources\application-example9.yml

```yaml
zuul:
  routes:
    client-a:
      path: /client/**
      url: forward:/client
```

（5）相同路径的加载规则

有一种特殊的情况，为一个映射路径指定多个 serviceId，那么它该加载哪个服务呢？如代码清单 7-14 所示：

代码清单7-14　ch7-3\ch7-3-zuul-server\src\main\resources\application-example4.yml

```
zuul:
  routes:
    client-b:
      path: /client/**
      serviceId: client-b
    client-a:
      path: /client/**
      serviceId: client-a
```

经过笔者测试，它总是会路由到 yml 配置文件后面那个服务，这里是 client-a 服务，也就是末尾那个服务。在 yml 解释器工作的时候，如果同一个映射路径对应多个服务，按照加载顺序，最末加载的映射规则会把之前的映射规则覆盖掉。

在选择映射规则的时候，只需要选择一种简单明了，后期维护方便，且自己能记住的方式就行了。

2. 路由通配符

此外，映射路径 /client/** 之后的 /** 也大有讲究，其还可配置为 /* 或者是 /?，规则如表 7-1 所示。

表 7-1　Zuul 的路由通配符映射规则

规则	释义	示例
/**	匹配任意数量的路径与字符	/client/add，/client/mul，/client/a /client/add/a，/client/mul/a/b
/*	匹配任意数量的字符	/client/add，/client/mul，/client/a
/?	匹配单个字符	/client/a，/client/b

我们在开发的时候要根据实际需要选择相应的通配符，选取的时候也要小心谨慎，有时候因为粗心把 /* 错当 /**，就可能导致一个很难排查的错误，会困扰你许久，若要避免这种情况的发生，在通晓其义的情况下，要做到仔细审慎。

7.3.2　功能配置

1. 路由前缀

在配置路由规则的时候，我们可以配置一个统一的代理前缀，如代码清单 7-15 所示：

代码清单7-15　ch7-3\ch7-3-zuul-server\src\main\resources\application-example1.yml

```
zuul:
  prefix: /pre    #使用prefix指定前缀
  routes:
    client-a: /client/**
```

下次我们在通过 Zuul 访问后端接口的时候就需要加上这个前缀了。注意，请求路径会变成 /pre/client/add，但是实际起作用的是 /client/add，也就是 Zuul 会把代理路径从请求路径中移除。可以使用 stripPrefix=false 来关闭这个功能，如代码清单 7-16 所示：

代码清单7-16 ch7-3\ch7-3-zuul-server\src\main\resources\application-example5.yml

```yaml
zuul:
    prefix: /pre
    routes:
        client-a:
            path: /client-a/**
            serviceId: client-a
            stripPrefix: false
```

关闭之后，请求路径是 /pre/client/add，实际起作用的还是 /pre/client/add，要注意一下，一般不选这个配置。

2. 服务屏蔽与路径屏蔽

有时候为了避免某些服务或者路径的侵入，可以将它们屏蔽掉，如代码清单 7-17 所示：

代码清单7-17 ch7-3\ch7-3-zuul-server\src\main\resources\application-example1.yml

```yaml
zuul:
    ignored-services: client-b         #忽略的服务，防服务侵入
    ignored-patterns: /**/div/**       #忽略的接口，屏蔽接口
    prefix: /pre
    routes:
        client-a: /client/**
```

加上 ignored-services 与 ignored-patterns 之后，Zuul 在拉取服务列表，创建映射规则的时候，就会忽略掉 client-b 服务与 /**/div/** 接口。

3. 敏感头信息

在构建系统的时候，使用 HTTP 的 header 传值是十分方便的，协议的一些认证信息默认也在 header，比如 Cookie，或者习惯把基本认证通过 BASE64 加密后放在 Authorization 里面。在我们内部系统没有什么问题，但是如果系统要和外部系统打交道，就可能会出现这些信息的泄露。幸运的是，在 Zuul 的配置里面可以指定敏感头，切断它和下层服务之间的交互，如代码清单 7-18 所示：

代码清单7-18 ch7-3\ch7-3-zuul-server\src\main\resources\application-example5.yml

```yaml
zuul:
    routes:
        client-a:
            path: /client/**
            sensitiveHeaders: Cookie,Set-Cookie,Authorization
            serviceId: client-a
```

4. 重定向问题

在笔者之前的一次实践中，客户端通过 Zuul 请求认证服务，认证成功之后重定向到一个欢迎页，但是笔者发现重定向的这个欢迎页的 host 变成了这个认证服务的 host，而不是 Zuul 的 host，如图 7-4 所示，直接暴露了认证服务的地址，我们可以在配置里面解决掉这个问题，

如代码清单 7-19 所示：

图 7-4　Zuul 的重定向问题

代码清单7-20　ch7-3\ch7-3-zuul-server\src\main\resources\application-example1.yml

```
zuul:
    add-host-header: true
    routes:
        client-a: /client/**
```

5. 重试机制

在生产环境中，由于各种各样的原因，可能会使一次请求偶然失败，考虑到某些业务的体验，不能通过有感知的操作来触发，这时候就会用到重试机制了，Zuul 可以配合 Ribbon（默认集成）来做重试，如代码清单 7-20 所示：

代码清单7-20　ch7-3\ch7-3-zuul-server\src\main\resources\application-example6.yml

```
zuul:
    retryable: true #开启重试

ribbon:
    MaxAutoRetries: 1 #同一个服务重试的次数(除去首次)
    MaxAutoRetriesNextServer: 1   #切换相同服务数量
```

当然，此功能要慎用，有一些接口要保证幂等性，一定要做好相关工作。

7.4　本章小结

本章主要讲解 Spring Cloud Zuul 的背景知识与基本用法，对于入门是十分有帮助的，也是学习 Zuul 的必经之路。通过本章的学习，你可以搭建一个基本的基于 Zuul 的工程，能够知道路由在微服务体系中的定位。在实际业务当中，往往需要使用各种各样的路由配置规则来达成目的，上文提到的只是有代表性的一部分，还有更多的方式需要读者自己去发现。在下一章，笔者将带你领略 Zuul 在实际企业级开发中的各种实用用法。

第 8 章 Chapter 8

Spring Cloud Zuul 中级篇

第 7 章讲述了 Zuul 的基本用法以及相关基本配置，至此我们能够基本搭建起一套可防止服务侵入的"网关–服务"微服务架构，但这在实际企业级开发中是远远不够的。在实际项目中，我们往往会在网关层涉及鉴权、限流、动态路由、文件上传、参数转换，以及做其他逻辑与业务处理，本章就围绕这些实用技巧展开，所有例子尽量做到开箱即用。熟悉本章后，你将能够搭建一套比较完善的微服务体系，对 Spring Cloud 微服务体系下的业务处理能力的处理也将提升一个层次。

8.1 Spring Cloud Zuul Filter 链

8.1.1 工作原理

Zuul 的核心逻辑是由一系列紧密配合工作的 Filter 来实现的，它们能够在进行 HTTP 请求或者响应的时候执行相关操作。可以说，没有 Filter 责任链，就没有如今的 Zuul，更不可能构成功能丰富的"网关"。本章后续实战技能大多可基于它完成，它是 Zuul 中最为开放与核心的功能。Zuul Filter 的主要特性有以下几点：

- Filter 的类型：Filter 的类型决定了此 Filter 在 Filter 链中的执行顺序。可能是路由动作发生前，可能是路由动作发生时，可能是路由动作发生后，也可能是路由过程发生异常时。
- Filter 的执行顺序：同一种类型的 Filter 可以通过 filterOrder() 方法来设定执行顺序。一般会根据业务的执行顺序需求，来设定自定义 Filter 的执行顺序。
- Filter 的执行条件：Filter 运行所需要的标准或条件。
- Filter 的执行效果：符合某个 Filter 执行条件，产生的执行效果。

Zuul 内部提供了一个动态读取、编译和运行这些 Filter 的机制。Filter 之间不直接通信，在请求线程中会通过 RequestContext 来共享状态，它的内部是用 ThreadLocal 实现的，当然你也可以在 Filter 之间使用 ThreadLocal 来收集自己需要的状态或数据。

Zuul 中不同类型 Filter 的执行逻辑核心在 com.netflix.zuul.http.ZuulServlet 类中定义，该类相关代码如下所示，流程图如图 8-1 所示。

```
try {
    ......
        try {
            preRoute();
        } catch (ZuulException e) {
            error(e);
            postRoute();
            return;
        }
        try {
            route();
        } catch (ZuulException e) {
            error(e);
            postRoute();
            return;
        }
        try {
            postRoute();
        } catch (ZuulException e) {
            error(e);
            return;
        }
} catch (Throwable e) {
    error(new ZuulException(e, 500, "UNHANDLED_EXCEPTION_" + e.getClass().
        getName()));
}
```

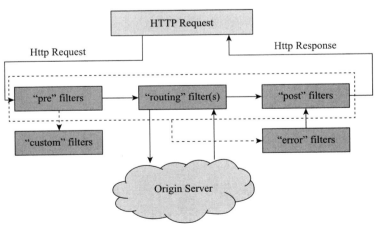

图 8-1　Zuul 请求生命周期官方原图

这张经典的官方流程图有些问题，其中 post Filter 抛错之后进入 error Filter，然后再进入

post Filter 是有失偏颇的。实际上 post Filter 抛错分两种情况：

1）在 post Filter 抛错之前，pre、route Filter 没有抛错，此时会进入 ZuulException 的逻辑，打印堆栈信息，然后再返回 status = 500 的 ERROR 信息。

2）在 post Filter 抛错之前，pre、route Filter 已有抛错，此时不会打印堆栈信息，直接返回 status = 500 的 error 信息。

也就是说，整个责任链流程终点不只是 post Filter，还可能是 error Filter，笔者这里重新整理了一下，如图 8-2 所示。

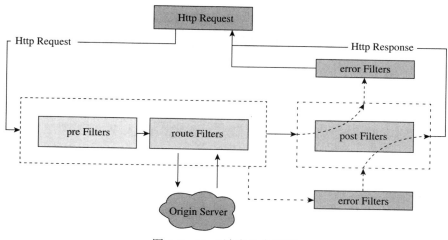

图 8-2　Zuul 请求生命周期

这样就比较直观地描述了 Zuul 关于 Filter 的请求生命周期。Zuul 中一共有四种不同生命周期的 Filter，分别是：

- pre：在 Zuul 按照规则路由到下级服务之前执行。如果需要对请求进行预处理，比如鉴权、限流等，都应考虑在此类 Filter 实现。
- route：这类 Filter 是 Zuul 路由动作的执行者，是 Apache HttpClient 或 Netflix Ribbon 构建和发送原始 HTTP 请求的地方，目前已支持 OkHttp。
- post：这类 Filter 是在源服务返回结果或者异常信息发生后执行的，如果需要对返回信息做一些处理，则在此类 Filter 进行处理。
- error：在整个生命周期内如果发生异常，则会进入 error Filter，可做全局异常处理。

在实际项目中，往往需要自实现以上类型的 Filter 来对请求链路进行处理，根据业务的需求，选取相应生命周期的 Filter 来达成目的。在 Filter 之间，通过 com.netflix.zuul.context. RequestContext 类来进行通信，内部采用 ThreadLocal 保存每个请求的一些信息，包括请求路由、错误信息、HttpServletRequest、HttpServletResponse，这使得一些操作是十分可靠的，它还扩展了 ConcurrentHashMap，目的是为了在处理过程中保存任何形式的信息，后面会对它进行具体讲解。

8.1.2 Zuul 原生 Filter

在讲 Zuul 原生 Filter 之前，先看一个小技巧。官方文档提到，Zuul Server 如果使用 @EnableZuulProxy 注解搭配 Spring Boot Actuator，会多两个管控端点。

1）/routes：返回当前 Zuul Server 中所有已生成的映射规则，加上 /details 可查看明细，如图 8-3、8-4 所示。

http://localhost:5555/actuator/routes


```
⊟{
  "/client/**":"client-a",
  "/client-a/**":"client-a"
}
```


图 8-3 Zuul 的 /routes 端点

http://localhost:5555/actuator/routes/details


```
⊟{
  "/client/**":⊟{
    "id":"client-a",
    "fullPath":"/client/**",
    "location":"client-a",
    "path":"/**",
    "prefix":"/client",
    "retryable":false,
    "customSensitiveHeaders":false,
    "prefixStripped":true
  },
  "/client-a/**":⊟{
    "id":"client-a",
    "fullPath":"/client-a/**",
    "location":"client-a",
    "path":"/**",
    "prefix":"/client-a",
    "retryable":false,
    "customSensitiveHeaders":false,
    "prefixStripped":true
  }
}
```


图 8-4 Zuul 的 /routes/details 端点

2）/filters：返回当前 Zuul Server 中所有已注册生效的 Filter，如图 8-5 所示。

http://localhost:5555/actuator/filters


```
⊟{
  "error":⊟[
    ⊟{
      "class":"org.springframework.cloud.netflix.zuul.filters.post.SendErrorFilter",
      "order":0,
      "disabled":false,
      "static":true
    }
  ],
  "post":⊟[
    ⊟{
      "class":"org.springframework.cloud.netflix.zuul.filters.post.SendResponseFilter",
      "order":1000,
      "disabled":false,
      "static":true
    }
  ],
  "pre":⊞[5],
  "route":⊞[3]
}
```


图 8-5 Zuul 的 /filters 端点

考虑到篇幅原因，pre 与 route 类型的 Filter 这里就不展开了。从返回数据可以清楚地看到所有已注册生效的 Filter 信息，包括：Filter 实现类路径、Filter 执行次序、是否被禁用、是否静态。根据返回的内容，将前面的图稍作扩展，即可得到 Zuul 内置 Filter 与生命周期的组合流程图，如图 8-6 所示。

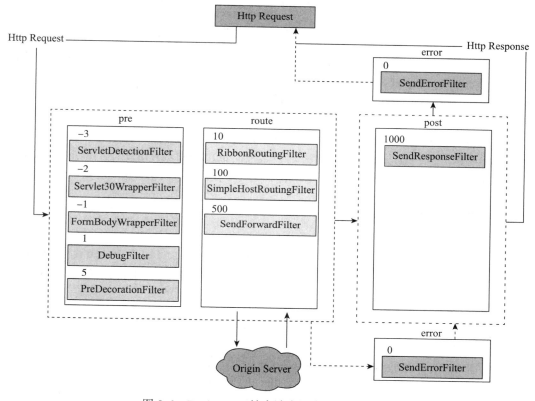

图 8-6　Zuul Filters 结合请求生命周期流程图

各内置 Filter 的详细说明见表 8-1。

表 8-1　Zuul 内置 Filter 信息

名　称	类　型	次　序	描　述
ServletDetectionFilter	pre	-3	通过 Spring Dispatcher 检查请求是否通过
Servlet30WrapperFilter	pre	-2	适配 HttpServletRequest 为 Servlet30RequestWrapper 对象
FormBodyWrapperFilter	pre	-1	解析表单数据并为下游请求重新编码
DebugFilter	pre	1	Debug 路由标识
PreDecorationFilter	pre	5	处理请求上下文供后续使用，设置下游相关头信息
RibbonRoutingFilter	route	10	使用 Ribbon、Hystrix 或者嵌入式 HTTP 客户端发送请求
SimpleHostRoutingFilter	route	100	使用 Apache Httpclient 转发请求

（续）

名 称	类 型	次 序	描 述
SendForwardFilter	route	500	使用 Servlet 转发请求
SendResponseFilter	post	1000	将代理请求的响应写入当前响应
SendErrorFilter	error	0	如果 RequestContext.getThrowable() 不为空，则转发到 error.path 配置的路径

以上是使用 @EnableZuulProxy 注解后安装的 Filter，如果使用 @EnableZuulServer 将缺少 PreDecorationFilter、RibbonRoutingFilter、SimpleHostRoutingFilter。

这些原生 Filter，不见得写得有多好，笔者曾经被其中一些处理弄得晕头转向，所以可以采取替代实现的方式，覆盖掉其原生代码，也可以采取禁用策略，语法如下：

3）zuul.<SimpleClassName>.<filterType>.disable=true。

比如要禁用 SendErrorFilter，在配置文件中添加 zuul.SendErrorFilter.error.disable=true 即可，笔者在生产环境中往往会这样使用，可以去除恼人的堆栈信息，使用自定义的异常处理器。

8.1.3 多级业务处理

本节这样命名是考虑到"网关类应用"的逻辑处理模型。在 Zuul 的 Filter 链体系中，我们可以把一组业务逻辑细分，然后封装到一个个紧密结合的 Filter 中，设置处理顺序，组成一组 Filter 链。这在一些业务场景下十分实用，以致除 Zuul 以外的网关中间件几乎都有类似的实现。

1. 实现自定义 Filter

在 Zuul 中实现自定义 Filter，继承 ZuulFilter 类即可，ZuulFilter 是一个抽象类，我们需要实现它的以下几个方法：

- String filterType()：使用返回值设定 Filter 类型，可以设置为 pre、route、post、error 类型。
- int filterOrder()：使用返回值设定 Filter 执行次序。
- boolean shouldFilter()：使用返回值设定该 Filter 是否执行，可以作为开关来使用。
- Object run()：Filter 里面的核心执行逻辑，业务处理在此编写。

代码清单 8-1 是一个最简 pre Filter 示例：

代码清单8-1 ch8-1\ch8-1-zuul-server\src\main\java\cn\springcloud\book\filter\FirstPreFilter.java

```java
public class FirstPreFilter extends ZuulFilter {

    @Override
    public String filterType() {
        return PRE_TYPE;
    }

    @Override
    public int filterOrder() {
        return 0;
    }
```

```
@Override
public boolean shouldFilter() {
    return true;
}

@Override
public Object run() throws ZuulException {
    System.out.println("这是第一个自定义Zuul Filter！");
    return null;
}
}
```

可以静态导入 FilterConstants 类里面的所有内容，它已经为我们定义好了一些常量信息。此外，应用在启动时，会把自实现的 Filter 加载进内存，但是仅仅有上述代码是不够的，还需要将它注入 Spring Bean 容器，我们只需要在启动主类加入如下代码即可完成一个 Filter 的完整配置：

```
@Bean
public FirstPreFilter firstPreFilter(){
    return new FirstPreFilter();
}
```

启动所有组件，访问 http://localhost:5555/client/mul?a=100&b=300，即可在控制台看到打印的特定内容，如图 8-7 所示。

2. 业务处理实战

在上一节我们已经知道了 Zuul 中 Filter 链的执行原理，也初步了解了一些内置原生 Filter 的功能，因为有了它们，Zuul 才能成为 Spring Cloud 生态系统中不可缺少的重要组件。但是 Zuul 作为一个"网关"，原始的功能往往不能满足实际业务需求，为了解决这个问题，官方为我们预留了 API，使得我们能够实现自定义业务处理，加入 Zuul 的逻辑流程。下面我们通过一个简单案例，来学习如何使用自定义 Filter 来处理具体业务。详细代码见 ch8-1。

图 8-7 FirstPreFilter 打印出的结果

这里来模拟一个业务需求，使用 SecondPreFilter 来验证是否传入 a 参数，使用 ThirdPreFilter 来验证是否传入 b 参数，最后在 PostFilter 里边统一处理返回内容。其流程图如图 8-8 所示。

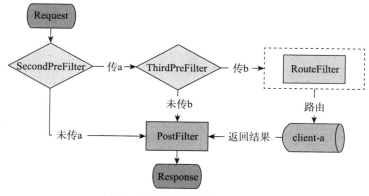

图 8-8 业务处理实战模型流程图

针对图 8-8 中的业务逻辑设定，核心类代码如代码清单 8-2 所示：

代码清单8-2　ch8-1\ch8-1-zuul-server\src\main\java\cn\springcloud\book\filter\SecondPreFilter.java

```java
public class SecondPreFilter extends ZuulFilter {

    @Override
    public String filterType() {
        return PRE_TYPE;
    }

    @Override
    public int filterOrder() {
        return 2;
    }

    @Override
    public boolean shouldFilter() {
        return true;
    }

    @Override
    public Object run() throws ZuulException {
        System.out.println("这是SecondPreFilter! ");
        //从RequestContext获取上下文
        RequestContext ctx = RequestContext.getCurrentContext();
        //从上下文获取HttpServletRequest
        HttpServletRequest request = ctx.getRequest();
        //从request尝试获取a参数值
        String a = request.getParameter("a");
        //如果a参数值为空则进入此逻辑
        if (null == a) {
            //对该请求禁止路由，也就是禁止访问下游服务
            ctx.setSendZuulResponse(false);
            //设定responseBody供PostFilter使用
            ctx.setResponseBody("{\"status\":500,\"message\":\"a参数为空！\"}");
            //logic-is-success保存于上下文，作为同类型下游Filter的执行开关
            ctx.set("logic-is-success", false);
            //到这里此Filter逻辑结束
            return null;
        }
        //设置避免报空
        ctx.set("logic-is-success", true);
        return null;
    }
}
```

SecondPreFilter 的执行次序设为 2，主要验证请求是否传入 a 参数。若没有传，则通过 ctx.setSendZuulResponse(false) 来禁止 route Filter 路由此次请求；通过 ctx.setResponseBody ("{\"status\":500,\"message\":\"a 参数为空！\"}") 来定制返回结果；使用 ctx.set("logic-is-success", false) 来关联 ThirdPreFilter，若是没有传入 a 参数，ThirdPreFilter 也就没有执行的必要了，如代码清单 8-3 所示。

代码清单8-3 ch8-1\ch8-1-zuul-server\src\main\java\cn\springcloud\book\filter\ThirdPreFilter.java

```java
public class ThirdPreFilter extends ZuulFilter {

    @Override
    public String filterType() {
        return PRE_TYPE;
    }

    @Override
    public int filterOrder() {
        return 3;
    }

    @Override
    public boolean shouldFilter() {
        RequestContext ctx = RequestContext.getCurrentContext();
        //从上下文获取logic-is-success值,用于判断此Filter是否执行
        return (boolean)ctx.get("logic-is-success");
    }

    @Override
    public Object run() throws ZuulException {
        System.out.println("这是ThirdPreFilter! ");
        //从RequestContext获取上下文
        RequestContext ctx = RequestContext.getCurrentContext();
        //从上下文获取HttpServletRequest
        HttpServletRequest request = ctx.getRequest();
        //从request尝试获取b参数值
        String b = request.getParameter("b");
        //如果b参数值为空则进入此逻辑
        if (null == b) {
            //对该请求禁止路由,也就是禁止访问下游服务
            ctx.setSendZuulResponse(false);
            //设定responseBody供PostFilter使用
            ctx.setResponseBody("{\"status\":500,\"message\":\"b参数为空! \"}");
            //logic-is-success保存于上下文,作为同类型下游Filter的执行开关,假定后续还有
                自定义Filter当设置此值
            ctx.set("logic-is-success", false);
            //到这里此Filter逻辑结束
            return null;
        }
        return null;
    }
}
```

ThirdPreFilter 的执行次序设为3,主要验证请求是否传入b参数。若没有传入,则通过 ctx.setSendZuulResponse(false) 来禁止 route Filter 路由此次请求;通过 ctx.setResponseBody ("{\"status\":500,\"message\":\"b参数为空! \"}") 来定制返回结果;使用 ctx.set("logic-is-success", false) 来关联下一级 Filter,如代码清单 8-4 所示。

代码清单8-4 ch8-1\ch8-1-zuul-server\src\main\java\cn\springcloud\book\filter\PostFilter.java

```java
public class PostFilter extends ZuulFilter {
```

```java
@Override
public String filterType() {
    return POST_TYPE;
}

@Override
public int filterOrder() {
    return 0;
}

@Override
public boolean shouldFilter() {
    return true;
}

@Override
public Object run() throws ZuulException {
    System.out.println("这是PostFilter! ");
    //从RequestContext获取上下文
    RequestContext ctx = RequestContext.getCurrentContext();
    //处理返回中文乱码
    ctx.getResponse().setCharacterEncoding("UTF-8");
    //获取上下文中保存的responseBody
    String responseBody = ctx.getResponseBody();
    //如果responseBody不为空，则说明流程有异常发生
    if (null != responseBody) {
        //设定返回状态码
        ctx.setResponseStatusCode(500);
        //替换响应报文
        ctx.setResponseBody(responseBody);
    }
    return null;
}
```

PostFilter 主要是用于检查有无定制 ResponseBody，以及设置响应字符集，避免中文乱码，此外还设定了 HTTP 状态码。

3. 测试

依次启动工程 ch8-1-eureka-server、ch8-1-zuul-server、ch8-1-client-a，测试如下。

1）调用 http://localhost:5555/client/mul?b=300，如图 8-9 所示。

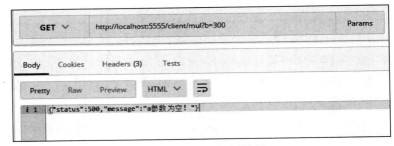

图 8-9　postman 调用结果一

ch8-1-zuul-server 控制台打印特定内容，如图 8-10 所示。

2）调用 http://localhost:5555/client/mul?a=100，结果如图 8-11 所示。

ch8-1-zuul-server 控制台打印结果，如图 8-12 所示。

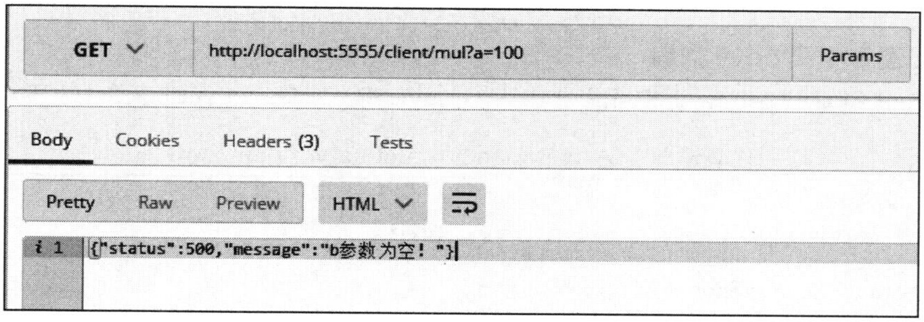

图 8-10　控制台打印结果一

图 8-11　postman 调用结果二

3）调用 http://localhost:5555/client/mul?a=100&b=300，结果如图 8-13 所示。

ch8-1-zuul-server 控制台打印特定内容，如图 8-14 所示。

这就是使用自定义 Zuul Filter 链处理业务的一个简单例子，RequestContext 中有许多非常实用的 API，我们可以结合自己的需求找到合适的解决方案，比如实战当中往往还使用 InputStream stream = ctx.getResponseDataStream() 获取下游服务返回的数据流，再做一些统一处理。更多功能需要读者去发现，这里就不一一列举了。

图 8-12　控制台打印结果二

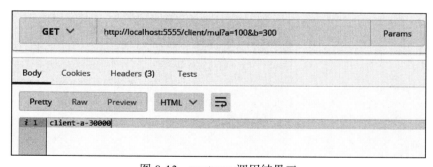

图 8-13　postman 调用结果三

8.1.4　使用 Groovy 编写 Filter

Groovy 语言是基于 JVM 的一门动态语言，它结合了 Python、Ruby 和 Smalltalk 的许多强大特性，能无缝引入 Java 代码与 Java 库，常常被用作 Java 的扩展语言来使用。它的语法与 Java 类似，书写起来比 Java 略为简洁，是一门很优秀的语言。

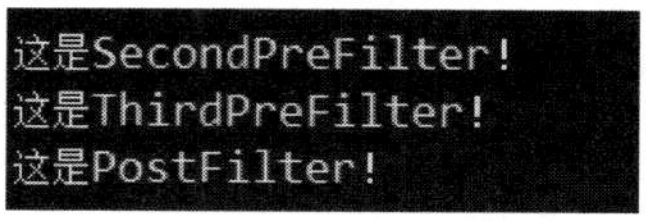

图 8-14　控制台打印结果三

Zuul 中提供 Groovy 的编译类 com.netflix.zuul.groovy.GroovyCompiler，结合 com.netflix.zuul.groovy.GroovyFileFilter 类，我们可以使用 Groovy 来编写自定义 Filter。也许到这里，你认为它在 Zuul 中还没有存在的必要，但是当你得知它可以不用编译（不用打进工程包），可以放在服务器上任意位置，可以任何时候修改由它编写的 Filter，且修改过后还不用重启服务的时候，你就会知道它有多实用了！下面我们来一起看一看这个令人兴奋的功能吧。

1. 为工程添加 Groovy 依赖

要在 Zuul 中使用 Groovy，首先需要引入 Groovy 相关依赖，如代码清单 8-5 所示。

代码清单8-5　ch8-1\ch8-1-zuul-server\pom.xml

```xml
<dependency>
    <groupId>org.codehaus.groovy</groupId>
    <artifactId>groovy-all</artifactId>
    <version>2.5.0-beta-2</version>
</dependency>
```

添加好依赖，我们就可以使用 Groovy 编写自定义 Filter 了。

2. 使用 Groovy 编写自定义 Filter：GroovyFilter.groovy

引入 Groovy 的依赖之后，我们就可以编写 Groovy 的 Filter 了，如代码清单 8-6 所示。

代码清单8-6　ch8-1\ch8-1-zuul-server\src\main\java\cn\springcloud\book\groovy\GroovyFilter.groovy

```groovy
class GroovyFilter extends ZuulFilter {

    @Override
    public String filterType() {
        return PRE_TYPE
    }

    @Override
    public int filterOrder() {
        return 10
    }

    @Override
    public boolean shouldFilter() {
        return true
    }

    @Override
    public Object run() throws ZuulException {
        HttpServletRequest request = RequestContext.currentContext.request as
            HttpServletRequest
        Iterator headerIt = request.getHeaderNames().iterator()
        while (headerIt.hasNext()) {
            String name = (String) headerIt.next()
            String value = request.getHeader(name)
            println("header: " + name + ":" + value)
        }
        println("This is Groovy Filter!")
```

```
        return null
    }
}
```

这里把自定义 Filter 设定为 PRE 类型的 Filter，执行次序为 10，shouldFilter 设置为 true，核心逻辑就是遍历请求的 Header 并打印出来，打印完之后再打印 This is Groovy Filter!。

3. 注册 GroovyFilter.groovy

在 ch8-1-zuul-server 工程的启动主类添加 Groovy 的加载器，如代码清单 8-7 所示。

代码清单8-7　ch8-1\ch8-1-zuul-server\src\main\java\cn\springcloud\book\ZuulServerApplication.java

```java
@Component
public static class GroovyRunner implements CommandLineRunner {
    @Override
    public void run(String... args) throws Exception {
        MonitoringHelper.initMocks();
        FilterLoader.getInstance().setCompiler(new GroovyCompiler());
        try {
            FilterFileManager.setFilenameFilter(new GroovyFileFilter());
            FilterFileManager.init(20, "/Users/Administrator/Desktop/groovy");
        } catch (Exception e) {
            throw new RuntimeException(e);
        }
    }
}
```

这里我将 GroovyFilter.groovy 放在 C:\Users\Administrator\Desktop\groovy 目录下，使用绝对路径每隔 20 秒扫描一次。

4. 测试

依次启动工程 ch8-1-eureka-server、ch8-1-zuul-server、ch8-1-client-a，访问 http://localhost:5555/client/mul?a=100&b=300，观察到控制台打印的结果，如图 8-15 所示。

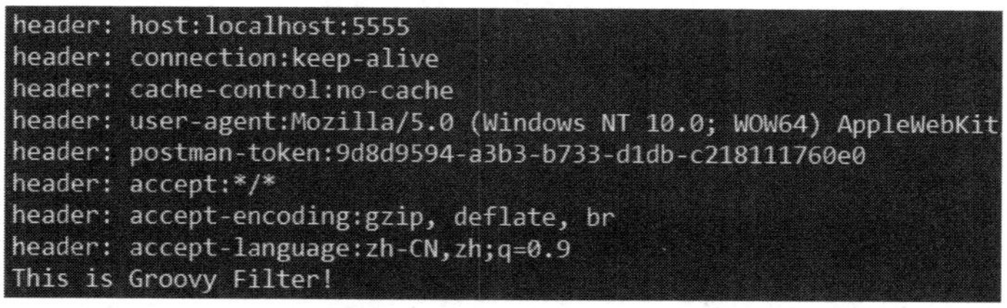

图 8-15　Groovy 测试控制台打印结果一

不启动服务，将 GroovyFilter.groovy 文件中的 println("This is Groovy Filter!") 改为 println("This is Groovy Filter!Modify!")，等待大约 20 秒，重复刚才操作，控制台输出如图 8-16 所示。

怎么样？是不是一个神器？不启动服务更新代码，是多少人梦寐以求的功能，而且效率很好，没有用到字节码生成库，且随处可放。既然这么好用，那就尽快实践起来吧！

```
header: host:localhost:5555
header: connection:keep-alive
header: cache-control:no-cache
header: user-agent:Mozilla/5.0 (Windows NT 10.0; WOW64) AppleWebKit
header: postman-token:8e54b789-b5fd-771d-2fac-4ae5b35f0948
header: accept:*/*
header: accept-encoding:gzip, deflate, br
header: accept-language:zh-CN,zh;q=0.9
This is Groovy Filter!Modify!
```

图 8-16　Groovy 测试控制台打印结果二

8.2　Spring Cloud Zuul 权限集成

8.2.1　应用权限概述

权限，是整个微服务体系乃至软件业永恒的话题，有资源的地方，就有权限约束。以往在构建单体应用的时候，比较流行的方式是使用 Apache Shiro，大家的印象都是 Apache Shiro 比 Spring Security 上手容易，学习成本相对较小，但是，到了 Spring Cloud 这里，面对成千上万的服务，而且服务之间无状态，难免显得力不从心，所以，Spring Cloud 没有选择它也是有原因的。在解决方案的选择上面，传统的譬如单点登录（SSO），或者分布式 Session，要么致使权限服务器集中化导致流量臃肿，要么需要实现一套复杂的存储同步机制，都不是最好的解决方案。作为 Spring Cloud 微服务体系流量前门的 Zuul，除去与它特性毫无相关的实现方式，比较好的有：

1. 自定义权限认证 Filter

由于 Zuul 对请求转发全程的可控性，我们可以在 RequestContext 的基础上做任何事情，如只需要在 8.1.3 节的基础上设置一个执行顺序靠前的 Filter，就可专门用于对请求特定内容做权限认证。如图 8-17 所示。

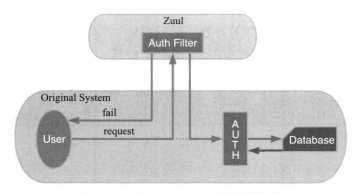

图 8-17　基于 Zuul Filter 的权限模型

这种方式的优点是实现灵活度高，可整合已有权限系统，对原始系统微服务化特别友好；缺点是需要开发一套新的逻辑，维护增加成本，而且也会使得调用链路变得紊乱。

2. OAuth2.0 + JWT

OAuth2.0 是业界对于"授权 – 认证"比较成熟的面向资源的授权协议。怎么理解这句话呢？举个例子，除了可以使用本站用户名与密码登录 Spring Cloud 中国社区，还可以使用第三方应用登录，比如：GitHub、QQ 等登录方式。第三方登录功能对用户十分有亲和力，而 Oauth2.0 就是用于定义 Spring Cloud 中国社区与用户之间的那个"授权层"的。Oauth2.0 的运行原理如图 8-18 所示。

在整个流程中，用户是资源拥有者，其关键还是在于客户端需要资源拥有者的授权，这个过程就相当于键入密码或者是其他第三方登录，触发了这个操作之后，客户端就可以向授权服务器

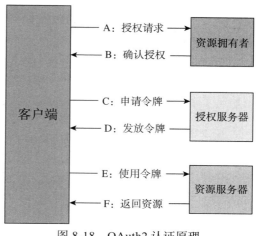

图 8-18　OAuth2 认证原理

申请 Token，拿到后再携带 Token 到资源所在服务器拉取相应资源。

JWT（JSON Web Token）是一种使用 JSON 格式来规约 Token 或者 Session 的协议。由于传统认证方式免不了会生成一个凭证，这个凭证可以是 Token 或者 Session，保存于服务端或者其他持久化工具中，这样一来，凭证的存取就变得十分麻烦，JWT 的出现打破了这一瓶颈，实现了"客户端 Session"的愿景。JWT 通常由三部分组成：

- Header 头部：指定 JWT 使用的签名算法。
- Payload 载荷：包含一些自定义与非自定义的认证信息。
- Signature 签名：将头部与载荷使用"."连接之后，使用头部的签名算法生成签名信息并拼装到末尾。

OAuth2.0+JWT 的意义就在于，使用 OAuth2.0 协议的思想拉取认证生成 Token，使用 JWT 瞬时保存这个 Token，在客户端与资源端进行对称或非对称加密，使得这个规约具有定时、定量的授权认证功能，从而免去 Token 存储所带来的安全或系统扩展问题。

8.2.2　Zuul+OAuth2.0+JWT 实战

我们知道，在服务注册到注册中心之后，是由 Zuul 来引导各个请求到内部敏感服务的，因此，在 Zuul 网关层实现与下游服务之间的鉴权至关重要。笔者基于 OAuth2.0 与 JWT 的思想，在 Zuul 上做了一次实践，示例代码见 ch8-2。

1. zuul-server 编写说明

zuul-server 中需要做的就是当请求接口时，判断是否登录鉴权，如果未登录，则跳转到 auth-server 的登录界面（这里使用的是 Spring Security OAuth 的默认登录界面，也可以重写相关代码定制页面），登录成功后 auth-server 颁发 jwt token，zuul-server 在访问下游服务时将 jwt token 放入 header 中即可。

pom.xml 需要引入对 OAuth2 与 Security 的支持，如代码清单 8-8 所示：

代码清单8-8　ch8-2\ch8-2-zuul-server\pom.xml

```xml
<dependency>
    <groupId>org.springframework.cloud</groupId>
    <artifactId>spring-cloud-starter-security</artifactId>
</dependency>
<dependency>
    <groupId>org.springframework.cloud</groupId>
    <artifactId>spring-cloud-starter-oauth2</artifactId>
</dependency>
```

编写核心配置文件 bootstrap.yml，提供对引入框架的支持，如代码清单 8-9 所示：

代码清单8-9　ch8-2\ch8-2-zuul-server\src\main\resources\bootstrap.yml

```yml
spring:
    application:
        name: zuul-server
server:
    port: 5555
eureka:
    client:
        serviceUrl:
            defaultZone: http://${eureka.host:127.0.0.1}:${eureka.port:8888}/eureka/
    instance:
        prefer-ip-address: true
zuul:
    routes:
        client-a:
            path: /client/**
            serviceId: client-a
security:
    basic:
        enabled: false
    oauth2:
        client:
            access-token-uri: http://localhost:7777/uaa/oauth/token #令牌端点
            user-authorization-uri: http://localhost:7777/uaa/oauth/authorize
            #授权端点
            client-id: zuul_server #OAuth2客户端ID
            client-secret: secret #OAuth2客户端密钥
        resource:
            jwt:
                key-value: springcloud123 #使用对称加密方式，默认算法为HS256
```

代码中定义了从 /client 到 client-a 服务的路由规则，比较重要的是 security 根下面的配置，验证授权端点为 http://localhost:7777/uaa/oauth/authorize，jwt token 的颁发地址为 http://localhost:7777/uaa/oauth/token，声明 jwt 头部签名算法为对称加密方式，默认加密算法为 HS256，密钥为 springcloud123，当然如果安全性要求更高，应该使用非对称加密方式，生成公钥与私钥放置于服务中。

编写核心启动主类 ZuulServerApplication.java，如代码清单 8-10 所示。

代码清单8-10 ch8-2\ch8-2-zuul-server\src\main\java\cn\springcloud\book\ZuulServerApplication.java

```java
@SpringBootApplication
@EnableDiscoveryClient
@EnableZuulProxy
@EnableOAuth2Sso
public class ZuulServerApplication extends WebSecurityConfigurerAdapter{

    public static void main(String[] args) {
        SpringApplication.run(ZuulServerApplication.class, args);
    }

    @Override
    protected void configure(HttpSecurity http) throws Exception {
        http
            .authorizeRequests()
            .antMatchers("/login", "/client/**")
            .permitAll()
            .anyRequest()
            .authenticated()
            .and()
            .csrf()
            .disable();
    }
}
```

重写 WebSecurityConfigurerAdapter 适配器的 configure(HttpSecurity http) 方法，声明需要鉴权的 url 信息。

2. auth-server 编写说明

整个示例的另一个核心就是 auth-server，它作为认证授权中心，会颁发 jwt token 凭证。pom.xml 只需要引入对 OAuth2 的支持，如代码清单 8-11 所示。

代码清单8-11 ch8-2\ch8-2-auth-server\pom.xml

```xml
<dependency>
    <groupId>org.springframework.cloud</groupId>
    <artifactId>spring-cloud-starter-oauth2</artifactId>
</dependency>
```

编写核心配置文件 bootstrap.yml，如代码清单 8-12 所示。

代码清单8-12 ch8-2\ch8-2-auth-server\src\main\resources\bootstrap.yml

```yaml
spring:
    application:
        name: auth-server
server:
    port: 7777
    servlet:
        contextPath: /uaa #web基路径
eureka:
    client:
        serviceUrl:
```

```yaml
      defaultZone: http://${eureka.host:127.0.0.1}:${eureka.port:8888}/eureka/
  instance:
    prefer-ip-address: true
```

这里说明一下，在 Spring Boot 2.0 之后，设定工程基路径需要使用 server.servlet.contextPath，而不是原来的 server. contextPath。

编写认证授权服务适配器类 OAuthConfiguration.java，如代码清单 8-13 所示。

代码清单8-13 ch8-2\ch8-2-auth-server\src\main\java\cn\springcloud\book\OAuthConfiguration.java

```java
@Configuration
@EnableAuthorizationServer
public class OAuthConfiguration extends AuthorizationServerConfigurerAdapter {

    @Autowired
    private AuthenticationManager authenticationManager;

    @Override
    public void configure(ClientDetailsServiceConfigurer clients) throws Exception {
        clients
            .inMemory()
            .withClient("zuul_server")
            .secret("secret")
            .scopes("WRIGTH", "read").autoApprove(true)
            .authorities("WRIGTH_READ", "WRIGTH_WRITE")
            .authorizedGrantTypes("implicit", "refresh_token", "password", "authorization_
                code");
    }

    @Override
    public void configure(AuthorizationServerEndpointsConfigurer endpoints)
            throws Exception {
        endpoints
            .tokenStore(jwtTokenStore())
            .tokenEnhancer(jwtTokenConverter())
            .authenticationManager(authenticationManager);
    }

    @Bean
    public TokenStore jwtTokenStore() {
        return new JwtTokenStore(jwtTokenConverter());
    }

    @Bean
    protected JwtAccessTokenConverter jwtTokenConverter() {
        JwtAccessTokenConverter converter = new JwtAccessTokenConverter();
        converter.setSigningKey("springcloud123");
        return converter;
    }
}
```

这个类主要用于指定客户端 ID、密钥，以及权限定义与作用域声明，指定 TokenStore 为 JWT，不同于以往将 TokenStore 指定为 Redis 或是其他持久化工具，主要的不同就是在这里。

编写启动配置类 AuthServerApplication.java，如代码清单 8-14 所示。

代码清单8-14 ch8-2\ch8-2-auth-server\src\main\java\cn\springcloud\book\AuthServerApplication.java

```java
@SpringBootApplication
@EnableDiscoveryClient
public class AuthServerApplication extends WebSecurityConfigurerAdapter {

    public static void main(String[] args) {
        SpringApplication.run(AuthServerApplication.class, args);
    }

    @Bean(name = BeanIds.AUTHENTICATION_MANAGER)
    @Override
    public AuthenticationManager authenticationManagerBean() throws Exception {
        return super.authenticationManagerBean();
    }

    @Override
    protected void configure(AuthenticationManagerBuilder auth) throws Exception {
        auth
            .inMemoryAuthentication()
            .withUser("guest").password("guest").authorities("WRIGTH_READ")
            .and()
            .withUser("admin").password("admin").authorities("WRIGTH_READ", "WRIGTH_WRITE");
    }

    @Bean
    public static NoOpPasswordEncoder passwordEncoder() {
      return (NoOpPasswordEncoder) NoOpPasswordEncoder.getInstance();
    }
}
```

声明用户 admin 具有读写权限，用户 guest 具有读权限；authenticationManagerBean() 方法用于手动注入 AuthenticationManager；passwordEncoder() 用于声明用户名和密码的加密方式，这个功能在 Spring Security 5.0 之前是没有的，需要注意一下。

3. client-a 编写说明

client-a 作为 zuul-server 的下游服务，需要的功能很简单，能够被注册发现，以及能够按照规则解析 jwt token 即可。

编写启动配置类 ClientAApplication.java，如代码清单 8-15 所示。

代码清单8-15 ch8-2\ch8-2-client-a\src\main\java\cn\springcloud\book\ClientAApplication.java

```java
@SpringBootApplication
@EnableDiscoveryClient
@EnableResourceServer
@RestController
public class ClientAApplication extends ResourceServerConfigurerAdapter {

    public static void main(String[] args) {
        SpringApplication.run(ClientAApplication.class, args);
```

```java
    }

    @RequestMapping("/test")
    public String test(HttpServletRequest request) {
        System.out.println("---------------header---------------");
        Enumeration headerNames = request.getHeaderNames();
        while (headerNames.hasMoreElements()) {
            String key = (String) headerNames.nextElement();
            System.out.println(key + ": " + request.getHeader(key));
        }
        System.out.println("---------------header---------------");
        return "hellooooooooooooooo!";
    }

    @Override
    public void configure(HttpSecurity http) throws Exception {
        http
        .csrf().disable()
        .authorizeRequests()
        .antMatchers("/**").authenticated()
        .antMatchers(HttpMethod.GET, "/test")
        .hasAuthority("WRIGTH_READ");
    }

    @Override
    public void configure(ResourceServerSecurityConfigurer resources) throws
        Exception {
        resources
        .resourceId("WRIGTH")
        .tokenStore(jwtTokenStore());
    }

    @Bean
    protected JwtAccessTokenConverter jwtTokenConverter() {
        JwtAccessTokenConverter converter = new JwtAccessTokenConverter();
        converter.setSigningKey("springcloud123");
        return converter;
    }

    @Bean
    public TokenStore jwtTokenStore() {
        return new JwtTokenStore(jwtTokenConverter());
    }
}
```

到这里相信大家很容易就能看出，它的配置和 auth-server 的配置其实相差无几：如反析 jwt token 配置，声明接口权限可见范围等。这里写了一个 /test 接口用于返回内容与打印 header 到控制台。

4. 测试

在测试之前，说明一下整个示例的流程，其流程图如图 8-19 所示。

依次启动工程 ch8-2-eureka-server、ch8-2-zuul-server、ch8-2-client-a、ch8-2-auth-server。

1）使用浏览器访问 http://localhost:5555/client/test 结果，如图 8-20 所示。

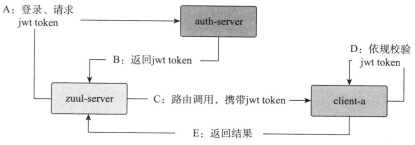

图 8-19　示例认证流程

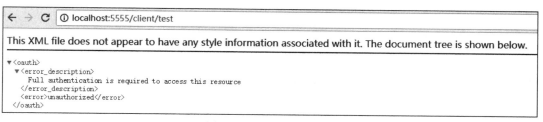

图 8-20　forbidden 页面

由于还未登录授权，这时候接口是调不通的。

2）使用浏览器访问 http://localhost:5555 结果，如图 8-21 所示。

跳转到 auth-server 的默认登录页面，使用用户名 admin 与密码 admin 登录。

3）使用浏览器再次访问 http://localhost:5555/client/test，结果如图 8-22 所示。

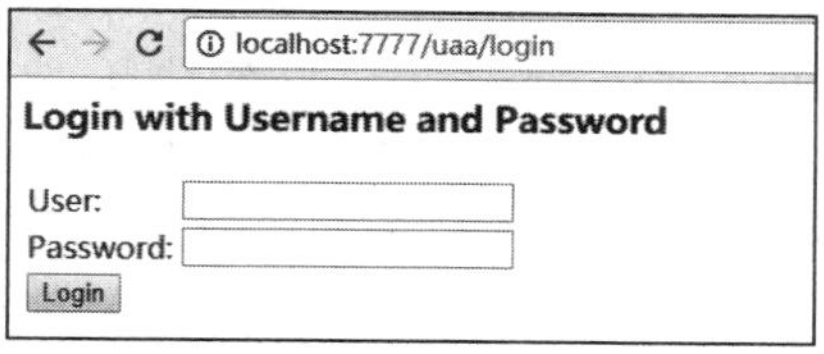

图 8-21　默认登录页面　　　　　　　　图 8-22　通过验证结果

调用接口成功，观察 client-a 服务控制台打印的结果，如图 8-23 所示。

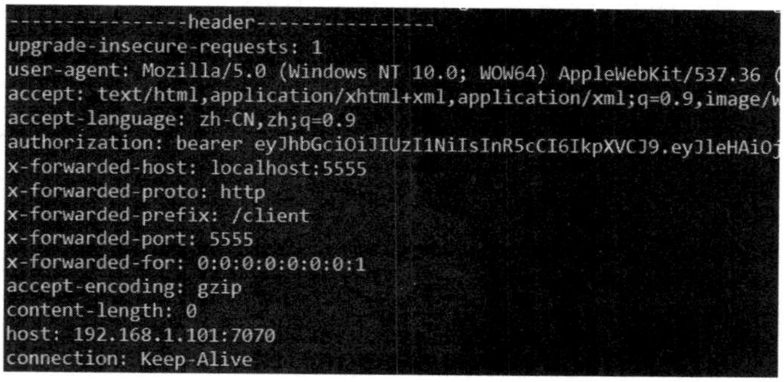

图 8-23　控制台打印结果

其中的 authorization 就是 jwt token，使用 BASE64 解码后为：

```
{
    "alg": "HS256",
    "typ": "JWT"
} {
    "exp": 1523734683,
    "user_name": "admin",
    "authorities": ["WRIGTH_WRITE", "WRIGTH_READ"],
    "jti": "fe85ea29-c9bd-4503-82a2-9cdd28272f97",
    "client_id": "zuul_server",
    "scope": ["WRIGTH", "read"]
}
FaJDhfvnnnVhOqUgvwBEfTeLQXmG3i3FbmsGINHPDHY
```

从其中可以清晰看到 jwt token 的头部、载荷、签名信息。

8.3 Spring Cloud Zuul 限流

构建一个自我修复型系统一直是各大企业进行架构设计的难点所在，在 Hystrix 中，我们可以通过熔断器来实现，通过某个阈值来对异常流量进行降级处理。其实，除对异常流量进行降级处理之外，我们也可以做一些其他操作来保护我们的系统免受"雪崩之灾"，比如：流量排队、限流、分流等。这一节我们基于 Zuul 来谈一谈限流策略的应用与实现。

8.3.1 限流算法

说到限流算法，脑袋中不自觉就想到了"漏桶"与"令牌桶"。诚然，两种限流的祖师级算法确有其独到之处，其他实现比如滑动时间窗或者三色速率标记法等，其实质还是"漏桶"与"令牌桶"的变种，要么是将"漏桶"容积换成了单位时间，要么是按规则将请求标记颜色进行处理，底层还是"令牌"的思想。所以，掌握"漏桶"与"令牌桶"算法原理，对理解其他限流算法有一定帮助。

1. 漏桶（Leaky Bucket）

漏桶的原型是一个底部有漏孔的桶，桶上方有一个入水口，水不断地流进桶内，桶下方的漏孔就会以一个相对恒定的速率漏水，在入大于出的情况下，桶在一段时间之后就会被装满，这时候多余的水就会溢出；而在入小于出的情况下，漏桶则不起任何作用。后来人们将这个经典模型运用在网络流量整形上面，通过漏桶算法的约束，突发流量可以被整形为一个规整的流量，如图 8-24 所示。

如图 8-24 所示，当我们的请求或者具有一定体量的数据流涌来的时候，在漏桶的作用下，流量被整形，不能满足要求的部分被削减掉，所以，漏桶算法能够强制限定流量速率。注意，在我们的应用中，这部分溢出的流量是可以被利用起来的，并非完全丢弃，我们可以把它们收集到一个队列里面，做流量排队，尽量做到合理利用所有资源。

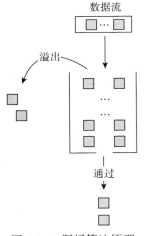

图 8-24　漏桶算法原理

2. 令牌桶（Token Bucket）

令牌桶算法和漏桶算法有点不一样，桶里面存放令牌，而令牌又是以一个恒定的速率被加入桶内，可以积压，可以溢出。当我们的数据流涌来时，量化请求用于获取令牌，如果取到令牌则放行，同时桶内丢弃掉这个令牌；如果不能取到令牌，请求则被丢弃，如图 8-25 所示。

由于令牌桶内可以存在一定数量的令牌，那么就可能存在一定程度的流量突发，这也是决定漏桶算法与令牌桶算法适用于不同应用场景的主要原因。

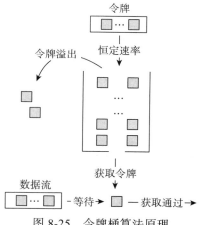

图 8-25　令牌桶算法原理

8.3.2　限流实战

在 Zuul 中实现限流最简单的方式是使用定义 Filter 加上相关限流算法，其中可能会考虑到 Zuul 的多节点部署，因为算法的原因，这时候需要一个 K/V 存储工具（推荐使用 Redis，充分利用 Redis 单线程的特性，可以有效避免多节点带来的一些问题）。当然如果 Zuul 是单节点应用，限流方式的选择就会广得多，完全可以将相关 prefix 放在内存之中，方便又快捷。本节示例代码见 ch8-3。

这里介绍一个开箱即用的工具，spring-cloud-zuul-ratelimit（https://github.com/marcosbarbero/spring-cloud-zuul-ratelimit），它是一位国外友人专门针对 Zuul 编写的限流库，提供多种细粒度策略：

- user：认证用户名或匿名，针对某个用户粒度进行限流。
- origin：客户机 ip，针对请求客户机 ip 粒度进行限流。
- url：特定 url，针对某个请求 url 粒度进行限流。
- serviceId：特定服务，针对某个服务 id 粒度进行限流。

以及多种粒度临时变量存储方式：

- IN_MEMEORY：基于本地内存，底层是 ConcurrentHashMap。
- REDIS：Redis 的 K/V 存储。
- CONSUL：Consul 的 K/V 存储。
- JPA：Spring Data JPA，基于数据库。
- BUKET4J：一个使用 Java 编写的基于令牌桶算法的限流库，其有四种模式，JCache、Hazelcast、Apache Ignite、Inifinispan，其中后面三种支持异步。

在选择粒度策略与存储方式时，应该结合自己的应用客观选取，比如 Zuul 如果需要多节点部署，那粒度临时变量存储方式就不能选择 MEMEORY。选择最适合自己应用的搭配，快速构建起自己的应用。

1. zuul-server 编写说明

截至目前，其最新版本是 LATEST 版本，可全面兼容 Spring Cloud Finchley，在 pom.xml

中需要引入相关依赖,如代码清单 8-16 所示。

代码清单8-16　ch8-3\ch8-3-zuul-server\pom.xml

```
<dependency>
    <groupId>com.marcosbarbero.cloud</groupId>
    <artifactId>spring-cloud-zuul-ratelimit</artifactId>
    <version>LATEST</version>
</dependency>
```

编写配置文件 bootstrap.yml,如代码清单 8-17 所示。

代码清单8-17　ch8-3\ch8-3-zuul-server\src\main\resources\bootstrap.yml

```
server:
    port: 5555
eureka:
    client:
        serviceUrl:
            defaultZone: http://${eureka.host:127.0.0.1}:${eureka.port:8888}/eureka/
    instance:
        prefer-ip-address: true
zuul:
    routes:
        client-a:
            path: /client/**
            serviceId: client-a
    ratelimit:
        key-prefix: springcloud-book #按粒度拆分的临时变量key前缀
        enabled: true #启用开关
        repository: IN_MEMORY #key存储类型,默认是IN_MEMORY本地内存,此外还有多种形式
        behind-proxy: true #表示代理之后
        default-policy: #全局限流策略,可单独细化到服务粒度
            limit: 2 #在一个单位时间窗口的请求数量
            quota: 1 #在一个单位时间窗口的请求时间限制
            refresh-interval: 3 #单位时间窗口
            type:
                - user #可指定用户粒度
                - origin #可指定客户端地址粒度
                - url #可指定url粒度
```

示例采用本地内存的存储方式,在 3 秒的时间窗口内不能有超过 2 次的接口调用。到这里,它的整合就算完成了,我们需要做的只是修改配置文件,定制我们需要的限流策略。eureka-server 与 client-a 服务不需要做任何修改。

2. 测试

依次启动工程 ch8-2-eureka-server、ch8-2-zuul-server、ch8-2-client-a。使用 postman 调用接口,如图 8-26 所示。

调用接口超限之后,返回如图 8-27 所示信息。

在时间窗阈值内访问接口,接口返回正确信息,一旦超限,后台会抛出 429 异常,接口返回上面的错误信息。

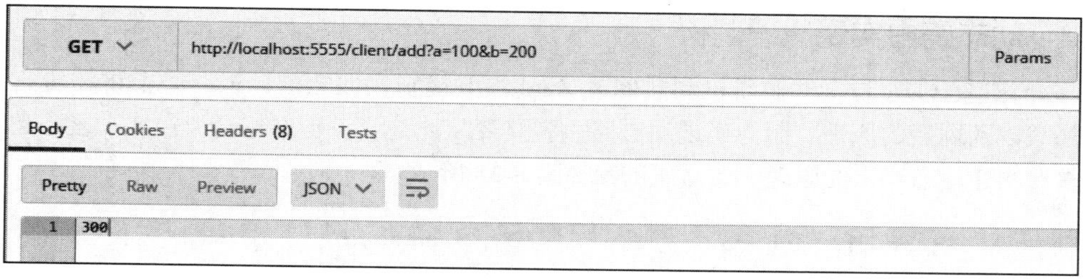

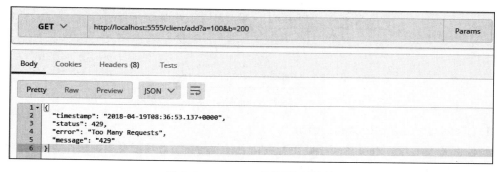

图 8-26　postman 调用正常结果

图 8-27　postman 调用限制结果

8.4　Spring Cloud Zuul 动态路由

8.4.1　动态路由概述

在第 7 章，我们详细介绍了 Zuul 的各种映射规则配置方式，这极大地增加了我们在构建应用时的选择余地，这些方式称为"静态路由（Static Routing）"。一般来说，我们在微服务构建前期就已经按照业务把各种映射关系制定好了，但是在后期迭代过程中，一个复杂的系统难免经历新服务的上线过程，这个时候我们不能轻易停掉线上某些映射链路，那么问题就来了，Zuul 是在启动的时候将配置文件中的映射规则写入内存，要新建映射规则，只能修改了配置文件之后再重新启动 Zuul 应用。能不能有一种方法，既能按需修改映射规则，又能使服务免于重启之痛呢？答案是有的，目前有两种解决方案实现"动态路由（Dynamic Routing）"：

- 结合 Spring Cloud Config + Bus，动态刷新配置文件。这种方式的好处是不用 Zuul 维护映射规则，可以随时修改，随时生效；唯一不好的地方是需要单独集成一些使用并不频繁的组件，Config 没有可视化界面，维护起规则来也相对麻烦。
- 重写 Zuul 的配置读取方式，采用事件刷新机制，从数据库读取路由映射规则。此种方式因为基于数据库，可轻松实现管理界面，灵活度较高。

通常采用第一种方式，这是 Spring Cloud 生态推崇的方式，但是也有它的局限性，有关于它的使用可以查看本书第 11～12 章，这里不再赘述。本节主要围绕第二种方式展开。本节示例代码见 ch8-4。

8.4.2 动态路由实现原理剖析

在介绍基于 DB 与事件刷新机制热修改 Zuul 映射规则实战之前，很有必要讲一讲这部分路由加载以及刷新的原理，因为实现动态路由就是要从此处入手，本节会通过 4 个核心类来使读者掌握构建动态路由的原理，类之间依赖如图 8-28 所示。

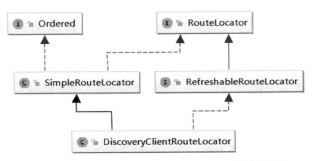

图 8-28　DiscoveryClientRouteLocator 类依赖图

1. DiscoveryClientRouteLocator

顾名思义，此类是 Zuul 中对于路由配置信息读取与新节点注册变更的操作类，本类依赖关系如图 8-28 所示，部分重要源码如下：

```
public class DiscoveryClientRouteLocator extends SimpleRouteLocator
        implements RefreshableRouteLocator {
...
    @Override
    protected LinkedHashMap<String, ZuulRoute> locateRoutes() {
        ...
    }

    @Override
    public void refresh() {
        doRefresh();
    }
...
}
```

locateRoutes() 方法继承自 SimpleRouteLocator 类并重写了规则，该方法主要的功能就是将配置文件中的映射规则信息包装成 LinkedHashMap<String, ZuulRoute>，键 String 是路径 path，值 ZuulRoute 是配置文件的封装类，以往所见的配置映射读取进来就是使用 ZuulRoute 来封装。refresh() 实现自 RefreshableRouteLocator 接口，添加刷新功能必须要实现此方法，doRefresh() 方法来自 SimpleRouteLocator 类。我们需要自定义路由配置加载器，以仿照它的实现。

2. SimpleRouteLocator

SimpleRouteLocator 是 DiscoveryClientRouteLocator 的父类，此类基本实现了 RouteLocator 接口，对读取的配置文件信息做一些基本处理，提供方法 doRefresh() 与 locateRoutes() 供子类实现刷新策略与映射规则加载策略，两个方法源码及签名如下：

```
public class SimpleRouteLocator implements RouteLocator, Ordered {

    /**
     * Calculate all the routes and set up a cache for the values. Subclasses can call
     * this method if they need to implement {@link RefreshableRouteLocator}.
     */
    protected void doRefresh() {
        this.routes.set(locateRoutes());
    }

    /**
     * Compute a map of path pattern to route. The default is just a static map from the
     * {@link ZuulProperties}, but subclasses can add dynamic calculations.
     */
    protected Map<String, ZuulRoute> locateRoutes() {
        LinkedHashMap<String, ZuulRoute> routesMap = new LinkedHashMap<>();
        for (ZuulRoute route : this.properties.getRoutes().values()) {
            routesMap.put(route.getPath(), route);
        }
        return routesMap;
    }
}
```

两个方法都是使用 protected 修饰，是为了让子类不用维护此类一些成员变量就能够实现刷新或者读取路由的功能。从注释也能看出，调用 doRefresh() 方法需要实现 RefreshableRouteLocator；locateRoutes() 方法，默认是一个静态的映射读取方法，方法签名说得很明白，如果需要动态读取加载映射，则需要子类重写此方法。

3. ZuulServerAutoConfiguration

在低版本的 Spring Cloud Zuul 中，这个类叫作 ZuulConfiguration，位于 org.springframework.cloud.netflix.zuul 包中，主要目的是注册各种过滤器、监听器以及其他功能。Zuul 在注册中心新增服务后刷新监听器也是在此注册的，底层是采用 Spring 的 ApplicationListener 来实现的，核心代码如下：

```
@Configuration
@EnableConfigurationProperties({ ZuulProperties.class })
@ConditionalOnClass(ZuulServlet.class)
@ConditionalOnBean(ZuulServerMarkerConfiguration.Marker.class)
public class ZuulServerAutoConfiguration {

    private static class ZuulRefreshListener
            implements ApplicationListener<ApplicationEvent> {

        @Autowired
        private ZuulHandlerMapping zuulHandlerMapping;

        private HeartbeatMonitor heartbeatMonitor = new HeartbeatMonitor();

        @Override
        public void onApplicationEvent(ApplicationEvent event) {
            if (event instanceof ContextRefreshedEvent
                    || event instanceof RefreshScopeRefreshedEvent
```

```
                    || event instanceof RoutesRefreshedEvent) {
                this.zuulHandlerMapping.setDirty(true);
            }
            else if (event instanceof HeartbeatEvent) {
                if (this.heartbeatMonitor.update(((HeartbeatEvent) event).getValue())) {
                    this.zuulHandlerMapping.setDirty(true);
                }
            }
        }
    }
}
```

由方法 onApplicationEvent(ApplicationEvent event) 可知，Zuul 会接收 3 种事件 ContextRefreshedEvent、RefreshScopeRefreshedEvent、RoutesRefreshedEvent 通知去刷新路由映射配置信息，此外心跳续约监视器 HeartbeatMonitor 也会触发这个动作。

4. ZuulHandlerMapping

此类是将本地配置的映射关系映射到远程的过程控制器，与事件刷新相关的代码如下：

```
public class ZuulHandlerMapping extends AbstractUrlHandlerMapping {
    private volatile boolean dirty = true;

    public void setDirty(boolean dirty) {
        this.dirty = dirty;
        if (this.routeLocator instanceof RefreshableRouteLocator) {
            ((RefreshableRouteLocator) this.routeLocator).refresh();
        }
    }
}
```

这个 dirty 属性很重要，它是用来控制当前是否需要重新加载映射配置信息的标记，在 Zuul 每次进行路由操作的时候都会检查这个值，如果为 true，就会触发配置信息的重新加载，同时再将其回设为 false。也由 setDirty(boolean dirty) 可知，启动刷新动作必须要实现 RefreshableRouteLocator 接口。

本节讲解了路由映射规则的加载原理以及 Zuul 的事件刷新方式。我们在构建动态路由的时候，只需要重写 SimpleRouteLocator 类的 locateRoutes() 方法，并且实现 RefreshableRouteLocator 接口的 refresh() 方法，再在内部调用 SimpleRouteLocator 类的 doRefresh() 方法，就可以构建起一个由 Zuul 内部事件触发的自定义动态路由加载器。如果不想使用内部事件触发配置更新操作，改为手动触发，可以重写 onApplicationEvent(ApplicationEvent event) 方法内部实现方式，事实上手动触发的控制性更好。

8.4.3 基于 DB 的动态路由实战

在上一小节，我们知道了动态路由的改造原理与方式，本节就做一个实战 demo，这里的 DB 暂且选用 MySQL，当然也可以选择其他持久化方式，目的是方便，易于管理，实际上选用 MongoDB 也是一种不错的选择。示例代码详见 ch8-4。

1. zuul-server 编写说明

这里 ORM 选用比较简单的 Spring JdbcTemplate，pom.xml 引入相关配置，如代码清单 8-18 所示。

代码清单8-18　ch8-4\ch8-4-zuul-server\pom.xml

```xml
<dependency>
    <groupId>mysql</groupId>
    <artifactId>mysql-connector-java</artifactId>
</dependency>
<dependency>
    <groupId>org.springframework.boot</groupId>
    <artifactId>spring-boot-starter-jdbc</artifactId>
</dependency>
```

编写工程配置文件 bootstrap.yml，如代码清单 8-19 所示。

代码清单8-19　ch8-4\ch8-4-zuul-server\src\main\resources\bootstrap.yml

```yaml
spring:
  datasource:
    url: jdbc:mysql://localhost:3306/test?useUnicode=true&characterEncoding=utf-8
    driver-class-name: com.mysql.jdbc.Driver
    username: root
    password: xxxxx

ribbon:
  eureka:
    enabled: true
```

与之前不同的是加入数据库连接配置，另外如果需要防止 Ribbon 从注册中心获取已注册服务列表进行负载均衡，影响动态路由功能，可以将 ribbon.eureka.enabled 设置为 false。

存储映射规则的数据库表设计，如图 8-29 所示。

id	path	service_id	url	strip_prefix	retryable	enabled	description
1	/baidu/**	(NULL)	http://www.baidu.com	1	0	1	重定向百度
2	/client/**	(NULL)	http://localhost:7070	1	0	1	url
3	/client-a/**	client-a	(NULL)	1	0	1	serviceId

图 8-29　数据库表设计

编写 DAO 类，从数据库读取路由配置信息，如代码清单 8-20 所示。

代码清单8-20　ch8-4\ch8-4-zuul-server\src\main\java\cn\springcloud\book\dao\PropertiesDao.java

```java
@Component
public class PropertiesDao {

    @Autowired
    private JdbcTemplate jdbcTemplate;

    private final static String SQL = "SELECT * FROM zuul_route WHERE enabled = TRUE";

    public Map<String, ZuulRoute> getProperties() {
        Map<String, ZuulRoute> routes = new LinkedHashMap<>();
```

```java
            List<ZuulRouteEntity> list = jdbcTemplate.query(SQL, new BeanPropertyRowMapper<>
                (ZuulRouteEntity.class));
            list.forEach(entity -> {
                if (StringUtils.isEmpty(entity.getPath())) return;
                ZuulRoute zuulRoute = new ZuulRoute();
                BeanUtils.copyProperties(entity, zuulRoute);
                routes.put(zuulRoute.getPath(), zuulRoute);
            });
            return routes;
        }
    }
```

读取数据库可用的路由配置信息，ZuulRouteEntity 是库表的映射实体类，这里就不展示了，然后把结果包装成 Map<String, ZuulRoute> 供自定义路由配置加载器使用。

编写自定义路由配置加载器，如代码清单 8-21 所示。

代码清单8-21 ch8-4\ch8-4-zuul-server\src\main\java\cn\springcloud\book\DynamicZuulRouteLocator.java

```java
public class DynamicZuulRouteLocator extends SimpleRouteLocator implements
    RefreshableRouteLocator {

    @Autowired
    private ZuulProperties properties;

    @Autowired
    private PropertiesDao propertiesDao;

    public DynamicZuulRouteLocator(String servletPath, ZuulProperties properties) {
        super(servletPath, properties);
        this.properties = properties;
    }

    @Override
    public void refresh() {
        doRefresh();
    }

    @Override
    protected Map<String, ZuulRoute> locateRoutes() {
        LinkedHashMap<String, ZuulRoute> routesMap = new LinkedHashMap<>();
        routesMap.putAll(super.locateRoutes());
        routesMap.putAll(propertiesDao.getProperties());
        LinkedHashMap<String, ZuulRoute> values = new LinkedHashMap<>();
        routesMap.forEach((key, value) -> {
            String path = key;
            if (!path.startsWith("/")) {
                path = "/" + path;
            }
            if (StringUtils.hasText(this.properties.getPrefix())) {
                path = this.properties.getPrefix() + path;
                if (!path.startsWith("/")) {
                    path = "/" + path;
                }
            }
```

```
            values.put(path, value);
        });
        return values;
    }
}
```

这个类是本次改造的核心类，locateRoutes() 方法从数据库加载配置信息，并且结合 Zuul 内部事件刷新机制，实际上每次心跳续约都会触发路由配置重新加载的操作，如果需要改为手动触发，可参考上一小节的描述。

最后是对自定义路由配置加载器的生效操作类，如代码清单 8-22 所示。

代码清单8-22   ch8-4\ch8-4-zuul-server\src\main\java\cn\springcloud\book\config\DynamicZuulConfig.java

```
@Configuration
public class DynamicZuulConfig {

    @Autowired
    private ZuulProperties zuulProperties;

    @Autowired
    private ServerProperties serverProperties;

    @Bean
    public DynamicZuulRouteLocator routeLocator() {
        DynamicZuulRouteLocator routeLocator = new DynamicZuulRouteLocator(
                serverProperties.getServlet().getServletPrefix(), zuulProperties);
        return routeLocator;
    }
}
```

2. 测试

依次启动工程 ch8-4-eureka-server、ch8-4-zuul-server、ch8-4-client-a。按照数据库的配置，使用 Chrome 调用接口，如图 8-30 所示。

8.5　Spring Cloud Zuul 灰度发布

8.5.1　灰度发布概述

灰度发布，是指在系统迭代新功能时的一种平滑过渡的上线发布方式。灰度发布是在原有系统的基础上，额外增加一个新版本，这个新版本包含我们需要待验证的新功能，随后用负载均衡器引入一小部分流量到这个新版本应用，如果整个过程没有出现任何差错，再平滑地把线上系统或服务一步步替换成新版本，至此完成了一次灰度发布，如图 8-31 所示。

这种发布方式由于可以在用户无感知的情况下完成产品的升级，在许多公司都有较为成熟的解决方案。对于 Spring Cloud 微服务生态来说，粒度一般是一个服务，往往通过使用某些带有特定标记的流量来充当灰度发布过程中的"小白鼠"，并且目前已经有比较好的开源项目来做这个事情。

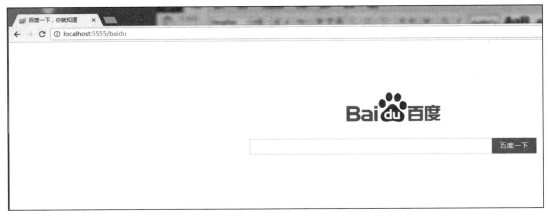

图 8-30 动态路由测试结果

8.5.2 灰度发布实战之一

灰度发布有很多种实现方式，本节要讲的是基于 Eureka 元数据（metadata）的一种方式。在 Eureka 里面，一共有两种元数据：

1）标准元数据：这种元数据是服务的各种注册信息，比如 ip、端口、服务健康信息、续约信息等，存储于专门为服务开辟的注册表中，用于其他组件取用以实现整个微服务生态。

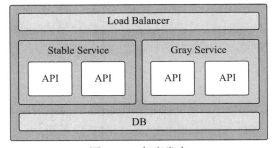

图 8-31 灰度发布

2）自定义元数据：自定义元数据是使用 eureka.instance.metadata-map.<key>=<value> 来配置的，其内部其实就是维护了一个 Map 来保存自定义元数据信息，可以配置在远端服务，随服务一并注册保存在 Eureka 注册表中，对微服务生态的任何行为都没有影响，除非我们知道其特定的含义。

本节我们基于 Eureka 的自定义元数据来完成灰度发布的操作。它的原理是通过获取 Eureka 实例信息，并鉴别元数据的含义，再分别进行路由规则下的负载均衡。示例代码详见 ch8-5。

1. zuul-server、client-a 编写说明

要实现这样的功能，首先要在 Zuul Server 引入一个开源项目包（github 地址：https://github.com/jmnarloch/ribbon-discovery-filter-spring-cloud-starter），如代码清单 8-23 所示。

代码清单8-23 ch8-5\ch8-5-zuul-server\pom.xml

```
<dependency>
```

```xml
        <groupId>io.jmnarloch</groupId>
        <artifactId>ribbon-discovery-filter-spring-cloud-starter</artifactId>
        <version>2.1.0</version>
</dependency>
```

接下来就是在 client-a 服务里设置 metadata 了,如代码清单 8-24 所示。

代码清单8-24 ch8-5\ch8-5-client-a\src\main\resources\bootstrap.yml

```yaml
server:
    port: 7070
spring:
    profiles: node1
    application:
        name: client-a
    instance:
        metadata-map:
            host-mark: running-host
---
server:
    port: 7071
spring:
    profiles: node2
    application:
        name: client-a
    instance:
        metadata-map:
            host-mark: running-host
---
server:
    port: 7072
spring:
    profiles: node3
    application:
        name: client-a
    instance:
        metadata-map:
            host-mark: gray-host
```

7070 与 7071 端口运行的是稳定的线上服务,我们将它们的 host-mark 设置成 running-host,需要上线的灰度服务端口为 7072,host-mark 为 gray-host,最后要达成的效果是:由于服务名称都为 client-a,但是在某一个值的作用下,部分请求被分发到 7070 与 7071 实例上,也就是 host-mark 为 running-host 的节点;另一部分则分发到 7072 实例,host-mark 为 gray-host 的节点。client-a 服务需要启动 3 个实例,启动类如代码清单 8-25 所示。

代码清单8-25 ch8-5\ch8-5-client-a\src\main\java\cn\springcloud\book\ClientAApplication.java

```java
@SpringBootApplication
@EnableDiscoveryClient
public class ClientAApplication {

    public static void main(String[] args) {
        SpringApplication.run(ClientAApplication.class, "--spring.profiles.
```

```
            active=node3");
    }
}
```

配合配置文件，每次启动将 node1 改为 node2 或 node3，就能启动多个服务实例。

以上配置都还是基础，最重要的是 Zuul Server 中的自定义过滤器编写，如代码清单 8-26 所示。

代码清单8-26 ch8-5\ch8-5-zuul-server\src\main\java\cn\springcloud\book\filter\GrayFilter.java

```java
public class GrayFilter extends ZuulFilter {

    @Override
    public Object run() throws ZuulException {
        HttpServletRequest request = RequestContext.getCurrentContext().getRequest();
        String mark = request.getHeader("gray_mark");
        if (!StringUtils.isEmpty(mark) && "enable".equals(mark)) {
            RibbonFilterContextHolder.getCurrentContext()
                .add("host-mark", "gray-host");
        } else {
            RibbonFilterContextHolder.getCurrentContext()
                .add("host-mark", "running-host");
        }
        return null;
    }
}
```

该过滤器的意思是，将 header 里面的 gray_mark 作为指标，如果 gray_mark 等于 enable 的话，就将该请求路由到灰度节点 gray-host 上，如果不等于或者没有这个指标，就路由到其他节点。RibbonFilterContextHolder 是该项目的一个核心类，它定义了基于 metadata 的一种负载均衡机制。

2. 测试

依次启动工程 ch8-5-eureka-server、ch8-5-zuul-server、ch8-5-client-a。但是需要修改 ch8-5-client-a 工程中 bootstrap.yml 中的 spring.profile=node 为 node2 或 node3，启动三次工程。使用 postman 调用接口：

1）header 不加 gray_mark，测试结果如图 8-32、8-33 所示。

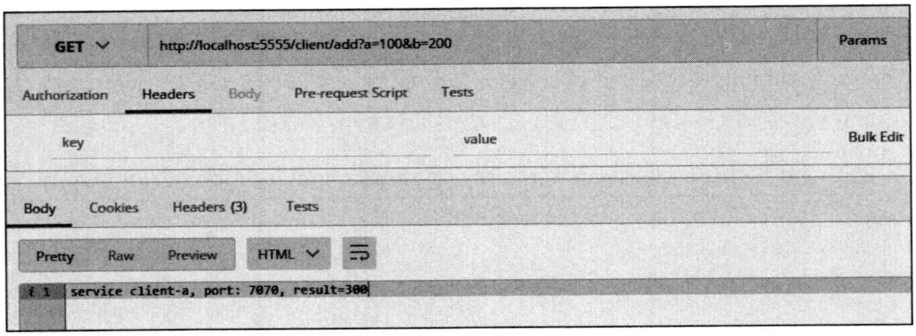

图 8-32　postman 测试结果一

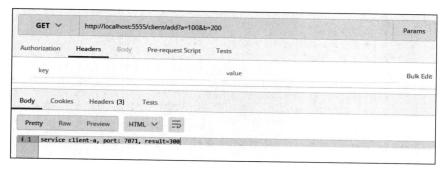

图 8-33　postman 测试结果二

可以看出，当没有在 header 中传 gray_mark 的时候，请求只会路由到端口为 7070 与 7071 的 client-a 服务上。

2）header 加上 gray_mark，并且值为 enable，如图 8-34 所示。

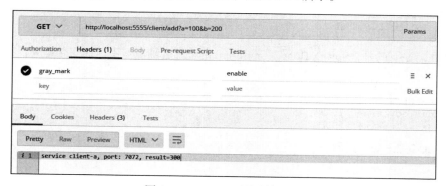

图 8-34　postman 测试结果三

此时无论请求多少次，都会路由到端口为 7072 的 client-a 服务。

8.6　Spring Cloud Zuul 文件上传

8.6.1　文件上传实战

文件上传的场景，相信很多开发者都会遇到，Zuul 作为一个网关中间件，自然也会面临文件上传的考验。Zuul 的文件上传功能是从 Spring Boot 承袭过来的，所以，也需要 Spring Boot 的相关配置。

1. zuul-server 编写说明

bootstrap.yml 中需要对上传文件的一些阈值做限定，以及对熔断或者负载均衡超时时间做改变，如代码清单 8-27 所示：

代码清单8-27　ch8-6\ch8-6-zuul-server\src\main\resources\bootstrap.yml

```
spring:
```

```yaml
  application:
    name: zuul-server
  servlet:   #spring boot2.0之前是http
    multipart:
      enabled: true       # 使用http multipart上传处理
      max-file-size: 100MB  # 设置单个文件的最大长度，默认1M，如不限制配置为-1
      max-request-size: 100MB # 设置最大的请求文件的大小，默认10M，如不限制配置为-1
      file-size-threshold: 1MB   # 当上传文件达到1MB的时候进行磁盘写入
      location: /   # 上传的临时目录
...
##### Hystrix默认超时时间为1秒，如果要上传大文件，为避免超时，稍微设大一点
hystrix:
  command:
    default:
      execution:
        isolation:
          thread:
            timeoutInMilliseconds: 30000
ribbon:
  ConnectTimeout: 3000
  ReadTimeout: 30000
```

需要注意一下的是，Spring Boot 对 multipart 的配置会随版本不同而变化挺大，在 1.4.x 版本之前：

```yaml
multipart:
  enabled: true
  max-file-size: 100MB
```

在 1.5.x 到 2.x 版本之间：

```yaml
spring:
  http:
    multipart:
      enabled: true
      max-file-size: 100MB
```

在 2.x 版本之后：

```yaml
spring:
  servlet:
    multipart:
      enabled: true
      max-file-size: 100MB
```

此外需要一个上传接口来接收前端传过来的文件，如代码清单 8-28 所示：

代码清单8-28 ch8-6\ch8-6-zuul-server\src\main\java\cn\springcloud\book\controller\ZuulUploadController.java

```java
@Controller
public class ZuulUploadController {

    @PostMapping("/upload")
    @ResponseBody
    public String uploadFile(@RequestParam(value = "file", required = true)
        MultipartFile file) throws IOException{
```

```
        byte[] bytes = file.getBytes();
        File fileToSave = new File(file.getOriginalFilename());
        FileCopyUtils.copy(bytes, fileToSave);
        return fileToSave.getAbsolutePath();
    }
}
```

这里在上传成功后返回了上传的文件的绝对路径，供测试使用。

2. 测试

依次启动工程 ch8-6-eureka-server、ch8-6-zuul-server。使用 postman 上传文件，如图 8-35 所示。

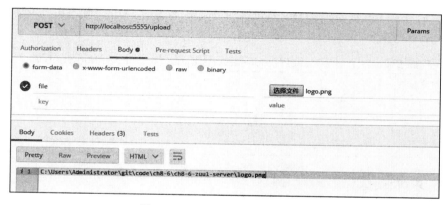

图 8-35　postman 文件上传结果

刷新工程后可以看到如图 8-36 所示的结果。

8.6.2　文件上传乱码解决

在 Finchley 之前的版本，上传中文名的文件会出现文件名乱码的情况，上传英文名的文件则不会，这是由于 Zuul 内部默认使用了 Spring MVC 来上传文件，这种方式对中文字符的处理有点瑕疵。如果要解决这个问题，我们可以改为使用 Zuul Servlet 来上传文件，当需

图 8-36　上传结果

要上传大文件的时候尤需如此，因为它带有一个缓冲区。我们只需要在请求路径前加上 /zuul 就可以使用 Servlet 了，官方对该问题的描述如图 8-37 所示。

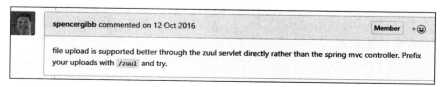

图 8-37　官方对乱码解决方案描述

比如将上一节的例子路径改为：http://localhost:5555/zuul/upload

8.7 Spring Cloud Zuul 实用小技巧

8.7.1 饥饿加载

Zuul 内部默认使用 Ribbon 来调用远程服务，所以由于 Ribbon 的原因，在我们部署好所有组件之后，第一次经过 Zuul 的调用往往会去注册中心读取服务注册表，初始化 Ribbon 负载信息，这是一种懒加载策略，但是这个过程是极其耗时的，尤其是服务过多的时候。为了避免这个问题，我们可以在启动 Zuul 的时候就饥饿加载应用程序上下文信息，开启饥饿加载只需添加配置，如代码清单 8-29 所示：

代码清单8-29　ch8-7\ch8-7-zuul-server\src\main\resources\bootstrap.yml

```yml
#开启饥饿加载
zuul:
    ribbon:
        eager-load:
            enabled: true
```

8.7.2 请求体修改

笔者曾经遇到过这样一种需求：在客户端对 Zuul 发送 post 请求之后，由于某些原因，在请求到下游服务之前，需要对请求体进行修改，常见的是对 form-data 参数的增减，对 application/json 的修改，对请求体做 UpperCase 等。在 Zuul 中可以很好地解决这种需求，通过新增一个 PRE 类型的 Filter 对请求体进行修改，如代码清单 8-30 所示：

代码清单8-30　ch8-7\ch8-7-zuul-server\src\main\java\cn\springcloud\book\filter\ModifyRequestEntityFilter.java

```java
@Configuration
public class ModifyRequestEntityFilter extends ZuulFilter {

    @Override
    public String filterType() {
        return PRE_TYPE;
    }

    @Override
    public int filterOrder() {
        return PRE_DECORATION_FILTER_ORDER + 1;// =6
    }
```

```java
@Override
public boolean shouldFilter() {
    return true;
}

@Override
public Object run() throws ZuulException {
    RequestContext ctx = RequestContext.getCurrentContext();
    HttpServletRequest request = ctx.getRequest();
    request.getParameterMap();
    Map<String, List<String>> requestQueryParams = ctx.getRequestQueryParams();
    if (requestQueryParams == null){
        requestQueryParams = new HashMap<>();
    }
    //这里添加新增参数的value,注意,只取list的0位
    ArrayList<String> arrayList = new ArrayList<>();
    arrayList.add("1wwww");
    requestQueryParams.put("test", arrayList);
    ctx.setRequestQueryParams(requestQueryParams);
    return null;
}
}
```

由于在 Zuul 中有 Filter（FormBodyWrapperFilter）会对请求体做封装，因此在编写此 Filter 的时候应当把它的执行次序放在该 Filter 之后，为了稳妥起见，把 ModifyRequestEntityFilter 的次序设置为 PRE 类型 Filter 的最后一级。例子 ModifyRequestEntityFilter 中做的事情就是对 form-data 新增一个参数 weight=140，在 shouldFilter() 方法中可以设定该 Filter 执行的特定条件。

8.7.3 使用 okhttp 替换 HttpClient

我们知道，在 Spring Cloud 中各个组件之间使用的通信协议都是 HTTP，而 HTTP 客户端使用的是 Apache 公司的 HttpClient，但是由于其难以扩展等诸多原因，已被许多技术栈弃用。Square 公司开发的 okhttp 正在逐渐被接受，它具有以下优点：

- 支持 SPDY，可以用于合并多个对于同一个 HOST 的请求。
- 使用连接复用机制有效减少资源消耗。
- 使用 GZIP 压缩减少数据量传输。
- 对响应的缓存，避免重复的网络请求。

在 Zuul 中使用 okhttp 替换 HttpClient，首先需要在 pom.xml 中增加 okhttp 的依赖包，如代码清单 8-31 所示：

代码清单8-31　ch8-7\ch8-7-zuul-server\pom.xml

```xml
<dependency>
    <groupId>com.squareup.okhttp3</groupId>
    <artifactId>okhttp</artifactId>
</dependency>
```

然后在配置文件中禁用 HttpClient 并开启 okhttp 即可，如代码清单 8-32 所示：

代码清单8-32　ch8-7\ch8-7-zuul-server\src\main\resources\bootstrap.yml

```yml
#禁用HttpClient并开启okhttp
ribbon:
    httpclient:
        enabled: false
    okhttp:
        enabled: true
```

8.7.4　重试机制

在 Spring Cloud 中有多种发送 HTTP 请求的方式可以与 Zuul 结合，Ribbon、Feign 或者 RestTemplate，但是无论选择哪种，都可能出现请求失败的情况，这在复杂的互联网环境是不可避免的。Zuul 作为一个网关中间件，在出现偶然请求失败时进行适当的重试是十分必要的，重试可以有效地避免一些突发原因引起的请求丢失。Zuul 中的重试机制是配合 Spring Retry 与 Ribbon 来使用的，因此需要在 pom.xml 引入 Spring Retry 的依赖包，如代码清单 8-33 所示：

代码清单8-33　ch8-7\ch8-7-zuul-server\pom.xml

```xml
<dependency>
    <groupId>org.springframework.retry</groupId>
    <artifactId>spring-retry</artifactId>
</dependency>
```

随后需要对重试进行具体配置，如代码清单 8-34 所示：

代码清单8-34　ch8-7\ch8-7-zuul-server\src\main\resources\bootstrap.yml

```yml
zuul:
    retryable: true #开启重试，D版之后默认为false，需要手动开启
#重试机制配置
ribbon:
    ConnectTimeout: 3000
    ReadTimeout: 60000
    MaxAutoRetries: 1 #对第一次请求的服务的重试次数
    MaxAutoRetriesNextServer: 1 #要重试的下一个服务的最大数量（不包括第一个服务）
    OkToRetryOnAllOperations: true
spring:
    cloud:
```

```yaml
    loadbalancer:
      retry:
        enabled: true #内部默认已开启，这里列出来说明这个参数比较重要
```

也可以对单个映射规则进行重试：zuul.routes.<route>.retryable=true

需要注意的是，在某些对幂等要求比较高的场景下，要慎用重试机制，因为如果没有相关处理的话，出现幂等问题是十分有可能的。

8.7.5　Header 传递

笔者之前在用 Zuul 构建网关的时候，遇到这样一个问题：在 Zuul 中对请求做了一些处理，需要把处理结果发给下游服务，但是又不能影响请求体的原始特性，这个问题该怎么解决好呢？后来在翻阅 Zuul 中一个重要的类 RequestContext 的时候，看到一个方法正好可以用来解决此问题，官方称之为 Header 的传递，如代码清单 8-35 所示：

代码清单8-35　ch8-7\ch8-7-zuul-server\src\main\java\cn\springcloud\book\filter\HeaderDeliverFilter.java

```java
@Configuration
public class HeaderDeliverFilter extends ZuulFilter {
...
    @Override
    public Object run() throws ZuulException {
        RequestContext context = RequestContext.getCurrentContext();
        context.addZuulRequestHeader("result", "to next service");
        return null;
    }
}
```

这样就可以动态增加一个 header 来传递给下游服务使用了，十分实用，这种使用方式与敏感头相反，需要注意传递信息的安全性。

8.7.6　整合 Swagger2 调试源服务

Swagger 是一个可视化的 API 测试工具，可以和应用完美融合，通过声明接口注解的方式，方便快捷地获取 API 调试界面，使得前后端可以很好地对接，极大地减少了后端人员编写接口文档的时间。

我们在 Spring Boot 构建的后端应用中常常看见它的身影，但是在 Spring Cloud 生态中有一个尴尬的事。由于 Zuul 的内部源服务可能有非常复杂的结构，抑或是 Zuul 中有鉴权功能，使得经过 Zuul 的 API 并不好调试。其实我们大可以在 Zuul 中整合 Swagger2 来实现对源服务 API 的调试。集成方式很简单，只需要在 Zuul 中加入对 Swagger2 的依赖，并且添加相关的配置即可，而源服务只需要像 Spring Boot 中整合 Swagger2 那样，配置 controller 扫描包路径即可。下面贴上 Zuul 中的核心配置，如代码清单 8-36 所示：

代码清单8-36　ch8-7\ch8-7-zuul-server\src\main\java\cn\springcloud\book\config\SwaggerConfig.java

```java
@Configuration
@EnableSwagger2
public class SwaggerConfig {

    @Autowired
    ZuulProperties properties;

    @Primary
    @Bean
    public SwaggerResourcesProvider swaggerResourcesProvider() {
        return () -> {
            List<SwaggerResource> resources = new ArrayList<>();
            properties.getRoutes().values().stream()
                    .forEach(route -> resources
                            .add(createResource(
                                    route.getServiceId(), route.getServiceId(), "2.0")));
            return resources;
        };
    }

    private SwaggerResource createResource(String name, String location, String version) {
        SwaggerResource swaggerResource = new SwaggerResource();
        swaggerResource.setName(name);
        swaggerResource.setLocation("/" + location + "/v2/api-docs");
        swaggerResource.setSwaggerVersion(version);
        return swaggerResource;
    }
}
```

效果页面如图 8-38 所示。

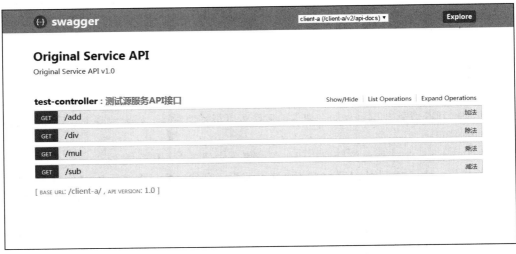

图 8-38　Zuul 中 Swagger2 生成的源服务接口

8.8 本章小结

本章主要讲解了 Spring Cloud Zuul 在应用层面上的一些实用技巧，包括：Filter 链的原理与使用、权限集成、应用限流、动态路由、灰度发布、文件上传以及一些小技巧。本章是在第 7 章的基础上进行扩展的，通过本章的学习，你将可以利用 Zuul 构建起一个功能较完善的网关应用。笔者还是提醒一下，Zuul 的 Filter 链非常重要，它是一个网关处理业务的灵魂所在，基于它还可以做更多的事情。本章的示例代码见代码部分 ch8-1 到 ch8-7。下一章，笔者将详细剖析 Zuul 的高级应用特性以及重要源码。

第 9 章

Spring Cloud Zuul 高级篇

第 7 章与第 8 章详述了当前各大企业对 Zuul 的实践经验，作为一个网关中间件，应付各种复杂场景是 Zuul 的使命，它整合的组件纷繁复杂，在我们受益于它的丰富功能的同时，许多"硬伤"也不得不面对。本章将详细介绍 Zuul 更深层次的知识，与上层负载均衡器的搭配，饱受诟病的性能问题，以及调优建议，最后是核心源码解析，进而了解它的工作原理，对每个请求的来龙去脉做到了然于胸。学习完本章后，相信你对 Zuul 的使用上会更加得心应手。

9.1　Spring Cloud Zuul 多层负载

9.1.1　痛点场景

在 Spring Cloud 微服务架构体系中，所有请求的前门的网关 Zuul 承担着请求转发的主要功能，对后端服务起着举足轻重的作用。当业务体量猛增之后，得益于 Spring Cloud 的横向扩展能力，往往加节点、加机器就可以使得系统支撑性获得大大提升，但是仅仅加服务而不加网关是会有性能瓶颈的，笔者在之前的实践经验是，单一 Zuul 的处理能力十分有限，因此扩张节点往往是服务连带 Zuul 一起扩张，然后再在请求上层加一层软负载，通常是使用 Nginx（Nginx 均分请求到 Zuul 负载层，"完美"地解决了问题），如图 9-1 所示。

但好景不长，在维持线上正常运行半个月后，不知什么原因，其中一台 Zuul 服务挂掉了，监控显示有四分之一的请求失败，原因很简单，从 Nginx 到 Zuul 其实没有什么关联性，如果服务宕掉，Nginx 也还是会把请求分过来，在 Nginx 没有采取相关应对措施的情况下，这是十分严重的问题。

9.1.2　解决方案

OpenResty 整合了 Nginx 与 Lua，实现了可伸缩的 Web 平台，内部集成了大量精良的 Lua

库、第三方模块以及多数的依赖项。能够非常快捷地搭建处理超高并发、扩展性极高的动态 Web 应用、Web 服务和动态网关。我们可以使用 Lua 脚本模块与注册中心构建一个服务动态增减的机制，通过 Lua 获取注册中心状态为 UP 的服务，动态地加入到 Nginx 的均衡列表中去，由于这种架构模式涉及了不止一个负载均衡器，我们称其为"多层负载"，如图 9-2 所示。

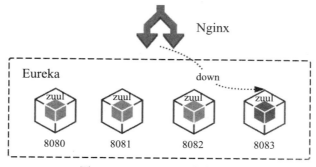

图 9-1　Nginx+Zuul 痛点场景

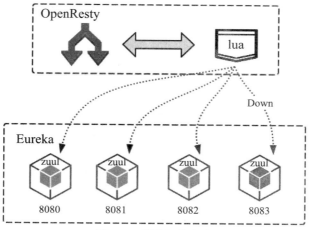

图 9-2　弹性解决方案

目前 Spring Cloud 中国社区针对这一场景开源了相关的 Lua 插件源码，GitHub 地址：https://github.com/SpringCloud/nginx-zuul-dynamic-lb。

相关典型配置如下：

```
http {
    #缓存区大小
    lua_shared_dict dynamic_eureka_balancer 128m;
    init_worker_by_lua_block {
        -- 引入Lua插件文件
        local file = require "resty.dynamic_eureka_balancer"
        local balancer = file:new({dict_name="dynamic_eureka_balancer"})
        -- Eureka地址列表
        balancer.set_eureka_service_url({"127.0.0.1:8888", "127.0.0.1:9999"})
        -- 配置初始被监听的服务名
```

```
            balancer.watch_service({"zuul", "client"})
        }
        upstream springcloud_cn {
            server 127.0.0.1:666;
            balancer_by_lua_block {
                -- 配置需要动态变化的服务名
                local service_name = "zuul"
                local file = require "resty.dynamic_eureka_balancer"
                local balancer = file:new({dict_name="dynamic_eureka_balancer"})
                --balancer.ip_hash(service_name) --IP Hash LB
                balancer.round_robin(service_name) --Round Robin LB
            }
        }
        server {
            listen       80;
            server_name  localhost;
            location / {
                proxy_pass  http://springcloud_cn/;
                proxy_set_header Host $http_host;
                proxy_set_header X-Real-IP $remote_addr;
                proxy_set_header X-Forwarded-For $proxy_add_x_forwarded_for;
                proxy_set_header X-Forwarded-Proto  $scheme;
            }
        }
    }
```

实现原理是使用 Lua 脚本定时根据配置的服务名与 Eureka 地址，去拉取该服务的信息，在 Eureka 里面提供 /eureka/apps/{serviceId} 端点，返回服务的注册信息，所以我们只需要取用状态为 "UP" 的服务，将它的地址加入 Nginx 负载列表即可。此项目使得 Nginx 与 Zuul 之间拥有一个动态感知能力，不用手动配置 Nginx 负载与 Zuul 负载，这样对于应用弹性扩展是极其友好的。

9.2 Spring Cloud Zuul 应用优化

9.2.1 概述

Zuul（这里指 Zuul1.0 版本，Zuul2.0 版本之后使用异步非阻塞模型）在给我们的微服务体系带来诸多便利的同时，也饱受着性能的争议，这一切还要从它的底层架构说起。Zuul 是建立在 Servlet 上的同步阻塞架构，所以在处理逻辑上面是和线程密不可分的，每一次请求都需要从线程池获取一个线程来维持 I/O 操作，路由转发的时候又需要从 HTTP 客户端获取线程来维持连接，这就会导致一个组件占用两个线程资源的情况。所以，在 Zuul 的使用中，对这部分的优化是很有必要的，一个好的优化体系会使得应用支撑的业务体量更大，也能最大化利用服务器资源。

在这里，笔者结合个人关于 Zuul 的实践经验，将对 Zuul 的优化分为以下几个类型。

❑ 容器优化：内置容器 Tomcat 与 Undertow 的比较与参数设置。

- 组件优化：内部集成的组件优化，如 Hystrix 线程隔离、Ribbon、HttpClient 与 OkHttp 选择。
- JVM 参数优化：适用于网关应用的 JVM 参数建议。
- 内部优化：一些内部原生参数，或者内部源码，以一种更恰当的方式重写它们。

本节就详细围绕这几种优化方式，给读者讲述更高效率运用 Zuul 的技巧。

9.2.2 容器优化

关于 Spring Boot 优化的文章，网上有很多，不过大部分都会提到把默认的内嵌容器 Tomcat 替换成 Undertow，那么 Undertow 是什么？Undertow 翻译为"暗流"，即平静的湖面下暗藏着波涛汹涌，所以 JBoss 公司取其意，为它的轻量级高性能容器命名。Undertow 提供阻塞或基于 XNIO 的非阻塞机制，它的包大小不足 1MB，内嵌模式运行时的堆内存占用只有 4MB 左右。要使用 Undertow 只需要在配置文件中移除 Tomcat，添加 Undertow 的依赖即可：

```xml
<dependency>
    <groupId>org.springframework.boot</groupId>
    <artifactId>spring-boot-starter-web</artifactId>
    <exclusions>
        <exclusion>
            <groupId>org.springframework.boot</groupId>
            <artifactId>spring-boot-starter-tomcat</artifactId>
        </exclusion>
    </exclusions>
</dependency>
<dependency>
    <groupId>org.springframework.boot</groupId>
    <artifactId>spring-boot-starter-undertow</artifactId>
</dependency>
```

Undertow 的主要参数配置如表 9-1 所示。

表 9-1　Undertow 参数说明

配　置　项	默　认　值	说　明
server.undertow.io-threads	Math.max(Runtime.getRuntime().availableProcessors(), 2);	设置 IO 线程数，它主要执行非阻塞的任务，它们会负责多个连接，默认设置每个 CPU 核心有一个线程。不要设置过大，如果过大，启动项目会报错：打开文件数过多
server.undertow.worker-threads	ioThreads * 8	阻塞任务线程数，当执行类似 Servlet 请求阻塞 IO 操作，Undertow 会从这个线程池中取得线程。它的值设置取决于系统线程执行任务的阻塞系数，默认值是 IO 线程数 *8
server.undertow.direct-buffers	取决于 JVM 最大可用内存大小（long maxMemory = Runtime.getRuntime().maxMemory();），小于 64MB 默认为 false，其余为 true	是否分配直接内存（NIO 直接分配的堆外内存）

(续)

配 置 项	默 认 值	说 明
server.undertow.buffer-size	最大可用内存 <64MB：512 字节 64MB <= 最大可用内存 <128MB：1024 字节 128MB < 最大可用内存：1024*16-20 字节	每块 buffer 的空间大小，空间越小利用越充分，不要设置太大，以免影响其他应用，合适即可
server.undertow.buffers-per-region	最大可用内存 <64MB：10 64MB <= 最大可用内存 <128MB：10 128MB <= 最大可用内存：20	每个区域分配的 buffer 数量，所以 pool 的大小是 buffer-size * buffers-per-region

关于阻塞任务线程数 worker-threads 的设置，这里提一下，其默认是 IO 线程数乘以 8，关于这个问题笔者曾经到 JBoss 的 Issue 咨询过，为什么是乘以 8，而不是 9 或者 10，又或者其他倍数？开发者给出的答案是：Performance testing showed that this was a reasonable default for most work loads。意思就是 8 是官方经过反复测试之后取的一个合理的默认值，对此，笔者有一点拙见，在《Java 虚拟机并发编程》中提到，关于 IO 密集型应用（我们的网关应用就是 IO 密集型应用）的最佳线程数设置，有一个计算公式：

$$最佳线程数 = CPU 核数 / (1 - 阻塞系数)$$

IO 密集型应用的阻塞系数一般在 0.8~0.9 之间，由于 io-threads 取的是 CPU 核数，worker-threads = io-threads * 8，代入公式得到阻塞系数等于 0.875，可以得知，JBoss 团队的测试环境应用线程的阻塞系数就为 0.875，这是一个比较中肯的值，在有条件的话，尽量根据测试得出的线程在 IO 操作上的时间与 CPU 密集任务所消耗的比值来确定，建议在 Zuul 下游是静态资源的情况下设置稍小，在与数据库交互的情况下设置稍大。

9.2.3 组件优化

在 Spring Cloud 微服务体系中，有一个容易被忽略，但是集成组件最多，功能最强大的组件——Zuul。Zuul 网关主要用于智能路由，同时也支持认证、区域和内容感知路由，将多个底层服务聚合成统一的对外 API。所以要更好地使用 Zuul，就免不了要对它集成的组件进行优化，使它可以更好地支撑我们的服务集群。

1. Hystrix

由于在 Zuul 中默认集成了 Hystrix 熔断器，使得网关应用具有弹性、容错的能力。但是如果使用缺省的配置，可能会遇到种种问题，其中最常见的问题就是当我们启动 Zuul 应用之后，第一次请求往往会失败，这很大程度上是因为预热导致的。因为在第一次请求的时候，Zuul 内部要初始化很多类信息，这是十分耗时的，而 Hystrix 就恰恰对这个时间不太买账，因此超过它的缺省超时时间 1000ms 后，就会返回错误信息。解决方式有两种，第一种是加大超时时间：

```
hystrix.command.default.execution.isolation.thread.timeoutInMilliseconds=5000
```

第二种是直接禁用掉 Hystrix 的超时时间：

```
hystrix.command.default.execution.timeout.enabled=false
```

Zuul 中关于 Hystrix 的配置还有一个很重要的点，那就是 Hystrix 的线程隔离模式的选择，包括线程池隔离模式（THREAD）或者信号量隔离模式（SEMAPHORE）。在网关中，对资源的使用是应该受到严格控制的，如果不加限制，会导致资源滥用，在恶劣的线上环境下就容易引起雪崩。两种模式的说明以及特定场景下的选择如表 9-2 所示。

表 9-2　Hystrix 两种线程隔离方式对比

	线程池模式（THREAD）	信号量模式（SEMAPHORE）
官方推荐	是	否
线程	与请求线程分离	与请求线程共用
开销	上下文切换频繁，较大	较小
异步	支持	不支持
应对并发量	大	小
适用场景	外网交互	内网交互

切换隔离模式的配置方式：

```
hystrix.command.default.execution.isolation.strategy=Thread|Semaphore
```

总结一下，当应用需要与外网交互，由于网络开销比较大与请求比较耗时，这时选用线程隔离策略，可以保证有剩余的容器（Tomcat & Undertow & Jetty）线程可用，而不会由于外部原因使得线程一直处于阻塞或等待状态，可以快速失败返回。但当我们的应用只在内网交互，并且体量比较大，这时使用信号量隔离策略就比较好，因为这类应用的响应通常会非常快（由于在内网），不会占用容器线程太长时间，使用信号量线程上下文就会成为一个瓶颈，可以减少线程切换的开销，提高应用运转的效率，也可以起到对请求进行全局限流的作用。

2. Ribbon

在 8.7.4 节讲过 Zuul 结合 Ribbon 实现超时重试的机制，这里再说明一下：

```
ribbon:
    ConnectTimeout: 3000
    ReadTimeout: 60000
    MaxAutoRetries: 1 #对第一次请求的服务的重试次数
    MaxAutoRetriesNextServer: 1 #要重试的下一个服务的最大数量（不包括第一个服务）
    OkToRetryOnAllOperations: true
```

配置当中的 ConnectTimeout 与 ReadTimeout 是当 HTTP 客户端使用 HttpClient 的时候生效的，这个超时时间最终会被设置到 HttpClient 中去。在设置的时候要结合 Hystrix 的超时时间来综合考虑，针对应用场景，设置太小会导致很多请求失败，设置太大会导致熔断功能控制性变差，所以需要经过压力测试得来。

9.2.4　JVM 参数优化

凡是 Java 应用，都免不了 JVM 调优这个过程。因为 JVM 构造的特殊性，加之多种可选的优化策略，使得应用的性能充满了不确定性，一次好的调优，宏观上可能会提升应用的吞吐量或者大大减小响应时间，反之如果不优化，应用不仅发挥不出最佳性能，可能还会因为某个区承受不住压力而使得应用挂掉，所以 JVM 调优是一个应用上线的必经之路。

对于 Zuul 来说，它起到的作用是"网关"，我们能够想到，网关最需要的就是吞吐量，所以优化应以此为切入点来综合考虑。这里推荐使用 Parallel Scavenge 收集器，注意它有一个参数 -XX:+UseAdaptiveSizePolicy，如果打开，JVM 会自动选择年轻代区大小和相应的 Survivor 区比例，以达到目标系统规定的最低响应时间或者 YGC 频率，官方建议在使用并行收集器的时候一直打开，但是笔者在压力测试中观察到效果并不理想，所以笔者的建议是将它关闭，改为 -XX:-UseAdaptiveSizePolicy，再根据实际情况调整 Eden 区与 Survivor 区的比例。这里还有一个参数比较管用 -XX:TargetSurvivorRatio，即 Survivor 区的对象利用率，默认是 50%，建议稍微加大，加大后 YGC 会看到比较明显的效果。另外，让垃圾对象尽量在新生代被回收掉，以免进入老年代触发 FGC（这也是一个技巧），这里使用的策略是在有限的堆空间下使用一个较大的新生代，并且 Eden 区也要比 Survivor 区大。老年代使用 Parallel Old 收集器，让网关应用彻底面向吞吐量。参数 -XX:+ScavengeBeforeFullGC，即 FGC 前先进行一次 YGC，推荐使用这个参数。

JVM 参数优化没有一个固定的值，在不同的压力环境下，不同的 JDK 版本下的选择也不一样，上文比较适合的是 JDK1.7 版本与 JDK1.8 版本，如果是 JDK1.9 版本之后，建议使用 G1。JVM 优化是一个繁琐的过程，这其中可能要反反复复折腾很多次，笔者的心得是，多尝试，在一定的理论基础上敢于使用参数，就一定能够调出一套适合你的应用的 JVM 参数。

最后，说一下 Spring Boot 以 jar 包模式启动时的 JVM 参数设置方法：java -JVM 参数 -jar application.jar

GC 收集器搭配（有连线的表示可以搭配使用），如图 9-3 所示。

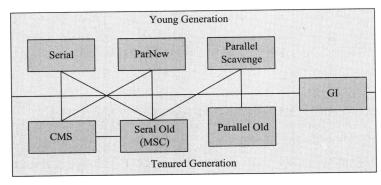

图 9-3　JVM 垃圾回收器搭配

9.2.5　内部优化

在官方文档中，Zuul 部分开篇讲了 zuul.max.host.connections 属性拆解成了 zuul.host.maxTotalConnections（服务 HTTP 客户端最大连接数）与 zuul.host.maxPerRouteConnections（每个路由规则 HTTP 客户端最大连接数），默认值分别为 200 与 20，如果使用 HttpClient 的时候

则有效，如果使用 OkHttp 则无效（替换方式见 8.7.3 节）。

前面讲过，Zuul 中有些 Filter 的设计并不是很合理，我们可以选择自实现替换掉它或者禁用：

zuul.<SimpleClassName>.<filterType>.disable=true

在 Zuul 中还有一个超时时间，使用 serviceId 映射与 url 映射的设置是不一样的，如果使用 serviceId 映射，ribbon.ReadTimeout 与 ribbon.SocketTimeout 生效；如果使用 url 映射，应该设置 zuul.host.connect-timeout-millis 与 zuul.host.socket-timeout-millis 参数。

9.3　Spring Cloud Zuul 原理 & 核心源码解析

9.3.1　工作原理与生命周期

在第 7 章开篇讲过，Zuul 本是由 Netflix 公司开发并开源，后经 Pivotal 公司整合进入 Spring Cloud 生态，使得我们可以通过引入相关依赖，添加一个简单的注解便可以创建一个 Zuul Server，如此便捷的操作与强大的封装让人想对它的运行原理一探究竟。

Zuul 的底层是由 Servlet 与一系列 Filter 组合而成，各个组件协同配合，让 Zuul 功能强大，且易于扩展。官方有一张框架图可以很好地描述 Zuul 的工作原理，如图 9-4 所示。

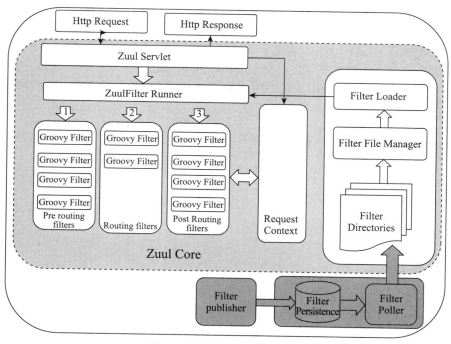

图 9-4　Zuul 官方生命周期图

由图 9-4 可以得知，ZuulServlet 通过 RequestContext 统管着由许多 Filter 组成的核心部件，所有操作都与 Filter 息息相关。请求、ZuulServlet 与 Filter 共同构建起了 Zuul 的运行时生命周期，如图 9-5 所示。

Zuul 的请求引入还是来自于 DispatcherServlet，然后交由 ZuulHandlerMapping（由 AbstractUrlHandlerMapping 继承得来，它是 Spring MVC 对 request 进行分发的一个重要的抽象类）处理由初始化得来的路由定位器 RouteLocater，为后续的请求分发做好准备，同时整合了基于事件从服务中心拉取服务列表的机制；进入 ZuulController，它的主要职责是初始化 ZuulServlet 以及继承 ServletWrappingController，通过重写 handleRequest 方法来将 ZuulServlet 引入生命周期，之后所有的请求都会经过 ZuulServlet；当请求进入 ZuulServlet 之后，第一次调用会初始化 ZuulRunner，非第一次调用就按 Filter 链 order 顺序执行；ZuulRunner 中将请求和响应初始化为 RequestContext，包装成 FilterProcessor 转化为调用 preRoute()、route()、postRoute() 和 error() 方法；最后便是我们的各个 Filter 执行逻辑了，原始请求在 Filter 链中经过种种变换，最后得到预期的结果。

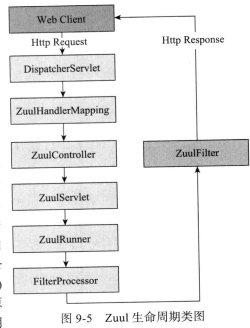

图 9-5　Zuul 生命周期类图

当我们在启动类中添加了 @EnableZuulProxy 或者 @EnableZuulServer 之后，底层发生了什么？各个版本的实现有些许不一样，当前是基于 Zuul 1.3.1 版本。打开第 1 个注解：

```
@EnableCircuitBreaker
@Target(ElementType.TYPE)
@Retention(RetentionPolicy.RUNTIME)
@Import(ZuulProxyMarkerConfiguration.class)
public @interface EnableZuulProxy {
}
```

打开第 2 个注解：

```
@Target(ElementType.TYPE)
@Retention(RetentionPolicy.RUNTIME)
@Documented
@Import(ZuulServerMarkerConfiguration.class)
public @interface EnableZuulServer {
}
```

从这里可以初步看出这两个注解的不同之处，@EnableZuulProxy 整合了 Hystrix 的断路器功能，而 @EnableZuulServer 并未整合。除此之外，不同之处就在于引入的 ZuulProxyMarkerConfiguration 与 ZuulServerMarkerConfiguration 了，它们的作用是什么呢？分别打开源码：

源码一：

```
@Configuration
public class ZuulProxyMarkerConfiguration {
```

```
    @Bean
    public Marker zuulProxyMarkerBean() {
        return new Marker();
    }

    class Marker {
    }
}
```

源码二：

```
@Configuration
public class ZuulServerMarkerConfiguration {
    @Bean
    public Marker zuulServerMarkerBean() {
        return new Marker();
    }

    class Marker {
    }
}
```

这里仅仅只是注入了两个 Bean，并无特别之处，查看注释之中的 @link，分别指向 ZuulProxyAutoConfiguration 与 ZuulServerAutoConfiguration，打开两者源码。

源码一：

```
@Configuration
@Import({ RibbonCommandFactoryConfiguration.RestClientRibbonConfiguration.class,
        RibbonCommandFactoryConfiguration.OkHttpRibbonConfiguration.class,
        RibbonCommandFactoryConfiguration.HttpClientRibbonConfiguration.class,
        HttpClientConfiguration.class })
@ConditionalOnBean(ZuulProxyMarkerConfiguration.Marker.class)
public class ZuulProxyAutoConfiguration extends ZuulServerAutoConfiguration {
    ...
}
```

源码二：

```
@Configuration
@EnableConfigurationProperties({ ZuulProperties.class })
@ConditionalOnClass(ZuulServlet.class)
@ConditionalOnBean(ZuulServerMarkerConfiguration.Marker.class)
public class ZuulServerAutoConfiguration {
    ...
}
```

原来它们是通过把 ZuulProxyMarkerConfiguration 与 ZuulServerMarkerConfiguration 的注入 Bean 当作开关（@ConditionalOnBean 的意思是，如果 Spring 上下文中存在注解中的 Bean，才创建当前 Bean），来开启两种初始化配置方式，两种注解方式决定了 Zuul 的功能，@EnableZuulServer 的底层初始化类 ZuulServerAutoConfiguration 的功能是：

- 初始化配置加载器。
- 初始化路由定位器。
- 初始化路由映射器。

- 初始化配置刷新监听器。
- 初始化 ZuulServlet 加载器。
- 初始化 ZuulController。
- 初始化 Filter 执行解析器。
- 初始化一些 Filter。
- 初始化 Metrix 监控。

而 @EnableZuulProxy 的底层初始化类 ZuulProxyAutoConfiguration 是由 ZuulServerAutoConfiguration 扩展得来，所以它具有 ZuulServerAutoConfiguration 的全部功能，并且新增了如下功能：

- 初始化服务注册、发现监听器。
- 初始化服务列表监听器。
- 初始化 Zuul 自定义 Endpoint。
- 初始化一些 ZuulServerAutoConfiguration 中没有的 Filter。
- 引入 HTTP 客户端的两种方式：HttpClient 与 OkHttp。

装载完成这些组件之后，Zuul 便工作了起来，形成了一个完整的生命周期。

9.3.2 Filter 装载与 Filter 链实现

1. Filter 装载

Zuul 中的 Filter 必须经过初始化装载，才能在请求中发挥作用，这其中的过程如图 9-6 所示。

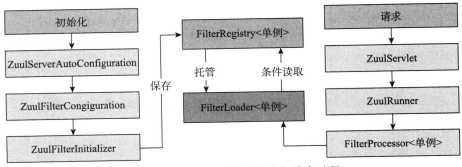

图 9-6 Zuul Filter 链初始化与请求过程

在初始化的时候，ZuulServerAutoConfiguration 中有一个配置类叫作 ZuulFilterConfiguration，它是 Zuul 的 Filter 初始化的入口：

```
@Configuration
protected static class ZuulFilterConfiguration {

    @Autowired
    private Map<String, ZuulFilter> filters;

    @Bean
```

```java
public ZuulFilterInitializer zuulFilterInitializer(
        CounterFactory counterFactory, TracerFactory tracerFactory) {
    FilterLoader filterLoader = FilterLoader.getInstance();
    FilterRegistry filterRegistry = FilterRegistry.instance();
    return new ZuulFilterInitializer(this.filters, counterFactory, tracer-
        Factory, filterLoader, filterRegistry);
}
```

它把 Filter 注入进来，获取 FilterRegistry（可以看成 Filter 的仓库）与 FilterLoader（FilterRegistry 与外界交互的媒介，可条件读取）的实例，注意，它们是单例对象，整个生命周期只存在一个实例，内部采用 ConcurrentHashMap 保存数据。然后将它们传入新构造的 ZuulFilterInitializer：

```java
@PostConstruct
public void contextInitialized() {
    log.info("Starting filter initializer");
    TracerFactory.initialize(tracerFactory);
    CounterFactory.initialize(counterFactory);
    for (Map.Entry<String, ZuulFilter> entry : this.filters.entrySet()) {
        filterRegistry.put(entry.getKey(), entry.getValue());
    }
}

@PreDestroy
public void contextDestroyed() {
    log.info("Stopping filter initializer");
    for (Map.Entry<String, ZuulFilter> entry : this.filters.entrySet()) {
        filterRegistry.remove(entry.getKey());
    }
    clearLoaderCache();
    TracerFactory.initialize(null);
    CounterFactory.initialize(null);
}

private void clearLoaderCache() {
    Field field = ReflectionUtils.findField(FilterLoader.class, "hashFiltersByType");
    ReflectionUtils.makeAccessible(field);
    @SuppressWarnings("rawtypes")
    Map cache = (Map) ReflectionUtils.getField(field, filterLoader);
    cache.clear();
}
```

这里用到了两个注解，@PostConstruct 标注的方法表明在 Bean 初始化之前就把 Filter 的信息保存入 FilterRegistry，@PreDestroy 标注的方法表示在 Bean 销毁的时候清空 FilterRegistry 与 FilterLoader。在 FilterLoader 中，可以通过 Filter 名、Filter 的 class、Filter 的类型来查询得到相应的 Filter。

在请求上下文走到 ZuulServlet 的时候，会检查 ZuulRunner 是否初始化。如果没有，则执行 init 方法初始化 ZuulRunner 对象；如果已初始化，则进入 service 方法执行 Filter 链的逻辑。ZuulRunner 中主要是对 RequestContext 的初始化，将请求上下文放入生命周期，以及获取 FilterProcessor 的单例对象，它是 Zuul 中 Filter 的解释执行器。最后在 FilterProcessor 中通过 FilterLoader 获取相应 Filter 并初始化。至此，Filter 从配置到装载等待执行的过程已经全部完成。

2. Filter 链实现

Filter 链的实现分为同类型与不同类型，同类型的 Filter 执行次序在 FilterLoader 与 FilterProcessor 中被定义。FilterProcessor 会通过 FilterLoader 获取一个类型的 Filter，这些 Filter 会按照 order 排序，然后在 FilterProcessor 中依次迭代执行。

不同类型 Filter 的执行次序在 ZuulServlet 中被定义：

```
try {
            try {
                preRoute();
            } catch (ZuulException e) {
                error(e);
                postRoute();
                return;
            }
            try {
                route();
            } catch (ZuulException e) {
                error(e);
                postRoute();
                return;
            }
            try {
                postRoute();
            } catch (ZuulException e) {
                error(e);
                return;
            }
        } catch (Throwable e) {
            error(new ZuulException(e, 500, "UNHANDLED_EXCEPTION_" + e.getClass().
                getName()));
        }
```

工作原理已在 8.1.1 节解析，这里不再赘述。

9.3.3 核心路由实现

在 8.4.2 节中对实现动态路由相关的路由原理做了一部分讲解，本节将详细地讲解 Zuul 中的核心功能——路由，是如何实现的。

路由部分有一个顶级接口叫作 RouteLocator，所有关于路由的功能都是由它实现而来，其中定义了三个基本方法：获取忽略的 path 集合、获取路由列表、根据 path 获取路由信息。它的实现类图如图 9-7 所示。

SimpleRouteLocator 是对它的一个基本实现，主要功能是对 Zuul Server 的配置文件中路由规则的维护，它实现了 Ordered 接口，可以对定位器优先级进行设置。Spring 是一个大量使用策略模式的框架，在策略模式下接口的实现类有一个优先级问题，Spring 就是通过 Ordered

接口来实现这个优先级的。配置文件的映射规则最终会被转换为一个 Map 结构，key 是路径，value 是对规则的封装：

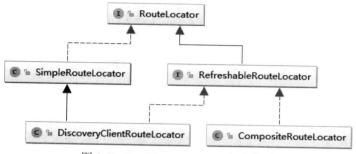

图 9-7　Zuul RouteLocator 实现类图

```
public static class ZuulRoute {

    private String id;

    private String path;

    private String serviceId;

    private String url;

    private boolean stripPrefix = true;

    private Boolean retryable;

    private Set<String> sensitiveHeaders = new LinkedHashSet<>();

    private boolean customSensitiveHeaders = false;
//省略get/set方法
}
```

RefreshableRouteLocator 扩展了 RouteLocator 接口，它在类 ZuulHandlerMapping 中才实质性生效，凡是实现了 RefreshableRouteLocator，都会被事件监听所刷新。

```
public class ZuulHandlerMapping extends AbstractUrlHandlerMapping {
    private volatile boolean dirty = true;

    public void setDirty(boolean dirty) {
        this.dirty = dirty;
        if (this.routeLocator instanceof RefreshableRouteLocator) {
            ((RefreshableRouteLocator) this.routeLocator).refresh();
        }
    }
}
```

DiscoveryClientRouteLocator 这个定位器的作用是整合配置文件与注册中心里的路由信息。它实现了 RefreshableRouteLocator，扩展了 SimpleRouteLocator，如果读者需要自定义路由加载规则，可以参照它的实现。

最后是 CompositeRouteLocator 定位器，这个定位器在 ZuulServerAutoConfiguration 中配置加载的时候，有一个注解很重要，即 @Primary，表示所有 RouteLocator 类型的 Bean 中，优先加载它，也就是说，所有的定位器都要到这里来装配，可以看作其他路由定位器的处理器。Zuul 通过它来将请求与路由规则进行关联，这个操作在 ZuulHandlerMapping 中：

```java
//省略部分代码
public class ZuulHandlerMapping extends AbstractUrlHandlerMapping {

    private final RouteLocator routeLocator;

    private final ZuulController zuul;

    public ZuulHandlerMapping(RouteLocator routeLocator, ZuulController zuul) {
        this.routeLocator = routeLocator;
        this.zuul = zuul;
        setOrder(-200);
    }

    @Override
    protected Object lookupHandler(String urlPath, HttpServletRequest request)
            throws Exception {
        if (this.dirty) {
            synchronized (this) {
                if (this.dirty) {
                    registerHandlers();
                    this.dirty = false;
                }
            }
        }
        return super.lookupHandler(urlPath, request);
    }

    private void registerHandlers() {
        Collection<Route> routes = this.routeLocator.getRoutes();
        else {
            for (Route route : routes) {
                registerHandler(route.getFullPath(), this.zuul);
            }
        }
    }
}
```

ZuulHandlerMapping 将映射规则交由 ZuulController 处理，ZuulController 又到 ZuulServlet 中处理，最后到达向异域或源服务发送 HTTP 请求的 route 类型的 Filter 中，默认有三种发送 HTTP 请求的 Filter：

- RibbonRoutingFilter：优先级是 10，使用 Ribbon、Hystrix 或者嵌入式 HTTP 客户端发送请求。
- SimpleHostRoutingFilter：优先级是 100，使用 Apache Httpclient 发送请求。
- SendForwardFilter：优先级是 500，使用 Servlet 发送请求。

9.4 本章小结

本章讲解了 Zuul 的一些高级应用场景以及核心原理解析。熟悉多层负载之后，可以结合 OpenResty 构建起一个从前端负载均衡到 Zuul 网关的弹性支撑系统；使用 Zuul 免不了要对它的某些特性进行优化，使我们的 Zuul 网关能够更好地工作；了解 Zuul 的核心功能实现（Filter 链与路由），可以对它进行二次开发，定制适合自己应用的独特功能。本书的 Zuul 篇到这里就告一段落了，相信会对目前大多数使用 Zuul 的公司有一定的参考意义，如今 Zuul 2.0 已经开源，笔者曾到官方提过 Issue 询问是否 Spring Cloud 会整合 Zuul 2.0，得到的答案是否定的。所以，Spring Cloud 新一代的网关中间件 Spring Cloud Gateway 必将大放异彩，详细内容请关注本书后面关于 Spring Cloud Gateway 的章节，想从 Zuul 迁移到 Spring Cloud Gateway，相信你会找到答案。

Chapter 10 第 10 章

Spring Cloud 基础综合案例

前面的章节分别介绍了 Spring Cloud 的组件和一些常用的场景和功能，这一章会把这些组件全部组合起来搭建一个简单的基础综合框架，即把这些组件整个串起来构成一个完整可用的 Spring Cloud 基础框架，并对其做一些简单的增强，希望能给广大开发者带来便利。

10.1 基础框架

10.1.1 搭建说明

现在利用常用的一些组件搭建一个基础框架，使用的组件包括 Eureka、Ribbon、Config、Zuul、Hystrix，完成一个用户信息管理的小服务，提供几个简单的接口来完成这些功能。

框架整体可采用前后端分离的架构，前台用 VUE 或者其他框架都可行，这里只实现后端框架。

后端框架包括注册中心 Eureka，配置中心 Spring Cloud Config，API 网关 Zuul，客户端负载均衡 Ribbon，断路器 Hystrix，同时后端包含两个业务服务，一个是用户服务 sc-user-service，一个是数据服务 sc-data-service。

从前端发起请求，根据接口获取相关的用户数据，如果用户服务有数据即刻返回，如果需要数据服务的数据，则调用数据服务的接口获取数据，组装后进行返回。

10.1.2 技术方案

技术方案的实现流程见图 10-1，用户从浏览器发起请求，经过浏览器，请求达到 Nginx，打开前台界面，由前台发起请求后台数据，当请求达到 Nginx 后，Nginx 对网关层进行负载，因为网关也需要做 HA，此时网关接收到请求后会根据请求路径进行动态路由，根据服务名发现是 UserService 中的服务，则从 Ribbon 中选择一台 UserService 的实例进行调用，由

UserService 返回数据，如果此时 UserService 需要使用第三方 DataService 的数据，则跟 Zuul 一样，选择一台 DataService 的实例进行调用，返回数据到前台即可渲染页面，流程结束。

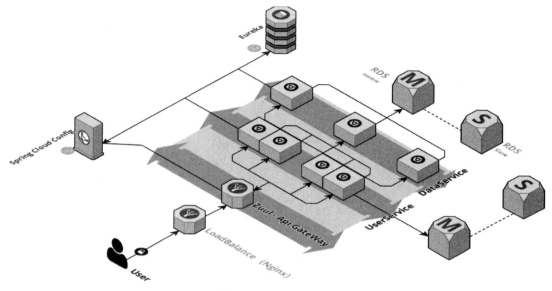

图 10-1　技术架构图

10.1.3　具体实现

1. 工程介绍

工程代码可见 ch10-1，具体工程结构如下，见表 10-1 所示。

表 10-1　项目工程介绍表

工　程　名	端　　口	描　　述
ch10-1	N/A	父工程
ch10-1-config-server	9090	配置中心
ch10-1-eureka-server	8761	注册中心
ch10-1-zuul-server	7777	API GateWay
ch10-1-hystrix-dashboard	9099	hystrix dashboard & Turbine
ch10-1-common	N/A	公共基础包，方便后台服务引用
ch10-1-user-service	9091	用户服务，对用户数据的操作
ch10-1-data-service	8099	数据服务，提供基础的数据

2. 业务接口

这里基于用户服务实现对相关数据的获取，写了 3 个简单接口，如代码清单 10-1 所示：

- 获取系统默认用户。
- 获取上下文用户。
- 根据用户名获取该用户的供应商数据。

代码清单10-1 ch10-1\ch10-1-user-service\src\main\java\cn\springcloud\book\user\controller\UserController.java

```java
/**
 * 获取配置文件中系统默认用户
 */
@GetMapping("/getDefaultUser")
public String getDefaultUser(){
    return userService.getDefaultUser();
}

/**
 * 获取上下文用户
 */
@GetMapping("/getContextUserId")
public String getContextUserId(){
    return userService.getContextUserId();
}

/**
 * 获取供应商数据
 */
@GetMapping("/getProviderData")
public List<String> getProviderData(){
    return userService.getProviderData();
}
```

3. 运行结果

1）按顺序启动 eureka、zuul-server、data-service、user-service，成功后浏览器访问 http://localhost:9091/getContextUserId，发现页面空白，控制台打印"the user is null, please access from gateway or check user info"，这说明拦截器起到了作用，对于没有用户信息这样不合法的请求进行了拦截，如代码清单 10-2 所示：

代码清单10-2 ch10-1\ch10-1-common\src\main\java\cn\springcloud\book\common\intercepter\UserContextInterceptor.java

```java
@Override
public boolean preHandle(HttpServletRequest request, HttpServletResponse respone,
    Object arg2) throws Exception {
    User user = new User(HttpConvertUtil.httpRequestToMap(request));
    if(StringUtils.isEmpty(user.getUserId()) && StringUtils.isEmpty(user.getUserName())) {
    log.error("the user is null, please access from gateway or check user info");
        return false;
    }
    UserContextHolder.set(user);
    return true;
}
```

2）从网关正确的访问 http://localhost:7777/sc-user-service/getContextUserId 这个地址，此时发现报错：

```
{"businessId":1,"exceptionType":"cn.springcloud.book.common.exception.BaseException","code":10001,"businessMessage":"the user is null, please check","codeEN":"AuthEmptyError"}
```

这是自定义了一个异常，没有传用户信息，因为这里在网关做了拦截，如果请求头里没有 x-customs-user 则鉴权不通过，如代码清单 10-3 所示：

代码清单10-3　ch10-1\ch10-1-zuul-server\src\main\java\cn\springcloud\book\filter\AuthFilter.java

```java
public final static String CONTEXT_KEY_USERID = "x-customs-user";
public static void authUser(RequestContext ctx) {
    HttpServletRequest request = ctx.getRequest();
    Map<String, String> header = httpRequestToMap(request);
    String userId = header.get(User.CONTEXT_KEY_USERID);
    if(StringUtils.isEmpty(userId)) {
        try {
BaseException BaseException = new BaseException(CommonError.AUTH_EMPTY_ERROR.getCode(),
            CommonError.AUTH_EMPTY_ERROR.getCodeEn(),
            CommonError.AUTH_EMPTY_ERROR.getMessage(),1L);
            BaseExceptionBody errorBody = new BaseExceptionBody(BaseException);
            ctx.setSendZuulResponse(false);
            ctx.setResponseStatusCode(401);
    ctx.setResponseBody(JSONObject.toJSON(errorBody).toString());
        } catch (Exception e) {
            logger.error("println message error",e);
        }
    }else {
        for (Map.Entry<String, String> entry : header.entrySet()) {
            ctx.addZuulRequestHeader(entry.getKey(), entry.getValue());
        }
    }
}
```

3）添加请求头 x-customs-user=spring，访问 http://localhost:7777/sc-user-service/getContextUserId，若请求成功，则返回正常的值，如图 10-2 所示，成功返回填写的上下文值为 spring。

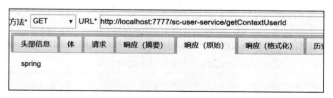

图 10-2　请求成功图

10.2　实战扩展

在实际项目开发中肯定会对基础框架做能力封装，这里也做了一些简单的封装，供大家参考，源代码包含在 ch10 里。

10.2.1 公共包（对象，拦截器，工具类等）

公司的框架一般会有一些值对象，拦截器，分页对象，权限等这些基础数据，并且其他的服务都是需要这些能力的，所以会抽取出这部分对象放入公共包里，供其他服务引用，在此建立了一个叫 common 的工程，然后其他的服务，如 user-service 和 data-service 都会用到它里面的一些组件和东西，这里面放的东西可以有这几部分，具体源码可参考 ch10-1\ch10-1-common。

10.2.2 用户上下文对象传递

在实际的项目开发工程中，很可能要对用户对象或者上下文对象进行传递，如用户对象，从 Zuul 网关一直到后面的微服务都需要用户对象，后面的微服务需要获取到用户 ID 后开展一些业务操作，所以需要此功能，实现的方案是：

1）在 Zuul 获取到用户信息，存入 header 头，后台服务进入方法前，获取到 header 进行组装 User 用户对象，后台服务通过 UserContextHolder 获取。

2）在后台服务之间相互调用时，增加拦截器，获取当前用户然后转换成 header 放入请求头，被调用服务拦截器拦截后解析到 header 放入上下文中，服务通过 UserContextHolder 获取。

其具体实现分为以下几个步骤：

1）在公共 SDK 里定义用户对象 User，如代码清单 10-4 所示：

代码清单10-4　ch10-1\ch10-1-common\src\main\java\cn\springcloud\book\common\vo\User.java

```
public final static String CONTEXT_KEY_USERID = "x-customs-user";
    /**
     * 用户ID
     */
    private String userId;

    private String userName;
```

2）增加三个拦截器：

a）FeignUserContextInterceptor：在使用 Feign 进行服务间调用时会拦截到请求，并将用户属性放到 header 里，如代码清单 10-5 所示：

代码清单10-5　ch10-1\ch10-1-common\src\main\java\cn\springcloud\book\common\intercepter\FeignUserContextInterceptor.java

```
public class FeignUserContextInterceptor implements RequestInterceptor {

    @Override
    public void apply(RequestTemplate template) {
        ServletRequestAttributes attributes = (ServletRequestAttributes) RequestContextHolder
                .getRequestAttributes();
        HttpServletRequest request = attributes.getRequest();
        Enumeration<String> headerNames = request.getHeaderNames();
```

```java
        if (headerNames != null) {
            while (headerNames.hasMoreElements()) {
                String name = headerNames.nextElement();
                String values = request.getHeader(name);
                template.header(name, values);
            }
        }
    }
}
```

b）RestTemplateUserContextInterceptor：在使用 RestTemplate 进行服务间调用时会拦截到请求，并将用户属性放到 header 里，如代码清单 10-6 所示：

代码清单10-6　ch10-1\ch10-1-common\src\main\java\cn\springcloud\book\common\intercepter\RestTemplateUserContextInterceptor.java

```java
public class RestTemplateUserContextInterceptor implements ClientHttpRequest-
    Interceptor {

    @Override
    public ClientHttpResponse intercept(HttpRequest request, byte[] body,
        ClientHttpRequestExecution execution)
            throws IOException {
        User user = UserContextHolder.currentUser();
        Map<String, String> headers = user.toHttpHeaders();
        for (Map.Entry<String, String> header : headers.entrySet()) {
            request.getHeaders().add(header.getKey(), header.getValue());
        }
        // 调用
        return execution.execute(request, body);
    }
}
```

c）UserContextInterceptor：进入 controller 控制器时会拦截到请求，从 header 头中解析出用户对象存入上下文中，方便服务里使用，如代码清单 10-7 所示。

代码清单10-7　ch10-1\ch10-1-common\src\main\java\cn\springcloud\book\common\intercepter\UserContextInterceptor.java

```java
public boolean preHandle(HttpServletRequest request, HttpServletResponse respone,
    Object arg2) throws Exception {
    User user = new User(HttpConvertUtil.httpRequestToMap(request));
    if(StringUtils.isEmpty(user.getUserId()) && StringUtils.isEmpty(user.get-
        UserName())) {
        log.error("the user is null, please access from gateway or check user info");
        return false;
    }
    UserContextHolder.set(user);
    return true;
}
```

3）增加 UserContextHolder 类，如代码清单 10-8 所示：

代码清单10-8　ch10-1\ch10-1-common\src\main\java\cn\springcloud\book\common\context\
UserContextHolder.java

```java
public class UserContextHolder {
    public static ThreadLocal<User> context = new ThreadLocal<User>();
    public static User currentUser() {
        return context.get();
    }
    public static void set(User user) {
        context.set(user);
    }
    public static void shutdown() {
        context.remove();
    }
}
```

Hystrix 并发策略，如代码清单 10-9 所示（这里只列出了主要的核心代码）：

代码清单10-9　ch10-1\ch10-1-common\src\main\java\cn\springcloud\book\common\context\
SpringCloudHystrixConcurrencyStrategy.java

```java
@Override
    public <T> Callable<T> wrapCallable(Callable<T> callable) {
        return new HystrixThreadCallable<>(callable, RequestContextHolder.getRe
            questAttributes(),HystrixThreadLocal.threadLocal.get());
    }
```

此类使用了 ThreadLocal 对象来保存用户信息，由于在线程池隔离的模式下，会导致前后线程传递对象丢失，这里使用自定义并发策略 HystrixConcurrencyStrategy 即可解决此问题，具体的详细使用方法和解释说明可见 10.3.3 节。

4）在配置类中注册这三个拦截器和并发策略，如代码清单 10-10 所示：

代码清单10-10　ch10-1\ch10-1-common\src\main\java\cn\springcloud\book\common\config\
CommonConfiguration.java

```java
    /**
     * 请求拦截器
     */
@Override
public void addInterceptors(InterceptorRegistry registry) {
    registry.addInterceptor(new UserContextInterceptor());
}

    /**
     * 创建Feign请求拦截器，在发送请求前设置认证的用户上下文信息
     */
    @Bean
    @ConditionalOnClass(Feign.class)
public FeignUserContextInterceptor feignTokenInterceptor(){
        return new FeignUserContextInterceptor();
}

    /**
```

```
 * RestTemplate拦截器
 * @return
 */
@LoadBalanced
@Bean
public RestTemplate restTemplate() {
    RestTemplate restTemplate = new RestTemplate();
    restTemplate.getInterceptors().add(new RestTemplateUserContextInterceptor());
    return restTemplate;
}

@Bean
public SpringCloudHystrixConcurrencyStrategy springCloudHystrixConcurrencyStrategy() {
    return new SpringCloudHystrixConcurrencyStrategy();
}
```

10.2.3　Zuul 的 Fallback 机制

在微服务里有 Hystrix 的 Fallback 机制，那么在微服务应用本身发生问题后，网关层能感知到这些错误并返回友好的提示吗？答案也是可以的，Zuul 提供了一个 Fallback 机制，可以在出现问题的时候进行统一处理，做起来也很简单，只要实现 FallbackProvider 接口，然后定义自己需要的错误码和错误信息即可。在老版本中是 ZuulFallbackProvider 接口，目前已经不建议使用了。

实现 FallbackProvider 接口，加上相应的处理逻辑，如代码清单 10-11 所示：

代码清单10-11　ch10-1\ch10-1-zuul-server\src\main\java\cn\springcloud\book\filter\ZuulFallback.java

```
@Component
public class ZuulFallback implements FallbackProvider{
    @Override
    public String getRoute() {
        return "*"; //可以配置指定的路由,值为serviceId,如sc-user-service
    }
    @Override
    public ClientHttpResponse fallbackResponse(String route, Throwable cause) {
        return new ClientHttpResponse() {
            @Override
            public HttpStatus getStatusCode() throws IOException {
                return HttpStatus.INTERNAL_SERVER_ERROR;
            }
            @Override
            public String getStatusText() throws IOException {
                return HttpStatus.INTERNAL_SERVER_ERROR.getReasonPhrase();
            }
            @Override
            public void close() {
            }
            @Override
            public InputStream getBody() throws IOException {
                //定义自己的错误信息
                return new ByteArrayInputStream(("microservice error").getBytes());
```

```java
        }
        @Override
        public HttpHeaders getHeaders() {
            HttpHeaders headers = new HttpHeaders();
            headers.setContentType(MediaType.APPLICATION_JSON);
            return headers;
        }
        @Override
        public int getRawStatusCode() throws IOException {
            // TODO Auto-generated method stub
            return HttpStatus.INTERNAL_SERVER_ERROR.value();
        }
    };
}
```

10.3 生产环境各组件参考配置

很多朋友不知道如何配置生产环境,这里给出了一份简单模板供大家参考,但由于每个业务项目的大小和业务本身不同,导致不能完全写完整,同时很多朋友也针对不同的组件进行了优化,如替换底层 httpClient 等,所以这里列出的是基础配置。

10.3.1 Eureka 推荐配置

1) Eureka 服务端配置:

```yaml
server:
    port: 8761
spring:
    application:
        name: sc-eurekaserver
eureka:
    client:
        serviceUrl:
            defaultZone: http://127.0.0.1:8761/eureka/,http://127.0.0.2:8761/eureka/
    instance:
        prefer-ip-address: true
    server:
        enable-self-preservation: false
        eviction-interval-timer-in-ms: 30000
```

Eureka 属性描述见表格 10-2。

表 10-2　Eureka 配置描述

属　　性	描　　述
eureka.client.serviceUrl.defaultZone	注册中心位置,高可用需要可配置多个地址,可根据启动参数传入
eureka.instance.prefer-ip-address	推荐大家都可以用 IP 来注册
eureka.server.enable-self-preservation	默认开启,关闭自我保护,如果规模较大,可以考虑不关闭
eureka.server.eviction-interval-timer-in-ms	主动失效检测间隔,默认 60s,可以设置短一点

2）Eureka 客户端配置：

```yaml
eureka:
  client:
    serviceUrl:
      defaultZone: http://127.0.0.1:8761/eureka/,http://127.0.0.2:8761/eureka/
  instance:
    prefer-ip-address: true
```

其他如心跳间隔时间的设置与上面服务端区别不大，推荐使用官方的方式，保持 30s 不变。

10.3.2 Ribbon 推荐配置

Ribbon 的配置如下，一般都配置为全局的，也可以配置为单个服务的，这里列了个全局的。对于是否重试，在真实的项目中发现，由于幂等性或网络不稳定等原因导致易出问题，默认都不进行重试，如果对于一些查询比较多的服务可以开启重试，这一般根据具体项目来定义，具体的配置可以看源代码的 DefaultClientConfigImpl 类。

```yaml
ribbon:
  ConnectTimeout: 2000          #全局请求连接的超时时间，默认2秒
  ReadTimeout: 5000             #全局请求的超时时间，默认5秒
  MaxAutoRetries: 0             #对当前实例的重试次数
  MaxAutoRetriesNextServer: 0   #切换下一个实例重试次数
  OkToRetryOnAllOperations: false  #对所有操作请求都进行重试
```

10.3.3 Hystrix 推荐配置

断路器的配置如下，源代码类的话可以查看 HystrixCommandProperties：

```yaml
hystrix:
  command:
    default:
      execution:
        isolation:
          thread:
            timeoutInMilliseconds: 10000
```

全局请求连接的超时时间默认为 1s，通常会调整这个值大小，推荐设为 10s，如果你的请求需要的时间过长则根据你的请求进行修改或对单个 command 设置单独时间，如果要设置线程池大小，可以看下面网关的配置。

```
hystrix.command.default.execution.isolation.thread.timeoutInMilliseconds
```

10.3.4 Zuul 推荐配置

Zuul 的配置如下：

```yaml
zuul:
  ribbonIsolationStrategy: THREAD
  threadPool:
    useSeparateThreadPools: true
    threadPoolKeyPrefix: zuulgateway
  host:
```

```yaml
            max-per-route-connections: 50
            max-total-connections: 300
            socket-timeout-millis: 5000
            connect-timeout-millis: 5000
hystrix:
    threadpool:
        default:
            coreSize: 20
            maximumSize: 50
            allowMaximumSizeToDivergeFromCoreSize: true
    command:
        default:
            execution:
                isolation:
                    thread:
                        timeoutInMilliseconds: 10000
```

Zuul 的属性描述见表 10-3。

表 10-3　Zuul 配置描述

属　性	描　述
hystrix.command.default.execution.isolation.thread.timeoutInMilliseconds	全局请求连接的超时时间，默认为 1s，通常会调整这个值大小，推荐改为 10s
hystrix.threadpool.default.coreSize	全局默认核心线程池大小，默认值为 10，一般会调大一点
hystrix.threadpool.default.maximumSize	全局默认最大线程池大小，默认值为 10，同样会调大一点
hystrix.threadpool.default.allowMaximumSizeToDivergeFromCoreSize	该属性允许配置 coreSize 和 maximumSize 生效，默认为 false
zuul.ribbonIsolationStrategy: THREAD	使用线程池隔离策略，默认 Zuul 使用信号量，这里改成线程池
zuul.threadPool.threadPoolKeyPrefix: THREAD	使用线程池隔离时每条线程的前缀
zuul.host.maxTotalConnections	目标主机的最大连接数，默认为 200
zuul.host.maxPerRouteConnections	每个路由可用的最大连接数，默认为 20
zuul.host.connect-timeout-millis zuul.host.socket-timeout-millis	如果你不是以服务名来路由而是使用 URL 来配置 Zuul 路由，则使用的是这两个超时配置

10.4　本章小结

本章主要是搭建了一个简单的基础综合案例，为之前的一些知识点做了一个小的 demo，并对工程做了简单增强，包括提供公共包和上下文传递、Zuul 的 Fallback 机制等，另外给出了一份生产环境的参考配置。总之讲解这些内容的初衷都是希望能给广大开发者带来一些便利。

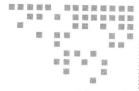

第 11 章　Spring Cloud Config 上篇

Spring Cloud Config 是 Spring Cloud 微服务体系中的中配置中心，是微服务中不可或缺的一部分，其能够很好地将程序中配置日益增多的各种功能的开关、参数的配置、服务器的地址，配置修改后实时生效，灰度发布，分环境、分集群管理配置等进行全面的集中化管理，有利于系统的配置管理和维护。本章节和下一章中将为大家讲解在 Spring Cloud 构建微服中如何使用 Spring Cloud Config 配置中心、扩展配置中心以及三方配置中心 Apollo 的接入。

11.1　Spring Cloud Config 配置中心概述

11.1.1　什么是配置中心

1. 配置中心的由来

在集中式开发时代，配置文件已经基本足够了，因为那时配置的管理通常不会成为一个很大的问题。但是在互联网时代，应用都是分布式系统，部署在 N 台服务器上，想要去线上一台台地重启机器肯定不靠谱，并且维护成本也很高，所以配置中心应运而生。配置中心被用作集中管理不同环境（Dev、Test、Stage、Prod）和不同集群配置，以及在修改配置后将实时动态推送到应用上进行刷新。

2. 配置中心概况

目前市面上有很多的配置中心，这里挑选主要的几个进行对比，如图 11-1 所示：

3. 配置中心应具备的功能

那么配置中心的定位是什么呢？它应该具备如下功能：

- Open API
- 业务无关性

	A	B	C	D	E	F
1	对比性	重要性	Spring Cloud Config	netflix archaius	ctrip apollo	disconf
2	功能特性					
3	静态配置管理	高	基于file	无	支持	支持
4	动态配置管理	高	支持	支持	支持	支持
5	统一管理	高	无,需要git、数据库等	无	支持	支持
6	多维度管理	中	无,需要git、数据库等	无	支持	支持
7	变更管理	高	无,需要git、数据库等	无	支持	无
8	本地配置缓存	高	无	无	支持	支持
9	配置更新策略	中	无	无	支持	支持
10	配置锁	中	无	不支持	不支持	不支持
11	配置校验	中	无	无	无	无
12	配置生效时间	高	重启生效、手动刷新	手动刷新生效	实时	实时
13	配置更新推送	高	需要手动触发	需要手动触发	支持	支持
14	配置定时拉取	高	无	无	支持	配置更新目前依赖事件驱动, client 重启或者Server推送操作
15	用户权限管理	中	无,需要git、数据库等	无	界面直接提供发布历史和回滚按钮	操作记录有落数据库,但无查询接口
16	授权、审核、审计	中	无,需要git、数据库等	无	支持	操作记录有落数据库,但无查询接口
17	配置版本管理	高	git	无	支持	支持
18	配置合规检测	高	不支持	不支持	支持(还需完善)	无
19	实例配置监控	高	需要结合spring admin	不支持	支持	支持,可以查看每个配置在那台机器上加载
20	灰度发布	中	不支持	不支持	支持	不支持部分更新
21	告警通知	中	不支持	不支持	支持邮件方式警告	支持邮件方式警告
22	统计报表	中	不支持	不支持	不支持	不支持
23	依赖关系	高	不支持	不支持	不支持	不支持
24	技术路线					
25	支持Spring Boot	高	原生支持	低	支持	与spring boot 无关
26	支持Spring Config	高	原生支持	低	支持	与spring boot 无关
27	客户端支持	低	java	java	java、.net	java
28	业务系统入侵性	高	入侵性弱	入侵性弱	入侵性弱	入侵性弱,支持注解和xml
29	可依赖组建	高				
30	可用性					
31	单点故障(SPOF)	高	支持HA部署	支持HA部署	支持HA部署	支出HA部署, 高可用由ZK提供
32	多数据中心部署	高	支持	支持	支持	支持
33	配置获取性能	高	unknow	unknow	unknow	unknow
34	易用性					
35	配置界面	中	无,需要git、数据库等操	无	统一界面(ng编写)	统一界面

图 11-1 多种配置中心对照图

- 配置生效监控
- 一致性 K-V 存储
- 统一配置实时推送
- 配合灰度与更新
- 配置全局恢复、备份与历史
- 高可用集群

具体如图 11-2 所示。

接下来看看实际的流转图,具体如图 11-3 所示。

接下来看看整个支撑体系,大致分为两类:

- 运维管理体系
- 开发管理体系

具体的规划如图 11-4 所示。

说了那么多,接下来让我们进入主题——Spring Cloud Config 配置中心。

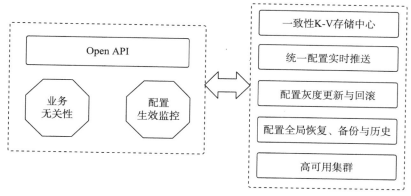

图 11-2 配置中心的功能图

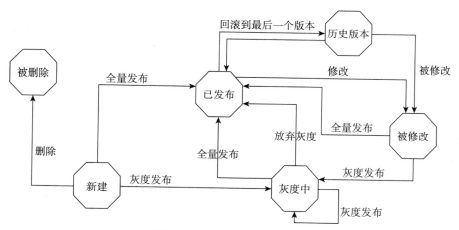

图 11-3 配置中心的流转图

11.1.2 Spring Cloud Config

1. Spring Cloud Config 概述

Spring Cloud Config 是一个集中化外部配置的分布式系统，由服务端和客户端组成。它不依赖于注册中心，是一个独立的配置中心。Spring Cloud config 支持多种存储配置信息的形式，目前主要有 jdbc、Vault、Native、svn、git，其中默认为 git。接下来开始学习 Spring Cloud Config 配置中心的构建和客户端的使用。

2. Git 版工作原理

在开始代码构建演示案例之前需要了解一下配置中心的工作原理，如图 11-5 所示。

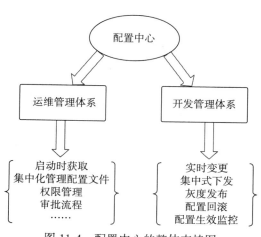

图 11-4 配置中心的整体支持图

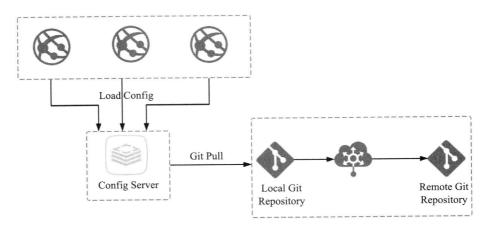

图 11-5 配置中心工作原理图

在图 11-5 中，配置客户端启动时会向服务端发起请求，服务端接收到客户端的请求后，根据配置的仓库地址，将 git 上的文件克隆到本地的一个临时目录中，这个目录是一个 git 的本地仓库目录，然后服务端再读取本地文件返回给客户端。这样做的好处是，当 git 服务器故障或者网络请求异常时，保证服务端仍然可以正常工作。

11.1.3　Spring Cloud Config 入门案例

1. Config Server 配置

（1）创建 Maven 工程

新建 maven 工程 ch11-1，入门案例代码如本书源码 ch11-1 所示。

（2）父级配置依赖

在 ch11-1 的 pom 中添加公共部分依赖，后续的子模块不用再添加相同的依赖，如代码清单 11-1 所示。

代码清单11-1　ch11-1 /pom.xml

```
<!--  利用传递依赖，公共部分  -->
    <dependencies>
        <dependency>
            <groupId>org.springframework.boot</groupId>
            <artifactId>spring-boot-starter-web</artifactId>
        </dependency>

        <dependency>
            <groupId>org.springframework.boot</groupId>
            <artifactId>spring-boot-starter-test</artifactId>
            <scope>test</scope>
        </dependency>

        <dependency>
            <groupId>org.springframework.boot</groupId>
            <artifactId>spring-boot-starter-actuator</artifactId>
```

```xml
        </dependency>
    </dependencies>
```

（3）创建工程模块

在工程 ch11-1 下面创建一个 module，命名为 ch11-1-config-server，代码如本书源码 ch11-1/ch11-1-config-server 所示。

（4）配置依赖

在 pom 中添加 config-servers 的依赖，具体如代码清单 11-2 所示。

代码清单11-2　code\ch11-1\ch11-1-config-server\pom.xml

```xml
<dependencies>

    <dependency>
        <groupId>org.springframework.boot</groupId>
        <artifactId>spring-boot-starter-actuator</artifactId>
    </dependency>

    <dependency>
        <groupId>org.springframework.cloud</groupId>
        <artifactId>spring-cloud-config-server</artifactId>
    </dependency>
</dependencies>
```

（5）编写主程序入口代码

创建 ConfigGitApplication 类作为 Config-Server 的启动类，并且添加相关的注解作为程序的入口，具体如代码清单 11-3 所示。

代码清单11-3　code\ch11-1\ch11-1-config-server\src\main\java\cn\springcloud\book\config\ConfigGitApplication.java

```java
@SpringBootApplication
@EnableConfigServer
public class ConfigGitApplication {

    public static void main(String[] args) {
        SpringApplication.run(ConfigGitApplication.class, args);
    }

}
```

在程序入口代码中的 @EnableConfigServer 注解是开启 Spring Cloud Config 的服务功能。

（6）配置文件配置

在 application.yml 中添加配置，具体如代码清单 11-4 所示。

代码清单11-4　code\ch11-1\ch11-1-config-server\src\main\resources\application.yml

```yml
spring:
  cloud:
    config:
      server:
```

```yaml
        git:
          uri: https://gitee.com/zhongzunfa/spring-cloud-config.git
          #username:
          #password:
          search-paths: SC-BOOK-CONFIG
  application:
    name: sc-config-git
server:
    port: 9090
```

这里解释一下相关配置，配置中的 uri 指的是 git 服务器的地址；username 和 password 指的是 git 访问的用户名和密码，如果设置访问需要用户名和密码，添加上对应的值即可；search-paths 这个属性值就比较厉害了，可以搜索 SC-BOOK-CONFIG 目录下所有满足条件的配置文件，用户可以根据需求添加多个目录，目录之间用逗号隔开，相关内容将在中级篇中进行详细讲解，到这里 config Server 就创建完成了。

接下来，在 git 仓库中创建文件夹目录 SC-BOOK-CONFIG，然后在该目录下创建三个文件，分别命名为 config-info-dev.yml、config-info-test.yml、config-info-prod.yml，在 config-info-dev.yml 中添加如下内容：

```yaml
cn:
    springcloud:
        book:
            config: I am the git configuration file from dev environment.
```

config-info-test.yml 与 config-info-prod.yml 文件中的内容与 config-info-dev.yml 的区别在于，dev environment 被相应地改成了 test environment 和 prod environment，三个文件如图 11-6 所示。

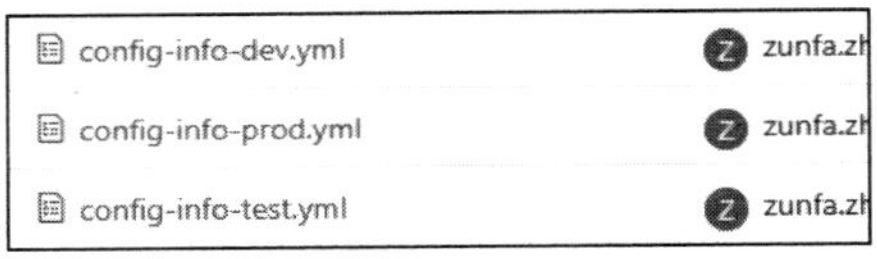

图 11-6　git 中的配置文件图

完成上述的工作之后，直接启动 ch11-1-config-server 工程，可以看到控制台信息如下：

```
main] s.w.s.m.m.a.RequestMappingHandlerMapping : Mapped "{[/{name}-{profiles}.
    json],methods=[GET]}" onto public org.springframework.http.ResponseEntity
main] s.w.s.m.m.a.RequestMappingHandlerMapping : Mapped "{[/{label}/{name}-
    {profiles}.yml || /{label}/{name}-{profiles}.yaml],methods=[GET]}" onto public
main] s.w.s.m.m.a.RequestMappingHandlerMapping : Mapped "{[/{name}/{profile}/
    {label}/**],methods=[GET],produces=[application/octet-stream]}" onto public
main] s.w.s.m.m.a.RequestMappingHandlerMapping : Mapped "{[/{name}/{profile}/
    {label}/**],methods=[GET]}" onto public java.lang.String org.springframework
main] s.w.s.m.m.a.RequestMappingHandlerMapping : Mapped "{[/{name}/{profile}/**],
    methods=[GET],
```

这里只是截取关键信息，从控制台信息里的 Mapped 中可以看到配置信息和 URL 的映射关系。下面先讲解一下官方给出的映射关系：

```
/{application}/{profile}[/{label}]
/{application}-{profile}.yml
/{label}/{application}-{profile}.yml
/{application}-{profile}.properties
```

```
/{label}/{application}-{profile}.properties
```

其中 application 是应用名，也可以理解成 git 上面的文件名，profile 指的是对应激活的环境名，例如 dev、test、prod 等，label 指的是 git 的分支，如果不写，默认的分支为 master。

使用网络访问工具访问地址：http://localhost:9090/config-info/dev/master。显示的结果如下：

```
{
    "name": "config-info",
    "profiles": [
        "dev"
    ],
    "label": "master",
    "version": "5339f7f6044dc1d0df45fdb138337d7da7cf061d",
    "state": null,
    "propertySources": [
        {
            "name": "https://gitee.com/zhongzunfa/spring-cloud-config.git/SC-
                BOOK-CONFIG/config-info-dev.yml",
            "source": {
                "cn.springcloud.book.config": "I am the git configuration file
                    from dev environment."
            }
        }
    ]
}
```

此时观察 config-server 控制台打印的信息可知，config-server 会在本地的临时目录下面克隆远程仓库中的配置信息，本地临时目录如图 11-7 所示：

```
[nio-9090-exec-1] o.s.c.c.s.e.NativeEnvironmentRepository : Adding property source:
    file:/C:/Users/ADMINI~1/AppData/Local/Temp/config-repo-4858102926375848355/SC-
    BOOK-CONFIG/config-info-dev.yml
```

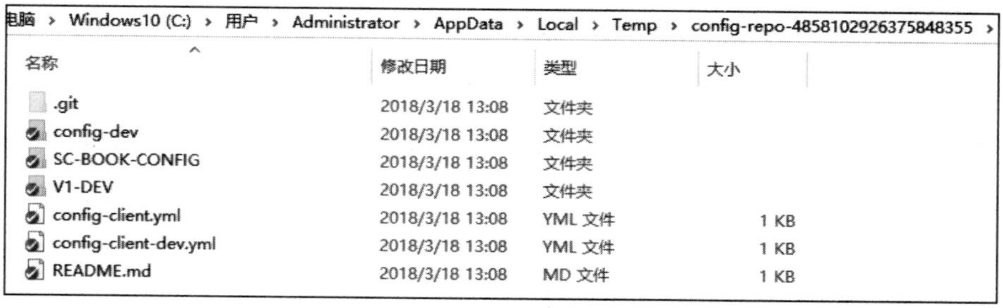

图 11-7 本地临时仓库图

到这里，config-server 就完成了，接下开始配置 config-client。

2. Config Client 配置

（1）创建工程模块

在工程 ch11-1 下面创建一个 module，命名为 ch11-1-config-client，代码如本书源码 ch11-

1/ch11-1-config-client 所示。

（2）配置依赖

在 pom 中添加 config-client 和 web 依赖，具体如代码清单 11-5 所示。

代码清单11-5　code\ch11-1\ch11-1-config-client\pom.xml

```xml
<dependencies>
    <dependency>
        <groupId>org.springframework.cloud</groupId>
        <artifactId>spring-cloud-config-client</artifactId>
    </dependency>
</dependencies>
```

spring-cloud-config-client 主要是客户端用来向服务端拉取配置所需依赖。

（3）编写主程序入口代码

创建 ClientConfigGitApplication 类并且添加对应的注解作为启动程序入口，具体如代码清单 11-6 所示。

代码清单11-6　code\ch11-1\ch11-1-config-client\src\main\java\cn\springcloud\book\config\client\ClientConfigGitApplication.java

```java
@SpringBootApplication
public class ClientConfigGitApplication {

    public static void main(String[] args) {
        SpringApplication.run(ClientConfigGitApplication.class, args);
    }
}
```

为了更好地观察拉取到的 git 上面的配置，这里需要创建一个 Controller 用于访问返回信息，同时还需要创建一个实体，用于注入远程配置上的信息。

创建实体类 ConfigInfoProperties，并且添加 @Component 作为一个组件，同时使用 @ConfigurationProperties 加载配置属性和指定配置前缀，具体如代码清单 11-7 所示。

代码清单11-7　code\ch11-1\ch11-1-config-client\src\main\java\cn\springcloud\book\config\client\config\ConfigInfoProperties.java

```java
@Component
@ConfigurationProperties(prefix = "cn.springcloud.book")
public class ConfigInfoProperties {

    private String config;
    // 省略 get set 方法
}
```

创建 ConfigClientController 类并且将 ConfigInfoProperties 对象注入使用属性值，具体如代码清单 11-8 所示。

代码清单11-8　code\ch11-1\ch11-1-config-client\src\main\java\cn\springcloud\book\config\client\controller\ConfigClientController.java

```java
@RestController
public class ConfigClientController {

    @Autowired
    private ConfigInfoProperties configInfoValue;

    @RequestMapping("/getConfigInfo")
    public String getConfigInfo(){
        return configInfoValue.getConfig();
    }
}
```

（4）配置文件配置

创建 application.yml 文件，添加如下内容，如代码清单11-9 所示。

代码清单11-9　ch11-1/ch11-1-config-client/src/main/resources/application.yml

```yaml
server:
    port: 9091

spring:
    application:
        name: ch11-1-config-client
```

再创建一个 bootstrap.yml 文件添加如下内容，如代码清单 11-10 所示。

代码清单11-10　ch11-1/ch11-1-config-client/src/main/resources/bootstrap.yml

```yaml
spring:
    cloud:
        config:
            label: master
            uri: http://localhost:9090
            name: config-info
            profile: dev
```

下面解释一下上述 bootstrap.yml 中的配置项。
- label 代表要请求哪个 git 分支，本案例中访问的是主分支。
- uri 代表请求的 config server 的地址。
- name 代表要请求哪个名称的远程文件，可以写多个，通过逗号隔开。
- profile 代表哪个分支的文件，例如 dev、test、prod 等。

到这里，读者可能会产生疑问，这些配置为什么要放在 bootrap.yml 里，而不放在 application.yml 中呢？这与 spring boot 的加载顺序有关，bootstrap.yml 文件会优先于 application.yml 加载，因此会去加载远程的配置文件信息，到此客户端中配置完成。

现在启动客户端，此时可以看到控制台信息中打印出配置中 git 的相关信息，比如 config-server 地址、label 请求地址的分支等，具体如下所示：

```
main] c.c.c.ConfigServicePropertySourceLocator : Fetching config from server
    at: http://localhost:9090
main] c.c.c.ConfigServicePropertySourceLocator : Located environment: name=config-
    info, profiles=[dev], label=master, version=5339f7f6044dc1d0df45fdb138337d7
    da7cf061d, state=null
main] b.c.PropertySourceBootstrapConfiguration : Located property source:
    CompositePropertySource {name='configService', propertySources=[MapPropert
    ySource {name='configClient'}, MapPropertySource {name='https://gitee.com/
    zhongzunfa/spring-cloud-config.git/SC-BOOK-CONFIG/config-info-dev.yml'}]}
```

接下来进行访问，输入如下访问地址：http://localhost:9091/getConfigInfo。

可以看到和 git 上配置的信息一致，具体如图 11-8 所示。

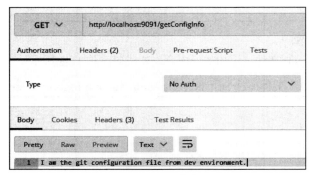

图 11-8　访问内容图

到此配置中心的基本使用就讲授完了，那么如果在远程 Git 仓库上修改配置文件属性会不会生效呢？让我们试验一下。修改 config-info-dev 里面的内容为 I am the git configuration file from dev environment. Add more info.，再次访问上面的地址，我们发现还是之前的配置内容，这时读者可能会想到是由于没有重启，重启一下看到信息果然为最新的。那这样的话，是不是修改配置都需要重启呢？这让人很难接受，在生产环境怎么能随意地进行重启呢！这样一来，是不是代表这个配置中心不够好用呢？当然不是！Spring Cloud 团队早就想到这些了，接下来我们将学习不用重启也能更新配置的方法。

11.2　刷新配置中心信息

11.2.1　手动刷新操作

本节将学习如何实现手动刷新配置，这样不用重启项目也可以获取到最新的配置信息，config server 还是使用之前的 ch11-1-config-server 的代码，不过是将其拷贝到 ch11-2 模块下重命名为 ch11-2-config-server，具体代码如本书源码 ch11-2-config-server 所示，此处就不详细列出。复制 ch11-1-config-client，重命名为 ch11-2-config-client-refresh，这里需要对 ch11-2-config-client-refresh 进行改造。

（1）配置依赖

在之前的依赖基础上添加下面的依赖，如代码清单 11-11 所示。

代码清单11-11　code\ch11-2\ch11-2-config-client-refresh\pom.xml

```xml
<dependency>
    <groupId>org.springframework.cloud</groupId>
    <artifactId>spring-cloud-config-client</artifactId>
</dependency>

<!-- 做简单的安全和端点开放 -->
<dependency>
    <groupId>org.springframework.boot</groupId>
    <artifactId>spring-boot-starter-security</artifactId>
</dependency>
```

第一个依赖是端点的访问依赖，第二个依赖是安全的依赖，需要进行权限过滤，不进行端点拦截。

（2）配置文件配置

在之前的基础上再创建 application.properties 文件来添加下面的内容，如代码清单11-12所示。

代码清单11-12　code\ch11-2\ch11-2-config-client-refresh\src\main\resources\application.properties

```
management.endpoints.web.exposure.include=*
management.endpoint.health.show-details=always
```

这里解释一下上面配置的作用。因为其 F 版的权限提高了，所以需要添加如下内容 management.endpoints.web.exposure.include=* 来表示包含所有端点的信息，默认情况下，只是打开了 info、health 的端点，细心的读者在上一个项目启动过程中的控制打印信息中可以发现，另外的配置 management.endpoint.health.show-details=always 总是表示详细信息的显示。

此外还需要修改一下 application.yml 文件中的端口，具体如下所示：

```
server:
    port: 9093
```

（3）添加安全配置

创建 SecurityConfiguration 类继承 WebSecurityConfigurerAdapter，具体如代码清单11-13所示。

代码清单11-13　code\ch11-2\ch11-2-config-client-refresh\src\main\java\cn\springcloud\book\config\client\config\SecurityConfiguration.java

```java
@Configuration
public class SecurityConfiguration extends WebSecurityConfigurerAdapter {
    @Override
    protected void configure(HttpSecurity http) throws Exception {
        http.csrf().disable();
    }
}
```

上述 configure 方法中的代码主要是用来关闭端点的安全校验。

（4）Controller 类的变更

在 ConfigClientController 上添加 @RefreshScope 注解，具体如代码清单11-14所示。

代码清单11-14　code\ch11-2\ch11-2-config-client-refresh\src\main\java\cn\springcloud\book\config\ConfigClientController

```
@RefreshScope
@RestController
public class ConfigClientController {

    @Autowired
    private ConfigInfoProperties configInfoValue;

    @RequestMapping("/getConfigInfo")
    public String getConfigInfo(){
        return configInfoValue.getConfig();
    }
}
```

这里需要对 ConfigInfoProperties 类进行改造，具体改造如下所示：

```
@Component
@RefreshScope
public class ConfigInfoProperties {

    @Value("${cn.springcloud.book.config}")
    private String config;
}
```

在类的头部添加了 @RefreshScope 注解，被 @RefreshScope 修饰的 Bean 都是延迟加载的，只有在第一次访问时才会被初始化，刷新 Bean 也是同理，下次访问时会创建一个新的对象。上述编写完成后分别启动项目 ch11-2-config-server 和 ch11-2-config-client-refresh。

访问地址：localhost:9093/configConsumer/getConfigInfo，会打印出如图 11-9 所示的信息。

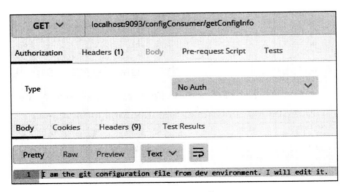

图 11-9　初始加载信息图

再修改一下远程仓库中的 config-info-dev.yml 配置信息，如图 11-10 所示。

I am the git configuration file from dev environment. I will edit it......

图 11-10　修改内容图

再访问地址 localhost:9093/configConsumer/getConfigInfo 发现内容没有改变，此时需要访

问一下端点进行刷新配置信息，端点访问地址为：localhost:9093/actuator/refresh，端点刷新信息如图 11-11 所示：

再次访问 localhost:9093/configConsumer/getConfigInfo，信息发生了变化。最终加载信息如图 11-12 所示。

到这里手动刷新配置就完成了，此时读者肯定会想，每次都要手动刷新，要是服务很多的话，当忘记或者遗漏时就会造成服务出错。那么有没有自动刷新配置信息的方法呢，答案是肯定的，在下一节中我们将进行交接自动刷新配置。

11.2.2 结合 Spring Cloud Bus 热刷新

首先介绍一下 Spring Cloud Config 结合 Spring Cloud Bus 进行刷新的整体流程图，具体如图 11-13 所示。

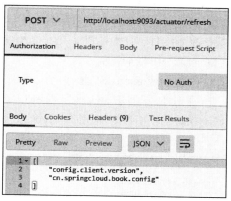

图 11-11 端点刷新信息图

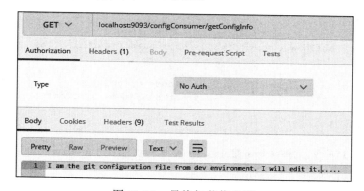

图 11-12 最终加载信息图

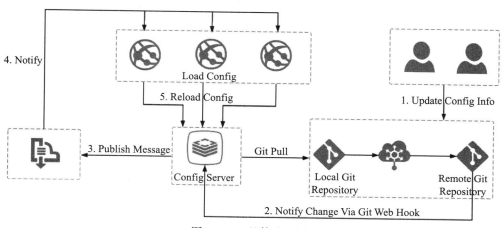

图 11-13 整体流程图

为了方便讲解，下文将 Spring Cloud Bus 简称为 Bus。

在图 11-13 中，用户更新配置信息时，检查到 Git Hook 变化，触发 Hook 配置地址的调用，Config Serve 接收到请求并发布消息，Bus 将消息发送到 config client，当 config client 接收到消息后会重新发送请求加载配置信息，大体流程就是这样。本节的案例使用 Rabbit MQ 作为消息中间件，这里不讲解 Rabbit MQ 的安装和使用方法，读者自己动手安装使用，接下来我们来实现一下以下案例。

（1）创建 maven 工程

在 code 下创建 maven 工程 ch11-3 并且在 ch11-3 的 pom 中添加公共部分依赖，这样后续的子模块就不用再添加相同的依赖了，如代码清单 11-15 所示。

代码清单11-15　ch11-3 /pom.xml

```xml
<!-- 利用传递依赖，公共部分   -->
<dependencies>
    <dependency>
        <groupId>org.springframework.boot</groupId>
        <artifactId>spring-boot-starter-web</artifactId>
    </dependency>
    <dependency>
        <groupId>org.springframework.boot</groupId>
        <artifactId>spring-boot-starter-actuator</artifactId>
    </dependency>

    <dependency>
        <groupId>org.springframework.cloud</groupId>
        <artifactId>spring-cloud-starter-bus-amqp</artifactId>
    </dependency>
</dependencies>
```

在工程 ch11-3 下面创建一个 module，命名为 ch11-3-config-server-bus，代码如本书源码 ch11-3/ch11-3-config-server-bus 所示。

（2）配置依赖

在 pom 中添加依赖信息，具体如代码清单 11-16 所示。

代码清单11-16　code\ch11-3\ch11-3-config-server-bus\pom.xml

```xml
<dependencies>
    <dependency>
        <groupId>org.springframework.cloud</groupId>
        <artifactId>spring-cloud-starter-bus-amqp</artifactId>
    </dependency>

    <dependency>
        <groupId>org.springframework.cloud</groupId>
        <artifactId>spring-cloud-config-server</artifactId>
    </dependency>

    <!-- 做简单的安全和端点开放 -->
    <dependency>
        <groupId>org.springframework.boot</groupId>
        <artifactId>spring-boot-starter-security</artifactId>
    </dependency>
</dependencies>
```

(3)编写主程序入口代码

创建启动类 ConfigGitApplication 并且添加相关注解,作为程序的启动类,具体如代码清单 11-17 所示。

代码清单11-17 code\ch11-3\ch11-3-config-server-bus\src\main\java\cn\springcloud\book\config\GitConfigServerApplication.java

```
@SpringBootApplication
@EnableConfigServer
public class GitConfigServerApplication {

    public static void main(String[] args) {
        SpringApplication.run(GitConfigServerApplication.class, args);
    }
}
```

除了添加上面的主类信息外,还需要添加权限开放的配置类信息,创建类 SecurityConfiguration.class 并且添加如下信息,具体如代码清单 11-18 所示。

代码清单11-18 code\ch11-3\ch11-3-config-server-bus\src\main\java\cn\springcloud\book\config\SecurityConfiguration.java

```
@Configuration
public class SecurityConfiguration extends WebSecurityConfigurerAdapter {

    @Override
    protected void configure(HttpSecurity http) throws Exception {
        http.csrf().disable();
    }
}
```

上述 configure 方法中的代码主要是关闭端点的安全校验。

(4)配置文件配置

配置 application.properties 文件,具体如代码清单 11-19 所示。

代码清单11-19 code\ch11-3\ch11-3-config-server-bus\src\main\resources\application.properties

```
management.endpoints.web.exposure.include=*
management.endpoint.health.show-details=always
```

相关解释同上一节,此处不再赘述。

配置 application.yml 文件,具体如代码清单 11-20 所示。

代码清单11-20 code\ch11-3\ch11-3-config-server-bus\src\main\resources\application.yml

```
spring:
  cloud:
    config:
      server:
        git:
          uri: https://gitee.com/zhongzunfa/spring-cloud-config.git
```

```yaml
          search-paths: SC-BOOK-CONFIG
  application:
    name: ch11-3-config-server-bus

  ## 配置rabbitMQ 信息
  rabbitmq:
        host: localhost
        port: 5672
        username: guest
        password: guest

server:
        port: 9090
```

解释一下配置文件中 rabbitMQ 节点的信息，host 是 rabbitMQ 安装的位置机器地址，port 是端口号，后面两项就是用户名和密码，这里使用安装完成后默认的用户名和密码。到此结合 Bus 的热刷新服务端就创建完成，接下来创建客户端程序。

（5）创建工程模块

同理在工程 ch11-3 下面创建一个 module，命名为 ch11-3-config-client-bus-refresh，代码如本书源码 ch11-3/ch11-3-config-client-bus-refresh 所示。

（6）配置依赖

在 pom 中添加 config-client 和安全相关的依赖信息，具体如代码清单 11-21 所示。

代码清单11-21　code\ch11-3\ch11-3-config-client-bus-refresh\pom.xml

```xml
<dependencies>
    <dependency>
        <groupId>org.springframework.cloud</groupId>
        <artifactId>spring-cloud-config-client</artifactId>
    </dependency>
    <!-- 做简单的安全和端点开放 -->
    <dependency>
        <groupId>org.springframework.boot</groupId>
        <artifactId>spring-boot-starter-security</artifactId>
    </dependency>
</dependencies>
```

这里添加 security 是为了访问端点信息，接下来创建启动程序入口类。

（7）编写主类程序入口代码

创建类 ClientConfigGitApplication，并且添加相关的注解作为程序启动类，具体如代码清单 11-22 所示。

代码清单11-22　code\ch11-3\ch11-3-config-client-bus-refresh\src\main\java\cn\spring-cloud\book\config\client\GitConfigClientApplication.java

```java
@SpringBootApplication
public class GitConfigClientApplication {

    public static void main(String[] args) {
        SpringApplication.run(GitConfigClientApplication.class, args);
```

```
        }
    }
```

除了创建主类外，还需要创建 controller 用于查看远程 git 上的信息，创建 ConfigClient-Controller.class 添加如下内容，具体如代码清单 11-23 所示。

代码清单11-23 code\ch11-3\ch11-3-config-client-bus-refresh\src\main\java\cn\springcloud\book\config\client\controller\ConfigClientController.java

```
@RefreshScope
@RestController
@RequestMapping("configConsumer")
public class ConfigClientController {

    @Autowired
    private ConfigInfoProperties configInfoValue;

    @RequestMapping("/getConfigInfo")
    public String getConfigInfo(){
        return configInfoValue.getConfig();
    }
}
```

此外 ConfigInfoProperties 和 SecurityConfiguration 类的代码同 ch11-2-config-client-refresh config 包下的一致，这里不再赘述，详细内容请看代码清单的对应信息。

代码清单 code\ch11-3\ch11-3-config-client-bus-refresh\src\main\java\cn\springcloud\book\config\client\config\ConfigInfoProperties.java

code\ch11-3\ch11-3-config-client-bus-refresh\src\main\java\cn\springcloud\book\config\client\config\SecurityConfiguration.java

（8）配置文件配置

在 bootstrap.yml 配置文件中添加内容，具体如代码清单 11-24 所示。

代码清单11-24 code\ch11-3\ch11-3-config-client-bus-refresh\src\main\resources\bootstrap.yml

```
spring:
  cloud:
    config:
      label: master
      uri: http://localhost:9090
      name: config-info
      profile: dev
```

在 application.yml 中添加如下配置内容，具体如代码清单 11-25 所示。

代码清单11-25 code\ch11-3\ch11-3-config-client-bus-refresh\src\main\resources\application.yml

```
server:
  port: 9095

spring:
```

```
    application:
        name: ch11-3-config-client-bus-refresh
```

此外 application.properties 同 h11-2-config-client-refresh config 配置相同，此处不再赘述，详情请看代码清单：code\ch11-3\ch11-3-config-client-bus-refresh\src\main\resources\application.properties。

到这里客户端和服务端的案例代码就编写完成了，启动服务端和客户端进行访问。访问地址为 localhost:9095/configConsumer/getConfigInfo。

看到第一次显示的信息如图 11-14 所示。

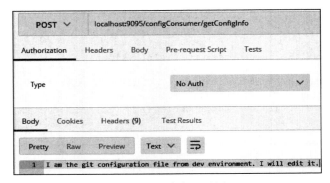

图 11-14　请求显示图

修改 git 上的配置信息，如图 11-15 所示。

```
cn:
  springcloud:
    book:
      config: I am the git configuration file from dev environment. I will edit it 2.
```

图 11-15　git 内容修改图

提交 git 修改的配置后，执行服务端的端点 bus-refresh 访问，地址为 http://localhost:9090/actuator/bus-refresh。

客户端的应用会收到如下信息：

```
c.c.c.ConfigServicePropertySourceLocator : Fetching config from server at:
    http://localhost:9090
c.c.c.ConfigServicePropertySourceLocator : Located environment: name=config-
    info, profiles=[dev], label=master, version=465f123168b96c54ac23a8d7ec18632
    36534f159, state=null
b.c.PropertySourceBootstrapConfiguration : Located property source: Composite-
    PropertySource {name='configService', propertySources=[MapPropertySource
    {name='configClient'}, MapPropertySource {name='https://gitee.com/zhong-
    zunfa/spring-cloud-config.git/SC-BOOK-CONFIG/config-info-dev.yml'}]}
o.s.cloud.bus.event.RefreshListener : Received remote refresh request. Keys
    refreshed [config.client.version, cn.springcloud.book.config]
o.s.a.r.c.CachingConnectionFactory : Attempting to connect to: [localhost:5672]
```

这里只是截取关键信息，在控制台打印的信息中可以看到刷新的 git 地址、文件以及刷新的 keys，再访问地址 localhost:9095/configConsumer/getConfigInfo，可见内容已经发生改变，具体如图 11-16 所示。

图 11-16　请求显示图 2

到此说明结合 bus 做热刷新成功。

当然我们不会每次都手动指定 config-server 这个请求刷新的地址，因此可以将地址配置在 WebHooks 上面，在提交（push）文件之后自动执行刷新的动作。

这里操作是在码云上执行的，点击码云上项目的 settings，在左侧导航栏中找到 WebHooks，具体的配置如图 11-17 所示。

图 11-17　WebHook 设置图

在点击 Add 添加规则，具体如图 11-18 所示。

图 11-18　WebHook 设置图 2

根据实际情况填写即可,如果读者使用不同的平台则以实际情况进行操作。

11.3 本章小结

本章主要讲解了配置中心的由来和配置中心应具备的功能,并且对 Spring Cloud Config 进行了案例讲解。在完成了服务端和客户端的基本配置、手动执行端点 /actuator/refresh 刷新客户端配置信息,以及自动刷新配置后,通过结合 Spring Cloud Bus 在服务端执行端点 /actuator/bus-refresh?destination=** 刷新所有的客户端,同时添加 webHook 配置对应的 Config Server 的回调地址,在 git 提交(pushs)后执行调用刷新。

第 12 章 Chapter 12

Spring Cloud Config 下篇

在第 11 章中已经介绍了 Spring Cloud Config 关于 Git 的基础用法和手动刷新以及借助 Spring Cloud Bus 做自动刷新的操作,本章主要介绍 Git 详细配置以及关系型数据库和非关系数据库存储实现配置中心,以及配置中心的扩展,包括:客户端自动刷新、客户端回退、安全认证、客户端高可用和服务端高可用等功能,此外还讲解了业界比较火的具有 portal 操作的三方配置中心 Apollo(携程),读者学会后可以很便捷地进行配置中心操作。

12.1 服务端 Git 配置详解与实战

12.1.1 Git 多种配置详解概述

本节将讲解更多 Config Server 关于 Git 的配置信息,主要的内容有:
- Git 中 URI 占位符
- 模式匹配和多个存储库
- 路径搜索占位符

12.1.2 Git 中 URI 占位符

Spring Cloud Config Server 支持占位符的使用,支持 {application}、{profile}、{label},这样的话就可以在配置 uri 的时候,通过占位符使用应用名称来区分应用对应的仓库然后进行使用,接下来让我们通过案例来说明。新建 maven 工程 ch12-1,案例代码如本书源 ch12-1 所示。

(1)父级配置依赖

在 ch12-1 的 pom 中添加公共部分依赖,后续的子模块便不用再添加相同的依赖,如代码清单 12-1 所示。

代码清单12-1　ch12-1/pom.xml

```xml
<dependencies>
    <dependency>
        <groupId>org.springframework.boot</groupId>
        <artifactId>spring-boot-starter-web</artifactId>
    </dependency>
    <dependency>
        <groupId>org.springframework.boot</groupId>
        <artifactId>spring-boot-starter-test</artifactId>
        <scope>test</scope>
    </dependency>
    <dependency>
        <groupId>org.springframework.boot</groupId>
        <artifactId>spring-boot-starter-actuator</artifactId>
    </dependency>
</dependencies>
```

在上述依赖中，主要是关于 web 和端点。接下来创建 config-server 工程。

（2）创建 config-server-placeholders

在 ch12-1 下创建 maven 工程，命名为 ch12-1-config-server-placeholders，并且在 pom 中添加相关依赖，具体如代码清单 12-2 所示。

代码清单12-2　ch12-1/ch12-1-config-server-placeholders/pom.xml

```xml
<dependencies>
    <dependency>
        <groupId>org.springframework.cloud</groupId>
        <artifactId>spring-cloud-config-server</artifactId>
    </dependency>
</dependencies>
```

上述依赖中主要是添加了 config-server 相关依赖。接下来创建加载远程配置的启动类。

（3）创建程序入口代码

创建名为 GitConfigServerApplication 的类，并且添加上相关依赖作为程序的主入口，具体如代码清单 12-3 所示。

代码清单12-3　ch12-1/ch12-1-config-server-placeholders/src/main/java/cn/springcloud/book/config/GitConfigServerApplication.java

```java
@SpringBootApplication
@EnableConfigServer
public class GitConfigServerApplication {

    public static void main(String[] args) {
        SpringApplication.run(GitConfigServerApplication.class, args);
    }

}
```

其中 @EnableConfigServer 表示开启配置中心功能。这里还需要配置 Git 配置信息，用于加载远程配置信息。

（4）配置文件配置

在 application.yml 中添加配置信息，具体如代码清单 12-4 所示。

代码清单12-4　ch12-1/ch12-1-config-server-placeholders/src/main/resources/application.yml

```
spring:
  cloud:
    config:
      server:
        git:
          uri: https://gitee.com/zhongzunfa/{application}
          search-paths: SC-BOOK-CONFIG
application:
  name: ch12-1-config-server-placeholders
server:
  port: 9090
logging:
  level:
    root: debug
```

这里需要将日志级别调到 dubug 级别，上述代码中还配置了 search-paths 这个路径。需要注意的是，uri 的最后为 {application}，设置为配置模式。到此 config-server 端就配置完成了，接下来创建客户端。

（5）创建 config-client-placeholders

在 ch12-1 下创建 maven 工程，命名为 ch12-1-config-client-placeholders，并且在 pom 中添加相关依赖，具体如代码清单 12-5 所示。

代码清单12-5　ch12-1/ch12-1-config-client-placeholders/pom.xml

```
<dependencies>
    <dependency>
        <groupId>org.springframework.cloud</groupId>
        <artifactId>spring-cloud-config-client</artifactId>
    </dependency>
</dependencies>
```

这里主要是添加了 config-client 依赖，用于加载 config-server 端的配置。接下来创建客户端应用启动类。

（6）创建入口程序

创建启动类 GitConfigClientApplication，添加启动注解。具体如代码清单 12-6 所示。

代码清单12-6　ch12-1/ch12-1-config-client-placeholders/src/main/java/cn/springcloud/book/config/client/GitConfigClientApplication.java

```
@SpringBootApplication
public class GitConfigClientApplication {

    public static void main(String[] args) {
        SpringApplication.run(GitConfigClientApplication.class, args);
    }

}
```

为了通过 controller 访问获取远程依赖的配置信息，创建 ConfigClientController 类，并且添加相关依赖，此外还将远程配置信息的配置类注入进来。具体如代码清单 12-7 所示。

代码清单12-7 ch12-1/ch12-1-config-client-placeholders/src/main/java/cn/springcloud/book/config/client/controller/ConfigClientController.java

```java
@RestController
public class ConfigClientController {
    @Autowired
    private ConfigInfoProperties configInfoValue;
    @RequestMapping("/getConfigInfo")
    public String getConfigInfo(){
        return configInfoValue.getConfig();
    }
}
```

上述代码清单中的 ConfigInfoProperties 是用来加载配置信息的，具体如代码清单 12-8 所示。

代码清单12-8 ch12-1/ch12-1-config-client-placeholders/src/main/java/cn/springcloud/book/config/client/config/ConfigInfoProperties.java

```java
@Component
@ConfigurationProperties(prefix = "cn.springcloud.book")
public class ConfigInfoProperties {
    private String config;   //省略getter setter方法
}
```

上面将其注解为一个组件，并且在 @ConfigurationProperties 中指明了前缀 cn.springcloud.book，接下配置启动需要的服务端地址、应用名称等。

（7）配置文件配置

在 application.yml 配置文件中添加应用名称和端口，具体代码清单 12-9 所示。

代码清单12-9 ch12-1/ch12-1-config-client-placeholders/src/main/resources/application.yml

```yaml
server:
   port: 9100
spring:
   application:
      name: ch12-1-config-client-placeholders
```

在 bootstrap.yml 中添加服务端的地址，具体如代码清单 12-10 所示。

代码清单12-10 ch12-1/ch12-1-config-client-placeholders/src/main/resources/bootstrap.yml

```yaml
spring:
   cloud:
      config:
         label: master
         uri: http://localhost:9090
         name: spring-cloud-config
         profile: dev
```

上述配置中指明了 name 为 spring-cloud-config。在服务端 ch12-1-config-server-placeholders

的配置中，uri 是 {application} 匹配模式，并且添加了 search-paths。

当客户端请求服务端仓库的连接地址的 uri 变成了 https://gitee.com/zhongzunfa/spring-cloud-config，连接到了 spring-cloud-config 仓库。

远程 git 上的配置如图 12-1 所示。

在 spring-cloud-config 仓 库 下 的 SC-BOOK-CONFIG 中创建了 spring-cloud-config.yml，pring-cloud-config.yml 中的内容如图 12-2 所示。

在上述工作准备完成后，启动程序。先启动服务端后再启动客户端，此时观察服务端会发现如下加载的信息：

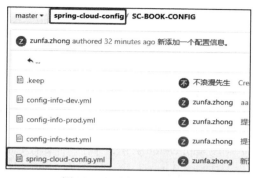

图 12-1　远程 git 配置图

图 12-2　spring-cloud-config.yml 内容图

```
m.m.a.RequestResponseBodyMethodProcessor  : Written [Environment [name=spring-
cloud-config, profiles=[dev], label=master, propertySources=[PropertySource
[name=https://gitee.com/zhongzunfa/spring-cloud-config/SC-BOOK-CONFIG/
spring-cloud-config.yml]], version=8a016385a2b492442a6d193a9014a77c3e728c97,
state=null]] as "application/json" using [org.springframework.http.converter.
json.MappingJackson2HttpMessageConverter@37637a24]
```

这里只是截取关键信息，想要看到信息，需要在配置文件上设置日志级别为 debug。

接下来访问地址：localhost:9100/getConfigInfo，会看到和上面 git 配置的一样的信息，具体如图 12-3 所示。

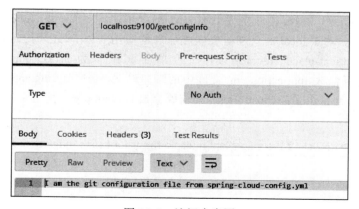

图 12-3　访问内容图

这里需要仓库名称和仓库下面的配置文件名称一致才可以，因为配置了 spring.cloud.config.name 默认占位符匹配的是 spring.application.name 的值。

12.1.3 模式匹配和多个存储库

在 application 和 profile 的名称上，Spring Cloud Server 还支持更复杂配置模式，可以使用通配符 {application}/{profile} 名称进行规则匹配，通过逗号分隔，接下来通过具体案例进行讲解。

创建 maven 工程 ch12-2，并且添加公共依赖，公共依赖同代码清单 12-1 所示。复制 ch12-1-config-server-placeholders 到 ch12-2 下并且重命名为 ch12-2-config-server-multiple-repositories，并修改 application.yml 文件，具体如代码清单 12-11 所示。

代码清单12-11　ch12-2/ch12-2-config-server-multiple-repositories/src/main/resources/application.yml

```yaml
spring:
  cloud:
    config:
      server:
        git:
          uri: https://gitee.com/zhongzunfa/spring-cloud-config
          search-paths: SC-BOOK-CONFIG
          repos:
            simple: https://gitee.com/zhongzunfa/simple
            special:
              pattern: special*/dev*,*special*/dev*
              uri: https://gitee.com/zhongzunfa/spring-cloud-config-special
            local:
              pattern: local*
              uri: /Users/zhongzunfa/all_test/spring-cloud-config
```

这里只是列出关键信息，如上配置中的 spring.cloud.config.server.uri（将 yml 写成 properties 形式）指明了默认的仓库地址，在使用 {application}/{profile} 匹配不上任何一个仓库时会使用默认的仓库进行匹配来获取信息。对于"spring-cloud-config-simples"匹配的是 spring-cloud-config-simples/*，需要注意的是其仅能匹配应用名称为 spring-cloud-config-simples 的所有 profile 配置。对于 local 的仓库将会匹配所有的应用名以 local 开头的 Profiles。同样复制 ch12-1-config-client-placeholders 到 ch12-2 下并且重命名为 ch12-2-config-client-multiple-repositories，并且修改 application.yml 文件中的端口改为 9101，spring.application.name 改为 ch12-2-config-client-multiple-repositories，还需要修改 bootstrap.yml 配置文件信息，具体如代码清单 12-12 所示。

代码清单12-12　ch12-2/ch12-2-config-client-multiple-repositories/src/main/resources/application.yml

```yaml
spring:
  cloud:
    config:
      label: master
      uri: http://localhost:9090
```

```
name: spring-cloud-config
profile: config-info
```

上述配置文件主要是修改 profile 的信息。到此就已经配置完成了。

读者可以启动服务端和客户端，再通过修改仓库信息进行验证是否和预期一样。这里不再进行试验。

12.1.4 路径搜索占位符

在前面的案例中已经有用到 searchPaths 参数，进行路径的搜索，可以根据路径和路径前缀等方式进行配置文件的获取，本节将详解其具体的用法。例如下面的代码片段配置：

```
spring.cloud.config.server.git.uri: https://gitee.com/zhongzunfa/spring-cloud-
    config.git
spring.cloud.config.server.git.search-paths: SC-BOOK-CONFIG, SC-CONFIG*
```

上述为 properties 格式，上述配置中的 SC-BOOK-CONFIG 是匹配当前路径下面所有的配置文件信息，SC-CONFIG* 表示的是配置以前 SC-CONFIG 前缀的文件夹，进行搜索所有配置文件。再看看使用占位符的例子，具体如下配置片段所示：

```
spring.cloud.config.server.git.uri: https://gitee.com/zhongzunfa/spring-cloud-
    config.git
spring.cloud.config.server.git.cloneOnStart: true search-paths: *{application}*
```

上述为 properties 格式，使用占位符的形式进行目录搜索，这样的话可以根据不同的项目对不同的配置文件进行路径配制，从而很好地划分文件。有一点需要注意的是，在引用上要加上 ** 字符否则不能识别，此外还可以使用 searchPaths: '{application}' 也就是使用单引号''形式。官方有具体的 bug 提问地址：

官方 bug 提问地址：https://github.com/spring-cloud/spring-cloud-config/issues/328。

这里对附加的一个属性进行一下讲解，cloneOnStart 表示的是当前服务端（config-server）仓库地址的合法性，表示其合法地将远程仓库克隆到本地。

Config-server 端关于 git 的配置到此就讲解完成了，接下来将讲解 config-server 的另一存储方式，即使用数据库进行存储，包括关系数据库和非关系数据库。

12.2 关系型数据库的配置中心的实现

12.2.1 Spring Cloud Config 基于 MySQL 的配置概述

Spring Cloud Config 提供了 jdbc 的方式，本节将通过实际案例来讲解其配置以及内部实现，本次案例主要使用的数据库是 MySQL。

在开始之前我们先简单了解一下架构，具体如图 12-4 所示。

图 12-4 中的大体请求流程是：config-client 请求 config-server，config-server 根据配置信息获取数据库中的表的相关配置。

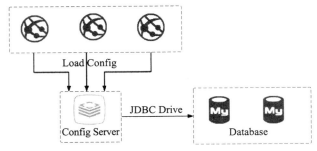

图 12-4 整体流程图

12.2.2 Spring Cloud Config 与 MySQL 结合案例

本案例中将使用 MySQL 代替 git 存储进行配置中心配置,接下来开始案例工程的创建。

1. 创建父工程

在工程 code 的根目录下创建 maven 工程,命名为 ch12-3,将子工程需要用到的共同依赖添加到 pom 中,公共依赖同 ch12-2 的 pom 一致,在这里不再重复赘述。

2. 创建 config-server-db

在工程 ch12-3 下创建一个 module,命名为 ch12-3-config-server-db,代码如本书源码 ch12-3/ch12-3-config-server-db 所示。

(1)配置依赖

在 ch12-3-config-server-db 的 pom 文件中添加 config-server 和 jdbc 以及 MySQL 相关依赖,具体如代码清单 12-13 所示。

代码清单12-13　code\ch12-3\ch12-3-config-server-db\pom.xml

```
<dependencies>
    <dependency>
        <groupId>org.springframework.cloud</groupId>
        <artifactId>spring-cloud-config-server</artifactId>
    </dependency>
    <dependency>
        <groupId>org.springframework.boot</groupId>
        <artifactId>spring-boot-starter-jdbc</artifactId>
    </dependency>
    <dependency>
        <groupId>mysql</groupId>
        <artifactId>mysql-connector-java</artifactId>
    </dependency>
</dependencies>
```

在这里使用 spring-jdbc 支持是因为 Spring Cloud Config 默认的配置方式,接下来创建应用的启动类。

(2)编写主程序入口代码

创建类 DbConfigServerApplication 添加相关注解,使其成为程序的启动入口,代码内容同代码清单 12-13 所示。详细代码清单读者可以查看：code\ch12-3\ch12-3-config-server-db\src\main\java\com\sc\book\config\DbConfigServerApplication.java 来启动相关配置。

（3）配置文件配置

在 application.yml 配置文件中添加激活 JDBC 配置、数据库链接地址的配置等，具体如代码清单 12-14 所示。

代码清单12-14　code\ch12-3\ch12-3-config-server-db\src\main\resources\application.yml

```yaml
server:
    port: 9090
spring:
    application:
        name: ch12-3-config-server-db
    cloud:
        config:
            server:
                jdbc:
                    sql: SELECT `KEY`, `VALUE` FROM PROPERTIES WHERE application =? AND profile =? AND lable =?
                label: master
            refresh:
                refreshable: none
    profiles:
        active: jdbc

## 数据配置
    datasource:
        url: jdbc:mysql://127.0.0.1:3306/spring-cloud?useUnicode=true&characterEncoding=UTF-8
        username: root
        password: 123456
        driver-class-name: com.mysql.jdbc.Driver
logging:
    level:
        org.springframework.jdbc.core: DEBUG
        org.springframework.jdbc.core.StatementCreatorUtils: Trace
```

先解释一下相关配置的信息：spring.cloud.config.server.jdbc.sql 是在调用时使用的 SQL，对于 SQL 上的参数 application、profile、label 同第 11 章的解释，spring.profiles.active=jdbc 使用的激活方式是 jdbc。

其中 spring.cloud.refresh.refreshable=none 是用来解决 DataSource 循环依赖问题，具体看官方 bug 解决地址如下：

官方 bug 解决地址：https://github.com/spring-cloud/spring-cloud-commons/issues/355。

（4）数据库脚本

接下来需要在 MySQL 数据库上创建名称为 spring-cloud 的数据库，再创建名称为 PROPERTIES 表，具体如代码清单 12-15 所示：

代码清单12-15　ch12-3/ch12-3-config-server-db/script/创建配置表.sql

```
-- 创建类型
```

```sql
CREATE TABLE `PROPERTIES` (
  `ID` int(11) NOT NULL AUTO_INCREMENT,
  `KEY` TEXT DEFAULT NULL,
  `VALUE` TEXT DEFAULT NULL,
  `APPLICATION` TEXT DEFAULT NULL,
  `PROFILE` TEXT DEFAULT NULL,
  `LABLE` TEXT DEFAULT NULL,
  PRIMARY KEY (`ID`)
) ENGINE=InnoDB AUTO_INCREMENT=3 DEFAULT CHARSET=utf8;
```

在创建好的表中插入数据，具体脚本如代码清单 12-16 所示。

代码清单12-16   ch12-3/ch12-3-config-server-db/script/插入数据.sql

```sql
INSERT INTO `spring-cloud`.`properties` (`ID`, `KEY`, `VALUE`, `APPLICATION`,
    `PROFILE`, `LABLE`) VALUES ('3', 'cn.springcloud.book.config', 'I am the mysql
    configuration file from dev environment.', 'config-info', 'dev', 'master');
INSERT INTO `spring-cloud`.`properties` (`ID`, `KEY`, `VALUE`, `APPLICATION`,
    `PROFILE`, `LABLE`) VALUES ('4', 'cn.springcloud.book.config', 'I am the mysql
    configuration file from test environment.', 'config-info', 'test', 'master');
INSERT INTO `spring-cloud`.`properties` (`ID`, `KEY`, `VALUE`, `APPLICATION`,
    `PROFILE`, `LABLE`) VALUES ('5', 'cn.springcloud.book.config', 'I am the mysql
    configuration file from prod environment.', 'config-info', 'prod', 'master');
```

上述脚本插入三个环境的数据分别是 dev、test、prod。

到这里基于数据库配置方式的 Server 端就完成了，接下开始配置 config-client。

3. 创建 config-client-db

在工程 ch12-3 下面创建一个 module，命名为 ch12-3-config-client-db，代码如本书源码 ch12-3/ch12-3-config-client-db 所示。

（1）配置依赖

在 ch12-3-config-client-db 的 pom 文件中添加 config-client 依赖，代码内容同代码清单 12-15 所示。

详细代码读者可以查看：ch12-3\ch12-3-config-client-db\pom.xml，接下来创建应用程序的启动类。

（2）编写主程序入口代码

创建 DbConfigClientApplication 类并且添加相关的注解信息，使其成为程序的启动入口类，具体代码内容同代码清单 12-6。

详细代码读者可查看：ch12-3/ch12-3-config-client-db/src/main/java/cn/springcloud/book/config/client/DbConfigClientApplication.java。

接下来创建配置实体类和 controller 访问类，配置实体 ConfigInfoProperties 和 ConfigClient-Controller 同代码清单 12-7 和代码清单 12-8 所述，这里不再赘述。

（3）配置文件配置

创建 application.yml 文件添加端口、应用名称、日志级别等配置，具体如代码清单 12-17 所示。

代码清单12-17　code\ch12-3\ch12-3-config-client-db\src\main\resources\application.yml

```yaml
spring:
  application:
    name: ch12-3-config-client-db
server:
  port: 9096
logging:
  level:
    root: INFO
```

再创建一个 bootstrap.yml 文件添加配置信息，具体如代码清单 12-18 所示。

代码清单12-18　code\ch12-3\ch12-3-config-client-db\src\main\resources\bootstrap.yml

```yaml
spring:
  cloud:
    config:
      label: master
      uri: http://localhost:9090
      name: config-info
      profile: dev
```

上述配置中主要是添加服务端的地址和请求的分支和名称等信息。到这里基于 jdbc 的客户端配置也完成了，启动服务端和客户端程序进行访问：localhost:9096/configConsumer/get-ConfigInfo，如图 12-5 所示。

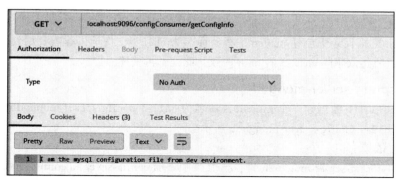

图 12-5　请求内容图

看到图 12-5 的内容，说明已经成功获取了数据库中的配置信息。使用 db 的好处是在封闭的环境内，不搭建 git 也可以使用配置中心。接下来讲解一下 Spring Cloud Config 和 MongoDB 的结合使用。

12.3　非关系型数据库的配置中心的实现

12.3.1　Spring Cloud Config 基于 MongoDB 的配置概述

Spring Cloud Config 并没有提供关于 MongoDB 的方式，但是目前 Spring Cloud 已经收录

了一个相关的在孵化器中，本节将通过实际案例讲解其配置以及内部实现，本次案例主要使用的非关系型数据库是 MongoDB 数据库。

在开始之前我们先简单了解一下整体架构，如图 12-6 所示。

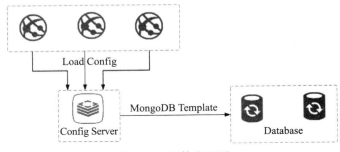

图 12-6　整体流程图

图 12-6 中的大体请求流程是：config-client 请求 config-server，config-server 根据配置信息获取 MongoDB 中的相关配置。

12.3.2　Spring Cloud Config MongoDB 案例

本案例中将使用 MongoDB 代替 Git 存储进行配置中心配置，这里不讲解 MongoDB 的安装方法，请读者参考相关资料自己安装使用，接下来开始案例工程的创建。

1. 创建父工程

在工程 code 根目录下创建 maven 工程，命名为 ch12-4，将子工程需要用到的共同依赖添加到 pom 中，公共依赖同 ch12-3 的 pom 一致，此处不再赘述。

2. 创建 config-server-mongod

在工程 ch12-4 下面创建一个 module，命名为 ch12-4-config-server-mongodb，代码如本书源码 ch12-4/ch12-4-config-server-mongodb 所示。

（1）配置依赖

在 ch12-4-config-server-mongodb 的 pom 文件中添加 config-sever 和 config-server-mongodb 依赖，具体如图代码清单 12-19 所示。

代码清单12-19　code\ch12-4\ch12-4-config-server-mongodb\pom.xml

```xml
<dependencies>
    <dependency>
        <groupId>org.springframework.cloud</groupId>
        <artifactId>spring-cloud-config-server</artifactId>
    </dependency>

    <dependency>
        <groupId>org.springframework.cloud</groupId>
        <artifactId>spring-cloud-config-server-mongodb</artifactId>
        <version>0.0.2.BUILD-SNAPSHOT</version>
    </dependency>
```

```
</dependencies>
```

上述依赖中添加了 MongoDB 的支持，这里使用的是 spring-cloud 孵化器中的支持。接下来创建程序入口类。

（2）编写主程序入口代码

创建 MongoDbConfigServerApplication 类，并且添加相关注解信息，具体如代码清单 12-20 所示。

代码清单12-20 code\ch12-4\ch12-4-config-server-mongodb\src\main\java\com\sc\boo\config\mongodb\MongoDbConfigServerApplication.java

```java
@SpringBootApplication
@EnableMongoConfigServer
public class MongoDbConfigServerApplication {

    public static void main(String[] args) {
        SpringApplication.run(MongoDbConfigServerApplication.class, args);
    }
}
```

其中 EnableMongoConfigServer 表示的是开启配置中心 MongoDB 的能力。接下来添加启动需要的配置信息。

（3）配置文件配置 application.yml

在 application.yml 文件中添加 MongoDB 的地址和端口信息等，具体如代码清单 12-21 所示。

代码清单12-21 code\ch12-4\ch12-4-config-server-mongodb\src\main\resources\application.yml

```yaml
server:
    port: 9090
spring:
    application:
        name: ch12-4-config-server-mongodb
    data:
        mongodb:
            uri: mongodb://localhost/springcloud
```

先解释一下相关配置的信息：spring.data.mongodb.uri 这个配置是用来指定 MongoDB 的数据库地址，其中最后面的 springcloud 是 MongoDB 的数据库，具体如图 12-7 所示。

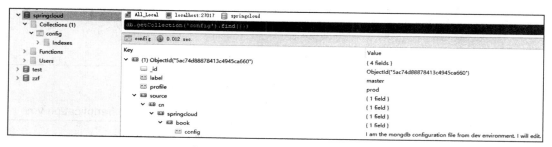

图 12-7 springcloud 数据库图

到此 server 的基本配置就完成了，还需要将相关数据存入 MongoDB 中，具体的数据如代码清单 12-22 所示。

代码清单12-22 ch12-4/ch12-4-config-server-mongodb/script/config.json

```
{ "label" : "master", "profile" : "prod", "source" : { "cn" : { "springcloud" : {
    "book" : { "config" : "I am the mongodb configuration file from dev environment.
    I will edit." } } } } }
```

接下来配置 MongoDB 版的 client 端。

3. 创建 config-client-mongodb

（1）创建工程模块

在工程 ch12-4 下面创建一个 module，命名为 ch12-4-config-client-mongodb，代码如本书源码 ch12-4/ch12-4-config-client-mongodb 所示。

（2）配置依赖

在 ch12-4-config-client-mongodb pom 中添加 config-client 依赖，代码内容同代码清单 12-25 所示。

详细代码读者可以查看：code\ch12-4\ch12-4-config-client-mongodb\pom.xml。配置依赖后创建程序启动类。

（3）编写主程序入口代码

创建程序启动类 MongoDbConfigClientApplication，并且添加上相关的注解成为启动程序入口，代码内容同代码清单 12-26 所示。

详细代码可以查看：code\ch12-4\ch12-4-config-client-mongodb\src\main\java\com\sc\boo\config\mongodb\MongoDbConfigClientApplication.java。

接下来创建配置实体类和 controller 访问类，配置实体类 ConfigInfoProperties 和 ConfigClientController 同代码清单 12-7 和代码清单 12-8 所述。

（4）配置文件配置 bootstrap.yml

在 bootstrap.yml 中添加服务端的访问地址，具体配置如代码清单 12-23 所示。

代码清单12-23 code\ch12-4\ch12-4-config-client-mongodb\src\main\resources\bootstrap.yml

```
spring:
  cloud:
    config:
      label: master
      uri: http://localhost:9090
      name: config
      profile: prod
```

此处还需要配置 application.yml 文件，具体如代码清单 12-24 所示。

代码清单12-24 ch12-4/ch12-4-config-client-mongodb/src/main/resources/application.yml

```
spring:
  application:
    name: ch12-4-config-client-mongodb
```

```
server:
    port: 9097
```

到此客户端的相关配置就完成了，接下来启动服务端和客户端进行案例测试。都启动好后，访问地址为 localhost:9097/configConsumer/getConfigInfo。访问结果和上述的 config.json 文件内容完全吻合，具体如图 12-8 所示。

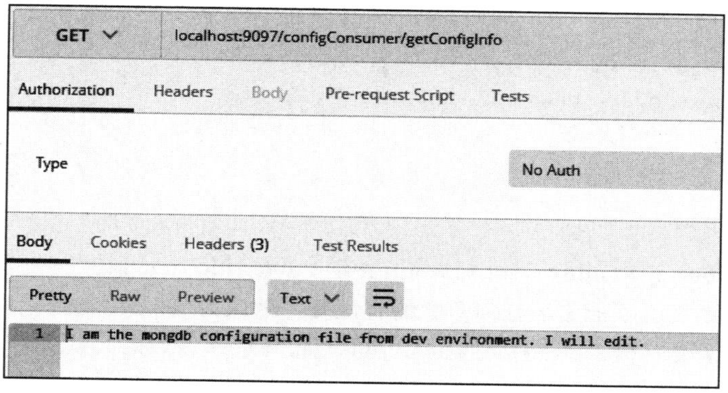

图 12-8　mongdb 配置访问结果图

可能有读者会想到，手动刷新和配置自动刷新对于 db 环境下是否同时支持呢？这是一个好问题，对于 db 操作来说，在自动刷新方面，肯定是做了界面化的配置和管理，当成功提交配置到 db 后，调用 config-server 的刷新接口即可实现和 git 的 webHook 一样的提交绑定执行功能。

12.4　Spring Cloud Config 使用技能

对于常常使用的功能小技能主要是指使用本地的参数覆盖远程的参数，在开发的时候经常会用到。

本地参数的覆盖远程参数

在某些时候需要使用当前系统的环境变量或者是应用本身设置的参数而不是使用远程拉取的参数，客户端可以进行如下配置。

```
spring:
    cloud:
        config:
            allowOverride: true
            overrideNone: true
            overrideSystemProperties: false
```

官方 bug 和解决方式如下。
官方 BUG 解决地址：

- https://github.com/spring-cloud/spring-cloud-config/issues/651。
- https://github.com/spring-cloud/spring-cloud-config/issues/359。

这里将源码关键的类的附上，具体如下所示。

```
@ConfigurationProperties("spring.cloud.config")
public class PropertySourceBootstrapProperties {

private boolean overrideSystemProperties = true;

private boolean allowOverride = true;
private boolean overrideNone = false;
// 省略getter setter 方法

}
```

解释上面三个属性的意思。
- overrideNone：当 allowOverride 为 true 时，overrideNone 设置为 true，外部的配置优先级更低，而且不能覆盖任何存在的属性源。默认为 false。
- allowOverride：标识 overrideSystemProperties 属性是否启用。默认为 true，设置为 false 意为禁止用户的设置。
- overrideSystemProperties：用来标识外部配置是否能够覆盖系统属性，默认为 true。

12.5 Spring Cloud Config 功能扩展

12.5.1 客户端自动刷新

在有些应用上面，不需要在服务端批量推送的时候，客户端本身需要获取变化参数的情况下，使用客户端的自动刷新能完成此功能。在工程 code 根目录下创建 maven 工程，命名为 ch12-5，将子工程需要用到的共同依赖添加到 pom 中，公共依赖同 ch12-3 的 pom 一致。此外 code\ch12-5\ch12-5-config-server 同 code\ch12-4\ch12-4-config-server 代码一致，接下来重点写一下 client 和 autoconfig 的代码。在工程 ch12-5 下面创建一个 module，命名为 ch12-5-config-client-refresh-autoconfig，代码如本书源码 ch12-5/ch12-5-config-client-refresh-autoconfig 所示。

（1）配置依赖

在 pom 中添加 config-client 和 autoconfigure 依赖，具体如代码清单 12-25 所示。

代码清单12-25 code\ch12-5\ch12-5-config-client-refresh-autoconfig\pom.xml

```xml
<dependencies>
    <dependency>
        <groupId>org.springframework.cloud</groupId>
        <artifactId>spring-cloud-config-client</artifactId>
    </dependency>
    <dependency>
        <groupId>org.springframework.boot</groupId>
        <artifactId>spring-boot-autoconfigure</artifactId>
    </dependency>
</dependencies>
```

（2）创建自动配置类

创建自动配置类 ConfigClientRefreshAutoConfiguration，并且添加相关注解，使其在 Spring boot 启动的时候将其加载，具体如代码清单 12-26 所示。

代码清单12-26 code\ch12-5\ch12-5-config-client-refresh-autoconfig\src\main\java\com\sc\book\config\ConfigClientRefreshAutoConfiguration.java

```java
@ConditionalOnClass(RefreshEndpoint.class)
@ConditionalOnProperty("spring.cloud.config.refreshInterval")
@AutoConfigureAfter(RefreshAutoConfiguration.class)
@Configuration
public class ConfigClientRefreshAutoConfiguration implements SchedulingConfigurer {

    private static final Log logger = LogFactory.getLog(ConfigClientRefreshAuto-
        Configuration.class);

    /**
     * 间隔刷新时间
     */
    @Value("${spring.cloud.config.refreshInterval}")
    private long refreshInterval;

    /**
     * 刷新的端点
     */
    @Autowired
    private RefreshEndpoint refreshEndpoint;

    @Override
    public void configureTasks(ScheduledTaskRegistrar scheduledTaskRegistrar) {

        final long interval = getRefreshIntervalInMilliseconds();

        logger.info(String.format("Scheduling config refresh task with %s second
            delay", refreshInterval));
        scheduledTaskRegistrar.addFixedDelayTask(new IntervalTask(new Runnable() {
            @Override
            public void run() {
                refreshEndpoint.refresh();
            }
        }, interval, interval));
    }

    /**
     * 以毫秒为单位返回刷新间隔。
     * @return
     */
    private long getRefreshIntervalInMilliseconds() {

        return refreshInterval * 1000;
    }

    /**
     * 如果没有在上下文中注册，则启用调度程序。
```

```
             */
            @ConditionalOnMissingBean(ScheduledAnnotationBeanPostProcessor.class)
            @EnableScheduling
            @Configuration
            protected static class EnableSchedulingConfigProperties {

            }
        }
```

在该类中,主要是注入了端点类,通过定时任务和刷新时间,进行配置请求刷新。

有一点需要说明的是,在 ConfigClientRefreshAutoConfiguration 是直接调用了端点的 refresh 方法,所以对于 F 版的安全机制不需要对端点进行打开也可以。接下来创建 spring.factories 文件添加自动配置类,也就是上述类,具体如代码清单 12-27 所示。

代码清单12-27 ch12-5\ch12-5-config-client-refresh-autoconfig\src\main\resources\META-INF\spring.factories

```
org.springframework.boot.autoconfigure.EnableAutoConfiguration=\
com.sc.book.config.ConfigClientRefreshAutoConfiguration
```

在工程 ch12-5 下面创建一个 module,命名为 ch12-5-config-client-auto-refresh,代码如本书源码 ch12-5/ch12-5-config-client-auto-refresh 所示。

(3)客户端配置依赖

在 ch12-5-config-client-auto-refresh 的 pom 中添加相关依赖,具体如代码清单 12-28 所示。

代码清单12-28 code\ch12-5\ ch12-5-config-client-auto-refresh\pom.xml

```xml
<dependencies>

    <dependency>
        <groupId>org.springframework.cloud</groupId>
        <artifactId>spring-cloud-config-client</artifactId>
    </dependency>

    <!-- 做简单的安全和端点开放 -->
    <dependency>
        <groupId>org.springframework.boot</groupId>
        <artifactId>spring-boot-starter-security</artifactId>
    </dependency>

    <dependency>
        <groupId>cn.springcloud.book</groupId>
        <artifactId>ch12-5-config-client-refresh-autoconfig</artifactId>
        <version>1.0-SNAPSHOT</version>
    </dependency>

</dependencies>
```

主要是将前面创建的 ch12-5-config-client-refresh-autoconfig 以依赖的形式添加上去。

(4)编写主程序入口

创建类 ClientConfigGitApplication 并且添加相关注解,使其成为应用启动入口程序,具体如代码清单 12-29 所示。

代码清单12-29 code\ch12-5\ch12-5-config-client-auto-refresh\src\main\java\cn\spring-cloud\book\config\ClientConfigGitApplication.java

```
@SpringBootApplication
public class ClientConfigGitApplication {

    public static void main(String[] args) {
        SpringApplication.run(ClientConfigGitApplication.class, args);
    }
}
```

此外还需要添加 controller 类用于观看配置信息，具体如代码清单 12-30 所示。

代码清单12-30 code\ch12-5\ch12-5-config-client-auto-refresh\src\main\java\cn\spring-cloud\book\config\controller\ConfigClientController.java

```
@RestController
@RequestMapping("configConsumer")
@RefreshScope
public class ConfigClientController {

    @Value("${cn.springcloud.book.config}")
    private String config;

    @RequestMapping("/getConfigInfo")
    public String getConfigInfo(){
        return config;
    }
}
```

下面还需要配置，然后启动配置。

（5）配置文件配置

在 application.yml 中添加刷新时间间隔和端口等配置，具体如代码清单 12-31 所示。

代码清单12-31 ch12-5\ch12-5-config-client-auto-refresh\src\main\resources\application.yml

```
server:
    port: 9012

spring:
    application:
        name: ch12-5-config-client-auto-refresh
    cloud:
        config:
            refreshInterval: 60
```

refreshInterval: 60 表示的意思是每 60 秒刷新配置信息。下面还需要配置 bootstrap.yml，具体如代码清单 12-32 所示。

代码清单12-32 code\ch12-5\ch12-5-config-client-auto-refresh\src\main\resources\bootstrap.yml

```
spring:
```

```
      cloud:
        config:
          label: master
          uri: http://localhost:9090
          name: config-info
          profile: dev
```

到此基本就完成了，接下来启动服务端和客户端，访问地址 http://localhost:9012/config-Consumer/getConfigInfo。得到的结果如图 12-9 所示。

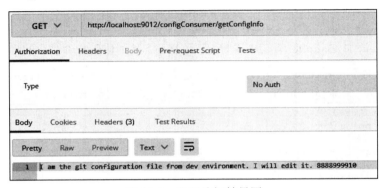

图 12-9　配置访问结果图

同时观察控制台每过 60 秒都会刷新，具体如下：

```
c.c.c.ConfigServicePropertySourceLocator : Fetching config from server at:
    http://localhost:9090
c.c.c.ConfigServicePropertySourceLocator : Located environment: name=config-
    info, profiles=[dev], label=master, version=86447836a46a8b5304c5d2a828038cb
    fd2db3f27, state=null
b.c.PropertySourceBootstrapConfiguration : Located property source: Composite-
    PropertySource {name='configService', propertySources=[MapPropertySource
    {name='configClient'}, MapPropertySource {name='https://gitee.com/zhong-
    zunfa/spring-cloud-config.git/SC-BOOK-CONFIG/config-info-dev.yml'}]}
```

这里只截取关键部分的信息，读者可以在操作的时候详细观察，当修改配置文件里面的内容，到了刷新时间后内容就会发生变更，如图 12-10 所示。

到此客户端刷新的案例就完成了，接下来将讲解客户端在特殊情况下的启动，比如 config-server 在服务网络中断等。

12.5.2　客户端回退功能

对于客户端回退功能可能有些读者会很疑问，什么时候会使用到这个功能呢？接下来就让我解释一下。

客户端配备回退机制，可以在出现网络中断时处理案例，或者配置服务因维护而关闭。当启用回退时，客户端适配器将"缓存"本地文件系统中的计算属性。要启用回退功能，只需指定存储缓存的位置即可。接下来介绍如何使用案例，对于服务端的使用方法如下。

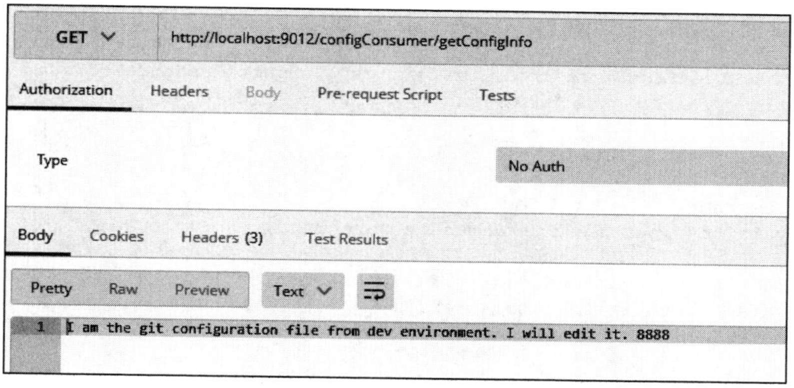

图 12-10 变更后的访问结果图

其中 ch12-6 的 pom 和 ch12-5 的一致，code\ch12-6\ch12-6-config-server 和 ch12-5/ch12-5-config-server 代码一致，这里重点讲解 ch12-6-config-client-fallback-autoconfig 里面的内容。在工程 ch12-6 下面创建一个 module，命名为 ch12-6-config-client-fallback-autoconfig，代码如本书源码 ch12-6/ch12-6-config-client-fallback-autoconfig 所示。

（1）配置依赖

在 ch12-6-config-client-fallback-autoconfig 的 pom 中添加如下依赖，具体如代码清单 12-33 所示。

代码清单12-33 code\ch12-6\ch12-6-config-client-fallback-autoconfig\pom.xml

```xml
<dependencies>
    <dependency>
        <groupId>org.springframework.cloud</groupId>
        <artifactId>spring-cloud-starter-config</artifactId>
    </dependency>
    <dependency>
        <groupId>org.springframework.security</groupId>
        <artifactId>spring-security-rsa</artifactId>
        <version>1.0.1.RELEASE</version>
    </dependency>
</dependencies>
```

其中 spring-security-rsa 这个依赖主要是用于当配置信息中存在敏感信息如用户名密码等缓存在本地的风险进行加密时用的。

（2）创建自动配置类

创建自动配置类 ConfigServerBootstrap，并且添加相关注解，使其在工程启动的时候进行加载，具体如代码清单 12-34 所示。

代码清单12-34 ch12-6\ch12-6-config-client-fallback-autoconfig\src\main\java\cn\springcloud\book\config\client\fallback\ConfigServerBootstrap.java

```
@Configuration
```

```java
@EnableConfigurationProperties
@PropertySource(value = {"configClient.properties", "file:${spring.cloud.config.
    fallbackLocation:}/fallback.properties"}, ignoreResourceNotFound = true)
public class ConfigServerBootstrap {

    public static final String FALLBACK_FILE_NAME = "fallback.properties";

    @Autowired
    private ConfigurableEnvironment environment;

    @Value("${spring.cloud.config.fallbackLocation:}")
    private String fallbackLocation;

    @Bean
    public ConfigClientProperties configClientProperties(){

        ConfigClientProperties clientProperties = new ConfigClientProperties (this.
            environment);
        clientProperties.setEnabled(false);
        return clientProperties;
    }

    @Bean
    public FallbackableConfigServicePropertySourceLocator fallbackableConfig-
        ServicePropertySourceLocator(){

        ConfigClientProperties client = configClientProperties();
        FallbackableConfigServicePropertySourceLocator fallbackableConfigService-
            PropertySourceLocator
                = new FallbackableConfigServicePropertySourceLocator(client,
                    fallbackLocation);
        return fallbackableConfigServicePropertySourceLocator;
    }
}
```

说明一下，spring.cloud.config.fallbackLocation 是指要回退配置文件的路径，file:${spring.cloud.config.fallbackLocation:}/fallback.properties 是回退本地配置文件所在的文件和名称。

configClient.properties 配置文件中信息具体如下：

```
spring.cloud.config.enabled=false
```

上面配置类中 new 的 FallbackableConfigServicePropertySourceLocator 类是用来创建本地回退文件的，具体如代码清单 12-35 所示。

代码清单12-35　ch12-6\ch12-6-config-client-fallback-autoconfig\src\main\java\cn\springcloud\book\config\client\fallback\FallbackableConfigServicePropertySourceLocator.java

```java
@Order(0)
public class FallbackableConfigServicePropertySourceLocator extends ConfigService-
    PropertySourceLocator {

    private boolean fallbackEnabled;
```

```java
    private String fallbackLocation;

@Autowired(required = false)
TextEncryptor textEncryptor;

public FallbackableConfigServicePropertySourceLocator(ConfigClientProperties
    defaultProperties, String fallbackLocation) {

    super(defaultProperties);
    this.fallbackLocation = fallbackLocation;
    this.fallbackEnabled = !StringUtils.isEmpty(fallbackLocation);
}

public PropertySource<?> locate(Environment environment){

    PropertySource<?> propertySource = super.locate(environment);
    if(fallbackEnabled){
        if(propertySource != null){
            storeLocally(propertySource);
        }
    }
    return propertySource;
}

private void storeLocally(PropertySource propertySource){

    StringBuilder sb = new StringBuilder();
    CompositePropertySource source = (CompositePropertySource)propertySource;
    for (String propertyName : source.getPropertyNames()) {
        Object value = source.getProperty(propertyName);
        if (textEncryptor != null)
            value = "{cipher}" + textEncryptor.encrypt(String.valueOf(value));
        sb.append(propertyName).append("=").append(value).append("\n");
    }

    System.out.println("file contents : " + sb.toString());
    saveFile(sb.toString());
}

private void saveFile(String contents){

    BufferedWriter output = null;
    File file = new File(fallbackLocation + File.separator + ConfigServer-
        Bootstrap.FALLBACK_FILE_NAME);
    try {
        if(!file.exists()){
            file.createNewFile();
        }
        output = new BufferedWriter(new FileWriter(file));
        output.write(contents);
    } catch (IOException e) {
        e.printStackTrace();
    } finally {
        if(output != null){
            try {
```

```
                    output.close();
                } catch (IOException e) {
                    System.out.print("Error" + e.getMessage());
                }
            }
        }
    }
}
```

上述代码主要是将加载到的远程配置文件在本地也创建一份，进行备份。

接下来是将 ConfigServerBootstrap 配置到 spring.factories 中，具体如代码清单 12-36 所示。

代码清单12-36 code\ch12-6\ch12-6-config-client-fallback-autoconfig\src\main\resources\META-INF\spring.factories

```
org.springframework.cloud.bootstrap.BootstrapConfiguration=\
cn.springcloud.book.config.client.fallback.ConfigServerBootstrap
```

在工程 ch12-6 下面创建一个 module，命名为 ch12-6-config-client-fallback，代码如本书源码 ch12-6/ch12-6-config-client-fallback 所示。

（3）客户端配置依赖

在 ch12-6-config-client-fallback 的 pom 中添加依赖，具体如代码清单 12-37 所示。

代码清单12-37 code\ch12-6\ch12-6-config-client-fallback\pom.xml

```xml
<dependencies>
    <dependency>
        <groupId>org.springframework.cloud</groupId>
        <artifactId>spring-cloud-config-client</artifactId>
    </dependency>

    <dependency>
        <groupId>cn.springcloud.book</groupId>
        <artifactId>ch12-6-config-client-fallback-autoconfig</artifactId>
        <version>1.0-SNAPSHOT</version>
    </dependency>
</dependencies>
```

上述的依赖，主要是将前面创建的自动配置工程 ch12-6-config-client-fallback-autoconfig 引入进来。接下来创建入口程序类。

（4）编写主程序入口

创建 ClientConfigGitApplication 类并且添加相关的注解作为启动程序入口，代码内容同代码清单 12-6 所示。

详细代码可查看：code\ch12-6\ch12-6-config-client-fallback\src\main\java\cn\springcloud\book\config\ClientConfigGitApplication.java。

此外还需要添加一个 controller 来显示加载的远程的配置内容，代码内容同代码清单 12-36 所示。

（5）配置文件配置

在 application.yml 文件中添加端口信息和应用名称，具体如代码清单 12-38 所示。

代码清单 12-38　code\ch12-6\ch12-6-config-client-fallback\src\main\resources\application.yml

```yaml
server:
    port: 9013
spring:
    application:
        name: ch12-6-config-client-fallback
```

在 bootstrap.yml 中添加回滚地址，具体如代码清单 12-39 所示。

代码清单 12-39　code\ch12-6\ch12-6-config-client-fallback\src\main\resources\bootstrap.yml

```yaml
spring:
    cloud:
        config:
            label: master
            uri: http://localhost:9090
            name: config-info
            profile: dev
            fallbackLocation: D:\\application_dev\\idea_workspace\\sc-book-
                local-config-test
```

fallbackLocation 指定了回退文件的路径，也就是当前服务所在的一个物理路径，到此编写就完成了。接下来分别启动 ch12-6-config-server 和 ch12-6-config-client-fallback，需要注意的是，在客户端启动过程中出现找不到系统文件的问题，因为指定的路径是在 ConfigServer-Bootstrap 配置上的，所以首次启动自然不存在这个文件，直接忽略就可以。启动完成后访问地址：localhost:9013/configConsumer/getConfigInfo。

可以看到如图 12-11 所示的结果。

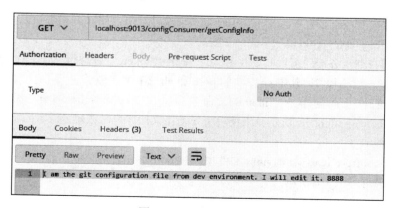

图 12-11　访问结果图

在图 12-11 中可以看到具体的远程配置内容，与此同时可以看看回退文件是否创建了文件，答案是肯定的，具体如图 12-12 所示。

图 12-12 退回文件报保存图

接下来关闭服务端和客户端,在单独启动客户端不启动服务的情况下查看客户端是否启动,经过测试是可以,客户端会尝试去连接远程,当远程失败后,再使用本地文件。

具体如图 12-13 所示:

```
: Fetching config from server at : http://localhost:9090
: Connect Timeout Exception on Url - http://localhost:9090. Will be trying the next url if available
: Could not locate PropertySource: I/O error on GET request for "http://localhost:9090/config-info/dev/master": Connection
: No active profile set, falling back to default profiles: default
```

图 12-13 客户端启动控制台信息图

这里就不具体截图了,读者手动尝试即可。到这里客户端的回退功能就完成了,这个功能也称之为客户端高可用的一部分,就是在服务端无法连接的情况下,客户端依然是可以用的。

接下来讲解配置安全认证。因为配置中心不是任何人知道地址就可以连接获取配置信息的,安全对于敏感的数据是很重要的。

12.5.3 客户端的安全认证机制 JWT

Spring Cloud Config 客户端使用 JWT 身份验证方法代替标准的基本身份验证。

这种方式需要对服务端和客户端都要改造。

授权三部曲:

1)客户端向服务端授权 Rest Controller 发送请求并且带上用户名和密码。

2)服务端返回 JWT Token。

3)客户端查询服务端的配置需要在 Header 中带上 token 令牌进行认证,在 code 的根目录下创建 module 工程 ch12-7,其中 ch12-7 的 pom 和 ch12-6 的一致。

在工程 ch12-7 下面创建一个 module,命名为 ch12-7-config-client-jwt,代码如本书源码 ch12-7/ch12-7-config-client-jw 所示,该模块主要是用来做客户端依赖的。

(1)配置依赖

在 ch12-7-config-client-jwt 的 pom 文件中添加 autoconfigure、configuration-processor 等依赖,具体如代码清单 12-40 所示。

代码清单12-40　code\ch12-7\ch12-7-config-client-jwt\pom.xml

```xml
<dependencies>
<dependency>
    <groupId>org.springframework.boot</groupId>
    <artifactId>spring-boot-autoconfigure</artifactId>
```

```xml
    </dependency>

    <dependency>
        <groupId>org.springframework.boot</groupId>
        <artifactId>spring-boot-configuration-processor</artifactId>
        <optional>true</optional>
    </dependency>

    <dependency>
        <groupId>org.springframework.cloud</groupId>
        <artifactId>spring-cloud-starter-config</artifactId>
    </dependency>
</dependencies>
```

接下来创建自动配置类。

（2）Config 配置类

创建 config 文件，具体如代码清单 12-41 所示。

代码清单12-41 code\ch12-7\ch12-7-config-client-jwt\src\main\java\cn\springcloud\book\config\jwt\config\ConfigClientBootstrapConfiguration.java

```java
@Configuration
@Order(Ordered.LOWEST_PRECEDENCE)
public class ConfigClientBootstrapConfiguration {

    private static Log logger = LogFactory.getLog(ConfigClientBootstrapConfiguration.
        class);

    @Value("${spring.cloud.config.username}")
    private String jwtUserName;

    @Value("${spring.cloud.config.password}")
    private String jwtPassword;

    @Value("${spring.cloud.config.endpoint}")
    private String jwtEndpoint;

    private String jwtToken;

    @Autowired
    private ConfigurableEnvironment environment;

    @PostConstruct
    public void init(){

        RestTemplate restTemplate = new RestTemplate();
        LoginRequest loginBackend = new LoginRequest();
        loginBackend.setUsername(jwtUserName);
        loginBackend.setPassword(jwtPassword);

        String serviceUrl = jwtEndpoint;
        Token token;

        try {
```

```java
            token = restTemplate.postForObject(serviceUrl, loginBackend, Token.class);
            if(token.getToken() == null){
                throw new Exception();
            }

            // 设置token
            setJwtToken(token.getToken());
        } catch (Exception e) {
            e.printStackTrace();
        }
    }

    public String getJwtToken() {
        return jwtToken;
    }

    public void setJwtToken(String jwtToken) {
        this.jwtToken = jwtToken;
    }

    @Bean
    public ConfigServicePropertySourceLocator configServicePropertySourceLocator
        (ConfigClientProperties configClientProperties){

        ConfigServicePropertySourceLocator configServicePropertySourceLocator =
            new ConfigServicePropertySourceLocator(configClientProperties);
        configServicePropertySourceLocator.setRestTemplate(customRestTemplate());
        return configServicePropertySourceLocator;
    }

    @Bean
    public ConfigClientProperties configClientProperties(){

        ConfigClientProperties clientProperties = new ConfigClientProperties
            (this.environment);
        clientProperties.setEnabled(false);
        return clientProperties;
    }

    private RestTemplate customRestTemplate(){

        Map<String, String> headers = new HashMap<>();
        headers.put("token", "Bearer:" + jwtToken);
        SimpleClientHttpRequestFactory requestFactory = new SimpleClientHttp-
            RequestFactory();
        requestFactory.setReadTimeout((60 * 1000 * 3) + 5000); // TODO 3m5s make
        RestTemplate template = new RestTemplate(requestFactory);
        if (!headers.isEmpty()) {
            template.setInterceptors(
                    Arrays.<ClientHttpRequestInterceptor> asList(new Generic-
                        RequestHeaderInterceptor(headers)));
        }

        return template;
    }
```

```java
    public static class GenericRequestHeaderInterceptor implements ClientHttpRequest-
        Interceptor{

        private final Map<String, String> headers;
        public GenericRequestHeaderInterceptor(Map<String, String> headers){
            this.headers = headers;
        }

        @Override
        public ClientHttpResponse intercept(HttpRequest httpRequest, byte[]
            bytes, ClientHttpRequestExecution clientHttpRequestExecution) throws
            IOException {

            headers.entrySet().stream().forEach(header -> {
                httpRequest.getHeaders().add(header.getKey(), header.getValue());
            });
            return clientHttpRequestExecution.execute(httpRequest, bytes);
        }
    }
}
```

@PostConstruct 注解是执行是在 servlet 构造函数和 Init 方法之间执行具体容器加载时序图如图 12-14 所示。

也就是说，在容器启动过程中会创建一个 restTemplate 对象将用户名和密码发送到 config-server 端进行认证，认证成功会返回 token，如果认证过程中用户名或者是密码出错将返回一个 401 认证失败。

其中 ${spring.cloud.config.username}、${spring.cloud.config.password} 等参数是配置在配置的客户端中，后面会进行讲解。需要创建 ConfigServicePropertySourceLocator 这个 bean 并且自定义一个 restTemplate 对象需要带上 token 信息，这就是上述代码中的 customRestTemplate() 方法。需要定义一个 ClientHttpRequestInterceptor 接口的实现类，也就是上述代码中的 GenericRequestHeaderInterceptor，主要用于拦截发送到 config-server 获取配置信息时，将 token 信息添加到 HttpServletRequest 的 Headers 中。

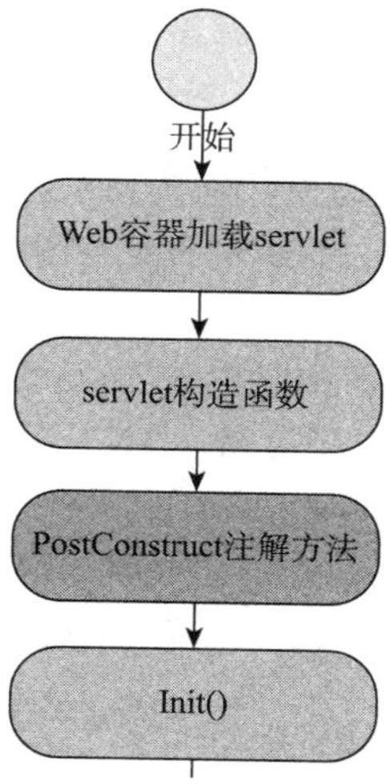

图 12-14 容器加载时序图

（3）创建实体信息

dto 实体类的编写，主要是用来传递用户信息。

首先创建 LoginRequest 实体类，具体如代码清单 12-42 所示。

代码清单12-42 code\ch12-7\ch12-7-config-client-jwt\src\main\java\cn\springcloud\book\config\jwt\dto\LoginRequest.java

```java
@JsonIgnoreProperties(ignoreUnknown = true)
@JsonInclude(JsonInclude.Include.NON_NULL)
public class LoginRequest {

    @JsonProperty
```

```
    private String username;
    @JsonProperty
    private String password;
    //省略getter setter toString方法
}
```

创建 token 实体类 Token.java，具体如代码清单 12-43 所示。

代码清单12-43　code\ch12-7\ch12-7-config-client-jwt\src\main\java\cn\springcloud\book\config\jwt\dto\Token.java

```
@JsonIgnoreProperties(ignoreUnknown = true)
@JsonInclude(JsonInclude.Include.NON_NULL)
public class Token {

    @JsonProperty
    private String token;
    //省略getter setter toString方法
}
```

（4）spring.factories 配置

此外需要创建 spring.factories 文件，将之前创建的 ConfigClientBootstrapConfiguration 类加入其中，具体如代码清单 12-44 所示。

代码清单12-44　code\ch12-7\ch12-7-config-client-jwt\src\main\resources\META-INF\spring.factories

```
org.springframework.cloud.bootstrap.BootstrapConfiguration=\
cn.springcloud.book.config.jwt.config.ConfigClientBootstrapConfiguration
```

到此客户端自动配置加载模块就完成了，接下来进行客户端编写和配置。

在工程 ch12-7 下面创建一个 module，命名为 ch12-7-config-client-auth-jwt，代码如本书源码 ch12-7/ch12-7-config-client-auth-jwt 所示。

（5）配置依赖

在 ch12-7-config-client-auth-jwt 的 pom 文件中添加依赖信息，具体如代码清单 12-45 所示。

代码清单12-45　code\ch12-7\ch12-7-config-client-auth-jwt\pom.xml

```xml
<dependencies>
    <dependency>
        <groupId>org.springframework.cloud</groupId>
        <artifactId>spring-cloud-config-client</artifactId>
    </dependency>

    <dependency>
        <groupId>cn.springcloud.book</groupId>
        <artifactId>ch12-7-config-client-jwt</artifactId>
        <version>1.0-SNAPSHOT</version>
    </dependency>
</dependencies>
```

上述内容主要是将之前创建好的工程 ch12-7-config-client-jwt 引入进来。接下来编写入口程序。

(6) 编写主入口程序

创建 ClientConfigGitApplication 类并且添加相关注解信息,作为程序入口启动类,代码内容同代码清单 12-6 所示。

详细代码可查看:code\ch12-7\ch12-7-config-client-auth-jwt\src\main\java\cn\springcloud\book\config\jwt\ClientConfigGitApplication.java。

此外还需要编写一个名为 ConfigClientController 的 controller 来查看加载 git 上的配置信息。具体代码同代码清单 12-30 所示。

(7) 配置文件配置

在 bootstrap.yml 中配置用户名和密码以及认证的请求地址,具体如代码清单 12-46 所示。

代码清单12-46 code\ch12-7\ch12-7-config-client-auth-jwt\src\main\resources\bootstrap.yml

```yaml
spring:
  cloud:
    config:
      label: master
      uri: http://localhost:9090
      name: config-info
      profile: dev
      username: root
      password: 123
      enabled: false
      endpoint: http://localhost:9090/auth
```

其中 password 和 username 是 config-server 端配置需要的认证用户信息,endpoint 是一个 http 地址也就是 config server 的访问验证授权的地址。

在 application.yml 文件中添加端口配置,具体如代码清单 12-47 所示。

代码清单12-47 code\ch12-7\ch12-7-config-client-auth-jwt\src\main\resources\application.yml

```yaml
server:
  port: 9014

spring:
  application:
    name: ch12-7-config-client-auth-jwt
```

到此客户端的编写就完成了,细心的读者应该已经发现,除了配置用户名和密码,请求地址外其他的部分都不用操作,这就是自动配置的好处。接下来编写 JWT 服务端,服务也是需要同步改造的。

在工程 ch12-7 下面创建一个 module,命名为 ch12-7-config-server-auth-jwt,代码如本书源码 ch12-7/ch12-7-config-server-auth-jwt 所示。

该模块的类的大体调用过程如图 12-15 所示。

(8) 配置依赖

在 ch12-7-config-server-auth-jwt 的 pom 文件中添加依赖信息,具体如代码清单 12-48 所示。

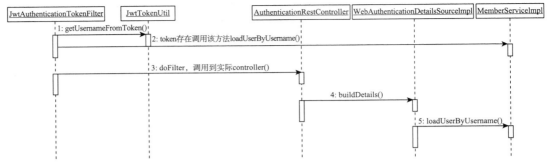

图 12-15 类调用时序图

代码清单12-48　code\ch12-7\ch12-7-config-server-auth-jwt\pom.xml

```xml
<dependencies>

    <dependency>
        <groupId>org.springframework.cloud</groupId>
        <artifactId>spring-cloud-config-server</artifactId>
    </dependency>

    <dependency>
        <groupId>org.springframework.boot</groupId>
        <artifactId>spring-boot-starter-security</artifactId>
    </dependency>

    <dependency>
        <groupId>io.jsonwebtoken</groupId>
        <artifactId>jjwt</artifactId>
        <version>0.9.0</version>
    </dependency>
    <dependency>
        <groupId>com.google.code.gson</groupId>
        <artifactId>gson</artifactId>
        <version>2.7</version>
    </dependency>
</dependencies>
```

在上述配置中添加。jjwt、gson 等相关认证和转化参数结构的依赖，接下来创建应用启动入口类。

（9）编写主程序入口

创建名为 ConfigGitApplication 类并且添加相关注解信息，使其成为程序的启动入口，代码内容同代码清单 12-3 所示。

详细代码可以查看：code\ch12-7\ch12-7-config-server-auth-jwt\src\main\java\cn\springcloud\book\config\ConfigGitApplication.java。

下面还需要创建认证请求的实体类。

（10）创建实体信息

创建 models 信息，创建 JwtAuthenticationRequest 类的具体代码如代码清单 12-49 所示。

代码清单12-49　code\ch12-7\ch12-7-config-server-auth-jwt\src\main\java\cn\springcloud\
book\config\models\JwtAuthenticationRequest.java

```java
public class JwtAuthenticationRequest implements Serializable {

    private String username;
    private String password;

    public JwtAuthenticationRequest() {
        super();
    }

    public JwtAuthenticationRequest(String username, String password) {
        this.setUsername(username);
        this.setPassword(password);
    }
}
```

该实体主要是用于传递用户名和密码。

创建 JwtAuthenticationResponse 类具体如代码清单 12-50 所示。

代码清单12-50　code\ch12-7\ch12-7-config-server-auth-jwt\src\main\java\cn\springcloud\
book\config\models\JwtAuthenticationResponse.java

```java
public class JwtAuthenticationResponse implements Serializable {

    private final String token;

    public JwtAuthenticationResponse(String token) {
        this.token = token;
    }
    //省略getter setter方法
}
```

该实体类主要是返回 token 信息。

创建 JwtUser 类具体如代码清单 12-51 所示。

代码清单12-51　code\ch12-7\ch12-7-config-server-auth-jwt\src\main\java\cn\springcloud\
book\config\models\JwtAuthenticationResponse.java

```java
public class JwtUser implements UserDetails {

    private final String username;
    private final String password;
    private final Collection<? extends GrantedAuthority> authorities;

    public JwtUser(String username, String password, Collection<? extends
        GrantedAuthority> authorities) {
        this.username = username;
        this.password = password;
        this.authorities = authorities;
    }
```

```java
    @JsonIgnore
    @Override
    public boolean isAccountNonExpired() {
        return true;
    }
    //省略getPassword、getUsername、getAuthorities、isEnabled、isAccountNonLocked、
      isCredentialsNonExpired等方法
}
```

该实体主要是 JWT 用户认证信息。

(11) Security 模块

创建 JWT 的 token 认证过滤器，具体代码如代码清单 12-52 所示。

代码清单12-52 code\ch12-7\ch12-7-config-server-auth-jwt\src\main\java\cn\springcloud\book\config\security\filters\JwtAuthenticationTokenFilter.java

```java
public class JwtAuthenticationTokenFilter extends UsernamePasswordAuthenticationFilter {

    @Autowired
    private UserDetailsService userDetailsService;

    @Autowired
    private JwtTokenUtil jwtTokenUtil;

    private final String tokenHeader = "token";

    @Override
    public void doFilter(ServletRequest request, ServletResponse response,
        FilterChain chain) throws IOException, ServletException {

        HttpServletRequest httpRequest = (HttpServletRequest) request;
        String authToken = httpRequest.getHeader(tokenHeader);
        String username = jwtTokenUtil.getUsernameFromToken(authToken);

        if (username != null && SecurityContextHolder.getContext().getAuthentication()
            == null) {
            UserDetails userDetails = this.userDetailsService.loadUserByUser-
                name(username);
            if(jwtTokenUtil.validateToken(authToken, userDetails)){
                UsernamePasswordAuthenticationToken auth = new UsernamePasswordAuthe
                    nticationToken(userDetails, null, userDetails.getAuthorities());
                auth.setDetails(new WebAuthenticationDetailsSource().buildDetails
                    (httpRequest));
                SecurityContextHolder.getContext().setAuthentication(auth);
            }
        }

        chain.doFilter(request, response);
    }
}
```

接下来创建 JWT 工具类，主要用于生成 JWT 的 token 和 token 验证，具体如代码清单 12-53 所示。

代码清单12-53 code\ch12-7\ch12-7-config-server-auth-jwt\src\main\java\cn\springcloud\
book\config\security\JwtTokenUtil.java

```java
@Component
public class JwtTokenUtil implements Serializable {

    private static final String CLAIM_KEY_USERNAME = "sub";
    private static final String CLAIM_KEY_AUDIENCE = "audience";
    private static final String CLAIM_KEY_CREATED = "created";

    private static final String AUDIENCE_UNKNOWN = "unknown";
    private static final String AUDIENCE_WEB = "web";

    private Key secret = MacProvider.generateKey();
    private Long expiration = (long) 120; // 2 minutes

    public String generateToken(JwtUser userDetails) {
        Map<String, Object> claims = new HashMap<>();
        claims.put(CLAIM_KEY_USERNAME, userDetails.getUsername());
        claims.put(CLAIM_KEY_AUDIENCE, AUDIENCE_WEB);
        claims.put(CLAIM_KEY_CREATED, new Date().getTime() / 1000);

        return generateToken(claims);
    }

    private String generateToken(Map<String, Object> claims) {
        return Jwts.builder().setClaims(claims).setExpiration(generateExpirationDate())
                .signWith(SignatureAlgorithm.HS512, secret).compact();
    }

    private Date generateExpirationDate() {
        return new Date(System.currentTimeMillis() + expiration * 1000);
    }

    public String getUsernameFromToken(String token) {
        if (token == null) {
            return null;
        }

        String username;
        try {
            final Claims claims = getClaimsFromToken(token);
            username = claims.getSubject();
        } catch (Exception e) {
            username = null;
        }

        return username;
    }

    private Claims getClaimsFromToken(String token) {
        Claims claims;

        final String tokenClean = token.substring(7); // remove "Bearer:"
        try {
            claims = Jwts.parser().setSigningKey(secret).parseClaimsJws (token-
```

```
                Clean).getBody();
        } catch (Exception e) {
            claims = null;
        }

        return claims;
    }

    //校验token的合法性
    public Boolean validateToken(String token, UserDetails userDetails) {
        JwtUser user = (JwtUser) userDetails;
        final String username = getUsernameFromToken(token);

        return (username.equals(user.getUsername()) && !isTokenExpired(token));
    }

    private Boolean isTokenExpired(String token) {
        final Date expiration = getExpirationDateFromToken(token);
        return expiration.before(new Date());
    }

    public Date getExpirationDateFromToken(String token) {
        Date expiration;
        try {
            final Claims claims = getClaimsFromToken(token);
            expiration = claims.getExpiration();
        } catch (Exception e) {
            expiration = null;
        }
        return expiration;
    }
}
```

该类主要是生产 token 的工具类，根据传递的用户信息来生产 token，或者是验证请求的 token 合法性。

创建 JWT 认证端点类——JwtAuthenticationEntryPoint 类，具体如代码清单 12-54 所示。

代码清单12-54　code\ch12-7\ch12-7-config-server-auth-jwt\src\main\java\cn\springcloud\book\config\security\JwtAuthenticationEntryPoint.java

```
@Component
public class JwtAuthenticationEntryPoint implements AuthenticationEntryPoint, Serializable {

    private static final long serialVersionUID = 3671302053846319657L;

    @Override
    public void commence(HttpServletRequest httpServletRequest, HttpServletResponse
        httpServletResponse, AuthenticationException e) throws IOException, Servlet-
        Exception {

        // 没有认证通过将添加401
        httpServletResponse.sendError(HttpServletResponse.SC_UNAUTHORIZED, "Unauthorized");
    }
}
```

在认证过程中,未能认证通过直接返回 401 状态码。创建一个认证账号的验证类——MemberServiceImpl 类,具体代码如代码清单 12-55 所示。

代码清单12-55 code\ch12-7\ch12-7-config-server-auth-jwt\src\main\java\cn\springcloud\book\config\security\MemberServiceImpl.java

```java
@Service("userDetailsService")
public class MemberServiceImpl implements UserDetailsService {

    private static final PasswordEncoder BCRYPT = new BCryptPasswordEncoder();

    // 这里的用户名和密码读者可以从数据库中获取
    @Value("${spring.security.user.name}")
    private String hardcodedUser;

    @Value("${spring.security.user.password}")
    private String password;

    @Override
    public JwtUser loadUserByUsername(String username) throws UsernameNotFoundException {
        // 对密码进行加密
        String hardcodedPassword = BCRYPT.encode(password);
        if (username.equals(hardcodedUser) == false) {
            throw new UsernameNotFoundException(String.format("No user found with username '%s'.", username));
        }else{

            SimpleGrantedAuthority simpleGrantedAuthority = new SimpleGrantedAuthority("ROLE_USER");
            List<GrantedAuthority> grantedAuthorityList = new ArrayList<GrantedAuthority>();
            grantedAuthorityList.add(simpleGrantedAuthority);
            return new JwtUser(hardcodedUser, hardcodedPassword, grantedAuthorityList);
        }
    }
}
```

创建一个将传递过来的对象数据封装到 JwtAuthenticationRequest 里面的 WebAuthenticationDetailsSourceImpl 对象,具体如代码清单 12-56 所示。

代码清单12-56 ch12-7/ch12-7-config-server-auth-jwt/src/main/java/cn/springcloud/book/config/security/WebAuthenticationDetailsSourceImpl.java

```java
@Component
public class WebAuthenticationDetailsSourceImpl implements AuthenticationDetailsSource<HttpServletRequest, JwtAuthenticationRequest> {

    @Override
    public JwtAuthenticationRequest buildDetails(HttpServletRequest request) {

        Gson gson = new Gson();
        String json = new String();
        String output = new String();
```

```
            BufferedReader br;
            StringBuffer buffer = new StringBuffer(16384);
            JwtAuthenticationRequest jwtAuthenticationRequest = new JwtAuthentication-
                Request();
            try {
                br = new BufferedReader(new InputStreamReader(request.getInputStream()));
                while ((output = br.readLine()) != null){
                    buffer.append(output);
                }

                json = buffer.toString();
                jwtAuthenticationRequest = gson.fromJson(json, JwtAuthenticationRequest.
                    class);
            } catch (IOException e) {
                e.printStackTrace();
            }

            return jwtAuthenticationRequest;
    }
}
```

该类将数据封装成 json 格式后返回到客户端。

（12）Controller 模块

客户端请求获取 token 的 controller 对象，具体如代码清单 12-57 所示。

代码清单12-57　code\ch12-7\ch12-7-config-server-auth-jwt\src\main\java\cn\springcloud\book\config\rest\AuthenticationRestController.java

```
@RestController
public class AuthenticationRestController {

    @Autowired
    private AuthenticationManager authenticationManager;

    @Autowired
    private JwtTokenUtil jwtTokenUtil;

    @Autowired
    private MemberServiceImpl userDetailsService;

    @Autowired
    private WebAuthenticationDetailsSourceImpl webAuthenticationDetailsSource;

    @RequestMapping(value = "/auth", method = RequestMethod.POST)
    public ResponseEntity<?> createAuthenticationToken(HttpServletRequest
        request){

        JwtAuthenticationRequest jwtAuthenticationRequest = webAuthentication-
            DetailsSource.buildDetails(request);
        UsernamePasswordAuthenticationToken authToken = new UsernamePasswordAuthenti-
            cationToken (jwtAuthenticationRequest.getUsername(), jwtAuthentication-
            Request.getPassword());
        authToken.setDetails(jwtAuthenticationRequest);
        Authentication authenticate = authenticationManager.authenticate(authToken);
```

```
                SecurityContextHolder.getContext().setAuthentication(authenticate);
                JwtUser userDetails = userDetailsService.loadUserByUsername(jwtAuthentica-
                    tionRequest.getUsername());
                final String token = jwtTokenUtil.generateToken(userDetails);

                return ResponseEntity.ok(new JwtAuthenticationResponse(token));
        }
}
```

该类是请求验证的 controller。

（13）Config 模块

创建 SecurityConfig 自动配置类，主要是 bean 的创建和安全认证的信息配置，具体如下代码清单 12-58 所示。

代码清单12-58 code\ch12-7\ch12-7-config-server-auth-jwt\src\main\java\cn\springcloud\book\config\config\SecurityConfig.java

```
@Configuration
@EnableWebSecurity
@EnableGlobalMethodSecurity(prePostEnabled = true)
public class SecurityConfig extends WebSecurityConfigurerAdapter {

    @Autowired
    private JwtAuthenticationEntryPoint unAuthorizedHandler;

    @Autowired
    private WebAuthenticationDetailsSourceImpl webAuthenticationDetailsSource;

    @Bean
    @ConditionalOnMissingBean(AuthenticationManager.class)
    public UsernamePasswordAuthenticationFilter usernamePasswordAuthenticationF
        ilter(AuthenticationManager authenticationManager)throws Exception{

        UsernamePasswordAuthenticationFilter usernamePasswordAuthenticationFilter
            = new UsernamePasswordAuthenticationFilter();
        usernamePasswordAuthenticationFilter.setAuthenticationManager(authentication-
            Manager);
        usernamePasswordAuthenticationFilter.setAuthenticationDetailsSource(web-
            AuthenticationDetailsSource);
        return usernamePasswordAuthenticationFilter;
    }

    @Bean
    public PasswordEncoder passwordEncoder(){
        return new BCryptPasswordEncoder();
    }

    @Bean
    @Override
    public AuthenticationManager authenticationManagerBean() throws Exception {
        return super.authenticationManagerBean();
    }
```

```java
    @Bean
    public JwtAuthenticationTokenFilter authenticationTokenFilter() throws Exception{
        JwtAuthenticationTokenFilter jwtAuthenticationTokenFilter = new Jwt-
            AuthenticationTokenFilter();
        jwtAuthenticationTokenFilter.setAuthenticationManager(authenticationManager());
        jwtAuthenticationTokenFilter.setAuthenticationDetailsSource(webAuthen-
            ticationDetailsSource);
        return jwtAuthenticationTokenFilter;
    }

    @Override
    protected void configure(HttpSecurity httpSecurity) throws Exception {
        httpSecurity
                .csrf().disable()
                .exceptionHandling().authenticationEntryPoint(unAuthorizedHandler)
                .and()
                .sessionManagement().sessionCreationPolicy(SessionCreationPolicy.
                    STATELESS)
                .and()
                .authorizeRequests()
                .antMatchers(HttpMethod.GET, "/").permitAll()
                .antMatchers("/auth/**").permitAll()
                .anyRequest().authenticated().and().formLogin()
                .authenticationDetailsSource(webAuthenticationDetailsSource)
                .permitAll();

        // 添加自定义的JWT的安全过滤的filter
        httpSecurity.addFilterBefore(authenticationTokenFilter(), UsernamePass-
            wordAuthenticationFilter.class);
        httpSecurity.headers().cacheControl();
    }
}
```

该类主要是进行安全认证和 token 的过滤。

到此服务端的部分也已经写完了。接下来启动服务端（ch12-7-config-server-auth-jwt）和客户端（ch12-7-config-client-auth-jwt）。

输入正确的密码，客户端启动过程中会出现如下信息：

```
c.c.c.ConfigServicePropertySourceLocator : Fetching config from server at : http://localhost:9090
c.c.c.ConfigServicePropertySourceLocator : Located environment: name=config-info, profiles=[dev], label=master, version=201917124912c018eb88adebf5b7d11d1c1c05bf, state=null
b.c.PropertySourceBootstrapConfiguration : Located property source: CompositePropertySource {name='configService', propertySources=[MapPropertySource {name='configClient'}, MapPropertySource {name='https://gitee.com/zhongzunfa/spring-cloud-config.git/SC-BOOK-CONFIG/config-info-dev.yml'}]}
```

服务同时也会出现如下信息：

```
o.s.c.c.s.e.NativeEnvironmentRepository : Adding property source: file:/C:/Users/ADMINI~1/AppData/Local/Temp/config-repo-1084479392871439955/SC-BOOK-
```

```
CONFIG/config-info-dev.yml
```

访问客户端的 controller，如图 12-16 所示，说明认证正常。

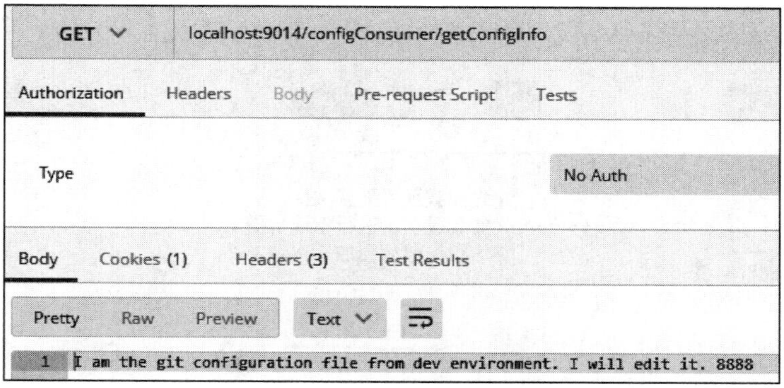

图 12-16　访问结果图

当使用错误的密码时，客户端启动会出现如下信息：

```
.ConfigurationPropertiesRebinderAutoConfiguration$$EnhancerBySpringCGLIB$$db61
    4e2d] is not eligible for getting processed by all BeanPostProcessors (for
    example: not eligible for auto-proxying)
org.springframework.web.client.HttpClientErrorException: 401 null
org.springframework.web.client.RestTemplate.doExecute(RestTemplate.java:730)
org.springframework.web.client.RestTemplate.execute(RestTemplate.java:686)
org.springframework.web.client.RestTemplate.postForObject(RestTemplate.
    java:437)
cn.springcloud.book.config.jwt.config.ConfigClientBootstrapConfiguration.init
    (ConfigClientBootstrapConfiguration.java:69)
```

这里只是截取关键信息，出现定义在代码中的 401 授权失败，读者可以自定义修改用户名或者密码进行验证。

当然读者也可以在关键地方打印日志，比如说打印 token 和传递进来的用户名和密码等，这样可以进行更好地观察。

12.6　高可用部分

对于线上的生产环境，通常对其都是有很高的要求，其中高可用是不可或缺的一部分，必须要保证服务是可用状态，才能保证系统更好地运行，这是业务稳定的保证。本节将讲解客户端高可用和服务端高可用问题，在此之前讲解的客户端回退功能其实就是客户端高可用的一种方式。

12.6.1　客户端高可用

对于客户端高可用的讲解，这里的方案主要还是用 file 的形式，前面讲解过一个回退的案例，两者思路大体一致。

这里解释一下，客户端高可用主要是解决当服务端不可用的情况下在客户端然可以正常启动。从客户端的角度出发，不是增加配置中心的高可用性，而是降低客户端对配置中心的依赖程度，从而提高整个分布式架构的健壮性。

客户端整体简要图示如图 12-17 所示：

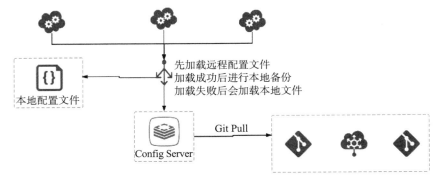

图 12-17　客户端加载配置高可用图

在 code 的根目录下面创建 module 工程 ch12-8，其中 ch12-8 的 pom 和 ch12-7 的一致，在工程 ch12-8 下面创建一个 module，命名为 ch12-8-config-client-high-availability-autoconfig，代码如本书源码 ch12-8/ch12-8-config-client-high-availability-autoconfig 所示。

（1）配置依赖

在 ch12-8-config-client-high-availability-autoconfig 的 pom.xml 文件中添加依赖，具体如代码清单所示 12-59 所示。

代码清单12-59　code\ch12-8\ch12-8-config-client-high-availability-autoconfig\pom.xml

```xml
<dependencies>
    <dependency>
        <groupId>org.springframework.cloud</groupId>
        <artifactId>spring-cloud-config-client</artifactId>
    </dependency>

</dependencies>
```

接下来创建自动配置类。

（2）自动加载属性配置类

创建配置属性加载 ConfigSupportProperties 类，具体如代码清单 12-60 所示。

代码清单12-60　code\ch12-8\ch12-8-config-client-high-availability-autoconfig\src\main\java\
　　　　　　　cn\springcloud\book\config\configuration\ConfigSupportConfiguration.java

```java
@Component
@ConfigurationProperties(prefix = ConfigSupportProperties.CONFIG_PREFIX)
public class ConfigSupportProperties {

    public static final String CONFIG_PREFIX = "spring.cloud.config.backup";
    private final String DEFAULT_FILE_NAME = "fallback.properties";
```

```java
    private boolean enable = false;
    private String fallbackLocation;

    public String getFallbackLocation() {
        return fallbackLocation;
    }

    public void setFallbackLocation(String fallbackLocation) {
        // 如果只是填写路径,那么就为其添加上一个默认的文件名
        if(fallbackLocation.indexOf(".") == -1){
            this.fallbackLocation = fallbackLocation + DEFAULT_FILE_NAME;
            return;
        }
        this.fallbackLocation = fallbackLocation;
    // 省略enable的getter和setter方法
}
```

创建配置 ConfigSupportConfiguration 类,该类主要的作用是判断远程加载信息是否可用,如果不能用将读取加载本地配置文件进行启动,具体如代码清单 12-61 所示。

代码清单12-61 code\ch12-8\ch12-8-config-client-high-availability-autoconfig\src\main\java\cn\springcloud\book\config\configuration\ConfigSupportConfiguration.java

```java
@Configuration
@EnableConfigurationProperties(ConfigSupportProperties.class)
public class ConfigSupportConfiguration implements ApplicationContextInitializer
    <ConfigurableApplicationContext>, Ordered {

    private final Logger LOGGER = LoggerFactory.getLogger(ConfigSupportConfiguration.
        class);
    private final Integer orderNum = Ordered.HIGHEST_PRECEDENCE + 11;

    @Autowired(required = false)
    private List<PropertySourceLocator> propertySourceLocators = Collections.
        EMPTY_LIST;

    @Autowired
    private ConfigSupportProperties configSupportProperties;

    @Override
    public void initialize(ConfigurableApplicationContext configurableApplicationContext) {

        if (!isHasCloudConfigLocator(this.propertySourceLocators)) {
            LOGGER.info("未启用Config Server管理配置");
            return;
        }

        LOGGER.info("检查Config Service配置资源");
        ConfigurableEnvironment environment = configurableApplicationContext.get-
            Environment();
        MutablePropertySources propertySources = environment.getPropertySources();
        LOGGER.info("加载PropertySources源: " + propertySources.size() + "个");

        if (!configSupportProperties.isEnable()) {
            LOGGER.warn("未启用配置备份功能,可使用{}.enable打开", ConfigSupportProperties.
```

```java
                CONFIG_PREFIX);
            return;
    }

    if (isCloudConfigLoaded(propertySources)) {
        PropertySource cloudConfigSource = getLoadedCloudPropertySource(property
            Sources);
        LOGGER.info("成功获取ConfigService配置资源");
        //备份
        Map<String, Object> backupPropertyMap = makeBackupPropertyMap(cloud
            ConfigSource);
        doBackup(backupPropertyMap, configSupportProperties.getFallbackLocation());

    } else {

        LOGGER.error("获取ConfigService配置资源失败");
        Properties backupProperty = loadBackupProperty(configSupportProperties.
            getFallbackLocation());
        if (backupProperty != null) {
            HashMap backupSourceMap = new HashMap<>(backupProperty);

            PropertySource backupSource = new MapPropertySource("backupSource",
                backupSourceMap);
            propertySources.addFirst(backupSource);
            LOGGER.warn("使用备份的配置启动: {}", configSupportProperties.getFall-
                backLocation());
        }
    }
}

@Override
public int getOrder() {
    return orderNum;
}

// 是否启用了Spring Cloud Config获取配置资源
private boolean isHasCloudConfigLocator(List<PropertySourceLocator> property-
    SourceLocators) {
    for (PropertySourceLocator sourceLocator : propertySourceLocators) {
        if (sourceLocator instanceof ConfigServicePropertySourceLocator) {
            return true;
        }
    }
    return false;
}

// 是否启用Cloud Config
private boolean isCloudConfigLoaded(MutablePropertySources propertySources) {
    if (getLoadedCloudPropertySource(propertySources) == null) {
        return false;
    }
    return true;
}

// 获取加载的Cloud Config 配置项
private PropertySource getLoadedCloudPropertySource(MutablePropertySources
    propertySources) {
```

```java
        if (!propertySources.contains(PropertySourceBootstrapConfiguration.
            BOOTSTRAP_PROPERTY_SOURCE_NAME)) {
            return null;
        }

        PropertySource propertySource = propertySources.get(PropertySourceBoot-
            strapConfiguration.BOOTSTRAP_PROPERTY_SOURCE_NAME);
        if (propertySource instanceof CompositePropertySource) {
            for (PropertySource<?> source : ((CompositePropertySource) propertySource).
                getPropertySources()) {
                if (source.getName().equals("configService")) {
                    return source;
                }
            }
        }

        return null;
    }

    // 生成备份的配置数据
    private Map<String, Object> makeBackupPropertyMap(PropertySource propertySource) {

        Map<String, Object> backupSourceMap = new HashMap<>();
        if (propertySource instanceof CompositePropertySource) {
            CompositePropertySource composite = (CompositePropertySource)
                propertySource;
            for (PropertySource<?> source : composite.getPropertySources()) {
                if (source instanceof MapPropertySource) {
                    MapPropertySource mapSource = (MapPropertySource) source;
                    for (String propertyName : mapSource.getPropertyNames()) {
                        // 前面的配置覆盖后面的配置
                        if (!backupSourceMap.containsKey(propertyName)) {
                            backupSourceMap.put(propertyName, mapSource.get-
                                Property(propertyName));
                        }
                    }
                }
            }
        }
        return backupSourceMap;
    }

    private void doBackup(Map<String, Object> backupPropertyMap, String filePath) {
        FileSystemResource fileSystemResource = new FileSystemResource(filePath);
        File backupFile = fileSystemResource.getFile();
        try {
            if (!backupFile.exists()) {
                backupFile.createNewFile();
            }
            if (!backupFile.canWrite()) {
                LOGGER.error("无法读写文件：{}", fileSystemResource.getPath());
            }

            Properties properties = new Properties();
            Iterator<String> keyIterator = backupPropertyMap.keySet().iterator();
            while (keyIterator.hasNext()) {
```

```
                String key = keyIterator.next();
                properties.setProperty(key, String.valueOf(backupPropertyMap.
                    get(key)));
            }

            FileOutputStream fos = new FileOutputStream(fileSystemResource.get-
                File());
            properties.store(fos, "Backup Cloud Config");
        } catch (IOException e) {
            LOGGER.error("文件操作失败: {}", fileSystemResource.getPath());
            e.printStackTrace();
        }
    }

    private Properties loadBackupProperty(String filePath) {
        PropertiesFactoryBean propertiesFactory = new PropertiesFactoryBean();
        Properties props = new Properties();
        try {
            FileSystemResource fileSystemResource = new FileSystemResource(filePath);
            propertiesFactory.setLocation(fileSystemResource);

            propertiesFactory.afterPropertiesSet();
            props = propertiesFactory.getObject();

        } catch (IOException e) {
            e.printStackTrace();
            return null;
        }

        return props;
    }
}
```

需要注意的是，启动顺序的设置，这是因为 Spring Cloud 使用的 PropertySourceBootstrap-Configuration 启动顺序为 private int order = -2147483638 + 10; 也就是 "HTGHEST PRECEDENCE = -2147483648;" 的值 order 的值越小越先加载。

所以上述的 orderNum 只要加上一个整数比其大即可，也就是一个比 10 大的整数即可。

（3）spring.factories 配置

此外需要创建 spring.factories 文件将 ConfigSupportConfiguration 类添加上去具体如代码清单 12-62 所示。

代码清单12-62 code\ch12-8\ch12-8-config-client-high-availability-autoconfig\src\main\resources\META-INF\spring.factories

```
org.springframework.cloud.bootstrap.BootstrapConfiguration=\
cn.springcloud.book.config.configuration.ConfigSupportConfiguration
```

到此客户端的自动配置加载模块就完成了，接下来将进行客户端编写和配置。

在工程 ch12-8 下面创建一个 module，命名为 ch12-8-config-client-high-availability，代码如本书源码 ch12-8/ch12-8-config-client-high-availability 所示。

（4）配置依赖

在 ch12-8-config-client-high-availability 的 pom 中添加依赖，具体如代码清单 12-63 所示。

代码清单12-63　ch12-8/ch12-8-config-client-high-availability/pom.xml

```xml
<dependencies>
    <dependency>
        <groupId>org.springframework.cloud</groupId>
        <artifactId>spring-cloud-config-client</artifactId>
    </dependency>

    <dependency>
        <groupId>cn.springcloud.book</groupId>
        <artifactId>ch12-8-config-client-high-availability-autoconfig</artifactId>
        <version>1.0-SNAPSHOT</version>
    </dependency>
</dependencies>
```

接下来创建应用的启动程序。

（5）主程序入口

创建类 ClientConfigGitApplication 并且添加相关的依赖，作为程序的入口程序，代码内容同代码清单 12-6。

详细代码可查看：ch12-8/ch12-8-config-client-high-availability/src/main/java/cn/springcloud/book/config/ClientConfigGitApplication.java。

创建 ConfigClientController 的 controller 类用于请求显示加载的配置信息，具体代码同代码清单 12-36 所示。

（6）配置文件配置

在 application.yml 文件配置中添加端口信息，具体如代码清单 12-64 所示。

代码清单12-64　code\ch12-8\ch12-8-config-client-high-availability\src\main\resources\application.yml

```yml
server:
    port: 9015

spring:
    application:
        name: ch12-8-config-client-high-availability
```

在 bootstrap.yml 文件配置中添加备份和回滚地址，具体如代码清单 12-65 所示。

代码清单12-65　code\ch12-8\ch12-8-config-client-high-availability\src\main\resources\bootstrap.yml

```yml
spring:
  cloud:
    config:
      label: master
      uri: http://localhost:9090
      name: config-info
      profile: dev
      backup:
        enable: true
```

```
              fallbackLocation: D:\\application_dev\\idea_workspace\\sc-book-
                  local-config-test\\fallback.properties
```

其中 enable 表示是否启动加载到远程配置信息进行本地备份，fallbackLocation 表示的是本地备份的路径，也可以是路径或者是路径加上文件名。到此客户端也编写完成。接下来编写服务端。

在工程 ch12-8 下面创建一个 module，命名为 ch12-8-config-server，代码如本书源码 ch12-8/ch12-8-config-server 所示。

（7）配置依赖

在 pom 中添加如下依赖，代码内容同代码清单 12-2 所示。

详细代码可查看：code\ch12-8\ch12-8-config-server\pom.xml。

配置依赖后，创建应用入口程序。

（8）主入口程序

创建 ConfigGitApplication 类并且添加相关的注解信息，使其成为入口程序。代码内容同代码清单 12-3。

详细代码可查看：code\ch12-8\ch12-8-config-server\src\main\java\cn\springcloud\book\config\ConfigGitApplication.java。

（9）配置文件配置

配置 application.yml 文件信息，具体如代码清单 12-66 所示。

代码清单12-66 code\ch12-8\ch12-8-config-server\src\main\resources\application.yml

```yaml
spring:
  cloud:
    config:
      server:
        git:
          uri: https://gitee.com/zhongzunfa/spring-cloud-config.git
          search-paths: SC-BOOK-CONFIG
  application:
    name: ch12-8-config-server
server:
  port: 9090
```

至此就全部编写完成了。接下来启动服务端（ch12-8-config-server）和客户端（ch12-8-config-client-high-availability），在客户端启动过程中会看到如下信息打印：

```
c.c.c.ConfigServicePropertySourceLocator : Fetching config from server at :
    http://localhost:9090
c.c.c.ConfigServicePropertySourceLocator : Located environment: name=config-
    info, profiles=[dev], label=master, version=201917124912c018eb88adebf5b7d11
    d1c1c05bf, state=null
b.c.PropertySourceBootstrapConfiguration : Located property source: Composite-
    PropertySource {name='configService', propertySources=[MapPropertySource
    {name='configClient'}, MapPropertySource {name='https://gitee.com/zhong-
    zunfa/spring-cloud-config.git/SC-BOOK-CONFIG/config-info-dev.yml'}]}
c.s.b.c.c.ConfigSupportConfiguration     : 成功获取ConfigService配置资源
```

这里只是截取关键信息，同时服务端可以看到如下信息：

```
2018-07-08 00:37:39.076  INFO 220812 --- [nio-9090-exec-1] o.s.c.c.s.e.Nativ
  eEnvironmentRepository   : Adding property source: file:/C:/Users/ADMINI~1/
  AppData/Local/Temp/config-repo-7903875733013216565/SC-BOOK-CONFIG/config-
  info-dev.yml
```

接下来通过访问工具访问地址：localhost:9015/configConsumer/getConfigInfo，具体如图 12-18 所示。

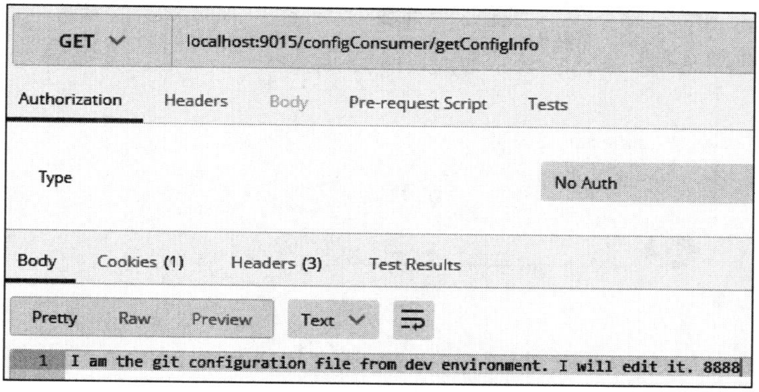

图 12-18　访问结果图

再将所有的服务都关闭，单独启动客户端，不启动服务端，看看客户端能够启动成功。通过验证，客户端依然可以启动成功的，客户端会打印出如下信息。

```
c.c.c.ConfigServicePropertySourceLocator : Fetching config from server at :
  http://localhost:9090
c.c.c.ConfigServicePropertySourceLocator : Connect Timeout Exception on Url -
  http://localhost:9090. Will be trying the next url if available
c.c.c.ConfigServicePropertySourceLocator : Could not locate PropertySource: I/
  O error on GET request for "http://localhost:9090/config-info/dev/master":
  Connection refused: connect; nested exception is java.net.ConnectException:
  Connection refused: connect
c.s.b.c.c.ConfigSupportConfiguration       : 获取ConfigService配置资源失败
c.s.b.c.c.ConfigSupportConfiguration       : 使用备份的配置启动: D:\\application_dev\\
  idea_workspace\\sc-book-local-config-test\\fallback.properties
```

这里只是截取关键信息，到此客户端的高可用就讲解完成了。有客户端高可用是不是也应该有服务端的高可用呢？答案是肯定的，接下我们将讲解服务端的高可用。

12.6.2　服务端高可用

同样的 Config Server 一样在生成环境下是要保证高可用的，本节将通过结合 eureka 注册中心的方式搭建 Config Server 的高可用，通过 ribbon 的负载均衡选择一个 Config Server 进行连接来获取配置信息，具体的 ribbon 策略如何操作的请查看第 5 章，此外对于 eureka 的高可用这里也不进行详解，详细关于 eureka 的高可用，请参考 eureka 相关章节。接下来先看一下整体的请求流程图，具体如图 12-19 所示。

在 code 的根目录下创建 mauen 工程 ch12-9，其中 ch12-9 的 pom 和 ch12-8 的一致。

1. eureka-server 服务端

在工程 ch12-9 下面创建一个 module，命名为 ch12-9-eureka-server，代码如本书源码 ch12-9/ch12-9-eureka-server 所示。

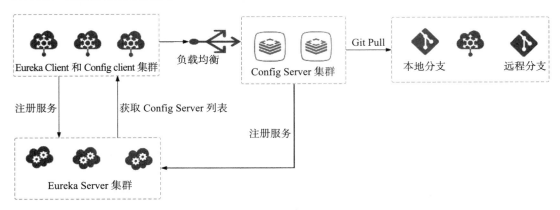

图 12-19　config 高可用流程图

（1）配置依赖

在 pom 中添加如下依赖，具体如代码清单 12-67 所示。

代码清单 12-67　code\ch12-9\ch12-9-eureka-server\pom.xml

```
<dependencies>
    <dependency>
        <groupId>org.springframework.cloud</groupId>
        <artifactId>spring-cloud-starter-netflix-eureka-server</artifactId>
    </dependency>
</dependencies>
```

接下来创建入口程序。

（2）主程序入口

创建类 EurekaServerApplication 并且添加相关注解，使其成为入口程序，具体如代码清单 12-68 所示。

代码清单 12-68　code\ch12-9\ch12-9-eureka-server\src\main\java\cn\springcloud\book\eureka\EurekaServerApplication.java

```
@SpringBootApplication
@EnableEurekaServer
public class EurekaServerApplication {

    public static void main(String[] args) {

        SpringApplication.run(EurekaServerApplication.class, args);
    }
}
```

还需要配置启动需要的配置信息。

（3）配置文件配置

在 application.yml 文件中添加 eureka 相关配置，具体如代码清单 12-69 所示。

代码清单12-69　ch12-9/ch12-9-eureka-server/src/main/resources/application.yml

```yaml
server:
    port: 8761

eureka:
    instance:
        hostname: localhost
    client:
        registerWithEureka: false
        fetchRegistry: false
        serviceUrl:
            defaultZone: http://${eureka.instance.hostname}:${server.port}/eureka/
    server:
        waitTimeInMsWhenSyncEmpty: 0
        enableSelfPreservation: false
```

到此 Eureka Sever 就编写完成了，更多关于 Eureka 相关配置请参考第 2 章。

2. config-server-high-availability 服务端

在工程 ch12-9 下面创建一个 module，命名为 ch12-9-config-server-high-availability，代码如本书源码 ch12-9/ch12-9-config-server-high-availability 所示。

（1）配置依赖

在 pom 中添加如下依赖，具体如代码清单 12-70 所示。

代码清单12-70　code\ch12-9\ch12-9-config-server-high-availability\pom.xml

```xml
<dependencies>
    <dependency>
        <groupId>org.springframework.cloud</groupId>
        <artifactId>spring-cloud-starter-netflix-eureka-client</artifactId>
    </dependency>

    <dependency>
        <groupId>org.springframework.cloud</groupId>
        <artifactId>spring-cloud-config-server</artifactId>
    </dependency>
</dependencies>

<build>
    <finalName>ch12-9-config-server-high-availability</finalName>
    <!--省略plugins部分信息-->
</build>
```

由于 config sever 需要注册到 Eureka Server 上，所以需要添加 Eureka Client 的依赖。

（2）主入口程序

创建 ConfigGitApplication 类并且添加相关注解信息，使其成为程序入口类，具体如代码清单 12-71 所示。

代码清单12-71 code\ch12-9\ch12-9-config-server-high-availability\src\main\java\cn\spring-cloud\book\config\ConfigGitApplication.java

```java
@SpringBootApplication
@EnableConfigServer
@EnableDiscoveryClient
public class ConfigGitApplication {

    public static void main(String[] args) {
        SpringApplication.run(ConfigGitApplication.class, args);

    }
}
```

其中 @EnableDiscoveryClient 表示的是启用服务发现功能。

（3）配置文件配置

application.yml 配置的具体代码如代码清单 12-72 所示。

代码清单12-72 code\ch12-9\ch12-9-config-server-high-availability\src\main\resources\application.yml

```yaml
spring:
  cloud:
    config:
      server:
        git:
          uri: https://gitee.com/zhongzunfa/spring-cloud-config.git
          search-paths: SC-BOOK-CONFIG
  application:
    name: ch12-9-config-server-high-availability
server:
  port: 9090

eureka:
  client:
    serviceUrl:
      defaultZone: http://localhost:8761/eureka/
```

从上面的配置文件中可以看到，这里只是比之前的配置多了一个 eureka 的配置项。
到此 config server 部分就编写完成了。

3. config-client 客户端

在工程 ch12-9 下面创建一个 module，命名为 ch12-9-config-client，代码如本书源码 ch12-9/ch12-9-config-client 所示。

（1）配置依赖

在 pom 中添加如下依赖配置，具体如代码清单 12-73 所示。

代码清单12-73 code\ch12-9\ch12-9-config-client\pom.xml

```xml
<dependencies>
```

```xml
<dependency>
    <groupId>org.springframework.cloud</groupId>
    <artifactId>spring-cloud-config-client</artifactId>
</dependency>
<dependency>
    <groupId>org.springframework.cloud</groupId>
    <artifactId>spring-cloud-starter-netflix-eureka-client</artifactId>
</dependency>

</dependencies>
```

该客户端需要连接 config 的同时将自己注册到 eureka 上。

（2）主入口程序

创建 ClientConfigGitApplication 类并且添加相关的注解信息，使其成为程序的入口类，代码内容同代码清单 12-6 所示。

详细代码可查看：code\ch12-9\ch12-9-config-client\src\main\java\cn\springcloud\book\config\ClientConfigGitApplication.java。

为了观察方便和断点等，需要添加一个 controller，具体代码同代码清单 12-36 所示。

下面还需要添加启动需要配置。

（3）配置依赖配置

在 application.yml 文件中添加配置，具体如代码清单 12-74 所示。

代码清单12-74   ch12-9/ch12-9-config-client/src/main/resources/application.yml

```yml
server:
    port: 9016

spring:
    application:
        name: ch12-9-config-client
```

配置 bootstrap.yml 文件，具体如代码清单 12-75 所示。

代码清单12-75   ch12-9/ch12-9-config-client/src/main/resources/bootstrap.yml

```yml
spring:
    cloud:
        config:
            label: master
            name: config-info
            profile: dev
            discovery:
                enabled: true
                service-id: ch12-9-config-server-high-availability
eureka:
    client:
        service-url:
            defaultZone: http://localhost:8761/eureka/
```

细心的读者看了这里的配置应该发现和之前的 client 配置有一些不一样，因为这里主要是去掉了 spring.cloud.config.uri 直接指向 Server 端地址的配置。

增加了最后的三个配置:
- spring.cloud.config.discovery.enabled。开启 Config 服务发现支持。
- spring.cloud.config.discovery.serviceId。指定 Server 端的 name,也就是 Server 端 spring.application.name 的值。
- eureka.client.service-url.defaultZone。指向注册中心的地址。

至此,config client 部分也编写完成了,首先通过 maven install 安装到本地,主要是用于命令行启动,读者可以选择自己喜欢的方式启动。可以直接 run 也可以在命令行下启动,先启动 eureka 在 main 下启动,接下来启动 config server。这里需要启动多个实例,所以命令行形式启动两个实例,默认的端口是 9090:

- java -jar ch12-9-config-server-high-availability.jar
- java -jar ch12-9-config-server-high-availability.jar --server.port=9091

接下来启动 config client,同样也是使用 java -jar ch12-9-config-client.jar 命令形式启动,此时观察打印的信息具体如图 12-20 所示。

图 12-20 client 启动打印信息图

读取的是 9090 端口 config-server 的配置信息,同时服务端打印信息如图 12-21 所示。

图 12-21 server 启动打印信息图

再启动一个客户端实例使用 9017 端口:java -jar ch12-9-config-client.jar --server.port=9017。观察看看是否已经负载了,很幸运这里遇上了如图 12-22 所示的情况。

这里读取的是 9091 启动实例上的配置信息,同时服务端打印信息如图 12-23 所示。

没有负载到也没关系,可以启动多个实例试试。或者直接关闭 9090 服务再启动客户端,再启动一个 config client 的实例,然后观察打印的信息,确定加载到的是 config server 另外的一个实例提供的配置信息,接下来验证一下目前启动好的部分是否可以正常使用,访问地址:localhost:9016/configConsumer/getConfigInfo。

```
2018-08-12 22:24:59.009  INFO 2952 --- [           main] c.c.c.ConfigServic
ePropertySourceLocator   : Fetching config from server at : http://192.168.31
.123:9091/
2018-08-12 22:25:02.417  INFO 2952 --- [           main] c.c.c.ConfigServic
ePropertySourceLocator   : Located environment: name=config-info, profiles=[d
ev], label=master, version=a7ebcdb3ea89251d5fba3d49eda07b59b7a6564b, state=
null
2018-08-12 22:25:02.418  INFO 2952 --- [           main] b.c.PropertySource
BootstrapConfiguration   : Located property source: CompositePropertySource {
name='configService', propertySources=[MapPropertySource {name='configClien
t'}, MapPropertySource {name='https://gitee.com/zhongzunfa/spring-cloud-con
fig.git/SC-BOOK-CONFIG/config-info-dev.yml'}]}
```

图 12-22　client 启动打印信息图 2 所示

```
2018-08-12 22:25:02.274  INFO 2881 --- [nio-9091-exec-4] f.a.AutowiredAnnotationBeanPostPr
ocessor   : JSR-330 'javax.inject.Inject' annotation found and supported for autowiring
2018-08-12 22:25:02.376  INFO 2881 --- [nio-9091-exec-4] o.s.c.c.s.e.NativeEnvironmentRepo
sitory    : Adding property source: file:/var/folders/z7/4brsh0w13zx0fnyc_3ct_k800000gn/T/co
nfig-repo-7990704829806327839/SC-BOOK-CONFIG/config-info-dev.yml
2018-08-12 22:25:02.376  INFO 2881 --- [nio-9091-exec-4] s.c.a.AnnotationConfigApplication
```

图 12-23　server 启动打印信息图 2

结果如图 12-24 所示。

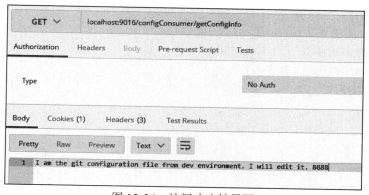

图 12-24　访问响应结果图

再访问 localhost:9017/configConsumer/getConfigInfo，结果如图 12-25 所示。

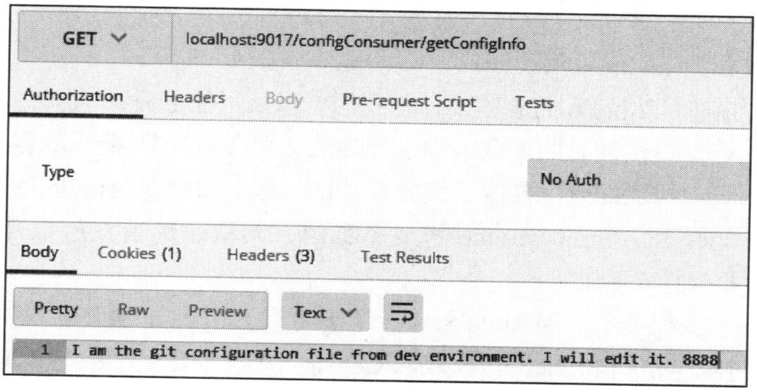

图 12-25　访问响应结果图

两个都可以访问，说明服务端的高可用没有问题，并且加载到的都是一样的配置信息。至此服务端的高可用就完成。可能有读者又回想，一直都是直接对配置文件进行操作，有没有直接基于 portal 的操作呢？对于界面化的操作当然是有的，接下来会讲解三方的配置中心和 spring cloud 的配置使用。

12.7 Spring Cloud 与 Apollo 配置使用

12.7.1 Apollo 简介

Apollo（阿波罗）是携程框架部门研发的开源配置管理中心，能够集中化管理应用不同环境、不同集群的配置，配置修改后能够实时推送到应用端，并且具备规范的权限、流程治理等特性。

Apollo 支持 4 个维度管理 Key-Value 格式的配置：application（应用）、environment（环境）、cluster（集群）、namespace（命名空间）。

12.7.2 Apollo 具备功能

目前 Apollo 提供了以下的特性。
- 统一管理不同环境、不同集群的配置
- 配置修改实时生效（热发布）
- 版本发布管理
- 灰度发布
- 权限管理、发布审核、操作审计
- 客户端配置信息监控
- 提供 Java 和 .Net 原生客户端
- 提供开放平台 API
- 部署简单

12.7.3 Apollo 总体架构模块

总体架构模块如图 12-26 所示。

图 12-26 简要描述了 Apollo 的总体设计，我们可以从下往上看：
- Config Service 提供配置的读取、推送等功能，服务对象是 Apollo 客户端。
- Admin Service 提供配置的修改、发布等功能，服务对象是 Apollo Portal（管理界面）。
- Config Service 和 Admin Service 都是多实例、无状态部署，所以需要将自己注册到 Eureka 中并保持心跳。
- 在 Eureka 上我们架了一层 Meta Server 用于封装 Eureka 的服务发现接口。
- Client 通过域名访问 Meta Server 获取 Config Service 服务列表（IP+Port），而后直接通过 IP+Port 访问服务，同时在 Client 侧会进行 load balance、错误重试。

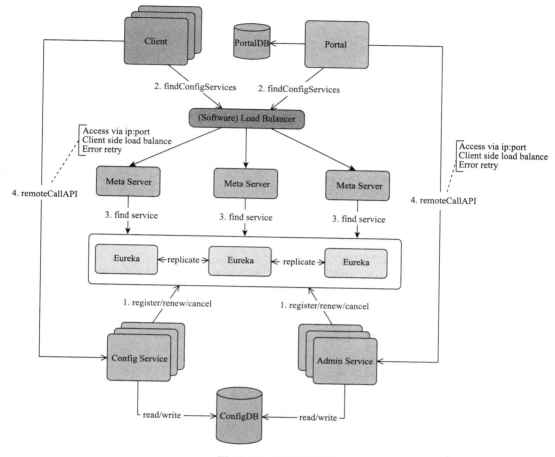

图 12-26　总体设计图

- Portal 通过域名访问 Meta Server 获取 Admin Service 服务列表（IP+Port），而后直接通过 IP+Port 访问服务，同时在 Portal 侧会进行 load balance、错误重试。

通过上述内容可以看到 Apollo 使用的注册中心也是 Eureka，此时读者会想到什么？是的！Apollo 可以和 Spring Cloud 搭建的微服务进行无缝集成。

12.7.4　客户端设计

客户端设计如图 12-27 所示。

图 12-27 简要描述了 Apollo 客户端的实现原理：

1）客户端和服务端保持了一个长连接，从而能第一时间获得配置更新的推送。
2）客户端还会定时从 Apollo 配置中心服务端拉取应用的最新配置。
- 这是一个 fallback 机制，为了防止推送机制失效导致配置不更新。
- 客户端定时拉取会上报本地版本，所以一般情况下，对于定时拉取的操作，服务端都会返回 304 - Not Modified。

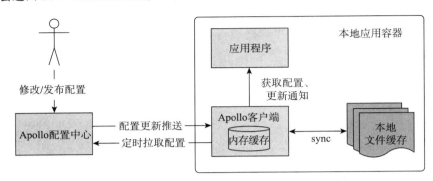

图 12-27　客户端设计图

- 定时频率默认为每 5 分钟拉取一次，客户端也可以通过在运行时指定 System Property: apollo.refreshInterval 来覆盖，单位为分钟。

3）客户端从 Apollo 配置中心服务端获取到应用的最新配置后，会保存在内存中。

4）客户端会把从服务端获取到的配置在本地文件系统缓存一份。在遇到服务不可用，或网络不通的时候，依然能从本地恢复配置。

5）应用程序从 Apollo 客户端获取最新的配置，订阅配置更新通知。

12.7.5　Apollo 运行环境方式

Apollo 客户端支持应用在不同的环境有不同的配置，所以 Environment 是另一个从服务器获取配置的重要信息。

Environment 可以通过以下 3 种方式的任意一个进行配置：

（1）通过 Java System Property

- 可以通过 Java 的 System Property env 来指定环境。
- 在 Java 程序启动脚本中，可以指定 -Denv=YOUR-ENVIRONMENT。
- 如果是运行 jar 文件，需要注意格式是 java -Denv=YOUR-ENVIRONMENT -jar xxx.jar。
- 注意 key 为全小写。

（2）通过操作系统的 System Environment

- 还可以通过操作系统的 System Environment ENV 来指定。
- 注意 key 为全大写。

（3）通过配置文件

- 通过配置文件来指定 env=YOUR-ENVIRONMENT。
- 对于 Mac/Linux，文件位置为 /opt/settings/server.properties。
- 对于 Windows，文件位置为 C:\opt\settings\server.properties。

到此 Apollo 具备的功能介绍就完成了，更多关于 Apollo 的详细介绍请参阅 Apollo 的 wiki 详细地址（https://github.com/ctripcorp/apollo/wiki），接下来让我们开始实战。

12.8　Spring Cloud 与 Apollo 结合使用实战

在开始和 Spring Cloud 结合使用之前，先将 Apollo 的配置服务端启动，接下来将介绍 Apollo 服务端的配置和启动。这里只是启动简单的单节点实例，如上面架构中所述 Config Service 和 Admin Service 都是多实例、无状态部署，所以需要将自己注册到 Eureka 中并保持心跳。

12.8.1　Apollo 环境的要求

Java 环境：
- Apollo 服务端：jdk 1.8+
- Apollo 客户端：jdk 1.7+

MySQL 环境：5.6.5+

12.8.2　Apollo 基础数据导入

在上述的环境都准备好后，需要到官方 GitHub 上下载源代码，具体地址为 https://github.com/ctripcorp/apollo/releases。

这里笔者使用的是 Apollo v1.0.0 版本，具体如图 12-28 所示。

图 12-28　Apollo v1.0.0 版本

Apollo 服务端运行需要两个数据库，ApolloPortalDB 和 ApolloConfigDB，在 MySQL 上创建这两个数据，笔者将数据库脚本和上述 Apollo 用到的服务放在工程 ch12-10/ch12-10-apollo-build-script/sql 中，具体如图 12-29 所示。

执行完成后数据库中就会出现下面的表，都是 Apollo 运行时需要的表，具体如图 12-30 所示。

读者可以使用数据库客户端工具执行。解压下载好的 apollo-adminservice-1.0.0-github.zip、apollo-

图 12-29　Apollo 数据库脚本结构图

configservice-1.0.0-github.zip、apollo-portal-1.0.0-github.zip，并且修改各自对应的 conf 目录下面的 application.yml 文件。其中主要修改的是数据库相关配置，这里以 apollo-configservice-1.0.0-github 为例进行演示。

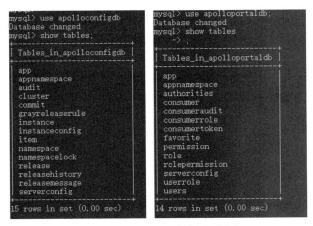

图 12-30　Apollo 数据库表图

解压后的目录，如图 12-31 所示。

图 12-31　apollo-configservice 解压目录

读者根据数据库地址，修改 application.yml 即可。

需要注意的是对于 script 目录下的 start.sh 脚本，需要修改其 log 路径，读者根据自己的实际情况修改即可。上述的三个工程配置文件修改完成后，分别在 apollo-adminservice、apollo-configservice、apollo-portal 工程 script 目录下执行 start.sh 启动。访问地址为 http://localhost:8070/。

正常会看到如图 12-32 所示的界面。

输入用户名 apollo，密码 admin 后登录，如图 12-33 所示。

创建一个项目，具体如图 12-34 所示。其中部门、项目负责人和项目管理员都使用默认的，这些可以添加后再选择。

上述的关键信息应用了 Id=123456789，一般用于应用获取配置使用，应用名使 sc-client-apollo。

创建后如图 12-35 所示。

新增配置信息，这里选择的是表格，创建情况如图 12-36 所示。

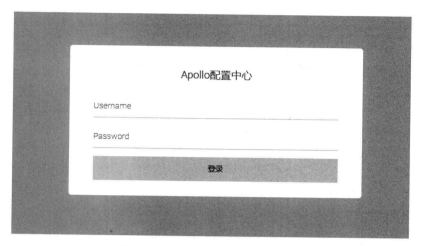

图 12-32　Apollo Portal 登录页面图

图 12-33　Apollo Portal 管理界面图

图 12-34　Apollo Portal 添加项目图

添加完成后如图 12-37 所示。

刚添加的是未发布的内容，要想向客户端发布，需要点击发布按钮将配置信息后发布出去，具体如图 12-38 所示。

306 ❖ 重新定义 Spring Cloud 实战

图 12-35　Apollo Portal 项目管理图

图 12-36　Apollo Portal 添加配置图

图 12-37　Apollo Portal 未发布配置图

图 12-38　Apollo Portal 发布配置图

点击发布会出现上述的提示，再点击发布，即可发布配置信息。Apollo 的服务端基本配置就完成了。接下来创建 spring cloud 项目信息来验证配置是否可用。在 code 根目录下创建 mauen 工程 ch12-10；其中 ch12-10 的 pom 和 ch12-9 的一致，并且 ch12-10-eureka-server 和 ch12-9-eureka-server 也一致。

在工程 ch12-10 下面创建一个 module，命名为 ch12-10-config-client-apollo，代码如本书源码 ch12-10/ch12-10-config-client-apollo 所示。

12.8.3　创建 config-client-apollo

（1）配置依赖

在 ch12-10-config-client-apollo 的 pom.xml 文件中添加相关依赖，具体如代码清单 12-76 所示。

代码清单 12-76　code\ch12-10\ch12-10-config-client-apollo\pom.xml

```xml
<dependencies>
    <dependency>
        <groupId>com.ctrip.framework.apollo</groupId>
        <artifactId>apollo-client</artifactId>
        <version>${apollo-client-version}</version>
    </dependency>
    <dependency>
        <groupId>org.springframework.cloud</groupId>
        <artifactId>spring-cloud-starter-netflix-eureka-client</artifactId>
    </dependency>
</dependencies>
```

这里引入了 Apollo 的客户端依赖。

（2）主入口程序

创建 ApolloClientApplication 类并且添加相关的注解信息，使其成为应用的入口程序，代码内容同代码清单 12-6 所示。

详细代码可查看：code\ch12-10\ch12-10-config-client-apollo\src\main\java\cn\springcloud\book\

config\ApolloClientApplication.java。

下面还需要添加开启 Apollo 的配置信息，具体如代码清单 12-77 所示。

代码清单12-77　ch12-10/ch12-10-config-client-apollo/src/main/java/cn/springcloud/book/
config/config/AppConfig.java

```
@Configuration
@EnableApolloConfig(value = "application", order = 10)
public class AppConfig {
}
```

查看信息的 controller，具体如代码清单 12-78 所示。

代码清单12-78　ch12-10/ch12-10-config-client-apollo/src/main/java/cn/springcloud/book/
config/controller/ConfigClientController.java

```
@RestController
@RequestMapping("configConsumer")
@RefreshScope
public class ConfigClientController {

    @Value("${config_info}")
    private String config;

    @RequestMapping("/getConfigInfo")
    public String getConfigInfo(){
        return config;
    }
}
```

（3）配置文件配置

在 application.yml 配置文件中添加 Eureka 和 Apollo 相关启动配置信息，具体如代码清单 12-79 所示。

代码清单12-79　code\ch12-10\ch12-10-config-client-apollo\src\main\resources\application.yml

```
server:
    port: 9018
spring:
    application:
        name: ch12-10-config-client-apollo
eureka:
    client:
        service-url:
            defaultZone: http://localhost:8761/eureka/
apollo:
    bootstrap:
        enabled: true
        namespaces: application
    meta: http://localhost:8080
```

meta 指向的地址是 apollo-configservice 服务对应的地址和端口。下面还需要创建 app.properties，存在应用的 id 需要和配置中心的 id 一致。

（4）app.properties

创建 app.properties 配置文件，具体如代码清单 12-80 所示。

代码清单12-80　code\ch12-10\ch12-10-config-client-apollo\src\main\resources\META-INF\
app.properties

```
app.id=123456789
```

到此代码和配置就完成了，启动项设置具体如图 12-39 所示。

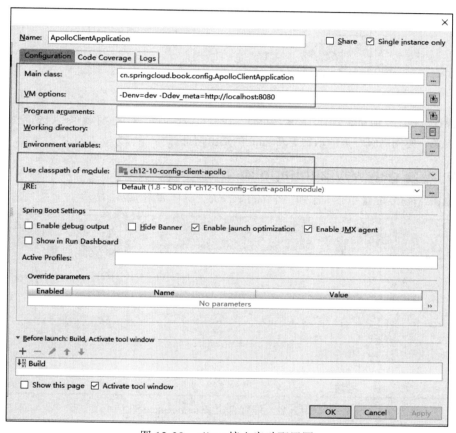

图 12-39　client 接入启动配置图

启动过程控制台会打印如下信息，如图 12-40 所示。

```
 : App ID is set to 123456789 by app.id property from /META-INF/app.properties
 : C:/opt/settings/server.properties does not exist or is not readable.
 : Environment is set to [dev] by JVM system property 'env'.
```

图 12-40　client 接入打印信息图

启动参数是通过 VM 参数设置进入的，启动成功访问地址：localhost:9018/configCon-sumer/

getConfigInfo。

具体结果如图 12-41 所示。

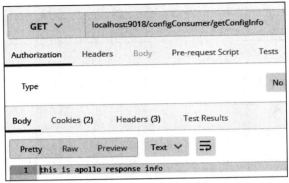

图 12-41　访问信息图

这样就得到了之前在配置中心配置的信息，读者可以往前翻看，这样就可以看到配置的信息。接下来，使用高级一点的动态更新 zuul 的路由信息。

12.8.4　创建 gateway-zuul-apollo

首先在 Apollo 配置中心中添加一个 namespace，具体如图 12-42 所示。

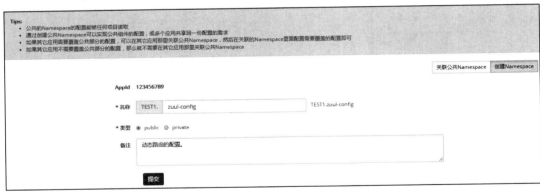

图 12-42　配置自定义 namespace 图

然后配置这个 namespace 的权限，如图 12-43 所示。

图 12-43　配置自定义 namespace 权限图

返回首页可以看到当前项目下存在两个 namespace，如图 12-44 所示。

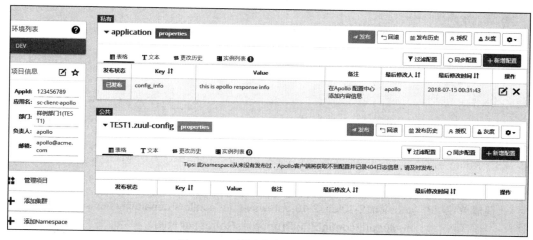

图 12-44　项目多个 namespace 图

接下来创建项目。在工程 ch12-10 下面创建一个 module，命名为 ch12-10-gateway-zuul-apollo，代码如本书源码 ch12-10/ch12-10-gateway-zuul-apollo 所示。

（1）配置依赖

在 ch12-10-gateway-zuul-apollo 的 pom.xml 文件中添加相关依赖，具体如代码清单 12-81 所示。

代码清单12-81　code\ch12-10\ch12-10-gateway-zuul-apollo\pom.xml

```xml
<dependencies>
    <dependency>
        <groupId>org.springframework.cloud</groupId>
        <artifactId>spring-cloud-starter-netflix-zuul</artifactId>
    </dependency>
    <dependency>
        <groupId>org.springframework.cloud</groupId>
        <artifactId>spring-cloud-starter-netflix-eureka-client</artifactId>
    </dependency>
    <dependency>
        <groupId>com.ctrip.framework.apollo</groupId>
        <artifactId>apollo-client</artifactId>
        <version>${apollo-client-version}</version>
    </dependency>
</dependencies>
```

（2）主入口程序

创建 ZuulServerApplication 类并且添加相关的注解信息，使其成为程序入口，具体如代码清单 12-82 所示。

代码清单12-82　code\ch12-10\ch12-10-gateway-zuul-apollo\src\main\java\cn\spring-cloud\book\zuul\ZuulServerApplication.java

```
@EnableApolloConfig
```

```java
@SpringBootApplication
@EnableDiscoveryClient
@EnableZuulProxy
public class ZuulServerApplication {

    public static void main(String[] args) {
        SpringApplication.run(ZuulServerApplication.class, args);
    }
}
```

其中 @EnableApolloConfig 表示的是开启 Apollo 配置，@EnableZuulProxy 表示的是启用 zuul 网关代理。

（3）Route 刷新

创建 zuul 路由规则刷新类 ZuulPropertiesRefresher，具体如代码清单 12-83 所示。

代码清单12-83 code\ch12-10\ch12-10-gateway-zuul-apollo\src\main\java\cn\springcloud\book\zuul\config\ZuulPropertiesRefresher.java

```java
@Component
public class ZuulPropertiesRefresher implements ApplicationContextAware {

    private static final Logger logger = LoggerFactory.getLogger(ZuulProperties-
        Refresher.class);
    private ApplicationContext applicationContext;

    @Autowired
    private RouteLocator routeLocator;

    @ApolloConfigChangeListener(value = "TEST1.zuul-config")
    public void onChange(ConfigChangeEvent changeEvent) {
        boolean zuulPropertiesChanged = false;
        for (String changedKey : changeEvent.changedKeys()) {
            if (changedKey.startsWith("zuul.")) {
                zuulPropertiesChanged = true;
                break;
            }
        }

        if (zuulPropertiesChanged) {
            refreshZuulProperties(changeEvent);
        }
    }

    private void refreshZuulProperties(ConfigChangeEvent changeEvent) {
        logger.info("Refreshing zuul properties!");

        this.applicationContext.publishEvent(new EnvironmentChangeEvent(change-
            Event.changedKeys()));

        this.applicationContext.publishEvent(new RoutesRefreshedEvent(routeLocator));

        logger.info("Zuul properties refreshed!");
    }
```

```java
    @Override
    public void setApplicationContext(ApplicationContext applicationContext)
            throws BeansException {
        this.applicationContext = applicationContext;
    }
}
```

@ApolloConfigChangeListener(value = "TEST1.zuul-config") 注解中的 value 参数默认是 application，因为自定义了 namespace 所以需要指定，监听服务端的配置下发，有配置更新时会调用 refreshZuulProperties() 方法进行参数刷新。

（4）配置文件配置

在 application.yml 配置文件中添加 zuul 路由规则和 Apollo 配置信息，具体如代码清单 12-84 所示。

代码清单12-84　ch12-10/ch12-10-gateway-zuul-apollo/src/main/resources/application.yml

```yaml
spring:
  application:
    name: ch12-10-gateway-zuul-apollo
server:
  port: 9019
eureka:
  client:
    serviceUrl:
      defaultZone: http://${eureka.host:127.0.0.1}:${eureka.port:8761}/eureka/
zuul:
  routes:
    client-apollo:
      path: /client/**
      serviceId: ch12-10-config-client-apollo
apollo:
  bootstrap:
    enabled: true
    namespaces: TEST1.zuul-config
  meta: http://localhost:8080
```

这里有一个比较明显配置了一个新的 namespace 的地址，就是上面在配置中心中创建的。

（5）app.properties

创建 app.properties 文件，具体如代码清单 12-85 所示。

代码清单12-85　code\ch12-10\ch12-10-gateway-zuul-apollo\src\main\resources\META-INF\app.properties

```
app.id=123456789
```

这样 zuul 的相关配置就完成了。接下来复制 ch12-10-config-client-apollo 命名为 ch12-10-config-client-apollo2。动态配置路由生效，主要变动 ConfigClientController 这个类的 config 获取信息改为 @Value({"$ config info new}")。

application.yml 将应用名改成 ch12-10-config-client-apollo2。

在配置中心的 application 的 namespace 中添加一个配置如图 12-45 所示。

图 12-45　添加配置信息图

启动项目，依次启动 ch12-10-eureka-server、ch12-10-config-client-apollo、ch12-10-config-client-apollo2、ch12-10-gateway-zuul-apollo。

访问 localhost:9019/client/configConsumer/getConfigInfo。

预期是可以访问的，因为在 zuul 的配置中有路由配置信息，所以实际也是可以访问的，如图 12-46 所示。

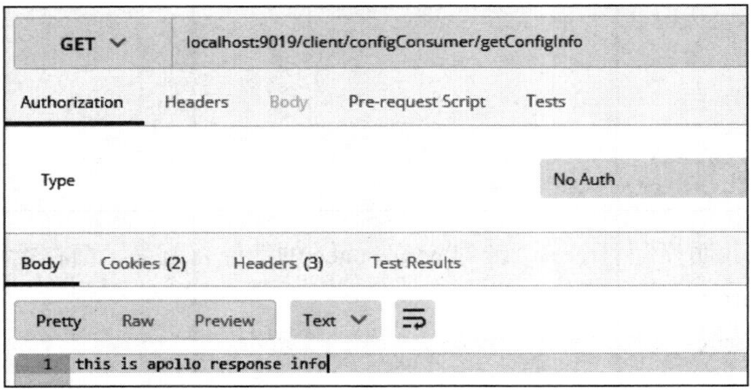

图 12-46　路由访问获取远程配置信息图

访问 localhost:9019/client_new/configConsumer/getConfigInfo。

预期如果不能访问，是因为没有配置相关的路由信息，具体如图 12-47 所示。

接下来，通过配置中心配置路由信息，再发布配置，如图 12-48 所示。

配置对应的路由实例，如图 12-49 所示。

发布配置后访问 localhost:9019/client_new/configConsumer/getConfigInfo，具体如图 12-50 所示。

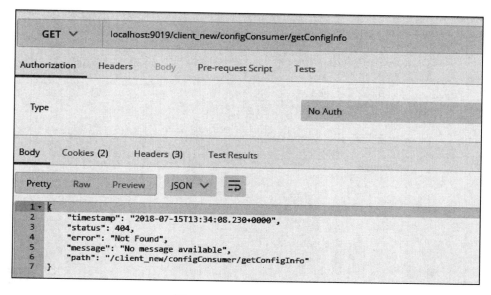

图 12-47 错误路由访问图

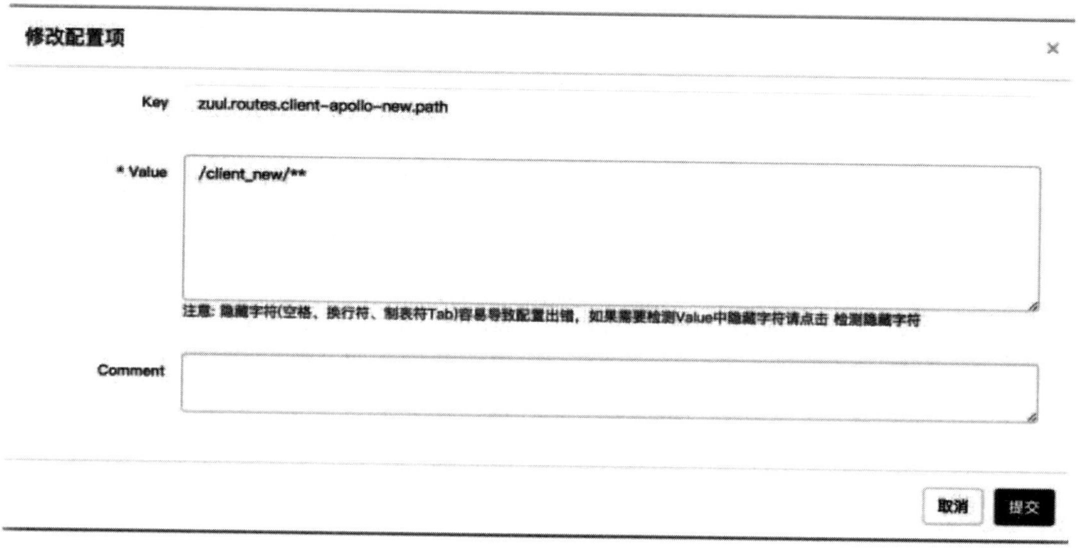

图 12-48 路由规则配置一图

可以正常访问，说明配置已经生效。到此 Spring Cloud 结合 Apollo 的基础案例使用就已经完成了，是不是感觉很流畅，直接进行界面操作不用编辑配置文件以及 git 之间的操作，就能完成配置下发的功能。关于审计等一系列好用的功能的详细使用方法请参考 Apollo 官方文档。

图 12-49　路由规则配置二图

图 12-50　新规则路由访问图

12.9　本章总结

本章主要讲解了 Spring Cloud Config 结合不同的存储（Git、MongoDB、MySQL）做配置中心，还对 Spring Cloud 的功能进行增强，主要增强的功能有客户自动刷新可配置化、客户端回退功能、客户端和服务结合改造的 JWT 安全认证机制等。客户端的高可用在服务端出现问题或者是网络出现异常情况下客户端依然可以启动和使用，服务端的高可用确保服务端不会出现单点故障带来灾难性问题。最后讲解携程开源配置中心（Apollo）和 Spring Cloud 的结合使用。通过上面的所有相关案例的学习，读者应该可以根据自己的实际情况来选择合适 Config 作为配置中心了。

第 13 章

Spring Cloud Consul 上篇

作为集群系统的灵魂，服务治理框架一直都是架构师的宠儿。随着微服务思想的普及，越来越多的服务治理框架如雨后春笋般冒了出来。除了前文所说的 Eureka，HashiCorp 公司的 Consul 也让诸多架构师青睐有加。本章将为读者介绍 Spring Cloud 的 Consul 组件：Spring Cloud Consul。

13.1 Consul 简介

13.1.1 什么是 Consul

Consul 是一个分布式高可用的服务网格（service mesh）解决方案，提供包括服务发现、配置和分段功能在内的全功能控制平面。这些功能中的每一个都可以根据需要单独使用，也可以一起使用以构建完整的服务网格。

Consul 是一个分布式高可用的系统服务发现与配置工具。简单来说，它跟 Eureka 的核心功能一样，但略有不同：

- Consul 使用 go 语言编写，以 HTTP 方式对外提供服务。
- Consul 支持多数据中心，这是它的一大特色。
- Consul 除了服务发现之外，还有一些别的功能。
- Consul 的一致性协议是 Raft。

当然，官方还是提供了 Consul 与其他框架的详细对比信息，有兴趣的读者可以看一下：https://www.consul.io/intro/vs/index.html。

13.1.2 Consul 能做什么

Consul 提供了以服务治理为核心的多种功能以满足分布式系统的需要，主要功能如下：

- 服务发现。有了 Consul，服务可以通过 DNS 或者 HTTP 直接找到它所依赖的服务。
- 健康检查。Consul 提供了健康检查的机制，从简单的服务端是否返回 200 的响应代码到较为复杂的内存使用率是否低于 90%。
- K/V 存储。应用程序可以根据需要使用 Consul 的 Key/Value 存储。Consul 提供了简单易用的 http 接口来满足用户的动态配置、特征标记、协调、leader 选举等需求。
- 多数据中心。Consul 原生支持多数据中心。这意味着用户不用为了多数据中心自己做抽象。

简单说来，Consul 可以作为服务治理组件和配置中心。当然，市场上还有很多其他类似功能的优秀框架，Consul 官方提供了对比信息，以便架构师们在做技术选型时可以尽快找到更适合自己的方案：https://www.consul.io/intro/vs/index.html

13.1.3　Consul 的安装

安装 Consul 比较简单，官方提供了可执行文件，读者朋友们可以去 https://www.consul.io/downloads.html 下载自己感兴趣的版本。本文以 1.2.0 版本为例。

1）我们将下载的 consul_1.2.0_linux_amd64.zip 解压到 /root/develop/consul/ 目录下。

2）添加 consul 到 PATH。

3）执行 consul -v；如果不报错，基本就算安装成功了。

```
[scbook@localhost root]$ consul -v
Consul v1.1.0
Protocol 2 spoken by default, understands 2 to 3 (agent will automatically use
    protocol >2 when speaking to compatible agents)
```

13.1.4　Consul 启动

Consul 集群默认需要至少三台 Consul 启动。如果只是想本地开发调试，可以使用开发者模式启动：

```
[scbook@localhost ~]$ consul agent -dev
==> Starting Consul agent...
==> Consul agent running!
           Version: 'v1.1.0'
           Node ID: '0ec0a993-f599-b8d8-642c-470801fc2f35'
         Node name: 'localhost.localdomain'
        Datacenter: 'dc1' (Segment: '<all>')
            Server: true (Bootstrap: false)
       Client Addr: [127.0.0.1] (HTTP: 8500, HTTPS: -1, DNS: 8600)
      Cluster Addr: 127.0.0.1 (LAN: 8301, WAN: 8302)
           Encrypt: Gossip: false, TLS-Outgoing: false, TLS-Incoming: false

==> Log data will now stream in as it occurs:

    2018/06/19 08:37:05 [DEBUG] agent: Using random ID "0ec0a993-f599-b8d8-
        642c-470801fc2f35" as node ID
    ……
```

13.1.5 Consul UI

Consul 默认是没有页面的，如果需要页面展示，可以加上 -ui 参数：

```
[scbook@localhost ~]$ consul agent -dev -ui
```

然后访问 http://localhost:8500，默认进入 ui 界面，如图 13-1 所示。

图 13-1　Consul 管理界面

13.1.6 Consul 实用接口

Consul 对外提供了丰富的 API，有运维人员喜欢的命令行接口[⊖]，也有开发人员喜欢的 HTTP 接口[⊖]。接下来我们介绍几个简单常用的接口。

1. Consul 管理命令

- consul members。查看当前 Consul 代理知道集群所有成员信息以及它们的状态：存活、离线、连接失败。
- consul monitor。持续打印当前 Consul 日志。这个命令很有用，因为 Consul 访问量比较大，所以生产环境一般不会保存日志，如果想查看实时日志，可以使用该命令。
- consul leave。退出集群。一般我们会使用这个命令而不是直接杀掉进程。

2. Consul 对外服务接口

- /v1/agent/members：列出集群所有成员及其信息。
- /v1/status/leader：显示当前集群 leader。
- /v1/catalog/services：显示目前注册的服务。
- /v1/kv/key：显示当前 key 对应的 value。

13.2　Spring Cloud Consul 简介

13.2.1　Spring Cloud Consul 是什么

Spring Cloud Consul 通过自动配置、对 Spring Environment 绑定和其他惯用的 Spring 模块

⊖　https://www.consul.io/docs/commands/index.html。
⊖　https://www.consul.io/api/index.html。

编程,为 Spring Boot 应用程序提供了 Consul 集成。只需要一些简单注解,你就可以快速启用和配置 Consul,并用它来构建大型分布式系统。

13.2.2　Spring Cloud Consul 能做什么

Spring Cloud Consul 作为 Spring Cloud 与 Consul 之间的桥梁,对二者都有良好的支持。
- 服务发现。实例可以向 Consul 注册服务,客户端可以使用 Spring bean 来发现服务提供方。
- 支持 Ribbon,客户端负载。
- 支持 Zuul,服务网关。
- 分布式配置中心,使用的是 Consul 的 K/V 存储。
- 控制总线,使用的是 Consul events。

13.2.3　Spring Cloud Consul 入门案例

接下来笔者将带领读者一起,从无到有创建一个 Spring Cloud Consul 项目。项目分成三个模块:consul-provider(服务提供方)、consul-consumer(服务消费方)和 consul-config(consul 配置)。其中 consul-provider 与 consul-consumer 模块有调用关系,而 consul-config 是利用 consul 作为配置中心的一个相对独立的案例。

案例代码如本书源码 ch13-1 所示。

(1) 创建 Maven 父级 pom 工程

在父工程里面配置好工程需要的父级依赖,目的是方便管理与简化配置,pom 文件见本书配套源码 ch13-1\pom.xml。

(2) 创建 consul-provider 模块

因为父工程中已经将依赖加进去了,consul-provider 模块的 pom 并不需要额外依赖,所以这里不展示 consul-provider 的 pom。

创建主程序入口代码。我们创建主程序入口代码 ConsulProviderApplication,提供程序启动入口、健康检查、对外接口等功能。如代码清单 13-1 所示。

代码清单13-1　ch13-1/ch13-1-consul-provider/src/main/java/cn/springcloud/book/consul/provider/ConsulProviderApplication.java

```java
// consul-server的启动主类,为了简化代码,我们将Controller代码放在主类中,实际工作中不建议
    这么做。
@RestController
@SpringBootApplication
public class ConsulProviderApplication {
    public static void main(String[] args) {
        SpringApplication.run(ConsulProviderApplication.class, args);
    }

    // 注意: 新版Spring Cloud Consul 默认注册健康检查接口为: /actuator/health
    @GetMapping("/actuator/health")
    public String health(){
        return "SUCCESS";
    }
```

```
    // 提供 sayHello 服务:根据对方传来的名字××,返回:hello,××
    @GetMapping("/sayHello")
    public String sayHello(String name){
        return "hello, " + name;
    }
}
```

（3）添加 consul-provider 配置信息

如代码清单 13-2 所示。

代码清单13-2   ch13-1/ch13-1-consul-provider/src/main/resources/bootstrap.yml

```yaml
server:
    port: 8081       # 因为本地启动,防止端口冲突
spring:
    application:
        name: consul-provider
    cloud:
        consul:
            host: 127.0.0.1      # consul 启动地址
            port: 8500           # consul 启动端口
```

（4）创建 consul-consumer 模块

这里我们用 Spring Cloud Openfeign 组件来作为服务调用组件,因此要在 pom.xml 中添加额外依赖。如代码清单 13-3 所示。

代码清单13-3   ch13-1/ch13-1-consul-consumer/pom.xml

```xml
<dependencies>
    <dependency>
        <groupId>org.springframework.cloud</groupId>
        <artifactId>spring-cloud-starter-openfeign</artifactId>
    </dependency>
</dependencies>
```

添加程序启动主类。我们创建主程序入口代码 ConsulProviderApplication,提供程序启动入口、健康检查、对外接口等功能。如代码清单 13-4 所示。

代码清单13-4   ch13-1/ch13-1-consul-consumer/src/main/java/cn/springcloud/book/consul/consumer/ConsulConsumerApplication.java

```java
/**
 * consul-consumer的启动主类
 * 为了简化代码,我们将 Controller 代码放在主类中,实际工作中不建议这么做
 */
@RestController
@SpringBootApplication
@EnableFeignClients
public class ConsulConsumerApplication {

    /** 调用 hello 服务*/
    @Autowired
    private HelloService helloService;
```

```java
@GetMapping("/actuator/health")
public String health(){
    return "SUCCESS";
}

/** 接收前端传来的参数，调用远程接口，并返回调用结果 */
@GetMapping("/hello")
public String hello(String name){
    return helloService.sayHello(name);
}

public static void main(String[] args) {
    SpringApplication.run(ConsulConsumerApplication.class,args);
}
}
```

使用 openfeign 作为远程调用组件。如代码清单 13-5 所示。

代码清单13-5 ch13-1/ch13-1-consul-consumer/src/main/java/cn/springcloud/book/consul/consumer/HelloService.java

```java
/** 使用 openfeign 组件，调用远程服务 */
@FeignClient("consul-provider")
public interface HelloService {
    @RequestMapping(value = "/sayHello",method = RequestMethod.GET)
    String sayHello(@RequestParam("name") String name);
}
```

添加 consul-consumer 配置信息。如代码清单 13-6 所示。

代码清单13-6 ch13-1/ch13-1-consul-consumer/src/main/resources/bootstrap.yml

```yaml
server:
    port: 8082    # 因为本地启动，防止端口冲突
spring:
    application:
        name: consul-consumer
    cloud:
        consul:
            host: 127.0.0.1      # consul 启动地址
            port: 8500           # consul 启动端口
```

（5）效果展示

1）先启动本地 consul 服务。

2）启动 consul-provider 和 consul-consumer。

3）先去 consul 的管理页面查看：http://localhost:8500（页面会自动重定向到：http://localhost:8500/ui/dc1/services），如果"Node Health"显示绿色对勾，即表示服务发布成功。如图 13-2 所示。

4）在浏览器地址栏输入：http://localhost:8082/hello?name=scbook 如果返回"hello, scbook"，即为运行成功。如图 13-3 所示。

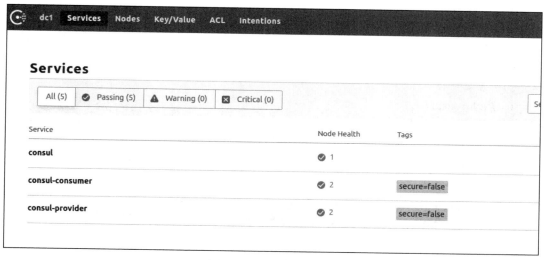

图 13-2 注册成功页面

（6）创建 consul-config 模块

因为父工程中已经将依赖加进去，consul-config 模块的 pom 并不需要额外依赖，所以这里不展示 consul-provider 的 pom。

图 13-3 运行结果展示

因为配置信息跟上文 consul-provider 和 consul-consumer 类似，所以这里不再展示 consul-config 的配置详情。

添加程序启动类。我们创建主程序入口代码 ConsulProviderApplication，提供程序启动入口、健康检查、对外接口等功能。如代码清单 13-7 所示。

代码清单13-7 ch13-1/ch13-1-consul-config/src/main/java/cn/springcloud/book/consul/config/ConsulConfigApplication.java

```
/**
 * consul-config启动类主类，为了简化代码，我们将Controller和配置代码放在主类中，实际工作中
   不建议这么做
 */
@SpringBootApplication
@RestController
@RefreshScope   // 参数值修改后自动刷新
public class ConsulConfigApplication {
    public static void main(String[] args) {
        SpringApplication.run(ConsulConfigApplication.class,args);
    }

    // 读取远程配置
    @Value("${foo.bar.name}")
    private String name;

    // 将配置展示在页面
    @GetMapping("/getName")
```

```
    public String getName(){
        return name;
    }
}
```

（7）效果展示

1）启动 consul。

2）浏览器地址栏进入：http://localhost:8500/ui/dc1/kv（或者等进入首页后，选择 Key/Value 菜单）。

3）点击 create 按钮，在"KEY OR FOLDER"栏输入：config/consul-config/foo.bar.name，在"VALUE"栏输入"scbook"，点击保存。如图 13-4 所示。

图 13-4　Consul 配置管理界面

4）启动 consul-config 模块，浏览器地址栏输入：http://localhost:8083/getName。

5）在 consul 页面中修改 key 对应的 value 值，再请求一次：http://localhost:8083/getName，可以看见值已经更新。如图 13-5 所示。

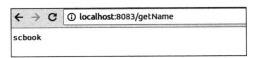

图 13-5　运行结果展示

13.3　本章小节

本章主要讲解 Consul 和 Spring Cloud Consul 的背景知识与基本用法。通过本章的学习，读者不但可以对 Consul 在微服务体系中的定位有一个大致的了解，还可以使用 Consul 和 Spring Cloud Consul 搭建一个基本的服务治理和配置中心实例。当然，在实际工作生产环境中，二者的搭建和使用方式要复杂得多。在下一章，笔者将带领大家踏上 SPring Cloud Consul 的企业级实战征程。

第 14 章

Spring Cloud Consul 下篇

在第 13 章中，我们学习了 Consul 和 Spring Cloud Consul 的相关背景知识与基础用法。但是实际工作中的情况远比一个简单的案例复杂得多，本章笔者就跟大家一起，循序渐进，探索 Spring Cloud Consul 的强大功能。

14.1 Spring Cloud Consul 深入

Spring Cloud Consul 是在 consul-api ⊖ 的基础上又封装了一层功能，使其跟现有 Spring Cloud 组件融合，达到开箱即用的目的。

14.1.1 Spring Cloud Consul 的模块介绍

围绕着 Consul 的核心功能，Spring Cloud Consul 也提供了相应的功能模块与之匹配。下面就为大家逐一介绍。

- Spring-cloud-consul-binder：对 Consul 的事件功能封装。
- Spring-cloud-consul-config：对 Consul 的配置功能封装。
- Spring-cloud-consul-core：基础配置和健康检查模块。
- Spring-cloud-consul-discovery：对 Consul 服务治理功能封装。

Consul 的事件功能比较弱化，应用比较多的是服务治理和配置功能，所以我们重点介绍 Spring Cloud Consul 的 discovery 和 config 模块。

14.1.2 Spring Cloud Consul Discovery

Spring-cloud-consul-discovery 在功能上分成两部分：服务注册和服务发现，如图 14-1 所示。

⊖ Consul-api 是 Consul 的一个 Java 客户端，Github 地址：https://github.com/Ecwid/consul-api。

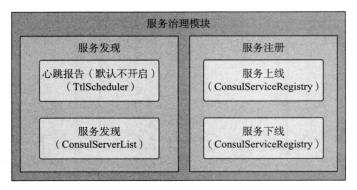

图 14-1　Spring Cloud Consul Discovery 主要功能分布

1. spring-cloud-consul-discovery 服务注册

服务启动时，会通过 ConsulServiceRegistry.register() 向 Consul 注册自身的服务。注册时，会告诉 Consul 以下信息。

- ID。服务 ID，默认是服务名 – 端口号。
- Name。服务名，默认是系统名称。
- Tags。给服务打的标签，默认是 [secure=false]。
- Address。服务地址，默认是主机名（如果没有，则返回 IP）。
- Port。服务端口，默认是服务的 web 端口。
- Check。健康检查信息，包括 Interval（健康检查间隔）和 HTTP（健康检查地址）。

一般情况下，我们不需要显式提供上述信息，spring-cloud-consul 会有默认值。但是在一些特殊业务场景中，可能就需要根据实际情况配置了。consul discovery 的常见配置如表 14-1 所示。

表 14-1　consul discovery 常见配置

配置项	修改方法（省略如下前缀：spring.cloud.consul.discovery）	适用场景
Address	prefer-ip-address=true	需要注册 IP 地址而非主机名
Address	ip-address=${HOST_IP}	如果服务部署在 docker 容器中，需要告诉 Consul 自己真实的地址
Port	port=${HOST_PORT}	如果服务部署在 docker 容器中，需要告诉 Consul 自己真实的端口号
Tags	tags=${GROUP},test	如果需要给相同服务进行分组，可以使用该配置给服务打不同标签，不同标签之间用英文逗号隔开
Check	health-check-interval=20s	Consul 默认对服务的健康检查时间是 10s，如果需要修改时长，可以修改此配置
Check	health-check-path=/health.json	默认健康检查路径是 /actuator/health，如果你没有使用 spring-cloud-actuator，或者由于其他原因想定制健康检查路径，可以修改此配置

接下来我们创建一个 consul-register 项目，来演示如何自定义注册信息。为了简化起见，

我们使用 ch13-1 的父级 pom。

（1）创建 consul-register 模块

添加程序启动主类，如代码清单 14-1 所示。

代码清单 14-1 ch14-1/ch14-1-consul-register/src/main/java/cn/springcloud/book/consul/register/ConsulRegisterApplication.java

```
@RestController
@SpringBootApplication
public class ConsulRegisterApplication {
    public static void main(String[] args) {
        SpringApplication.run(ConsulRegisterApplication.class,args);
    }
    /**
    这里我们不使用默认的健康检测，而是使用自己定义的接口
    */
    @GetMapping("/health")
    public String health(){
        return "SUCCESS";
    }
}
```

添加 consul-register 配置信息，如代码清单 14-2 所示。

代码清单 14-2 ch14-1/ch14-1-consul-register/src/main/resources/bootstrap.yml

```
spring:
  application:
    name: consul-register
  cloud:
    consul:
      host: 127.0.0.1          # consul 启动地址
      port: 8500               # consul 启动端口
      discovery:
        prefer-ip-address: true        # 优先使用 IP 注册
        ip-address: 127.0.0.1          # 假装部署在 docker中,指定了宿主机 IP
        port: 8080                     # 假装部署在 docker中,指定了宿主机端口
        health-check-interval: 20s     # 健康检查间隔时间为 20s
        health-check-path: /health     # 自定义健康检查路径
        tags: ${LANG},test             # 指定服务的标签，用逗号隔开
```

（2）效果展示

- 启动本地 Consul 服务。
- 启动 consul-register。
- 在浏览器地址栏输入：http://localhost:8500/ui/dc1/services，可以看到 consul-register 注册成功，且健康检查路径和端口均已生效，如图 14-2 所示。

> **注意** Tags 一栏会有一个 "secure=true"，这个是 Spring Cloud Consul 默认为我们加上的，表示是否是 HTTPS 协议。它取自配置 spring.cloud.consul.discovery.scheme，默认值是 http。如果你的服务提供的是 https 的服务时，需要配置该值为 https。它的作用是告诉客户端，调用服务方接口时需要哪种协议。

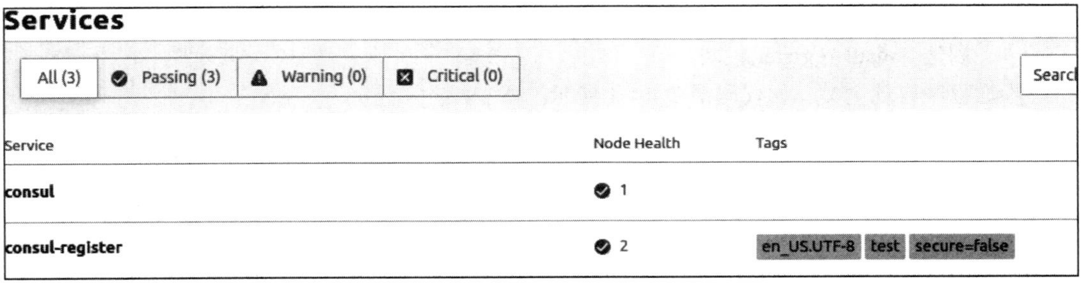

图 14-2 应用注册成功页面

2. Spring-cloud-consul-discovery 服务发现

Spring Cloud Consul 提供了两种方式提供服务发现功能：Ribbon 和 DiscoveryClient。

如果客户端使用了 Feign 或者 @LoadBalanced，那么默认使用的是 ConsulServerList.java 提供的服务发现逻辑。如果客户端只想独立使用服务发现功能，那么可以直接使用 DiscoveryClient。

上文中，我们提到了使用 Consul 的 tags 功能将服务分组，本节我们就用上面说的两种方式分别调用服务方功能。我们复用之前的父级 pom 配置，创建四个子模块：consul-provider-tag-1、consul-provider-tag-2、consul-consumer-ribbon、consul-consumer-discoveryclient。

（1）创建 consul-provider-tag-1 模块

添加程序启动类，如代码清单 14-3 所示。

代码清单14-3 ch14-2/ch14-2-consul-provider-tag-1/src/main/java/cn/springcloud/book/consul/provider/tag1/ConsulProviderTag1Application.java

```java
@SpringBootApplication
@RestController
public class ConsulProviderTag1Application {
    public static void main(String[] args) {
        SpringApplication.run(ConsulProviderTag1Application.class, args);
    }
    /** 此处省略健康检查接口
     * 提供 sayHello 服务:根据对方传来的名字 XX, 返回:hello XX, I am tag1
     */
    @GetMapping("/sayHello")
    public String sayHello(String name){
        return "hello," + name + ". I am tag1";
    }
}
```

添加程序配置文件，如代码清单 14-4 所示。

代码清单14-4 ch14-2/ch14-2-consul-provider-tag-1/src/main/resources/bootstrap.yml

```
server:
    port: 8082
spring:
```

```yaml
    application:
      name: consul-provider
    cloud:
      consul:
        host: 127.0.0.1        # consul 启动地址
        port: 8500             # consul 启动端口
        discovery:
          tags: tag1           # 指定服务的标签，用逗号隔开
```

（2）创建 consul-provider-tag-2 模块

tag-2 模块基本跟 tag-1 一样，只需要将启动类和配置中的 tag1 改为 tag2 即可，这里不再赘述。

（3）创建 consul-consumer-ribbon 模块

在 consumer-ribbon 模块中，我们建立两个 rest 接口，一个通过 feign 的方式访问 consul-provider，一个通过 RestTemplate 的方式访问 consul-provider。

接下来是添加 pom 依赖，因为需要用到 openfeign，所以 pom 中要添加对它的依赖，如代码清单 14-5 所示。

代码清单14-5　ch14-2/ch14-2-consul-consumer-ribbon/pom.xml

```xml
<dependencies>
    <dependency>
        <groupId>org.springframework.cloud</groupId>
        <artifactId>spring-cloud-starter-openfeign</artifactId>
    </dependency>
</dependencies>
```

添加程序入口启动类，如代码清单 14-6 所示。

代码清单14-6　ch14-2/ch14-2-consul-consumer-ribbon/src/main/java/cn/springcloud/book/consul/consumer/ribbon/ConsulConsumerRibbonApplication.java

```java
@SpringBootApplication
@RestController
@EnableFeignClients
public class ConsulConsumerRibbonApplication {
    @Autowired
    private HelloService helloService;      // hello 服务提供方
    @Autowired
    private RestTemplate restTemplate;

    /** 创建 RestTemplate Bean,并用 @LaodBalanced 注解*/
    @LoadBalanced
    @Bean
    public RestTemplate restTemplate() {
        return new RestTemplate();
    }

    /** 接收前端传来的参数，使用 feign的方式调用远程接口，并返回调用结果 */
    @GetMapping("/hello1")
    public String hello1(String name){
        return helloService.sayHello(name);
    }
```

```
/** 接收前端传来的参数，使用 restTemplate的方式调用远程接口，并返回调用结果 */
@GetMapping("/hello2")
public String hello2(String name){
    return restTemplate.getForObject("http://consul-provider/sayHello?name=
     "+name,String.class);
}
// 省略main方法和健康检查接口
```

添加 HelloService 服务接口，如代码清单 14-7 所示。

代码清单14-7 ch14-2/ch14-2-consul-consumer-ribbon/src/main/java/cn/springcloud/book/consul/consumer/ribbon/HelloService.java

```
/** 使用 openfeign 组件，调用远程服务 */
@FeignClient("consul-provider")
public interface HelloService {
    @RequestMapping(value = "/sayHello",method = RequestMethod.GET)
    String sayHello(@RequestParam("name") String name);
}
```

添加程序配置项，如代码清单 14-8 所示。

代码清单14-8 ch14-2/ch14-2-consul-consumer-ribbon/src/main/resources/bootstrap.yml

```
server:
    port: 8083              # 因为本地启动，防止端口冲突
spring:
    application:
        name: consul-consumer-ribbon
    cloud:
        consul:
            host: 127.0.0.1     # consul 启动地址
            port: 8500          # consul 启动端口
            discovery:
                server-list-query-tags:     # 注意 server-list-query-tags是一个 map
                    consul-provider: tag1   # 在调用consul-provider 服务时，使用
                                            tag1 对应的实例
```

（4）创建 consul-consumer-discoveryclient 模块

该模块使用 discoveryclient 注入的方式，手动去 Consul 中获取服务列表。这里需要说明的是，ConsulDiscoveryClient 中不支持根据服务方名称自定义 tag，后文我们会讲到如何自己加上这个功能。

discoveryclient 模块不需要引入额外组件，可以使用父级 pom，这里省略模块 pom 的展示。添加程序启动类，如代码清单 14-9 所示。

代码清单14-9 ch14-2/ch14-2-consul-consumer-discoveryclient/src/main/java/cn/springcloud/book/consul/consumer/discoveryclient/ConsulConsumerDiscoveryclientApplication.java

```
@SpringBootApplication
```

```
@RestController
public class ConsulConsumerDiscoveryclientApplication {
    @Autowired   // ConsulDiscoveryClient 会在程序启动时,初始化为DiscoveryClient实例
    private DiscoveryClient discoveryClient;

    // 这里只举例获取服务方信息,不去请求服务方接口
    @GetMapping("/getServer")
    public List<ServiceInstance> getServer(String serviceId){
        return discoveryClient.getInstances(serviceId);
    }

    // 省略main方法和健康检查接口
}
```

添加配置信息,如代码清单 14-10 所示。

代码清单14-10　ch14-2/ch14-2-consul-consumer-discoveryclient/src/main/resources/bootstrap.yml

```
server:
    port: 8084                    # 因为本地启动,防止端口冲突
spring:
    application:
        name: consul-consumer-discoveryclient
    cloud:
        consul:
            host: 127.0.0.1       # consul 启动地址
            port: 8500            # consul 启动端口
```

效果展示如下:

❏ 启动本地 Consul 服务。
❏ 依次启动 consul-provider-tag1、consul-provider-tag2、consul-consumer-ribbon、consul-consumer-discoveryclient。
❏ 打开 Consul 管理页面,查看服务注册情况,如图 14-3 所示。

Service	Port	Tags
consul	8300	
consul-consumer-discoveryclient	8084	secure=false
consul-consumer-ribbon	8083	secure=false
consul-provider	8081	tag1 secure=false
consul-provider	8082	tag2 secure=false

图 14-3　应用注册 tag 明细

❏ 请求 ribbon 模块的 feign 接口。在浏览器地址栏输入: http://localhost:8083/hello1?name=scbook,查看返回信息是否为 "hello,scbook. I am tag1"。

- 请求 ribbon 模块的 RestTemplate 接口。在浏览器地址栏输入：http://localhost:8083/hello2?name=scbook，查看返回信息是否为"hello,scbook. I am tag1"，如图 14-4 所示。

- 请求 discoveryclient 模块接口。在浏览器地址栏输入：http://localhost:8084/getServer，查看返回的服务方信息是否只有 consul-provider-tag1 的信息，如图 14-5 所示。

图 14-4　运行结果展示

```
① localhost:8084/getServer?serviceId=consul-provider
[
  {
    "serviceId": "consul-provider",
    "host": "192.168.44.151",
    "port": 8081,
    "secure": false,
    "metadata": {
      "tag1": "tag1",
      "secure": "false"
    },
    "uri": "http://192.168.44.151:8081",
    "scheme": null
  },
  {
    "serviceId": "consul-provider",
    "host": "192.168.44.151",
    "port": 8082,
    "secure": false,
    "metadata": {
      "tag2": "tag2",
      "secure": "false"
    },
    "uri": "http://192.168.44.151:8082",
    "scheme": null
  }
]
```

图 14-5　运行结果展示

 这里对于 consul-consumer-ribbon 的验证并不严谨，因为 ribbon 会对两个 provider 做负载。如果要确认我们的 tag 配置是生效的，则需要将 consul-provider-tag1 服务停止后，再一次请求 hello 接口，看是否会返回 tag2 的结果。限于篇幅的原因，我们这里不作赘述，有兴趣的读者可以接着实验。

14.1.3　Spring Cloud Consul Config

第 13 章我们演示了 Spring Cloud Consul Config 获取和刷新配置的简单用法。细心的读者可能会问，Spring Cloud Consul Config 与 Consul 是通过 HTTP 进行交互的，那配置刷新是如何做到的呢？另外，示例中只有一条配置，可是实际工作中的配置可能有成百上千条，难道我们要一个个在 Consul 页面中添加吗？别着急，这两个问题会在本节一一进行解答。

1. Spring Cloud Consul Config 配置刷新原理

我们知道，Spring Cloud Consul 是通过 HTTP 的方式跟 Consul 交互，那配置是如何实时生效的呢？答案其实很简单，那就是配置并没有实时生效。org.springframework.cloud.consul.config.ConfigWatch 中有一个定时方法 watchConfigKeyValues()，它默认每秒执行一次（可以通过 spring.cloud.consul.config.watch.delay 自定义），去 Consul 中获取最新的配置信息，一旦配置发生改变，Spring 通过 ApplicationEventPublisher 重新刷新配置。Consul Config 组件就是通过这种方式，达到配置"实时生效"的目的。那客户端如何得知配置被更新过了呢，答案在 Consul 返回的数据里。Consul 会给每一项配置加一个"consulIndex"属性，类似于版本号，如果配置更新，它就会自增。Spring Cloud Consul Config 就是通过缓存"consulIndex"来判断配置是否发生改变。

2. Spring Cloud Consul Config 高级配置

Consul 只支持用 K/V 的方式进行配置，那怎么让 Consul 支持同时配置多条的方式呢？难道要给 Consul 增加一个导入功能吗？聪明的读者马上就能想到，K/V 不仅可以代表一条配置，还可以代表一个应用的配置，我们将应用名作为 K，V 中用来存放它所有的配置，这样就可以达到同时配置多条的效果。

Spring Cloud Consul Config 就是这样，通过将 yml 或者 properties 放在 V 中来实现配置的批量操作。

接下来，我们创建一个 ch14-3 项目，里面包含一个 consul-config-customize 模块。该模块使用 Consul 的配置功能，将整个配置以 yml 的方式放在 Consul 中。

（1）创建父级 pom 依赖

在父工程里面配置好工程需要的父级依赖，目的是方便管理与简化配置。如代码清单 14-11 所示。

代码清单14-11　ch14-3\pom.xml

```xml
<dependencies>
    <!-- 因为只用到了Spring Cloud Consul的配置功能，这里不全部引入-->
    <dependency>
        <groupId>org.springframework.cloud</groupId>
        <artifactId>spring-cloud-starter-consul-config</artifactId>
    </dependency>
    <dependency>
        <groupId>org.springframework.boot</groupId>
        <artifactId>spring-boot-starter-web</artifactId>
    </dependency>
</dependencies>
```

（2）创建 consul-config-customize 模块

这里我们使用父级的 pom 依赖，当前模块 pom 配置省略，详见代码清单：ch14-3/ch14-3-consul-config-customize/pom.xml。

创建项目启动类，如代码清单 14-12 所示。

代码清单14-12 ch14-3/ch14-3-consul-config-customize/src/main/java/cn/springcloud/book/
consul/config/customize/ConsulCustomizeApplication.java

```java
@SpringBootApplication
@RestController
public class ConsulCustomizeApplication {
    public static void main(String[] args) {
        SpringApplication.run(ConsulCustomizeApplication.class,args);
    }
    // 读取远程配置
    @Value("${foo.bar.name}")
    private String name;
    // 将配置展示在页面
    @GetMapping("/getName")
    public String getName(){
        return name;
    }
}
```

添加 application.yml 配置信息，如代码清单 14-13 所示。

代码清单14-13 ch14-3/ch14-3-consul-config-customize/src/main/resources/application.yml

```yaml
server:
    port: 8081          # 这里使用 8081 端口，Consul中配置8082端口，验证生效配置
spring:
    profiles:
        active: test    # 指定启动时的 profiles
```

添加 bootstrap.yml 配置信息，如代码清单 14-14 所示。

代码清单14-14 ch14-3/ch14-3-consul-config-customize/src/main/resources/bootstrap.yml

```yaml
spring:
    application:
        name: consul-config-customize
    cloud:
        consul:
            config:
                format: yaml                    # Consul中 Value 配置格式为 yaml
                prefix: configuration           # Consul中配置文件目录为 configuration, 默认为 config
                default-context: app            # 去该目录下查找缺省配置,默认为 application
                profile-separator: ':'          # profiles配置分隔符,默认为 ','
                data-key: data                  # 如果指定配置格式为 yaml 或者 properties, 则需要该值作为key,默认为 data
```

效果展示如下。

1）启动本地 Consul 服务。

2）在 consul 管理页面中增加配置。

key 值为：configuration/consul-config-customize:dev/data。

value 值如下：

```
foo:
    bar:
        name: scbook-dev
server:
    port: 8082
```

key 值为：configuration/consul-config-customize:test/data

value 值如下：

```
foo:
    bar:
        name: scbook-test
server:
    port: 8083
```

3）启动 consul-config-customize，查看启动端口。启动日志如图 14-6 所示。

图 14-6　启动端口生效日志

4）在浏览器地址栏输入：http://localhost:8083/getName，查看生效字段。运行结果如图 14-7 所示。

5）修改 Consul 中的配置，然后重新查看生效字段是否已经刷新，这里不做演示。

图 14-7　运行结果

14.2　Spring Cloud Consul 功能重写

Spring Cloud Consul 提供了很多方便实用的功能供我们使用，但是面对五花八门的需求，我们还是希望可以重写它的原有逻辑。本节笔者就以服务发现为例，跟大家一起踏上重写 Spring Cloud Consul 的征程。

14.2.1　重写 ConsulDiscoveryClient

在前文，我们留下了一个问题还未解决：ConsulDiscoveryClient 并不支持自定义 tag。一般来说，ConsulServerList 和 ConsulDiscoveryClient 虽然面对的需求不同，但是实现的功能都是一样的，那就是根据条件查找服务。可能 Spring 认为 ConsulDiscoveryClient 是为一些框架型的功能准备的，用户完全可以拿到服务列表后自行筛选所需的数据，但是笔者认为 ConsulServerList 已经有现成的方法来支持该功能，在 ConsulDiscoveryClient 中加上它是举手之劳的事情，加完后还能实现两个类功能上的统一，不易带来误解和错觉。

但凡需要重写 Spring Cloud 的功能，我们第一步要看的都是该功能如何初始化。

如果在该类的初始化之前，有一个 @ConditionalOnMissingBean[⊖]注解，那么事情就已经成功了一半。

很幸运，ConsulDiscoveryClient 就是这样一个类：

```
// 省略注解信息
public class ConsulDiscoveryClientConfiguration {
    // 省略其他Bean的初始化代码
    @Autowired
    private ConsulClient consulClient;
    @Bean
    @ConditionalOnMissingBean
    public ConsulDiscoveryClient consulDiscoveryClient(ConsulDiscoveryProperties
        discoveryProperties) {
        return new ConsulDiscoveryClient(consulClient, discoveryProperties);
    }
}
```

接下来，我们创建 ch14-4 项目，里面有三个模块：consul-override-provider-tag1、consul-override-provider-tag2 和 consul-override-consumer。两个 provider 提供相同的服务名：consul-provider，consumer 通过 DiscoveryClient 指定访问 tag2 服务。

其中，父级 pom 和子模块 pom 均与前文服务发现项目配置相同，consul-override-provider-tag1-2 与 ch14-2 项目下 consul-provider-tag1-2 相同，因此文中均不展示。

（1）创建 consul-override-consumer 模块

添加服务启动程序，如代码清单 14-15 所示

代码清单14-15 ch14-4/ch14-4-consumer-override-consumer/src/main/java/cn/springcloud/book/consul/consumer/discoveryclient/ConsulConsumerDiscoveryclient-Application.java

```
@SpringBootApplication
@RestController
public class ConsulConsumerDiscoveryclientApplication {

    @Autowired   // ConsulDiscoveryClient 会在程序启动时,初始化为DiscoveryClient实例
    private DiscoveryClient discoveryClient;

    @Bean
    @Order(Ordered.HIGHEST_PRECEDENCE) // 保证优先被加载
    public MyConsulDiscoveryClient ConsulDiscoveryClient(ConsulClient client,
        ConsulDiscoveryProperties properties){
        return new MyConsulDiscoveryClient(client,properties);
    }
    // 这里只举例获取服务方信息,不去请求服务方接口
    @GetMapping("/getServer")
    public List<ServiceInstance> getServer(String serviceId){
        return discoveryClient.getInstances(serviceId);
    }
```

⊖ 该注解详情可以参考 :https://docs.spring.io/spring-boot/docs/current/api/org/springframework/boot/autoconfigure/condition/ConditionalOnMissingBean.html。

```
        // 省略main()方法和健康检查接口
    }
```

添加自定义 DiscoveryClient 实现类：MyConsulDiscoveryClient，如代码清单 14-16 所示。

代码清单14-16 ch14-4/ch14-4-consumer-override-consumer/src/main/java/cn/springcloud/book/consul/consumer/discoveryclient/MyConsulDiscoveryClient.java

```java
public class MyConsulDiscoveryClient implements DiscoveryClient {
    // 主体代码跟ConsulDiscoveryClient一致，因此相同代码省略
    private void addInstancesToList(List<ServiceInstance> instances, String serviceId,
                        QueryParams queryParams) {
        String aclToken = properties.getAclToken();
        Response<List<HealthService>> services;
        if (StringUtils.hasText(aclToken)) {
            services = client.getHealthServices(serviceId,
                    getTag(serviceId),    // 此处由获取默认tag改为获取指定tag
                    this.properties.isQueryPassing(), queryParams, aclToken);
        }
        else {
            services = client.getHealthServices(serviceId,
                    getTag(serviceId),    // 此处由获取默认tag改为获取指定tag
                    this.properties.isQueryPassing(), queryParams);
        }
        // 省略重复代码
    }
    // 添加获取tag的方法，该方法在 ConsulServerList中已经存在，直接复制过来
    private String getTag(String serviceId) {
        return this.properties.getQueryTagForService(serviceId);
    }
}
```

添加服务配置信息，如点清单 14-17 所示。

代码清单14-17 ch14-4/ch14-4-consumer-override-consumer/src/main/resources/bootstrap.yml

```yaml
spring:
    application:
        name: consul-override-consumer
    cloud:
        consul:
            host: 127.0.0.1       # consul 启动地址
            port: 8500            # consul 启动端口
            discovery:
                server-list-query-tags:    # 注意server-list-query-tags是一个 map
                    consul-provider: tag1  # 在调用consul-provider服务时，使用tag1
                                           # 对应的实例
```

（2）效果展示

1）分别启动 consul-override-provider-tag1、consul-override-provider-tag2、consul-override-consumer 三个应用。

2）在浏览器地址栏输入：http://localhost:8083/getServer?serviceId=consul-provider，我们

可以发现，仅能获取 tag1 的服务信息。运行结果如图 14-8 所示。

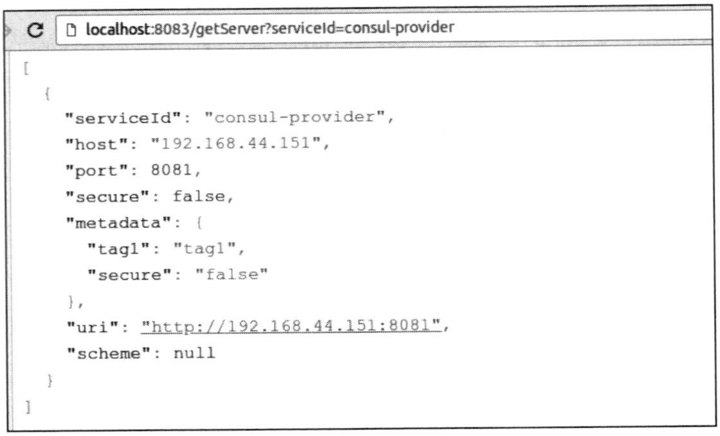

```
C   localhost:8083/getServer?serviceId=consul-provider
[
  {
    "serviceId": "consul-provider",
    "host": "192.168.44.151",
    "port": 8081,
    "secure": false,
    "metadata": {
      "tag1": "tag1",
      "secure": "false"
    },
    "uri": "http://192.168.44.151:8081",
    "scheme": null
  }
]
```

图 14-8　运行结果展示

14.2.2　重写 ConsulServerList

看到标题，肯定有读者有疑问，既然已经学习了重写 ConsulDiscoveryClient，那么重写跟他相同功能的 ConsulServerList 肯定是大同小异，有必要再拿出来说吗？

答案是：有必要。

上一节我们讲了，重写 Spring Cloud 的核心逻辑，首先要看它是否允许外部实现注入。那有读者不免更疑惑了：ConsulServerList 是 ServerList 的实现，而且在 ConsulRibbonClientConfiguration 中，它的初始化方法也确实有 @ConditionalOnMissingBean 注解㊀，那重写岂不是更加容易了？我们可以试下，案例实现如下。

我们创建一个 ch14-5 的项目，里面有三个模块：consul-provider1、consul-provider2、consul-consumer。其中，consul-consumer 中分别调用 provider1 和 provider2。

简单起见，我们只展示 consul-consumer 的核心代码，其他代码参见代码清单：ch14-5。

（1）创建 consul-consumer 模块

添加 pom 依赖，如代码清单 14-18 所示。

代码清单14-18　ch14-3/ch14-3-consul-config-customize/pom.xml

```xml
<dependencies>
    <dependency>
        <groupId>org.springframework.cloud</groupId>
        <artifactId>spring-cloud-starter-openfeign</artifactId>
    </dependency>
</dependencies>
```

添加程序启动程序，如代码清单 14-19 所示。

㊀ 有兴趣的读者可以翻阅源码验证一下。

代码清单14-19 ch14-5/ch14-5-consul-override-consumer/src/main/java/cn/springcloud/book/consul/override/consumer/consulserverlist/ConsulConsumerApplication.java

```java
@SpringBootApplication
@RestController
@EnableFeignClients
public class ConsulConsumerApplication {
    @Autowired
    private Hello1Service hello1Service; // consul-provider1 服务提供方
    @Autowired
    private Hello2Service hello2Service; // consul-provider2 服务提供方

    /** 接收前端传来的参数，使用feign的方式调用 consul-provider2 远程接口，并返回调用结果 */
    @GetMapping("/hello1")
    public String hello1(String name){
        return hello1Service.sayHello(name);
    }
    /** 接收前端传来的参数，使用 feign的方式调用 consul-provider2 远程接口，并返回调用结果 */
    @GetMapping("/hello2")
    public String hello2(String name){
        return hello2Service.sayHello(name);
    }

    // 使用自定义的 ConsulServerList
    // 这里的 config 没有使用注入的方式，因为启动时会报没有 IClientConfig Bean的错误
    @Bean
    public ServerList<?> ribbonServerList(ConsulClient client, ConsulDiscoveryProperties
        properties) {
        MyConsulServerList serverList = new MyConsulServerList(client, properties);
        IClientConfig config = new DefaultClientConfigImpl();
        serverList.initWithNiwsConfig(config);
        return serverList;
    }
    public static void main(String[] args) {
        SpringApplication.run(ConsulConsumerApplication.class,args);
    }
    @GetMapping("/actuator/health")
    public String health(){
        return "SUCCESS";
    }
}
```

添加服务提供方接口，如代码清单 14-20 所示。

代码清单14-20 ch14-5/ch14-5-consul-override-consumer/src/main/java/cn/springcloud/book/consul/override/consumer/consulserverlist/Hello1Service.java、Hello2Service.java

```java
/** 使用 openfeign 组件，调用 consul-provider1 远程服务 */
@FeignClient("consul-provider1")
public interface Hello1Service {
    @RequestMapping(value = "/sayHello",method = RequestMethod.GET)
    String sayHello(@RequestParam("name") String name);
}
/** 使用 openfeign 组件，调用 consul-provider2 远程服务 */
@FeignClient("consul-provider2")
```

```
public interface Hello2Service {
    @RequestMapping(value = "/sayHello",method = RequestMethod.GET)
    String sayHello(@RequestParam("name") String name);
}
```

添加程序配置信息，如代码清单 14-21 所示：

代码清单14-21　ch14-5/ch14-5-consul-override-consumer/src/main/resources/bootstrap.yml

```yml
server:
    port: 8081
spring:
    application:
        name: consul-provider2
    cloud:
        consul:
            host: 127.0.0.1      # consul 启动地址
            port: 8500           # consul 启动端口
```

添加自定义的 ConsulServerList，如代码清单 14-22 所示。

代码清单14-22　ch14-5/ch14-5-consul-override-consumer/src/main/java/cn/springcloud/book/consul/override/consumer/consulserverlist/MyConsulServerList.java

```java
public class MyConsulServerList extends AbstractServerList<ConsulServer> {
    // 省略无关代码，只在这里打印一句提示
    private List<ConsulServer> getServers() {
        ○ ○ ○ ○
        System.out.println("===== 自定义服务发现逻辑 =====");
        String tag = getTag(); // null is ok
        Response<List<HealthService>> response = this.client.getHealthServices(
            this.serviceId, tag, this.properties.isQueryPassing(),
            createQueryParamsForClientRequest(), this.properties.getAclToken());
        if (response.getValue() == null || response.getValue().isEmpty()) {
            return Collections.emptyList();
        }
        return transformResponse(response.getValue());
    }
}
```

（2）效果展示

分别启动 consul-provider1、consul-provider2 和 consul-consumer 等本地服务。

在浏览器地址栏输入 http://localhost:8083/hello2?name=scbook，我们发现报错了，日志显示找不到服务。如图 14-9 所示。

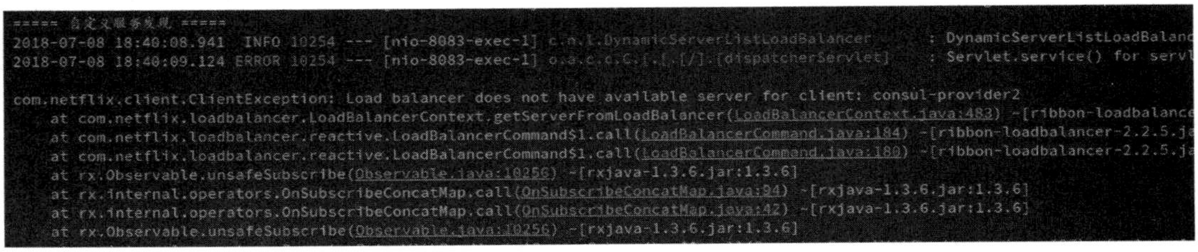

图 14-9　运行结果展示

经过如上示例,我们发现,单纯的自定义 ServerList 的接口并不能达到重写 Consul-ServerList 的目的。那么问题在哪儿呢?经过 debug 我们很容易得出,是因为 ConsulServerList 的 serviceId 属性为 null 导致的。这个属性很让人费解,它表示服务方的名称,但是却作为 ConsulServerList 的成员变量。那如果我依赖了多个服务,岂不是会出问题?由此我们很容易联想到,是不是 Spring Cloud Consul 为每个服务都创建了一个 ConsulServerList 实例呢?

经过源码追踪,我们可以发现,ConsulServerList 并不是在启动的时候初始化,而是在服务调用时,通过 Ribbon 来初始化的。

- Feign 通过 serviceId 去 Ribbon 中获取服务端配置。
- Ribbon 根据 serviceId 去缓存中找是否存在这个名称的。AnnotationConfigApplication-Context 实例,如果有就立即返回,如果没有就创建一个。而创建 AnnotationConfig-ApplicationContext 的过程,就是 ConsulServerList 初始化的过程。

AnnotationConfigApplicationContext 跟 ConsulServerList 是如何关联的呢?答案是 @Ribbon-Client 注解,具体可以参考 RibbonClientConfigurationRegistrar 源码,如下所示:

```
protected AnnotationConfigApplicationContext createContext(String name) {
// name 亦即服务名
    AnnotationConfigApplicationContext context = new AnnotationConfigApplicationContext();
    ......
    // 重新注册 RibbonClientConfiguration,这里的defaultConfigType就是RibbonClient-
       Configuration.class
    context.register(PropertyPlaceholderAutoConfiguration.class,this.default-
       ConfigType);
    .....
    context.refresh(); // 刷新 RibbonClientConfiguration, 其下的 serverList 也被重
       新初始化
    return context;
}
```

所以重写 ConsulServerList 的过程比较麻烦:要么重新写一套类似的 spring-cloud-consul-discovery 源码,要么就从源头的 RibbonClientConfiguration 开始直到 ConsulServerList,均改成自己的。

这里我们演示一下第二种方式。

在 ch14-5 项目下,创建新的模块:consul-override-consumer-new,主体代码跟上一节中的 consul-override-consumer 一样,这里我们新加几个类。

(1) 添加 MyRibbonConsulAutoConfiguration,如代码清单 14-23 所示。

代码清单14-23 ch14-5/ch14-5-consul-override-consumer-new/src/main/java/cn/spring-cloud/book/consul/override/consumer/newconsulserverlist/MyRibbonConsulAutoConfiguration.java

```
@Configuration
@ConditionalOnConsulEnabled
@ConditionalOnBean(SpringClientFactory.class)
@AutoConfigureAfter(RibbonAutoConfiguration.class)
@ConditionalOnExpression("${spring.cloud.consul.ribbon.enabled}==false")
@RibbonClients(defaultConfiguration = ConsulRibbonClientConfiguration.class)
```

```
public class MyRibbonConsulAutoConfiguration {
    /** 该类主要是将原有入口取代，因此它的生效逻辑刚好跟RibbonConsulAutoConfiguration相反：
     * 当 spring.cloud.consul.ribbon.enabled 为 false 时，重写逻辑生效
     */
}
```

（2）添加 MyConsulRibbonClientConfiguration，如代码清单 14-24 所示。

代码清单14-24　ch14-5/ch14-5-consul-override-consumer-new/src/main/java/cn/springcloud/book/consul/override/consumer/newconsulserverlist/MyConsulRibbonClientConfiguration.java

```
@Configuration
public class MyConsulRibbonClientConfiguration {
    // 该类主要是在此处有改动，将ServerList生效的实现改为 MyServerList
    @Bean
    @ConditionalOnMissingBean
    public ServerList<?> ribbonServerList(IClientConfig config, ConsulDiscoveryProperties
        properties) {
        MyConsulServerList serverList = new MyConsulServerList(client, properties);
        serverList.initWithNiwsConfig(config);
        return serverList;
    }
    // 其他逻辑没变，此处省略
}
```

添加配置信息，如代码清单 14-25 所示。

代码清单14-25　ch14-5/ch14-5-consul-override-consumer-new/src/main/resources/bootstrap.yml

```
server:
    port: 8084
spring:
    application:
        name: consul-override-consumer-new
    cloud:
        consul:
            host: 127.0.0.1        # consul 启动地址
            port: 8500             # consul 启动端口
            ribbon:
                enabled: false     # 此处配置很重要，为 true 时走原有逻辑，为 false 时走
                                     重写逻辑
```

（3）添加 META-INF，如代码清单 14-26 所示。

代码清单14-26　ch14-5/ch14-5-consul-override-consumer-new/src/main/resources/META-INF/spring.factories

```
org.springframework.boot.autoconfigure.EnableAutoConfiguration=\
cn.springcloud.book.consul.override.consumer.myconsultool.MyRibbonConsulAutoConfiguration
```

效果展示如下。

- 依次启动 consul-provider1、consul-provider2、consul-consumer 等本地 Consul 服务。
- 在浏览器地址栏输入 http://localhost:8084/hello2?name=scbook，即可返回 provider 传回的信息。运行结果如图 14-10 所示。

```
← → C   ① localhost:8084/hello2?name=scbook
hello,scbook. I am provider2
```

图 14-10　运行结果

因为 spring 加载顺序的问题，这里 consul 的复写逻辑不能跟 main 方法在同一包或子包，而且需要使用 META-INF 的方式来加载 MyRibbonConsulAutoConfiguration。否则配置会不生效。

总结

本节只是演示了一种重写逻辑的思路，对于正常工作中的场景，原生的 ConsulServerList 已经提供了足够的功能，并不提倡大家重写它的逻辑。而且本文的重写方式并不优雅，主要是笔者在工作中遇到了一些历史遗留问题，才不得不这么做。

可能有读者会有疑问：为什么 Spring Cloud Consul 不能提供更方便的方式让我们重写它的 ConsulServerList 呢？答案是：Ribbon。因为 Ribbon 需要支持对每一个服务单独配置它的服务请求参数，比如超时时间、负载算法重试策略等，所以它不得不为每一个服务单独实例化一个服务发现逻辑。

14.3　常见问题排查

14.3.1　版本兼容的那些坑

我们前文提起，Spring Cloud Consul 组件是基于 ecwid 的 consul-api 包封装过的，而 consul-api 则是对 Consul 的 HTTP 接口封装。因为参与方比较多，加上 Consul 这两年迅速崛起，迭代很快，各个组件之间的版本兼容问题就成了架构升级的一大难题。

1. 健康检查的兼容问题

Spring Cloud Consul 也接入了 acturator 的健康检查，实现为 ConsulHealthIndicator。在 1.1.2.RELEASE 版本之前，ConsulHealthIndicator 的实现是这样的：

```
public class ConsulHealthIndicator extends AbstractHealthIndicator {
    // 这里只做演示，省略无关代码
    @Override
    protected void doHealthCheck(Health.Builder builder) throws Exception {
        try {
            Response<Self> self = consul.getAgentSelf();
```

```
            Config config = self.getValue().getConfig();
            // 获取consul返回的advertiseAddress
            builder.up().withDetail("advertiseAddress", config.getAdvertise-
                Address());
        }
        catch (Exception e) {
            builder.down(e);
        }
    }
}
```

然而，Consul 在 1.0.0 版本之后，就不返回该字段了，导致健康检查报错：Data must not be null。

这也就意味着，如果你的 Consul 在 1.0.0 版本之前，Spring Cloud Consul 在 1.1.2.RELEASE 版本之前，想要升级 Consul，首先要做的是升级 Spring Cloud Consul。

2. consul-api 的兼容问题

上面那个例子比较极端，一旦出现带来的问题是灾难性的。不过好在涉及的 Spring 版本比较老，最近两年遇到这个问题的可能性不大。接下来这个例子就距离我们就比较近了：

Consul 在 1.0.0 版本后，将一些接口（/agent/check/pass，/agent/service/deregister）由 GET 方法改成 PUT，这个 bug 在 consul-api 的 1.3.0 版本才得到解决，对应到 Spring Cloud Consul 已经是 2.0.0.M1 版本了。

当然，只要产品在迭代，软件的兼容问题就会一直存在，技术选型不能因噎废食。基础架构的升级在任何时候，都是一件充满挑战的事情。上文列举的两个兼容问题是笔者工作中遇到的比较头疼的事情，这里列举出来给大家作为前车之鉴，尽可能避免出现类似困扰。

14.3.2 Spring Cloud Consul 的一些问题

我们前文讲到关于 ConsulServerList 支持 tag 而 ConsulDiscoveryClient 不支持的情况。虽然不能肯定 Spring 这么做是否是有意为之，但是对于有切实需求的开发者来说多多少少是一个遗憾。更加遗憾的是，Spring 类似的遗憾还有很多，限于篇幅，我们再举一个例子。

Spring Cloud Consul 的异常信息不完整。

笔者在工作中，经常遇到 Spring Cloud Consul 打印的异常堆栈中，message 为 null 的情况，导致排查问题异常困难。后来发现，是 Spring Cloud Consul 对 consul-api 自定义的 OperationException 异常没有做特殊处理导致的。

当 Consul 的 http 响应代码为非 200 时，consul-api 会抛出 OperationException。如下所示：

```
@Override
public Response<Map<String, List<String>>> getCatalogServices(QueryParams
    queryParams, String token) {
        // 省略无关代码
        if (rawResponse.getStatusCode() == 200) {
            Map<String, List<String>> value = GsonFactory.getGson().fromJson
                (rawResponse.getContent(),
```

```
            new TypeToken<Map<String, List<String>>>() {
            }.getType());
        return new Response<Map<String, List<String>>>(value, rawResponse);
    } else {
        throw new OperationException(rawResponse);
    }
}
```

而 OperationException 会对 Consul 的响应做一些自定义处理，如下所示：

```
public final class OperationException extends ConsulException {

    private final int statusCode;
    private final String statusMessage;
    private final String statusContent;

    public OperationException(int statusCode, String statusMessage, String
        statusContent) {
        super("OperationException(statusCode=" + statusCode + ", statusMessage='"
            + statusMessage + "', statusContent='" + statusContent + "')");
        this.statusCode = statusCode;
        this.statusMessage = statusMessage;
        this.statusContent = statusContent;
    }

    public OperationException(RawResponse rawResponse) {
        this(rawResponse.getStatusCode(), rawResponse.getStatusMessage(), rawResponse.
            getContent());
    }
}
```

Spring Cloud Consul 在调用 consul-api 接口时，有些代码会简单地使用 Exception 捕获，然后打印 log 日志，导致 OperationException 的属性丢失。

ConsulCatalogWatch.catalogServicesWatch 的处理方式如下所示：

```
@Scheduled(fixedDelayString = "${spring.cloud.consul.discovery.catalogServices-
    WatchDelay:30000}")
public void catalogServicesWatch() {
    try {
        // 省略无关代码
        Response<Map<String, List<String>>> response = consul
            .getCatalogServices(new QueryParams(properties
                .getCatalogServicesWatchTimeout(), index));
    } catch (Exception e) {
        log.error("Error watching Consul CatalogServices", e);
    }
}
```

在 ConsulWatch、ConsulPropertySourceLocator、ConsulHealthIndicator 和 ConsulCatalogWatch 中均使用这种处理方式。虽然 Consul 并不经常报错，但是一旦出现异常，排查异常原因就是一件很麻烦的事情，因为服务方返回的数据被吞掉了。

Spring Cloud 社区一直比较活跃，相信在不久的将来，这个 Bug 会被修复。这里也提醒广大读者朋友，接口调用时，要正确处理接口中的自定义异常，不然很容易丢失异常数据。

14.4 本章小节

本章开头讲解了 Spring Cloud Consul 的一些底层知识和高级用法，紧接着就以实战方式演示了如何重写 Spring Cloud Consul 的服务发现功能。ConsulServerList 作为 Spring Cloud Consul 的功能核心，值得每一个使用它的人去阅读源码。本章的最末尾，笔者提及了自己在工作中遇到的一个 Spring Cloud Consul 的小 Bug，旨在提醒大家，Spring 作为 Java 领域最著名的开源组织，在快速迭代的过程中，也会有不足之处。崇拜但是不迷信权威，我们才能在领略编程之美的同时，给自己带来编程的成就感。

第 15 章

Spring Cloud 认证和鉴权

在本章将介绍 Spring Cloud 中认证和鉴权的相关知识点，介绍几种常见的认证和鉴权方案，以及 UAA、OAUTH2、JWT 等相关实际案例，关于 JWT 和 OAUTH2 如果有兴趣可以参考 Zuul 的第 8 章，里面做了相关的论述。

15.1 微服务安全与权限

我们在做系统时，权限问题一直是我们的核心问题之一，从原来的单体变成现在的微服务架构，安全和权限的架构一直在不停地演进和变化，在微服务的场景下，与传统的单体应用鉴权已经有很大不同，服务拆分后，服务间调用的授权和用户的认证，都需要进行重新考虑和规划，老的单体应用我相信很多朋友已经做过很多次了，在微服务的架构下，一个典型的权限相关的架构就变成了这样，如图 15-1 所示。

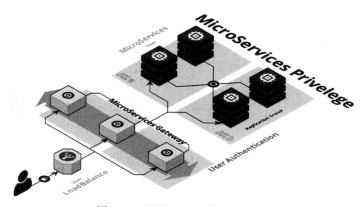

图 15-1　微服务典型权限架构图

我们发送请求通过负载均衡软件，负载均衡发到微服务网关上，网关进行用户的认证，解析出用户的基本信息，通过认证后，携带用户本身的标识到后台微服务，微服务获得标识后进行相关的鉴权，这样就能形成一个完整的权限链路。

15.2 Spring Cloud 认证与鉴权方案

我们首先来简单解释下，认证和鉴权这 2 个名称的含义。
- 认证：通俗来讲，就是要验证这个用户是谁。
- 鉴权：通俗来讲，就是要了解这个用户能做什么事。

15.2.1 单体应用下的常用方案

如图 15-2 所示，我们在传统的单体应用中，一般会写一个固定的认证和鉴权的包，里面包含了很多认证和鉴权的类，当用户发起请求时我们可以利用 session 的方式，把用户存入 session 并生成一个 sessionId，之后返回客户端。客户端可以存在 cookie 里，从而在后续的请求中顺利通过验证，其他的方案在这里不再做过多论述，我们主要来讲一下微服务架构下的权限方案。

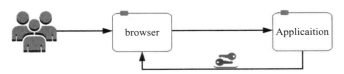

图 15-2　单体应用下认证和鉴权方案图

15.2.2 微服务下 SSO 单点登录方案

如图 15-3 所示，我们传统的做法就是通过 SSO 来实现，但在微服务架构情况下，应用被拆分，单体应用按照规则拆分成很多小的服务，这种情况下，如果对每个服务都进行每个用户的 SSO 动作，那么每个服务里都会做用户的认证和鉴权，可能保存用户信息或者每个用户都会和鉴权服务打交道，这些情况都将会带来非常大的网络消耗和性能损耗，也有可能会造成数据不一致，所以不太建议用这种方案。

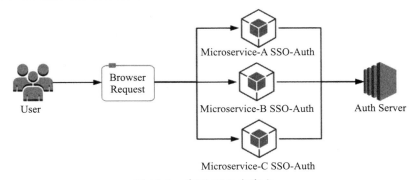

图 15-3　常用 SSO 方案之一

15.2.3 分布式 Session 与网关结合方案

分布式 Session 方案在使用上很广泛，业界很多项目都使用了这样的方案。这种方式的步骤，可以参见图 15-4。

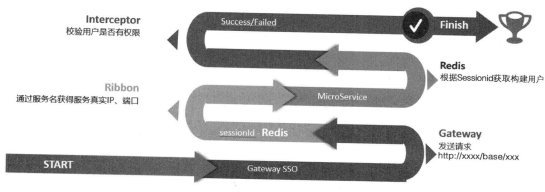

图 15-4　分布式 Session 方案图

1）用户在网关进行 SSO 登录，进行用户认证，检查用户是否存在和有效。
2）如果认证通过，则将用户的信息或数据存储在第三方部件中，如 MySQL、Redis。
3）后端微服务可以从共享存储中拿到用户的数据。

很多场景下，这种方案是推荐的，因为很方便同时可以做扩展，也可以保证高可用的方案。但是这种方案的缺点是依赖于第三方部件，且这些部件需要做高可用，并且需要增加安全的控制，所以对于实现起来有一定复杂度。

15.2.4 客户端 Token 与网关结合方案

客户端 Token 与网关结合方案如图 15-5 所示。

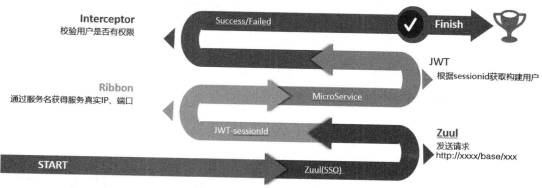

图 15-5　客户端 Token 与网关结合

它的实现步骤和方案如下。
1）客户端持有一个 Token，通常可用 JWT 或者其他加密的算法实现自己的一种 Token，然后通过 Token 保存了用户的信息。

2）发起请求并携带 Token，Token 传到网关层后，网关层进行认证和校验。

3）校验通过，携带 Token 到后台的微服务中，可以进行具体的接口或者 url 验证。

4）如果要涉及用户的大量数据存放，则 Token 就有可能不太适合，或者和上面的分布式 Session 一样，使用第三方部件来存储这些信息，这种方案也是业界很常用的方案，但对于 Token 来说，它的注销会有一定的麻烦，需要在网关层进行 Token 的注销。

15.2.5　浏览器 Cookie 与网关结合方案

这种方式和上面的方式类似，但不同的是我们会把用户的信息存在 Cookie 里，然后通过网关来解析 Cookie，从而获得用户相关的信息，这种方式在一些老系统做改造时遇到得比较多，适合作为老系统改造时采取的方案，因为很多系统需要继承，这时 Cookie 在别的系统中也是同样通用的。

15.2.6　网关与 Token 和服务间鉴权结合

图 15-6 所示，我们都知道网关适合做认证和鉴权，但在安全层面，我们要求更严格的权限，对于有些项目而言，本身网络跟外部隔离，加上其他的安全手段，所以只要求在网关层做认证和鉴权就行，对于后台服务之间的调用，不需要做微服务之间权限的控制。但某些时候我们还是需要对服务与服务之间的调用进行鉴权，知道某个用户是否有权限调用某个接口，这些都要进行鉴权。

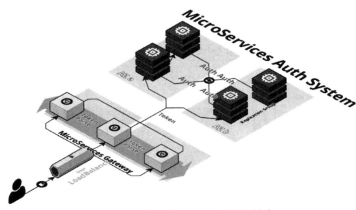

图 15-6　客户端 Token 与网关结合

这时的方案如下。

1）在 Gateway 网关层做认证，通过对用户校验后，传递用户的信息到 header 中，后台微服务在收到 header 后进行解析，解析完后查看是否有调用此服务或者某个 url 的权限，然后完成鉴权。

2）从服务内部发出的请求，在出去时进行拦截，把用户信息保存在 header 里，然后传出去，被调用方获取到 header 后进行解析和鉴权。

15.3　Spring Cloud 认证鉴权实战案例

这个案例我们利用 spring cloud gateway + JWT + 服务间鉴权实现，如果你使用的 Zuul 网关，可以参见 Zuul 的第 8 章，有 Zuul+OAUTH2+JWT 实战可以参考。这个案例我们采用前后端分离的架构，我们利用 Spring Cloud Gateway 实现用户的认证，解析 JWT 后传递用户信息到后端服务，后端服务根据用户和路由信息进行鉴权，具体的架构图参见图 15-7。

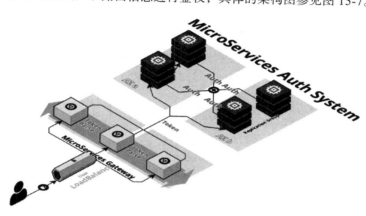

图 15-7　Spring Cloud Gateway + JWT + Auth

15.3.1　创建 Spring Cloud Gateway 及关联信息

1）引入网关依赖和客户端注册依赖，如代码清单 15-1 所示。

代码清单15-1　ch15-1\ch15-1-gateway\pom.xml

```
<dependency>
        <groupId>org.springframework.cloud</groupId>
        <artifactId>spring-cloud-starter-gateway</artifactId>
</dependency>
<dependency>
        <groupId>org.springframework.cloud</groupId>
        <artifactId>spring-cloud-starter-netflix-eureka-client</artifactId>
</dependency>
```

2）引入 JWT 和相关依赖，我们使用 io.jsonwebtoken 提供的类和方法来使用 JWT，如代码清单 15-2 所示。

代码清单15-2　ch15-1\ch15-1-gateway\pom.xml

```
<dependency>
        <groupId>io.jsonwebtoken</groupId>
        <artifactId>jjwt</artifactId>
        <version>0.7.0</version>
</dependency>
```

3）增加 JWT 生成和验证解析的方法，根据用户名生成 JWT，验证的时候同样传入 JWT

进行验证，如代码清单 15-3 所示。

代码清单15-3 ch15-1\ch15-1-gateway\src\main\java\cn\springcloud\book\gateway\filter\JwtUtil.java

```java
    public static String generateToken(String user) {
        HashMap<String, Object> map = new HashMap<>();
        map.put("id", new Random().nextInt());
        map.put("user", user);
        String jwt = Jwts.builder()
                .setSubject("user info").setClaims(map)
                .signWith(SignatureAlgorithm.HS512, SECRET)
                .compact();
        String finalJwt = TOKEN_PREFIX + " " +jwt;
        return finalJwt;
    }

    public static Map<String,String> validateToken(String token) {
        if (token != null) {
            HashMap<String, String> map = new HashMap<String, String>();
            Map<String,Object> body = Jwts.parser()
                    .setSigningKey(SECRET)
                    .parseClaimsJws(token.replace(TOKEN_PREFIX, ""))
                    .getBody();
            String id =  String.valueOf(body.get("id"));
            String user = (String) (body.get("user"));
            map.put("id", id);
            map.put("user", user);
            if(StringUtils.isEmpty(user)) {
                throw new PermissionException("user is error, please check");
            }
            return map;
        } else {
            throw new PermissionException("token is error, please check");
        }
    }
```

4）添加 GlobalFilter，所有的请求会经过此 filter，然后对 JWT Token 进行解析校验，并转换成系统内部的 Token，并把路由的服务名也加入 header，送往接下来的路由服务里，方便进行鉴权，如代码清单 15-4 所示。

代码清单15-4 ch15-1-gateway\src\main\java\cn\springcloud\book\gateway\filter\AuthFilter.java

```java
    @Override
    public Mono<Void> filter(ServerWebExchange exchange, GatewayFilterChain chain) {
        Route gatewayUrl = exchange.getRequiredAttribute(ServerWebExchangeUtils.
            GATEWAY_ROUTE_ATTR);
        URI uri = gatewayUrl.getUri();
        ServerHttpRequest request = (ServerHttpRequest)exchange.getRequest();
        HttpHeaders header = request.getHeaders();
        String token = header.getFirst(JwtUtil.HEADER_AUTH);
        Map<String,String> userMap = JwtUtil.validateToken(token);
        ServerHttpRequest.Builder mutate = request.mutate();
        if(userMap.get("user").equals("admin") || userMap.get("user").equals
            ("spring") || userMap.get("user").equals("cloud")) {
            mutate.header("x-user-id", userMap.get("id"));
```

```java
                mutate.header("x-user-name", userMap.get("user"));
                mutate.header("x-user-serviceName", uri.getHost());
            }else {
                throw new PermissionException("user not exist, please check");
            }
        ServerHttpRequest buildReuqest = mutate.build();
         return chain.filter(exchange.mutate().request(buildReuqest).build());
    }
```

15.3.2 核心的公共工程 core-service

1）服务拦截器，在进入控制器之前进行校验，如代码清单 15-5 所示。

增加 core 工程，提供拦截器，用于微服务之间调用时进行鉴权。

增加 RestTemplate 拦截器，用于调用时传递上下文信息。

代码清单15-5 ch15-1\ch15-1-core-service\src\main\java\cn\springcloud\book\common\intercepter\UserContextInterceptor.java

```java
@Override
    public boolean preHandle(HttpServletRequest request, HttpServletResponse
        respone, Object arg2) throws Exception {
        User user = getUser(request);
        UserPermissionUtil.permission(user);
        if(!UserPermissionUtil.verify(user,request)) {
            respone.setHeader("Content-Type", "application/json");
            String jsonstr = JSON.toJSONString("no permisson access service,
                please check");
            respone.getWriter().write(jsonstr);
            respone.getWriter().flush();
            respone.getWriter().close();
            throw new PermissionException("no permisson access service, please check");
        }
        UserContextHolder.set(user);
        return true;
    }

    @Override
    public void afterCompletion(HttpServletRequest request, HttpServletResponse
        respone, Object arg2, Exception arg3)
            throws Exception {
        UserContextHolder.shutdown();
    }
```

增加具体的校验类和方法，需要注意的是，我们这里采用的鉴权方式是给了不同用户可以访问服务的列表，再根据请求路由的服务名来校验该用户是否有访问这个服务的权限。这里可以根据项目自己的需求和业务来定制不同的校验逻辑，如查询数据库获取具体的菜单 url 或者角色校验等，如代码清单 15-6 所示。

代码清单15-6 ch15-1\ch15-1-core-service\src\main\java\cn\springcloud\book\common\util\UserPermissionUtil.java

```java
/**
```

```java
 * 模拟权限校验，可以根据自己项目需要定制不同的策略,如查询数据库获取具体的菜单url或者角色等
 * @param user
 */
public static boolean verify(User user,HttpServletRequest request){
    String url = request.getHeader("x-user-serviceName");
    if(StringUtils.isEmpty(user)) {
        return false;
    }else {
        List<String> str = user.getAllowPermissionService();
        for (String permissionService : str) {
            if(url.equalsIgnoreCase(permissionService)) {
                return true;
            }
        }
        return false;
    }
}

/**
 * 模拟权限赋值，可以根据自己项目需要定制不同的策略,如查询数据库获取具体的菜单url或者角色等
 * @param user
 */
public static void permission(User user){
    if(user.getUserName().equals("admin")) {
        List allowPermissionService = new ArrayList();
        allowPermissionService.add("client-service");
        allowPermissionService.add("provider-service");
        user.setAllowPermissionService(allowPermissionService);
    }else if(user.getUserName().equals("spring")) {
        List allowPermissionService = new ArrayList();
        allowPermissionService.add("client-service");
        user.setAllowPermissionService(allowPermissionService);
    } else {
        List allowPermissionService = new ArrayList();
        user.setAllowPermissionService(allowPermissionService);
    }
}
```

2）增加上下文持有对象，如代码清单15-7所示。

代码清单15-7 ch15-1\ch15-1-core-service\src\main\java\cn\springcloud\book\common\interceper\UserContextHolder.java

```java
public class UserContextHolder {
    public static ThreadLocal<User> context = new ThreadLocal<User>();
    public static User currentUser() {
        return context.get();
    }
    public static void set(User user) {
        context.set(user);
    }
    public static void shutdown() {
        context.remove();
    }
}
```

3）增加 RestTemplate 拦截器。

拦截请求后传递上下文信息和服务名到 header 中，如代码清单 15-8 所示。

代码清单15-8　ch15-1\ch15-1-core-service\src\main\java\cn\springcloud\book\common\
intercepter\RestTemplateUserContextInterceptor.java

```java
@Override
public ClientHttpResponse intercept(HttpRequest request, byte[] body, Client-
    HttpRequestExecution execution)
        throws IOException {
    User user = UserContextHolder.currentUser();
    request.getHeaders().add("x-user-id",user.getUserId());
    request.getHeaders().add("x-user-name",user.getUserName());
    request.getHeaders().add("x-user-serviceName",request.getURI().getHost());
    return execution.execute(request, body);
}
```

4）把拦截器加载到程序里，如代码清单 15-9 所示。

代码清单15-9　ch15-1\ch15-1-core-service\src\main\java\cn\springcloud\book\common\
config\CommonConfiguration.java

```java
/**
 * 请求拦截器
 */
@Override
    public void addInterceptors(InterceptorRegistry registry) {
    registry.addInterceptor(new UserContextInterceptor());
}

/***
 * RestTemplate 拦截器，在发送请求前设置鉴权的用户上下文信息
 * @return
 */
@LoadBalanced
    @Bean
    public RestTemplate restTemplate() {
        RestTemplate restTemplate = new RestTemplate();
        restTemplate.getInterceptors().add(new RestTemplateUserContextInterceptor());
        return restTemplate;
}
```

15.3.3　服务提供方工程 provider-service

1）增加核心工程的依赖，如代码清单 15-10 所示。

代码清单15-10　ch15-1\ch15-1-provider-service\pom.xml

```xml
<dependency>
        <groupId>cn.springcloud.book</groupId>
        <artifactId>ch15-1-core-service</artifactId>
        <version>0.0.1-SNAPSHOT</version>
    </dependency>
```

2）增加数据接口供客户端调用。

```
@GetMapping("/provider/test")
    public String test(HttpServletRequest request) {
        System.out.println("---------------success access provider service---------------");
        return "success access provider service!";
    }
```

15.3.4 客户端工程 client-service

1）在客户端增加核心工程的依赖，如代码清单 15-11 所示。

代码清单15-11　ch15-1\ch15-1-client-service\pom.xml

```
<dependency>
        <groupId>cn.springcloud.book</groupId>
        <artifactId>ch15-1-core-service</artifactId>
        <version>0.0.1-SNAPSHOT</version>
    </dependency>
```

2）增加接口及调用其他服务的接口，如代码清单 15-12 所示。

代码清单15-12　ch15-1\ch15-1-client-service\src\main\java\cn\springcloud\book\controller\TestController.java

```
@RequestMapping("/test")
    public String test(HttpServletRequest request) {
        System.out.println("---------------success access test method!---------------");
        Enumeration headerNames = request.getHeaderNames();
        while (headerNames.hasMoreElements()) {
            String key = (String) headerNames.nextElement();
            System.out.println(key + ": " + request.getHeader(key));
        }
        return "success access test method!!";
    }

    @RequestMapping("/accessProvider")
    public String accessProvider(HttpServletRequest request) {
        String result = rest.getForObject("http://provider-service/provider/test",
            String.class);
        return result;
    }
```

15.3.5 运行结果

我们依次启动：eureka-service、gateway、client-service 和 provider-service 这四个工程。如果不是 admin、spring 和 cloud 这三个用户则代表系统不存在该用户，如果鉴权时没有某个服务的访问权限，这两种情况我们都将拒绝访问。

1）为了方便测试，我们在 Gateway 里增加一个接口，传入姓名用于获取 Token，模拟客户端获取到 Token 的动作，如代码清单 15-13 所示。

代码清单15-13　ch15-1\ch15-1-gateway\src\main\java\cn\springcloud\book\gateway\filter\
　　　　　　　　TokenController.java

```
@GetMapping("/getToken/{name}")
public String get(@PathVariable("name") String name)  {
    return JwtUtil.generateToken(name);
}
```

2）访问 http://localhost:9001/getToken/abcd 获取 Token：

```
Bearer eyJhbGciOiJIUzUxMiJ9.eyJpZCI6LTI2NDM2Mjc3MiwidXNlciI6ImFiY2QifQ.SgcPPpPir5
   jlTu0wuOf073FVn6KmNjVeh82PXqVYISkN7ioI_NnzE6xQEn05521p3IE8Znc_DddUMBREp-O_rQ
```

添加 header 的 Authorization 信息值为此 JWT Token，访问 client-service 的测试接口 http://localhost:9001/CLIENT-SERVICE/test，由于名称是 abcd，所以结果如图 15-8 所示，提示系统没有该用户。

图 15-8　不存在的用户

3）访问 http://localhost:9001/getToken/cloud，使用 cloud 用户获取 Token，修改 Authorization，发起调用，由于 cloud 用户没有权限访问任何一个服务，所以结果如图 15-9 所示，表示没有权限访问。

图 15-9　没有权限访问服务

4）访问 http://localhost:9001/getToken/spring，使用 spring 用户获取 Token，修改 Authorization，发起调用，结果成功如图 15-10 所示。

图 15-10　成功访问服务

5）上面我们使用了 JWT 进行了认证，我们现在访问另外一个接口 http://localhost:9001/CLIENT-SERVICE/accessProvider，使 client 端调用 provider 的服务进行鉴权。

使用 spring 用户生成的 JWT Token 访问：http://localhost:9001/CLIENT-SERVICE/accessProvider，结果如图 15-11 所示，由于服务之间调用鉴权不通过导致访问失败。

图 15-11　服务之间鉴权不通过

6）使用 admin 用户的 JWT Token 访问：http://localhost:9001/CLIENT-SERVICE/accessProvider，结果如图 15-12 所示，表示访问成功。

图 15-12　服务鉴权通过

并且我们在 provider-service 里打印出了上下文的用户，如图 15-13 所示。

案例总结：

这里我们使用了 Spring Cloud Gateway+JWT 来实现简单的鉴权和认证，如果要使用 Zuul+JWT+OAUTH2 的方式，大家可以参考第 8 章，里面详细介绍了 Zuul 的用法和结合案例。

图 15-13　控制台打印用户

15.4　本章小结

本篇介绍了在微服务架构中常见的认证和鉴权方案以及案例，大家可以参考这些内容来实现自己的权限认证和鉴权。对于微服务架构来说，没有一种完全适合所有场景的方案，所以需要根据自己的业务场景和需求来确定具体的做法。

第 16 章 Chapter 16

Spring Cloud 全链路监控

在传统的 SOA 架构体系中，系统调用层级不多，调用关系也不复杂，一旦出现问题，根据异常信息可以很快定位到问题模块并进行排查。但是在微服务的世界里，实例数目成百上千，实例之间的调用关系几乎是网状结构，靠人力去监控和排查问题已经不太可能。这种情况下，一个完善的调用链路监控框架对于运维和开发来说，都是不可或缺的。

16.1 全链路监控概述

16.1.1 链路监控的原理来源

微服务架构下，服务按照不同的维度进行拆分，一次请求可能会涉及多个服务，并且有可能是由不同的团队开发，可能使用不同的编程语言来实现，有可能布在了几千台服务器上，横跨多个不同的数据中心。因此，就需要一些可以帮助理解系统行为、用于分析性能问题的工具，以便发生故障的时候，能够快速定位和解决问题。

这些工具就是 APM（Application Performance Management），其中最出名的是谷歌公开的论文提到的 Dapper。

Dapper 论文中对实现一个分布式跟踪系统提出了如下需求。

- ❏ 性能低损耗：分布式跟踪系统对服务的性能损耗应尽可能做到可以忽略不计，尤其是对性能敏感的应用不能产生损耗。
- ❏ 对应用透明：即要求尽可能用非侵入的方式来实现跟踪，尽可能做到业务代码的低侵入，对业务开发人员应该做到透明化。
- ❏ 可伸缩性：是指不能随着微服务和集群规模的扩大而使分布式跟踪系统瘫痪。
- ❏ 跟踪数据可视化和迅速反馈：即要有可视化的监控界面，从跟踪数据收集、处理到结果的展现尽量做到快速，这样就可以对系统的异常状况做出快速反应。

❑ 持续监控：即要求分布式跟踪系统必须是 7×24 小时工作的，否则将难以定位到系统偶尔抖动的行为。

论文中的一个基础案例具体如图 16-1 所示。

这里解释一下，调用流程：A～E 分别表示五个服务，用户发起一次请求到 A，然后 A 分别发送 RPC 请求到 B 和 C，B 处理请求后返回，C 还要发起两个 RPC 请求到 D 和 E。更多关于 Dapper 的内容请查看完整的论文。

Spring Cloud Sleuth 是 Spring Cloud 的分布式调用链解决方案，它从 Dapper、Zipkin、HTrace 中借鉴了很多思路。Sleuth 对于大部分用户来说都是透明的，系统间的交互信息都能被自动采集。用户可以通过日志文件获取链路数据，也可以将数据发给远程服务进行统一收集展示。

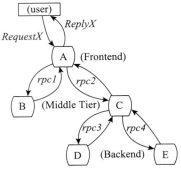

图 16-1　Dapper 论文案例图解

术语

❑ Span：基本工作单元。比如，发送一次 RPC 请求就是一个新的 Span。Span 通过一个 64 位的 ID 标识，还包含有描述、事件时间戳、标签、调用它的 Span 的 ID、处理器 ID（一般为 IP 地址）。注意：第一个 Span 是 root span，它的 ID 值和 trace 的 ID 值一样。
❑ Trace：一系列 Span 组成的树状结构。简而言之，就是一次调用请求。
❑ Annotation：标注，用来描述事件的实时状态。事件有如下几种状态。
  ○ cs：Client Sent。客户端发起请求，它表示一个 Span 的开始。
  ○ sr：Server Received。服务方接收到请求并开始处理，它减去 cs 的时间就是网络延迟时间。
  ○ ss：Server Sent。它表示请求处理完成，将响应数据返回给客户端。它减去 sr 的时间就是服务方处理时间。
  ○ cr：Client Received。它表示客户端收到服务方的返回值，是当前 span 结束的信号。它减去 cs 的时间就是一次请求的完整处理时间。

16.1.2　Sleuth 原理介绍

根据前文的描述，我们知道，Sleuth 通过 Trace 定义一次业务调用链，根据它的信息，我们能知道有多少个系统参与了该业务处理。而系统间的调用顺序和时间戳信息，则是通过 Span 来记录的。Trace 和 Span 的信息经过整合，就能知道该业务的完整调用链。二者的详细流转情况，如图 16-2 和图 16-3 所示：

16.1.3　Brave 和 Zipkin

Brave[⊖] 是一个用于捕捉分布式系统之间调用信息的工具库，然后将这些信息以 Span 的形式发送给 Zipkin。

⊖　https://github.com/openzipkin/brave

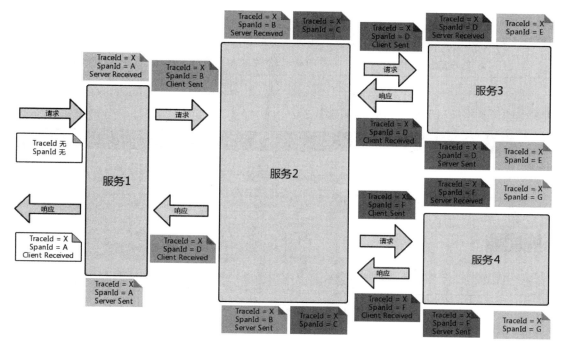

图 16-2 一次业务处理中，Trace 和 Span 的流转图

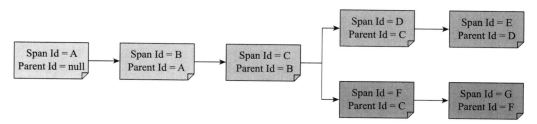

图 16-3 各个 Span 之间的父子关系图

从 2.0.0 版本开始，Sleuth 不再自己存储上下文信息，而是使用 Brave 作为调用链工具库，并且遵循 Brave 的命名和标记惯例。

> **注意** 如果你想要沿用老版本使用方式，可以将 spring.sleuth.http.legacy.enabled 属性设置为 true。

Zipkin [⊖] 是一个基于 Google Dapper 论文设计的分布式跟踪系统，它收集系统的延时数据并提供展示界面，以便用户排查问题。

Zipkin 的部署很简单，去官网下载所需的版本，然后直接 java -jar zipkin.jar 即可。

⊖ https://github.com/openzipkin/zipkin

注意
1）Zipkin 还支持源码部署和 Docker 镜像启动。
2）Zipkin 支持各种持久化方式，默认是数据存储在内存中。
3）关于 Zipkin 的更多信息，有兴趣的读者可以参考官方文档，这里不做过多介绍。

部署后，默认的启动端口为 9411。我们在浏览器地址栏输入 http://localhost:9411/，即可看到 zipkin 数据展示页面。如图 16-4 所示。

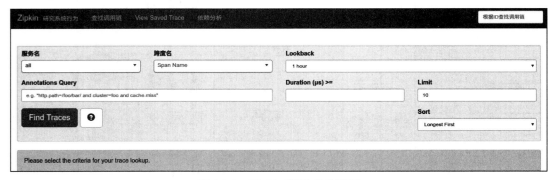

图 16-4　Zipkin 部署成功页面

16.2　Sleuth 基本用法

下面我们就用一个简单的案例来演示 Sleuth 是如何使用的。该案例包括一个父工程 ch16-1 和两个子模块：sleuth-consumer、sleuth-provider。其中，sleuth-consumer（端口号：8081）会使用 Feign、RestTemplate、新建子线程的方式请求 sleuth-provider（端口号：8082）。

（1）创建父工程：ch16-1

创建父依赖 pom，相关源码参考代码清单 16-1。

代码清单16-1　ch16-1/pom.xml

```
<dependencies>
    <dependency>
        <groupId>org.springframework.cloud</groupId>
        <artifactId>spring-cloud-starter-sleuth</artifactId>
    </dependency>
    <dependency>
        <groupId>org.springframework.cloud</groupId>
        <artifactId>spring-cloud-starter-openfeign</artifactId>
    </dependency>
    <dependency>
        <groupId>org.springframework.boot</groupId>
        <artifactId>spring-boot-starter-web</artifactId>
    </dependency>
</dependencies>
```

（2）创建子模块 sleuth-consumer

Sleuth-consumer 继承父工程的 pom，本身没有增加依赖。受限于篇幅，我们这里省略了

pom，启动主类。

创建配置类 ConsumerConfiguration。相关源码参考代码清单 16-2。

代码清单16-2 ch16-1/ch16-1-sleuth-consumer/src/main/java/cn/springcloud/book/sleuth/consumer/ConsumerConfiguration.java

```java
// 配置类,用于注册RestTemplate和ExecutorService;
@Configuration
public class ConsumerConfiguration {
    @Autowired
    BeanFactory beanFactory;
    @Bean
    public RestTemplate restTemplate(){
        return new RestTemplate();
    }

    @Bean  // 简单起见,我们注册固定大小的线程池
    public ExecutorService executorService(){
        ExecutorService executorService = Executors.newFixedThreadPool(2);
        return new TraceableExecutorService(this.beanFactory, executorService);
    }
}
```

（3）创建用户交互接口 ConsumerController

相关源码参考代码清单 16-3。

代码清单16-3 ch16-1/ch16-1-sleuth-consumer/src/main/java/cn/springcloud/book/sleuth/consumer/ConsumerController.java

```java
// 用户交互接口,分别使用feign、restTemplate、新线程的方式请求服务端
@RestController
public class ConsumerController {
    private static final Logger log = LoggerFactory.getLogger(ConsumerController.class);
    @Autowired
    private HelloService helloService;

    @Autowired
    private RestTemplate restTemplate;

    @Autowired
    private ExecutorService executorService;

    @GetMapping("/helloByFeign")
    public String helloByFeign(String name){
        log.info("client sent. Feign 方式, 参数: {}",name);

        String result = helloService.sayHello(name);

        log.info("client received. Feign 方式, 结果: {}",result);
        return result;
    }

    @GetMapping("/helloByRestTemplate")
    public String helloByRestTemplate(String name){
```

```
        log.info("client sent. RestTemplate方式, 参数: {}",name);

        String url = "http://localhost:8082/sayHello?name="+name;
        String result = restTemplate.getForObject(url,String.class);

        log.info("client received. RestTemplate方式, 结果: {}",result);
        return result;
    }

    @GetMapping("/helloByNewThread")
    public String hello(String name) throws ExecutionException, InterruptedException {
        log.info("client sent. 子线程方式, 参数: {}",name);

        Future future = executorService.submit(() -> {
            log.info("client sent. 进入子线程, 参数: {}",name);
            String result = helloService.sayHello(name);
            return result;
        });
        String result = (String) future.get();
        log.info("client received. 返回主线程, 参数: {}",result);
        return result;
    }
}
```

（4）创建 feign 调用接口 HelloService

相关源码参考代码清单 16-4。

代码清单16-4　ch16-1/ch16-1-sleuth-consumer/src/main/java/cn/springcloud/book/sleuth/consumer/HelloService.java

```
@FeignClient(name = "sleuth-provider",url = "localhost:8082")
public interface HelloService {
    @RequestMapping("/sayHello")
    String sayHello(@RequestParam("name")String name);
}
```

（5）创建子模块 sleuth-provider

Sleuth-provider 继承父工程的 pom，本身没有增加依赖。受限于篇幅，我们这里省略了 pom，启动主类。

（6）创建对外服务接口 ProviderController

相关源码参考代码清单 16-5。

代码清单16-5　ch16-1/ch16-1-sleuth-provider/src/main/java/cn/springcloud/book/sleuth/provider/ProviderController.java

```
// sleuth-provider 对外服务接口
@RestController
public class ProviderController {
    private static final Logger log = LoggerFactory.getLogger(ProviderController.class);
    @GetMapping("/sayHello")
    public String hello(String name){
        log.info("server received. 参数: {}",name);
```

```
        String result = "hello, "+name;
        log.info("server sent. 结果: {}",result);
        return result;
    }
}
```

效果展示

分别启动 sleuth-provider 和 sleuth-consumer。

在浏览器地址栏输入：http://localhost:8081/helloByFeign?name=scbook，日志如下（已简化）。

```
[sleuth-consumer,5497172f127eb5a5,5497172f127eb5a5,false] -- client Feign , 参数: scbook
[sleuth-consumer,5497172f127eb5a5,5497172f127eb5a5,false] -- client Feign , 结果 hello, scbook
[sleuth-provider,5497172f127eb5a5,eb933f187a7c127b,false]  --server received. 参数: scbook
[sleuth-provider,5497172f127eb5a5,eb933f187a7c127b,false]  -- server sent. 结果: hello, scbook
```

在浏览器地址栏输入：http://localhost:8081/helloByRestTemplate?name=scbook，日志如下（已简化）。

```
[sleuth-consumer,79329c1890c68346,79329c1890c68346,false]-client,RestTemplate参数: scbook
[sleuth-consumer,79329c1890c68346,79329c1890c68346,false]-client,RestTemplate 结果: hello, ..
[sleuth-provider,79329c1890c68346,d6f3a5b96bbf53f5,false]-server received. 参数: scbook
[sleuth-provider,79329c1890c68346,d6f3a5b96bbf53f5,false]- server sent. 结果: hello, scbook
```

在浏览器地址栏输入：http://localhost:8081/helloByNewThread?name=scbook，日志如下（已经简化）。

```
[sleuth-consumer,54e60bfce1a8563d,54e60bfce1a8563d,false]-client sent. 子线程方式, 参数: scbook
[sleuth-consumer,54e60bfce1a8563d,54406d3fc3c2c78f,false]- client sent. 进入子线程, 参数: scbook
[sleuth-consumer,54e60bfce1a8563d,54e60bfce1a8563d,false]-client received. 结果: hello, scbook
[sleuth-provider,54e60bfce1a8563d,99d0d707a2736233,false]-server received. 参数: scbook
[sleuth-provider,54e60bfce1a8563d,99d0d707a2736233,false]-server sent. 结果: hello, scbook
```

由上面的案例我们可以看出，引入了 spring-cloud-sleuth 之后，首先我们的日志组件可以自动打印 Span 信息，然后 Span 信息不仅可以随着 feign、restTemplate 往服务端传递，还可以在父子线程之间传递。相信读者朋友肯定是很好奇它是怎么做到的，接下来我们就一起探究一下，Sleuth 如何做 Span 信息传递的。

16.2.1 Sleuth 对 Feign 的支持

Feign 提供了 feign.Client 接口以便开发者自定义远程调用功能。

Sleuth 就是使用了 TracingFeignClient 实现 Feign 接口，然后在执行 http 调用前，在 header 中添加 Span 信息。

```
final class TracingFeignClient implements Client {

@Override public Response execute(Request request, Request.Options options)
        throws IOException {
    Map<String, Collection<String>> headers = new HashMap<>(request.headers());
    Span span = handleSend(headers, request, null);
```

```
        Response response = null;
        Throwable error = null;
        try (Tracer.SpanInScope ws = this.tracer.withSpanInScope(span)) {
            return response = this.delegate.execute(modifiedRequest(request, headers),
                options);
        }catch (IOException | RuntimeException | Error e) {
            error = e;
            throw e;
        }
        finally {
            handleReceive(span, response, error);
        }
    }
```

16.2.2　Sleuth 对 RestTemplate 的支持

在上面的例子中,我们演示了使用 RestTmeplate 跟客户端交互时,Span 信息也能从客户端传递给服务端。不过细心的读者或许会发现,我们使用的 RestTemplate 并不是临时初始化的,而是在启动时注册成一个 bean。那临时初始化的 RestTemplate 行不行呢?答案是否定的。我们先看下官方文档[○]:

 你必须将 RestTemplate 注册成为一个 bean,这样我们定义的拦截器(interceptors)才能注入进去。如果你使用 new 的方式创建一个 RestTemplate 实例,我们的拦截器就失效了。

从文档中,我们不难看出,Sleuth 是使用拦截器的方式对 RestTemplate 做了定制。这个拦截器是 brave.spring.web.TracingClientHttpRequestInterceptor,Sleuth 对它做了一层封装,为 LazyTracingClientHttpRequestInterceptor。RestTemplate 在执行 execute 的时候,request 会经过拦截器的处理,添加上 Span 信息。这就解释了为什么我们 new RestTemplate() 不行,把它注册成 bean 就立即可以了。

```
public final class TracingClientHttpRequestInterceptor implements ClientHttpRequest-
    Interceptor {
    // 省略非核心代码
    @Override
    public ClientHttpResponse intercept(HttpRequest request, byte[] body,
            ClientHttpRequestExecution execution) throws IOException {
        Span span = handler.handleSend(injector, request.getHeaders(), request);
        ClientHttpResponse response = null;
        Throwable error = null;
        try (Tracer.SpanInScope ws = tracer.withSpanInScope(span)) {
            return response = execution.execute(request, body);
        } catch (IOException | RuntimeException | Error e) {
            error = e;
            throw e;
        } finally {
            handler.handleReceive(response, error, span);
        }
    }
}
```

○　http://cloud.spring.io/spring-cloud-sleuth/single/spring-cloud-sleuth.html#_http_client_integration。

16.2.3 Sleuth 对多线程的支持

Sleuth 提供了 LazyTraceExecutor、TraceableExecutorService 和 TraceableScheduledExecutorService 三种多线程实现，它们都可以在你创建新的任务时新建一个 Span。也就是说，如果你想在多线程场景下使用 Sleuth，必须使用它提供的多线程实现，而不是自己去初始化线程池。

16.3 Sleuth 深入用法

16.3.1 TraceFilter

对于提供 HTTP 接口的服务方来说，我们很容易想到，它接收客户端 Span 信息的方式是 Filter。没错，Sleuth 通过 Brave 的 TracingFilter 达到获取 Span 信息的目的。

如果你想对 Span 信息有一些自定义的修改，比如增加 tag 或者响应头信息，那么只需要注册一个你自己的 Filter 就可以做到。

 你的 Filter 优先级要比 TracingFilter 的优先级要低，不然你无法拿到 TracingFilter 处理之后的信息。

16.3.2 Baggage

Baggage 是存储在 Span 的上下文中的一组 Key/Value 键值对，跟 traceId 和 spanId 不同，它不是必选项。"Baggage"翻译成中文是"行李"，意即我们可以把一些信息像行李一样，挂在 Sleuth 中，由 Sleuth 帮我们沿着调用链路一路往下传递。

毫无疑问，Baggage 是一个非常有用的功能，它相当于 Sleuth 暴露的一个功能接口，通过它，你可以让你的数据跟着 Sleuth 一起往后接连传递。

Baggage 的一个典型应用场景就是登录信息的传递。

16.3.3 案例

上面两小节我们对 TraceFilter 和 Baggage 做了一个简单的介绍，本节我们通过项目 ch16-2 来演示二者的实战用法。

项目 ch16-2 下有两个模块：consumer 和 provider，其中 consumer 使用自定义的 Filter 获取前端传来的 sessionId，放入 Baggage 中，通过 feign 调用的方式将 sessionId 传递给 provider。

（1）创建 Maven 父级 pom 工程

在父工程里面配置好工程需要的父级依赖，目的是方便管理与简化配置。相关源码参考代码清单 16-6。

代码清单16-6　ch16-2\pom.xml

```
<dependencies>
    <dependency>
        <groupId>org.springframework.cloud</groupId>
        <artifactId>spring-cloud-starter-sleuth</artifactId>
```

```xml
    </dependency>
    <dependency>
        <groupId>org.springframework.cloud</groupId>
        <artifactId>spring-cloud-starter-openfeign</artifactId>
    </dependency>
    <dependency>
        <groupId>org.springframework.boot</groupId>
        <artifactId>spring-boot-starter-web</artifactId>
    </dependency>
</dependencies>
```

（2）创建 sleuth-consumer 模块

sleuth-consumer 模块中复用父级 pom，篇幅所限，这里省略程序入口主类、feign 调用接口 HelloService、对外测试接口 ConsumerController 和 pom 的展示，读者可以在本书附录源码 ch16-2/ch16-2-sleuth-consumer 中找到相关代码。

1）创建 SessionFilter

自定义自己的过滤器 SessionFilter，将自定义信息放入 Span 中。相关源码参考代码清单 16-7。

代码清单16-7　ch16-2/ch16-2-sleuth-consumer/src/main/java/cn/springcloud/book/sleuth/consumer/SessionFilter.java

```java
/**
 * 自定义过滤器，
 * 获取当前的SessionId，放入Baggage中
 * 注意，因为不是所有的请求都需要往后传递，所以会对一些请求跳过执行
 */
@Component
@Order(TraceWebServletAutoConfiguration.TRACING_FILTER_ORDER + 1)
public class SessionFilter extends GenericFilterBean {
    private Pattern skipPattern = Pattern.compile(SleuthWebProperties.DEFAULT_
        SKIP_PATTERN);

    @Override
    public void doFilter(ServletRequest request, ServletResponse response,
        FilterChain filterChain) throws IOException, ServletException {
        if (!(request instanceof HttpServletRequest) || !(response instanceof
            HttpServletResponse)) {
            throw new ServletException("Filter just supports HTTP requests");
        }
        HttpServletRequest httpRequest = (HttpServletRequest) request;
        boolean skip = skipPattern.matcher(httpRequest.getRequestURI()).matches();

        if (!skip) {
            // 将 SessionId 放到 Baggage中
            ExtraFieldPropagation.set("SessionId", httpRequest.getSession().getId());
        }
        filterChain.doFilter(request, response);
    }
}
```

2）创建配置文件

创建模块的启动配置文件 bootstrap.yml。相关源码参考代码清单 16-8。

代码清单16-8　ch16-2/ch16-2-sleuth-consumer/src/main/resources/bootstrap.yml

```yaml
server:
    port: 8081
spring:
    application:
        name: sleuth-consumer
    sleuth:
        baggage-keys:   # 注意，Sleuth2.0.0之后，baggage的 key 必须在这里配置才能生效
            - SessionId
```

（3）创建 sleuth-provider 模块

sleuth-provider 模块中复用父级 pom，篇幅所限，这里省略程序入口主类和配置文件，其中 provider 启动端口号为 8082。读者可以参考本书附录源码：ch16-2/ch16-2-sleuth-consumer。

创建对外服务接口 ProviderController

相关源码参考代码清单 16-9。

代码清单16-9　ch16-2/ch16-2-sleuth-provider/src/main/java/cn/springcloud/book/sleuth/provider/ProviderController.java

```java
/**
 * sleuth-provider 对外服务接口
 */
@RestController
public class ProviderController {
    @GetMapping("/sayHello")
    public String hello(String name){
        return "hello, "+name+",SessionId is "+ ExtraFieldPropagation.get("SessionId");
    }
}
```

（4）效果展示

1）分别启动 sleuth-consumer 和 sleuth-provider。
2）在浏览器地址栏输入：http://localhost:8081/hello?name=scbook。
3）页面显示 "hello, scbook,SessionId is ***" 即为请求成功。
到此 Sleuth 相关部分内容就讲解完了。
接下来开始学习两款非入侵性的 APM——SkyWalking 和 Pinpoint。

16.4　Spring Cloud 与 SkyWalking

16.4.1　Skywalking 概述

SkyWalking 创建与 2015 年，用于提供分布式追踪功能。从 5.x 开始，项目进化为一个功能齐全的 APM 系统，被用于追踪、监控和诊断分布式系统，特别是使用微服务架构、云原生或容器技术。那么 SkyWalking 具体有什么功能和特性呢？接下来让我一一讲解。

16.4.2 SkyWalking 提供主要功能

SkyWalking 提供主要功能如下：
- 分布式追踪和上下文传输。
- 应用、实例、服务性能指标分析。
- 根源分析。
- 应用拓扑分析。
- 应用和服务依赖分析。
- 慢服务检测。
- 性能优化。

16.4.3 SkyWalking 主要特性

SkyWalking 主要特性如下。
- 多语言探针或者类库：
  - Java 自动探针，追踪和监控程序时，不需要修改源码。
  - 社区提供的语言探针：.NET Core、Node.js。
- 多种后端存储：ElasticSearch、H2。
  - 支持 OpenTrancing：Java 自动探针和 OpenTracing API 协同工作。
- 轻量级、完善的后端聚合和分析功能。
- 现代化 Web UI。
- 日志集成。
- 应用、实例和服务的告警。
- 支持接受其他跟踪器数据格式：
  - Zipkin JSON、Thrift、Protobuf v1 和 v2 格式，由 OpenZipkin 库提供支持。
  - Jaeger 采用 Zipkin Thrift 或 JSON v1/v2 格式。

16.4.4 SkyWalking 整体架构

对于整体架构主要由四部分组成 collector、agent、web、storage，具体如图 16-5 所示。

下面对于图 16-5 中的整体架构进行讲解。从上到下是应用级别的接入，可以使用 SDK 形式的接入，也使用非入侵性的 Agent 形式接入，agent 将数据转化成 SkyWalking Trace 数据协议，通过 HTTP 或者是 gRPC 发送到 collector，collector 对收集到的数据进行分析和聚合，最后存储到 ElasticSearch 或者 H2 中，一般情下 H2 用于测试，在图 16-5 右边的 UI 是通过 HTTP+GrahQL 进行数据获取展示，大体的流程就是这样。接下来讲解一下 Spring Cloud 与 SkyWalking 实战操作。

16.5 Spring Cloud 与 Skywalking 实战

本次实战的案例使用两个服务（service-a、service-b）、Zuul 网关、Eureka 作为注册中心，案例的调用流程是这样的，首先通过访问 Zuul 网关然后再转发到 service-a，service-a 通过 feign 远程调用 service-b，最后返回内容，具体的调用流程图，如图 16-6 所示。

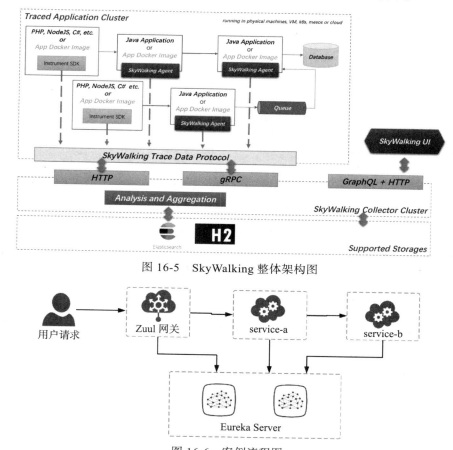

图 16-5　SkyWalking 整体架构图

图 16-6　案例流程图

接下来开始编写上述谈到的服务。先创建父级工程，将需要的公共依赖放在其 pom 中。

16.5.1　父工程创建

新建 maven 工程 ch16-3，案例代码如本书源 ch16-3 所示。

父级配置依赖

在 ch16-3 的 pom 中添加公共部分依赖，后续的子模块不用再添加相同的依赖，如代码清单 16-10 所示。

代码清单16-10　ch16-3/pom.xml

```
<dependencies>
    <dependency>
        <groupId>org.springframework.boot</groupId>
```

```xml
        <artifactId>spring-boot-starter-web</artifactId>
    </dependency>

    <dependency>
        <groupId>org.springframework.boot</groupId>
        <artifactId>spring-boot-starter-test</artifactId>
        <scope>test</scope>
    </dependency>

    <dependency>
        <groupId>org.springframework.boot</groupId>
        <artifactId>spring-boot-starter-actuator</artifactId>
    </dependency>
</dependencies>
```

主要是添加 web 以及测试和端点的依赖。创建 Eureka 注册中心工程。

16.5.2　创建 eureka-server-skywalking 工程

eureka-server-skywalking 工程主要是作为本次案例的服务注册中心。

（1）创建工程模块

在工程 ch16-3 下面创建一个 module，命名为 ch16-3-eureka-server-skywalking，代码如本书源码 ch16-3/ ch16-3-eureka-server-skywalking 所示。

（2）配置依赖

在 ch16-3-eureka-server-skywalking 工程的 pom 中添加 eureka-server 的依赖，具体如代码清单 16-11 所示。

代码清单16-11　ch16-3/ch16-3-eureka-server-skywalking/pom.xml

```xml
<dependencies>
    <dependency>
        <groupId>org.springframework.cloud</groupId>
        <artifactId>spring-cloud-starter-netflix-eureka-server</artifactId>
    </dependency>
</dependencies>
```

在配置好 pom 后创建入口程序。

（3）编写主程序入口代码

创建 EurekaServerApplication 类作为启动类，并且添加相关的注解作为程序的入口，具体如代码清单 16-12 所示。

代码清单16-12　ch16-3/ch16-3-eureka-server-skywalking/src/main/java/cn/springcloud/ book/eureka/EurekaServerApplication.java

```java
@SpringBootApplication
@EnableEurekaServer
public class EurekaServerApplication {

    public static void main(String[] args) {
```

```
        SpringApplication.run(EurekaServerApplication.class, args);
    }
}
```

其中 @EnableEurekaServer 是开启 Eureka 服务注册的能力，还需要配置启动需要的参数，具体在 application.yml 中配置。

（4）配置文件配置

在 application.yml 中添加启动的相关配置如代码清单 16-13 所示。

代码清单16-13 ch16-3/ch16-3-eureka-server-skywalking/src/main/resources/application.yml

```
server:
    port: 8761
eureka:
    instance:
        hostname: localhost
    client:
        registerWithEureka: false
        fetchRegistry: false
        serviceUrl:
            defaultZone: http://${eureka.instance.hostname}:${server.port}/eureka/
    server:
        waitTimeInMsWhenSyncEmpty: 0
        enableSelfPreservation: false
```

到此服务注册中心就编写完成了，更多相关 Eureka 的信息请到 Eureka 相关章节中观看，接下来创建 Zuul 网关工程。

16.5.3 创建 zuul-skywalking

zuul-skywalking 工程主要是作为本次案例的网关，用途是做服务转发。

（1）创建工程模块

在工程 ch16-3 下面创建一个 module，命名为 ch16-3-zuul-skywalking，代码如本书源码 ch16-3/ch16-3-zuul-skywalking 所示。

（2）配置依赖

在 ch16-3-zuul-skywalking 工程的 pom 中添加 eureka-client 和 Zuul 的依赖，具体如代码清单 16-14 所示。

代码清单16-14 ch16-3/ch16-3-eureka-server-skywalking/pom.xml

```xml
<dependencies>
    <dependency>
        <groupId>org.springframework.cloud</groupId>
        <artifactId>spring-cloud-starter-netflix-zuul</artifactId>
    </dependency>
    <dependency>
        <groupId>org.springframework.cloud</groupId>
        <artifactId>spring-cloud-starter-netflix-eureka-client</artifactId>
    </dependency>
```

```
</dependencies>
```

添加依赖完成后，再创建 Zuul 的启动类。

（3）编写主程序入口代码

创建 ZuulServerApplication 类作为启动类，并且添加相关的注解作为程序的入口，具体如代码清单 16-15 所示。

代码清单16-15　ch16-3/ch16-3-zuul-skywalking/src/main/java/cn/springcloud/book/zuul/ZuulServerApplication.java

```
@SpringBootApplication
@EnableDiscoveryClient
@EnableZuulProxy
public class ZuulServerApplication {

    public static void main(String[] args) {
        SpringApplication.run(ZuulServerApplication.class, args);
    }
}
```

其中类上的 @EnableDiscoveryClient 和 @EnableZuulProxy 注解分别是开启服务发现和 Zuul 代理转发能力。还需要创建配置文件，用于添加 Eureka 注册地址和路由规则等。

（4）配置文件配置

在 application.yml 中添加启动的相关配置如代码清单 16-16 所示。

代码清单16-16　ch16-3/ch16-3-zuul-skywalking/src/main/resources/application.yml

```
spring:
  application:
    name: ch16-3-zuul-skywalking
server:
  port: 9020
eureka:
  client:
    serviceUrl:
      defaultZone: http://${eureka.host:127.0.0.1}:${eureka.port:8761}/eureka/

zuul:
  routes:
    service-a:
      path: /client/**
      serviceId: ch16-3-service-a

ribbon:
  eureka:
    enabled: true
  ReadTimeout: 30000
  ConnectTimeout: 30000
  MaxAutoRetries: 0
  MaxAutoRetriesNextServer: 1
  OkToRetryOnAllOperations: false
```

```yaml
hystrix:
  threadpool:
    default:
      coreSize: 1000 ##并发执行的最大线程数，默认为10
      maxQueueSize: 1000 ##BlockingQueue的最大队列数
      queueSizeRejectionThreshold: 500 ## 即使maxQueueSize没有达到，达到queue-
                                          SizeRejectionThreshold该值后，请求
                                          也会被拒绝
  command:
    default:
      execution:
        isolation:
          thread:
            timeoutInMilliseconds: 120001
```

上述规则中添加了 Ribbon 和 Hystrix 相关配置是为了在首次请求时不会出现超时情况，具体的详细配置解释请看第 8 章。这里的路由规则只是配置了 service-a 的，关于 service-a 的具体代码后面会有详细的讲解。

到这里 Zuul 也编写完成了，更多关于 Zuul 的配置和详细讲解请查看第 8 章。接下来创建上述谈到的 service-a 工程。

16.5.4　创建 service-a

service-a 工程是本次案例链路调用的一个工程，会在此工程中通过远程调用的方式来调用 service-b，service-b 工程后面会进行详细讲解。

（1）创建工程模块

在工程 ch16-3 下面创建一个 module，命名为 ch16-3-service-a，代码如本书源码 ch16-3/ch16-3-service-a 所示。

（2）配置依赖

在 ch16-3-service-a 工程的 pom 中添加 eureka-client 和 openfeign 远程调用需要的依赖，具体如代码清单 16-17 所示。

代码清单16-17　ch16-3/ch16-3-service-a/pom.xml

```xml
<dependencies>
    <dependency>
        <groupId>org.springframework.cloud</groupId>
        <artifactId>spring-cloud-starter-netflix-eureka-client</artifactId>
    </dependency>
    <dependency>
        <groupId>org.springframework.cloud</groupId>
        <artifactId>spring-cloud-starter-openfeign</artifactId>
    </dependency>
</dependencies>
```

添加依赖后，再创建应用程序的启动类。

（3）编写主程序入口代码

创建 ZuulServerApplication 类作为启动类，并且添加相关的注解作为程序的入口，具体如

代码清单 16-18 所示。

> **代码清单16-18**　ch16-3/ch16-3-service-a/src/main/java/cn/springcloud/book/AService-Application.java

```
@SpringBootApplication
@EnableFeignClients
@EnableDiscoveryClient
public class AServiceApplication {

    public static void main(String[] args) {

        SpringApplication.run(AServiceApplication.class, args);
    }

}
```

其中类上的 @EnableFeignClients 注解表示的是开启 Feign 远程调用的功能，接下来配置启动需要的配置。

（4）配置文件配置

在 application.yml 中添加启动的相关配置如代码清单 16-19 所示。

> **代码清单16-19**　ch16-3/ch16-3-service-a/src/main/resources/application.yml

```
erver:
    port: 9021
spring:
    application:
        name: ch16-3-service-a
eureka:
    client:
        service-url:
            defaultZone: http://localhost:8761/eureka/
```

上述配置就是 Eureka 注册地址和基本的配置，为了进行远程调用，还需要创建远程调用接口和请求的 controller 类。

（5）创建 SkyFeignSerivece 远程调用接口

创建用于远程调用 service-b 的服务接口，通过 Feign 的形式发起调用，具体如代码清单 16-20 所示。

> **代码清单16-20**　ch16-3/ch16-3-service-a/src/main/java/cn/springcloud/book/service/SkyFeign-Serivece.java

```
@FeignClient(name = "ch16-3-service-b")
public interface SkyFeignSerivece {

    // 调用service-b serviceName 传递过去的是service-a
    @RequestMapping(value = "/getSendInfo", method = RequestMethod.GET)
    String getSendInfo(@RequestParam("serviceName") String serviceName);

}
```

上述类中使用 @FeignClient(name = "ch16-3-service-b") 标注为调用 service-b 服务的远程调用接口，更多关于 Feign 的内容请参考第 4 章。接下来创建接受用户发起请求的 controller 接受类。

创建类 SkyController，并且添加上 springmvc 的相关注解使其具备接受能力。

具体如代码清单 16-21 所示。

代码清单16-21 ch16-3/ch16-3-service-a/src/main/java/cn/springcloud/book/controller/SkyController.java

```java
@RestController
@RequestMapping("skyController")
public class SkyController {

    @Autowired
    private SkyFeignSerivece skyFeignSerivece;

    @RequestMapping(value = "/getInfo")
    public String getInfo(){
        return skyFeignSerivece.getSendInfo("service-a");
    }
}
```

到此 service-a 服务也创建完成，最后创建 service-b 工程也是整个案例中调用链的最后请求应用。

16.5.5 创建 service-b

service-b 工程是本次案例链路最后调用的工程，会在此工程中返回调用信息，主要是提示从 service-a 调用 service-b。

（1）创建工程模块

在工程 ch16-3 下面创建一个 module，命名为 ch16-3-service-b，代码如本书源码 ch16-3/ch16-3-service-b 所示。

（2）配置依赖

在 ch16-3-service-b 工程的 pom 中添加 eureka-client，具体如代码清单 16-22 所示。

代码清单16-22 ch16-3/ch16-3-service-b/pom.xml

```xml
<dependencies>
    <dependency>
        <groupId>org.springframework.cloud</groupId>
        <artifactId>spring-cloud-starter-netflix-eureka-client</artifactId>
    </dependency>
</dependencies>
```

依赖配置完成后，需创建应用的启动类 BServiceApplication，该类中的代码同 AServiceApplication，这里不再赘述，上文中知道 service-a 要调用 service-b，因此需要创建一个 controller 类，具体如代码清单 16-23 所示。

代码清单16-23　ch16-3/ch16-3-service-b/src/main/java/cn/springcloud/book/controller/SkySecondController.java

```java
@RestController
public class SkySecondController {

    @RequestMapping(value = "/getSendInfo", method = RequestMethod.GET)
    String getSendInfo(@RequestParam("serviceName") String serviceName){

        return  serviceName + " ----> " + "service-b";
    }
}
```

该 controller 是将调用方传递过来的参数和本身拼接的一个 service-b 字符串返回回去。还需要配置启动需要的参数信息。

（3）配置文件配置

在 application.yml 中添加启动的相关配置如代码清单 16-24 所示。

代码清单16-24　ch16-3/ch16-3-service-b/src/main/resources/application.yml

```yaml
server:
    port: 9022

spring:
    application:
        name: ch16-3-service-b

eureka:
    client:
        service-url:
            defaultZone: http://localhost:8761/eureka/
```

到此 service-b 就编写完成了，同时也是本次使用案例的所有工程编写完成，接下来构建 SkyWalking 的基础环境。

16.5.6　SkyWalking Collector 基础环境安装

为了讲解方便，下文将 elasticsearch 简称为 es。
SkyWalking5.x 需要的基础环境基础需要如下：

- JDK6+（被监控的应用程序运行在 jdk6 及以上版本）
- JDK8+（SkyWalking collector 和 WebUI 部署在 jdk8 及以上版本）
- Elasticsearch 5.x（集群模式或不使用）

在 SkyWalking5.x 开始使用 es 作为存储，所以先下载 es，本案例使用的 es 是 5.6.10 版本。将下载好的 es 解压，解压后的结构如图 16-7 所示。

其中 bin 中是启动等命令，config 是 es 的启动配置，里面的配置需要和 collector 中的配置信息对应上。config 中的目录结构具体如图 16-8 所示。

这里主要是对 elasticsearch.yml 的配置。首先将 elasticsearch.yml 中的 cluster.name 配置为 CollectorDBCluster，这个名字可以是其他的，只要和后面的 Collector 的配置一致即可。

图 16-7　es 解压图

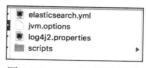

图 16-8　config 目录结构

elasticsearch.yml 具体配置位置如图 16-9 所示：

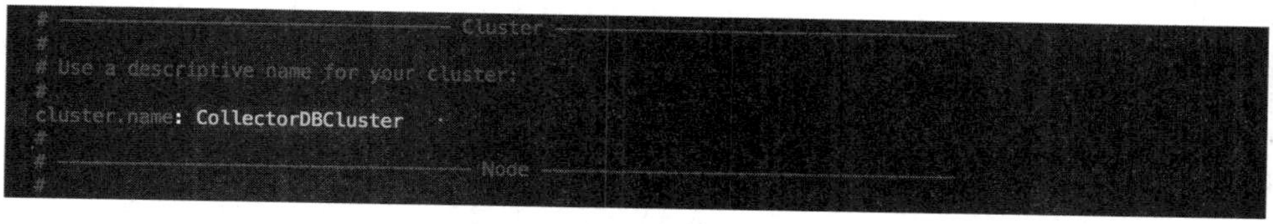

图 16-9　es 配置图

这里不对 es 展开讲解，更多关于 es 配置的相关解释请读者查阅相关资料。

配置完成后启动 es，es 在 bin 启动相关的脚本，读者根据当前使用的系统选择对应的启动即可。笔者使用的是 mac OS 所以进入 bin 目录执行命令：./elasticsearch。

启动成功会看如图 16-10 所示。

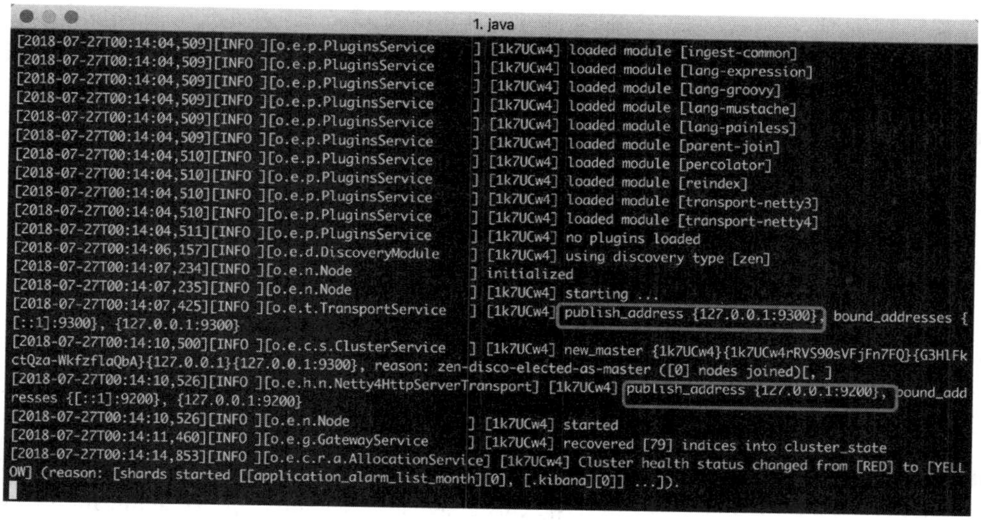

图 16-10　es 启动示意图

es 相关操作完成后，需要下载 SkyWalking 的相关资料，从 Apache 官方网站下载当前 SkyWalking 发布的版本。具体下载地址：http://skywalking.apache.org/downloads/。

本次案例下载到的是 5.0.0-beta2 版本，具体如图 16-11 所示。

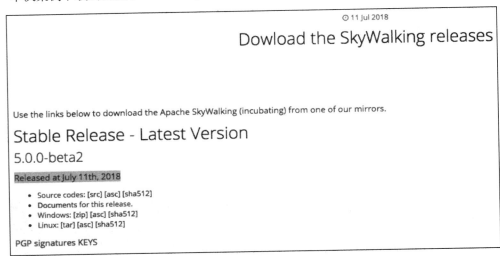

图 16-11　SkyWalking 下载图示

读者根据系统情况下载对应的版本即可，下载后进行解压，解压后的目录结构如图 16-12 所示。

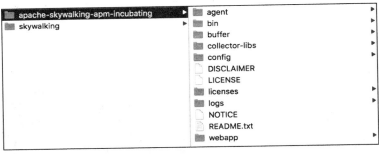

图 16-12　SkyWalking 解压目录结构

这里解释一下目录中的结构，agent 是探针相关，后面会有详细的讲解，bin 中存放的是 collectorService 和 webappService 的启动脚本，startup 是可以同时启动前面两个脚本的合并命令。config 是 collector 的相关配置信息。log 是存放启动 collector 和 web 后生成的日志文件，webapp 是存放 SkyWalking 展示 UI 的 jar 和配置文件。SkyWalking 中默认使用 8080、10800、11800、1280 这些端口，如果要修改端口，则应该在 SkyWalking 的 config 目录下的 application.yml 文件和 webapp 下面的 webapp.yml 配置文件中修改。在本案例中使用默认的配置。上文中提到 elaticsearch.yml 中配置的 cluster.name 要和 config 目录中 application.yml 文件中的 clusterName 对应上。application.yml 具体配置如图 16-13 所示。

接下来启动 colloector 和 web，进入下载好的 SkyWalking 目录下的 bin 目录执行命令：./startup.sh。具体如图 16-14 所示。

图 16-13　SkyWalking application.yml 配置图

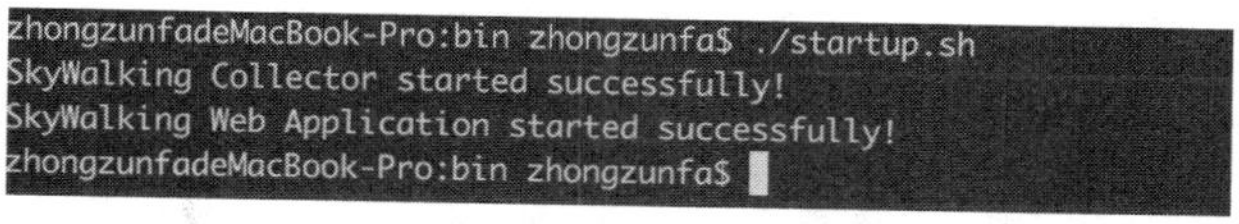

图 16-14　SkyWalking 启动 collector 和 web 进程图

都启动成功后，访问地址 localhost:8080，具体如图 16-15 所示。

图 16-15　SkyWalking 登录页面

默认用户名 / 密码是 admin/admin，登录后可以看到如图 16-16 所示页面。

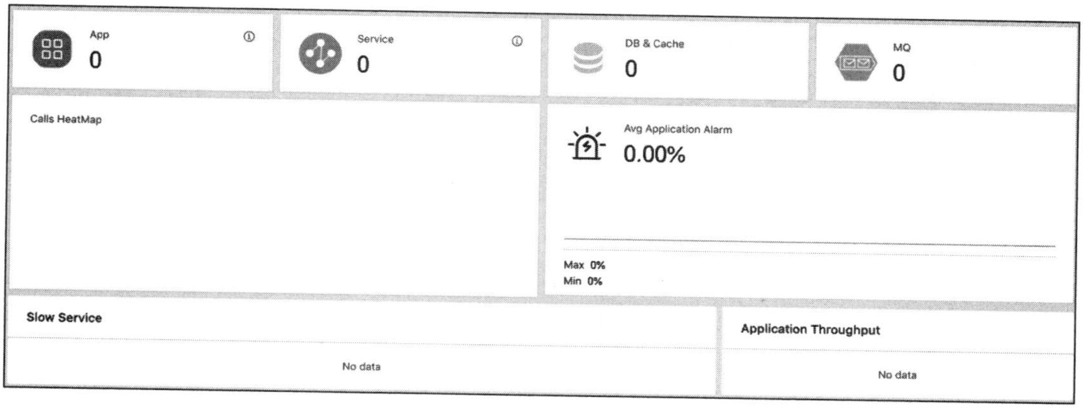

图 16-16　SkyWalking 首页图

其中 web（UI 进程）通过 127.0.0.1:10800 访问本地 collector，无须额外配置。
接下来使用 agent 启动应用进行观察。

16.5.7　使用 Agent 启动服务和监控查看

这里创建四个目录分别为 service-a、service-b、service-eureka、service-zuul。

主要放被监控的 jar 和 agent，每个应用使用一个对应的 agent 进行启动。

具体如图 16-17 所示。

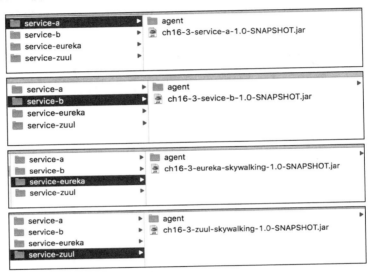

图 16-17　agent 和 jar 共存目录

其中 agent 是之前下载的 SkyWalking 中的 agent 目录，修改 agent/config 目录左面的 agent.config 文件，这里只修改 agent.application_code 这项配置，这个代表应用。每个配置文件对应改为：agent.application_code=service-a、agent.application_code=service-b、agent.application_code=service-zuul、agent.application_code=service-eureka。

而服务应用是我们之前编写好的，通过 maven 命令 maven install 打包出来的。接下来通过下面的命令启动要监控的服务，具体如下所示：

```
-- 启动Eureka
java -javaagent:/Users/zhongzunfa/all_test/skywalking/service-eureka/agent/
    skywalking-agent.jar -jar ch16-3-eureka-skywalking-1.0-SNAPSHOT.jar
-- 启动Zuul
java -javaagent:/Users/zhongzunfa/all_test/skywalking/service-zuul/agent/
    skywalking-agent.jar -jar ch16-3-zuul-skywalking-1.0-SNAPSHOT.jar
-- 启动B服务
java -javaagent:/Users/zhongzunfa/all_test/skywalking/service-b/agent/
    skywalking-agent.jar -jar ch16-3-sevice-b-1.0-SNAPSHOT.jar
-- 启动A服务
java -javaagent:/Users/zhongzunfa/all_test/skywalking/service-a/agent/skywalking-
    agent.jar -jar ch16-3-service-a-1.0-SNAPSHOT.jar
```

命令从上到下依次执行即可。监控的应用都执行完成后，查看 UI 管理界面，具体如图 16-18 所示。

图 16-18　应用和服务数图

这样就会出现发布的应用数和服务数，接下来操作调用，完成如图 16-6 所示调用流程观察。在 postman 上面访问地址：localhost:9020/client/skyController/getInfo。
具体如图 16-19 所示。

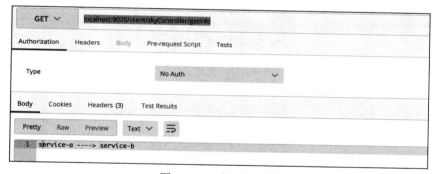

图 16-19　访问结果图

此时观察 SkyWalking 首页最下面会出现如图 16-20 所示。

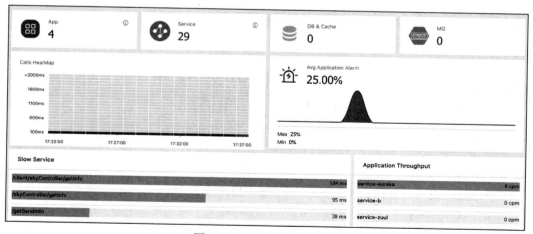

图 16-20　链路调用展示图

在图最下面出现一个调用过程，更详细的调用可以查看左侧菜单中的 applicaiotn、service、topology、Trace 和 Alarm 等。

首先看一下 application 的调用，具体如图 16-21 所示。

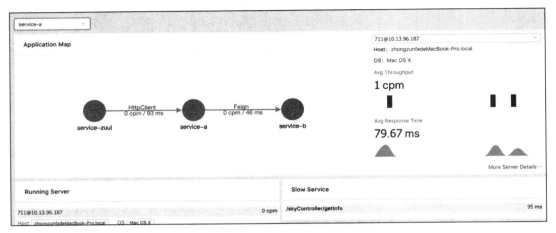

图 16-21　application 调用展示

其中左上角可以选择不同的应用观察，从图 16-21 中可以看到主机信息、平均通过率、平均响应时间等。接下来看 service 展示的情况，具体如图 16-22 所示。

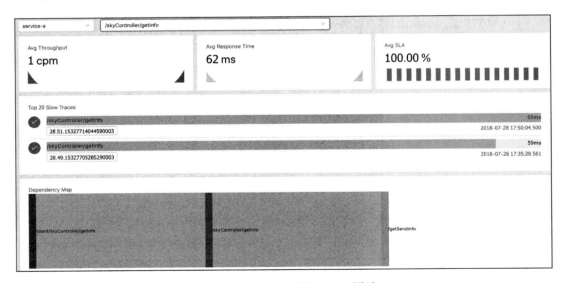

图 16-22　service-a 的 service 图示

同样也是有相应的指标信息。接下来看看整个拓扑图情况。

此时刷新 toplogy 可能会先看到一个 service-eureka 的节点。重复刷新可能会出现 serivce-a 指向 service-b 的请求，再请求一次，再刷新可能会出现 user-->service-zuul-->service-a-->

service-b 的调用，为什么会出现这样的现象呢？

解释一下，延迟出现的原因和调用顺序有关系，调用是顺序开始，完成是倒序完成的，然后在每个应用中完成后，发送监控的信息到 collector，然后 collector 才能依次输出采集的结果。所以就出现上面的情况。

最终会出现完整的拓扑图，如图 16-23 所示。

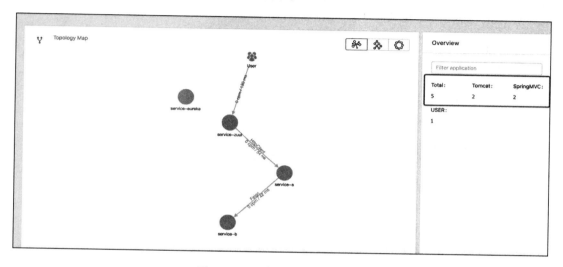

图 16-23　整体调用链头拓扑图

从图上看是 httpclient 插件生效了，变向实现了 Zuul 的监控。

并且和开始介绍的图 16-6 的请求一样，说明本次的链路调用监控吻合预期的结果。是不是很令人激动？整个调用流程都可以监控，比如微服务在调用过程中出现的延迟和各种问题都可以借助其进行很好地分析。更多关于 SkyWalking 的使用情况请到官方网站查看使用说明。官方地址：https://github.com/apache/incubator-skywalking

16.5.8　总结

最后总结一下整个案例的启动顺序。

1）启动 elaticsearch。
2）启动 skywakling collector 和 UI 两个进程。
3）使用 agent 启动 jar（下面的从 Eureka ---> Zuul -----> service-a -----> service-b）。
4）服务之间发起调用。
5）观察链路调用的结果图。

这里强调是因为启动顺序是相互依赖的，如果随意启动就可能会出现莫名其妙的问题，这个需要注意。到这里，Spring Cloud 和 SkyWalking 的结合案例就完成了，接下来讲解另外一个 APM 竞品——Pinpoint。

16.6 Spring Cloud 与 Pinpoint

16.6.1 Pinpoint 概述

Pinpoint 是一个分析大规模分布式系统的平台，并提供处理大量跟踪数据的解决方案。它自 2012 年 7 月开发，并于 2015 年 1 月 9 日作为开源项目推出。

Pinpoint 是由韩国人编写的著名的 APM 系统。

1. Pinpoint 特点

Pinpoint 的特点如下：

- 分布式事务跟踪，跟踪跨分布式应用的消息。
- 自动检测应用拓扑，帮助你搞清楚应用的架构。
- 水平扩展以便支持大规模服务器集群。
- 提供代码级别的可见性以便轻松定位失败点和瓶颈。
- 使用字节码增强技术，添加新功能而无须修改代码。

2. Pinpoint 的优势

Pinpoint 的优势如下：

- 非入侵式：使用字节码增强技术，添加新功能而无须修改代码。
- 资源消耗：对性能的影响最小（资源使用量增加约 3%）。

16.6.2 Pinpoint 架构模块

Pinpoint 主要包含如下 4 个模块：

- HBase（用于存储数据）。
- Pinpoint Collector（部署在 Web 容器上）。
- Pinpoint Web（部署在 Web 容器上）。
- Pinpoint Agent（附加到用于分析的 Java 应用程序）。

为了更好地说明 Pinpoint 中架构的模块，需要看详细的架构图，具体如图 16-24 所示。

解析图 16-24，大体的流程是这样的：首先通过 agent 收集调用应用的数据，将数据发送到 collector，collector 通过处理和分析数据最后存储到 HBase 中，可以通过 Pinpoint Web UI 查看已经分析好的调用分析数据。具体的收集数结构信息下面会进行详细讲解。

16.6.3 Pinpoint 的数据结构

在 Pinpoint 中，数据结构的核心由 Span、Trace 和 TraceId 组成。

- Span：RPC（远程过程调用）跟踪的基本单位。它表示 RPC 到达时处理的工作，包含跟踪数据。为确保代码级可见性，Span 将子项标记为 SpanEvent，作为数据结构。每个 Span 包含一个 TraceId。
- Trace：一系列跨度；它由相关的 RPC（Spans）组成。同一跟踪中的跨距共享相同的 TransactionId。Trace 通过 SpanIds 和 ParentSpanIds 排序为分层树结构。

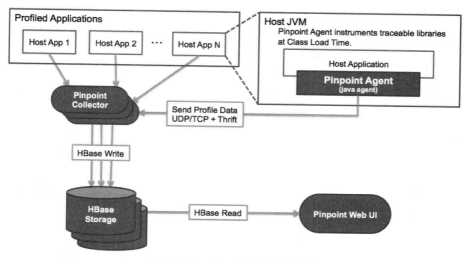

图 16-24　pinpoint 整体架构图

- TraceId：由 TransactionId、SpanId 和 ParentSpanId 组成的密钥集合。TransactionId 表示消息 ID，SpanId 和 ParentSpanId 都表示 RPC 的父子关系。
  - TransactionId（TxId）：来自单个事务的分布式系统发送/接收的消息的 ID；它必须在整个服务器组中是全局唯一的。
  - SpanId：接收 RPC 消息时处理的作业的 ID；它是在 RPC 到达节点时生成的。
  - ParentSpanId（pSpanId）：生成 RPC 的父 span 的 SpanId。如果节点是事务的起始点，则不会有父跨度（对于这些情况，我们使用值 −1 来表示跨度是事务的根跨度）。

从图 16-25 中能够比较直观地说明这些 ID 结构之间的关系。

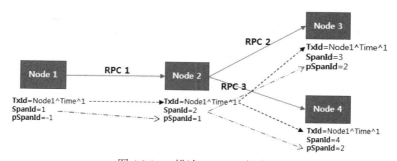

图 16-25　描述 TraceId 行为图

从图 16-25 中可以很好地描述一个 Trace 的过程。接下来看看 Pinpoint 的兼容性。

16.6.4　Pinpoint 兼容性

下面对 Pinpoint 涉及的组件和存储方面的中间件进行兼容对比。
Pinpoint 运行需要的 Java 版本，具体如表 16-1 所示。

表 16-1 Pinpoint 需要 Java 版本

Pinpoint Version	Agent	Collector	Web
1.0.x	6-8	6+	6+
1.1.x	6-8	7+	7+
1.5.x	6-8	7+	7+
1.6.x	6-8	7+	7+
1.7.x	6-8	8+	8+
1.8.x	6-8 9+(Experimental)	8+	8+

HBase 的兼容版本，具体如表 16-2 所示。

表 16-2 HBase 兼容版本

Pinpoint Version	HBase 0.94.x	HBase 0.98.x	HBase 1.0.x	HBase 1.1.x	HBase 1.2.x
1.0.x	yes	no	no	no	no
1.1.x	no	not tested	yes	not tested	not tested
1.5.x	no	not tested	yes	not tested	not tested
1.6.x	no	not tested	not tested	not tested	yes
1.7.x	no	not tested	not tested	not tested	yes
1.8.x	no	not tested	not tested	not tested	yes

Agent - Collector 兼容版本，具体如表 16-3 所示。

表 16-3 Agent 和 Collector 兼容版本

Agent Version	Collector 1.0.x	Collector 1.1.x	Collector 1.5.x	Collector 1.6.x	Collector 1.7.x	Collector 1.8.x
1.0.x	yes	yes	yes	yes	yes	yes
1.1.x	not tested	yes	yes	yes	yes	yes
1.5.x	no	no	yes	yes	yes	yes
1.6.x	no	no	not tested	yes	yes	yes
1.7.x	no	no	no	no	yes	yes
1.8.x	no	no	no	no	no	yes

Flink 兼容版本，具体如表 16-4 所示。

表 16-4 Flink 兼容版本

Pinpoint Version	flink 1.3.X	flink 1.4.X
1.7.x	yes	no

从上述的几点中我们已经基本上对 Pinpoint 有了一个很好的了解，接下来继续讲解 Spring Cloud 和 Pinpoint 的结合使用。

16.7 Spring Cloud 与 Pinpoint 实战

笔者将下文中需要的 Pinpoint 的 Hbase 初始化脚本等放在工程 ch16-3/ch16-3-script/script 中，读者可以直接使用，这样就不用去源代码中找了。接下来讲解 Pinpoint 基础环境的构建。

16.7.1 Pinpoint 基础环境

本次案例使用 Pinpoint 的 1.7.x 系列版本，使用的基础环境如下。

- java version: jdk1.6、jdk1.7、jdk1.8
    - 环境变量 JAVA_6_HOME 设置为 JDK 6 home 目录
    - 环境变量 JAVA_7_HOME 设置为 JDK 7+ home 目录
    - 环境变量 JAVA_8_HOME 设置为 JDK 8+ home 目录
- hbase：1.2.6
- maven：3.5（官方要求：3.2+）

上述配置多个 JDK 的环境变量的原因是 Pinpoint 的源码在 Maven 打包编译需要。下载 Pinpoint 的源代码，可以使用 Git 下载克隆源代码打包，具体如下命令所示：

- git clone https://github.com/naver/pinpoint.git
- cd pinpoint
- mvn install -Dmaven.test.skip=true

或者是通过地址：https://github.com/naver/pinpoint/archive/master.zip
下载源代码，再执行打包。需要注意几点。

- 需要将克隆下来或者下载 zip 的代码切到 1.7.x 版本上。
- 下载和打包过程时间比较长，因为要下载很多的 jar 文件。
- maven 编译打包过程需要同时安装多个版本 jdk，1.7.x 版本需要同时存在 JDK6、JDK7 和 JDK8。

笔者建议初级使用者，避免不必要的打包过程的问题，请前往下面的地址下载最新 release 发布出来的包，这样就可以直接使用，不用进行上面繁杂的操作。Pinpoint release 下载地址：https://github.com/naver/pinpoint/releases/。

本次案例下载最新 release 的 1.7.3，如图 16-26 所示。

下载 agent、collector、web 即可进行案例操作。

Pinpoint 存储还需要 Hbase，接下来安装 Hbase。

首先到 Apache 官方下载 Hbase，地址为：http://apache.mirror.cdnetworks.com/hbase/。

访问地址会看到如图 16-27 所示。

点击 1.2.6.1 进去再下载即可，具体如图 16-28 所示。

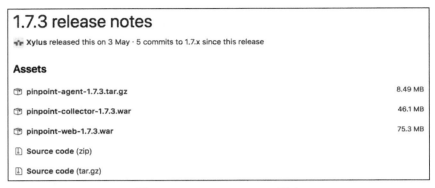

图 16-26　Pinpoint release 版本

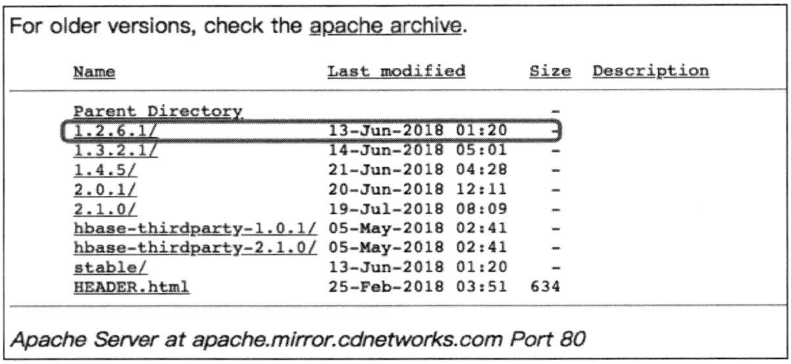

图 16-27　Hbase 版本选择

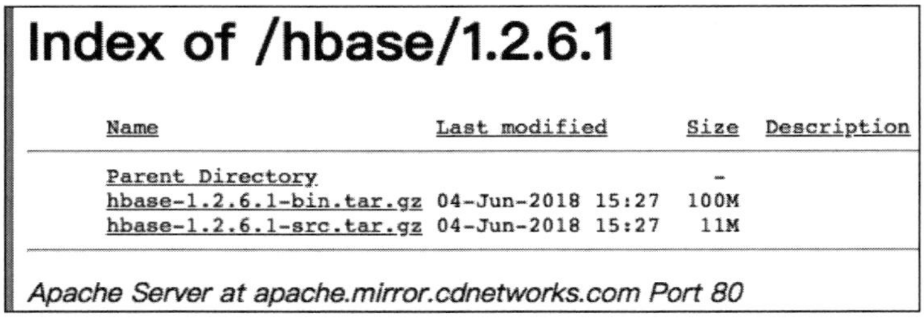

图 16-28　Hbase 下载图

将下载好的 Hbase 解压，解压后的目录结构如图 16-29 所示。

其中 bin 主要是启动等脚本，config 是配置相关，logs 主要是启动相关日志。

修改 config 中的 hbase-env.sh 的 JAVA_HOME，具体配置如图 16-30 所示。

图 16-30 中的配置读者可以根据自己 Java home 的路径进行配置。

配置完成后，进入 bin 目录中使用命令：./start-hbase.sh。

图 16-29　Hbase 解压目录结构

启动 Hbase。启动后验证是否启动成功，使用功能 jps 命令，如果看到 "HMaster" 的进程，说明已经启动成功了，如图 16-31 所示。

图 16-30　hbase-env.sh 配置

接下来将 Pinpoint 官方提供的 Hbase 初始化脚本导入，下面使用的脚本在工程 ch16-3/ch16-3-script/script 中，读者也可以在 Pinpoint 的源代码中找。进入 Hbase 的 bin 目录下执行脚本：./hbase shell / Users/zhongzunfa/all_test/pinpoint/hbase_init/init-hbase.txt。

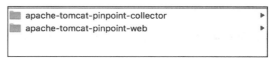

图 16-31　Hbase 验证启动图

最后的目录是笔者用于存放 Hbase 的初始化脚本的，读者根据自己的目录情况进行替换即可。到此基础环境就准备完成了。接下来部署 Collector 和 web 浏览界面。

16.7.2　Collector 和 Web 部署

Pinpoint 的 Collector 和 web 都是 war 包，这里使用 Tomcat 进行部署。笔者这里下载的是 tomcat8。解压两份并且命名为：apache-tomcat-pinpoint-collector、apache-tomcat-pinpoint-web，具体如图 16-32 所示。

图 16-32　Tomcat 重命名图

将 apache-tomcat-pinpoint-collector 中的 config 目录下的 server.xml 端进行修改。

将 8005 修改为 18005，8080 改为 18080，8443 改为 18443，8009 改为 18009。

将 apache-tomcat-pinpoint-web 中 config 目录下的 server.xml 端口进行修改。

将 8005 修改为 28005，8080 改为 28080，8443 改为 28443，8009 改为 28009。

将 apache-tomcat-pinpoint-collector 的 web/webapps/ROOT 目录下的文件清空并且将 pinpoint-collector-1.7.3.war 复制到下面。将其解压，解压后的目录结构如图 16-33 所示。

再进入 bin 目录中，执行启动脚本：./startup.sh。

同时观察启动日志，看是否启动成功，使用命令：tail -f ../logs/catalina.out。

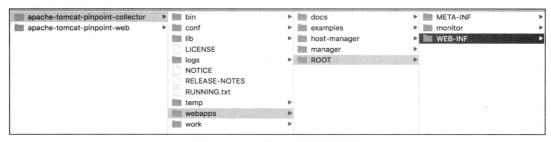

图 16-33　collector 解压部署图

同样将 apache-tomcat-pinpoint-web 的 web/webapps/ROOT 目录下的文件清空并将其复制到下面。将其解压，解压后的目录如图 16-34 所示。

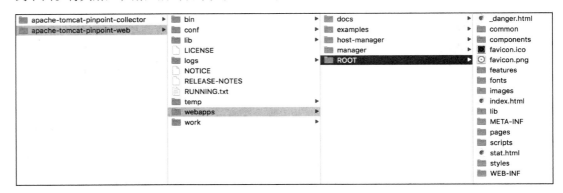

图 16-34　web 解压部署图

再进入 apache-tomcat-pinpoint-web 下的 bin 目录中执行启动脚本：./startup.sh。

同时观察启动日志，看是否启动成功，使用命令：tail -f ../logs/catalina.out。

当 collector 和 web 都启动成功后，在浏览器上访问地址：http://localhost:28080，初次访问后可以看到如图 16-35 所示。

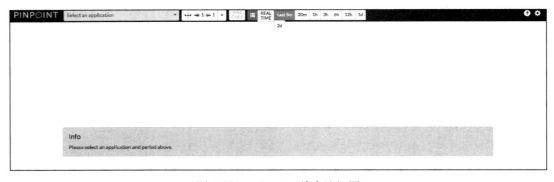

图 16-35　pinpoint 首次访问图

到此，collector 和 web 就部署完成了。接下来部署应用，使用 agent 方式启动。

16.7.3　Agent 启动应用

本次使用的应用案例同 16.2 节中的案例代码，同样创建四个目录分别为：service-a、service-b、service-zuul、service-eureka，将对应的应用复制到目录中，此外还需要将 Pinpoint 的 agent 复制进去并且将其解压，具体如图 16-36 所示。

这里讲解一下 agent 中的 pinpoint.config 配置文件。

- 配置 pinpoint.config：profiler.collector.ip=127.0.0.1 是指 pinpoint-collector 的地址，如果是同一服务器，则不用修改。其他默认。

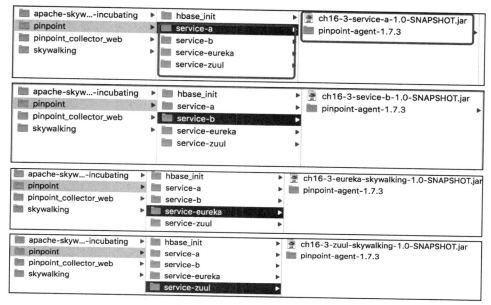

图 16-36　agent 和应用图

- 安装 pinpoint-collector 启动后，自动就开启了 9994、9995、9996 的端口了，这里默认即可。如果有端口需求，要去 pinpoint-collector 的配置文件（"pinpoint-collector/webapps/ROOT/WEB-INF/classes/pinpoint-collector.properties"）中，修改这些端口。此外里面还有很多的配置项，包括插件的启用等。

Pinpoint 通过将下面的三行加到 JVM 启动脚本中就可以轻易地为应用启用 Pinpoint agent 功能。具体如下所示：

```
-javaagent:$AGENT_PATH/pinpoint-bootstrap-$VERSION.jar
-Dpinpoint.agentId=<Agent's UniqueId>
-Dpinpoint.applicationName=<The name indicating a same service (AgentId collection)>
```

其中 -javaagent 冒号后面的参数是 agent 所在位置，-Dpinpoint.agentId 表示的是 agent 的唯一标识，-Dpinpoint.applicationName 表示的是应用名称。agent 有两种启动方式，一种是使用 Tomcat 的方式启动，另一种是 Spring Boot jar 包的方式启动。

方式一：修改 tomcat 目录下的 bin/catalina.sh，在 Control Script for the CATALINA Server 加入以下三行代码：

```
CATALINA_OPTS="$CATALINA_OPTS -javaagent: you agent path
CATALINA_OPTS="$CATALINA_OPTS -Dpinpoint.agentId= you agent id
CATALINA_OPTS="$CATALINA_OPTS -Dpinpoint.applicationName= you applicaiton name
```

第一行指的是 pinpoint-bootstrap-x.x.x.jar 的位置；

第二行的 agentId 必须唯一，标志一个 jvm；

第三行的 applicationName 表示同一种应用，同一个应用的不同实例应该使用不同的 agentId 与相同的 applicationName。

方式二：SpringBoot 启动

```
java -javaagent:you agent path/pinpoint-bootstrap-x.x.x.jar
-Dpinpoint.agentId=you agentId
-Dpinpoint.applicationName=you application name
-jar youjar.jar
```

接下来，通过下面的四个启动脚本来启动应用：

```
-- 启动eureka
java -javaagent:/Users/zhongzunfa/all_test/pinpoint/service-eureka/pinpoint-
    agent-1.7.3/pinpoint-bootstrap-1.7.3.jar -Dpinpoint.agentId=service-eureka
    -Dpinpoint.applicationName=service-eureka -jar ch16-3-eureka-skywalking-
    1.0-SNAPSHOT.jar

-- 启动zuul
java   -javaagent:/Users/zhongzunfa/all_test/pinpoint/service-zuul/pinpoint-
    agent-1.7.3/pinpoint-bootstrap-1.7.3.jar -Dpinpoint.agentId=service-zuul
    -Dpinpoint.applicationName=service-zuul -jar ch16-3-zuul-skywalking-1.0-
    SNAPSHOT.jar

-- 启动service-b
java   -javaagent:/Users/zhongzunfa/all_test/pinpoint/service-b/pinpoint-
    agent-1.7.3/pinpoint-bootstrap-1.7.3.jar -Dpinpoint.agentId=service-b
    -Dpinpoint.applicationName=service-b -jar ch16-3-sevice-b-1.0-SNAPSHOT.jar

-- 启动service-a
java   -javaagent:/Users/zhongzunfa/all_test/pinpoint/service-a/pinpoint-
    agent-1.7.3/pinpoint-bootstrap-1.7.3.jar -Dpinpoint.agentId=service-a
    -Dpinpoint.applicationName=service-a -jar ch16-3-service-a-1.0-SNAPSHOT.jar
```

在终端中执行上述命令后，再查看 webUI 的显示情况，具体请看 UI 浏览指标。

16.7.4　UI 浏览指标

启动四个 agent 后，再看 UI，如图 16-37 所示。

接下来测试应用间的访问调用，通过 postmant 访问地址：localhost:9020/skyController/getInfo，具体如图 16-38 所示。

此时再去 Pinpoint 的 UI 上查看服务器图，具体如图 16-39 所示。

图 16-39 显示的调用链图和图 16-6 调用流程一致，需要注意的是 Outbound 选择 4 也就是 1-4 的形式输出图。再访问一下 localhost:9020/client/skyController/getInfo，刷新 UI 会看到线上由之前的 1 变成了 2，具体如图 16-40 所示。

线上面的数字代表的是访问的次数，其中图 16-40 右边部分，最上面的部分表示的是成功率和失败率情况，中间部分表示的是响应时间，最下面是加载使用时间。还可以通过检查器查看更多的信息。

检查器（Inspector）：这里以 service-zuul 为例，Timeline 显示的是请求的时间段，information 指定当前节点启动的信息，包含应用名、agentId、启动时间等，具体如图 16-41 所示。

Heap 信息使用情况，具体如图 16-42 所示。

第 16 章　Spring Cloud 全链路监控　◆　395

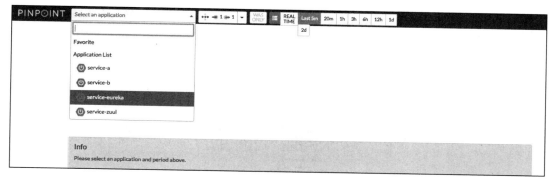

图 16-37　webUI 显示应用图

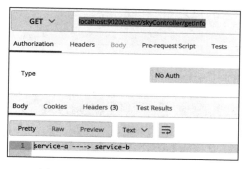

图 16-38　应用访问响应结果图

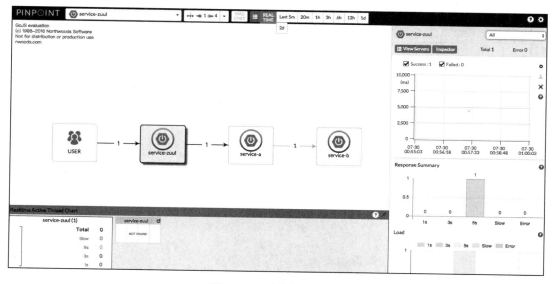

图 16-39　应用调用链图

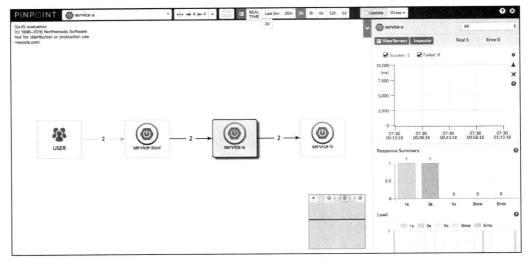

图 16-40 应用调用链图 2

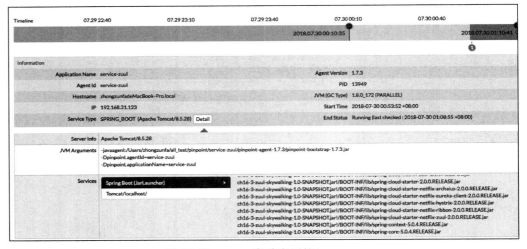

图 16-41 agent 启动应用的 inspector

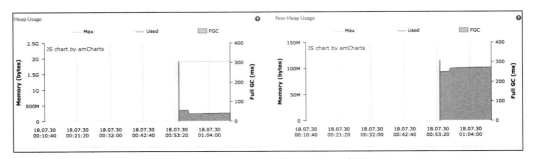

图 16-42 agent 启动应用的 Heap 情况

系统 CPU 等使用情况，具体如图 16-43 所示。

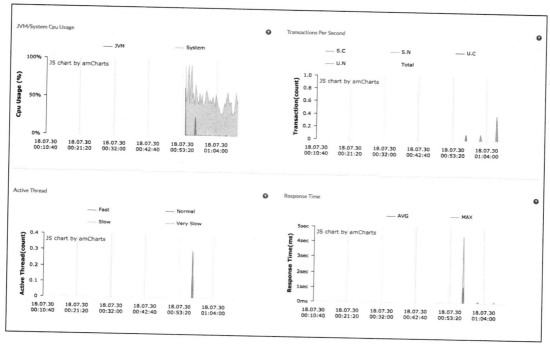

图 16-43　CPU 使用率和线程情况图

更多 Pinpoint 的信息，请读者进行试验来观看结果，这里不再一一复述。到此 Spring Cloud 和 Pinpoint 结果使用就完成了。更多关于 Pinpoint 信息请读者前往 Pinpoint 官网查看，地址为：https://github.com/naver/pinpoint。

16.7.5　总结

这里最后总结一下整个案例启动顺序：

1）启动 Hbase
2）启动 Collector
3）启动 web
4）启动 agent（下面的从 eureka → zuul → service-a → service-b）
5）应用调用
6）查看 top 分析图

这里强调启动顺序是因为其是相互依赖的，如果随意启动可能会出现莫名其妙的问题。到此 Spring Cloud 和 APM 相关的操作就完成了。APM 可以很好地帮助我们理解系统行为，也是用于分析性能问题的工具，以便发生故障的时候，能够快速定位和解决问题。

16.8 本章总结

本章讲解 Sleuth 的基础用法和 Sleuth 的高级用法，还讲解了零入侵链路追踪 SkyWalking 和 Pinpoint 的基本概述以及 Spring Cloud 结合实战，从中讲述了微服务架构下，APM 可以帮助我们更好地理解系统行为，也是用于分析性能问题的工具，以便发生故障的时候，能够快速定位和解决问题。

第 17 章 Chapter 17

Spring Cloud Gateway 上篇

本书的第 7~9 章介绍了什么是 Zuul 以及 Zuul 的工作原理和实战用法，本章将系统全面地介绍 Spring Cloud 生态体系中的第二代网关，即 Spring Cloud Gateway。本章将会介绍什么是 Spring Cloud Gateway、为什么会出现 Spring Cloud Gateway，以及 Spring Cloud Gateway 的工作原理和实战用法。

17.1 Spring Cloud Gateway 概述

17.1.1 什么是 Spring Cloud Gateway

Spring Cloud Gateway 是 Spring 官方基于 Spring 5.0、Spring Boot 2.0 和 Project Reactor 等技术开发的网关，Spring Cloud Gateway 旨在为微服务架构提供简单、有效且统一的 API 路由管理方式，如图 17-1 所示。Spring Cloud Gateway 作为 Spring Cloud 生态系统中的网关，目标是替代 Netflix Zuul，其不仅提供统一的路由方式，并且还基于 Filter 链的方式提供了网关基本的功能，例如：安全、监控/埋点、限流等。

17.1.2 Spring Cloud Gateway 的核心概念

网关提供 API 全托管服务，丰富的 API 管理功能，辅助企业管理大规模的 API，以降低管理成本和安全风险，包括协议适配、协议转发、安全策略（WAF）、防刷、流量、监控日志等功能。一般来说，网关对外暴露的 URL 或者接口信息，我们统称为路由信息。如果研发过网关中间件，或者使用或了解过 Zuul 的人，会知道网关的核心肯定是 Filter 以及 Filter Chain（Filter 责任链）。Spring Cloud Gateway 也具有路由和 Filter 的概念。下面介绍一下 Spring Cloud Gateway 中最重要的几个概念。

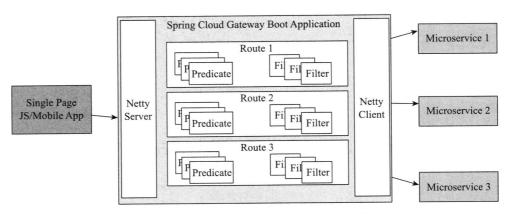

图 17-1　Spring Cloud Gateway 所有对外请求的入口

- 路由（route）。路由是网关最基础的部分，路由信息由一个 ID、一个目的 url、一组断言工厂和一组 Filter 组成。如果路由断言为真，则说明请求的 url 和配置的路由匹配。
- 断言（Predicate）。Java 8 中的断言函数。Spring Cloud Gateway 中的断言函数输入类型是 Spring 5.0 框架中的 ServerWebExchange。Spring Cloud Gateway 中的断言函数允许开发者去定义匹配来自于 Http Request 中的任何信息，比如请求头和参数等。
- 过滤器（filter）。一个标准的 Spring webFilter。Spring Cloud Gateway 中的 Filter 分为两种类型的 Filter，分别是 Gateway Filter 和 Global Filter。过滤器 Filter 将会对请求和响应进行修改处理。

17.2　Spring Cloud Gateway 的工作原理

　　Spring Cloud Gateway 的核心处理流程如图 17-2 所示，Gateway 的客户端会向 Spring Cloud Gateway 发起请求，请求首先会被 HttpWebHandlerAdapter 进行提取组装成网关的上下文，然后网关的上下文会传递到 DispatcherHandler。DispatcherHandler 是所有请求的分发处理器，DispatcherHandler 主要负责分发请求对应的处理器，比如将请求分发到对应 RoutePredicate-HandlerMapping（路由断言处理映射器）。路由断言处理映射器主要用于路由的查找，以及找到路由后返回对应的 FilteringWebHandler。FilteringWebHandler 主要负责组装 Filter 链表并调用 Filter 执行一系列的 Filter 处理，然后把请求转到后端对应的代理服务处理，处理完毕之后，将 Response 返回到 Gateway 客户端。

　　在 Filter 链中，通过虚线分割 Filter 的原因是，过滤器可以在转发请求之前处理或者接收到被代理服务的返回结果之后处理。所有的 Pre 类型的 Filter 执行完毕之后，才会转发请求到被代理的服务处理。被代理的服务把所有请求处理完毕之后，才会执行 Post 类型的过滤器。

> 注意　在配置路由的时候，如果不指定端口的话，http 默认设置端口为 80，https 默认设置端口为 443。Spring Cloud Gateway 的启动容器目前只支持 Netty。

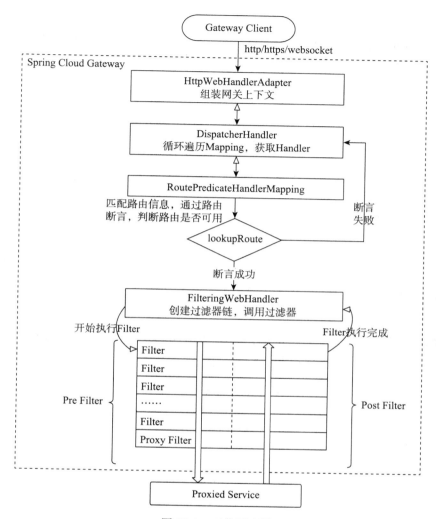

图 17-2　工作原理图

17.3　Spring Cloud Gateway 入门案例

作为网关来说，网关最重要的功能就是协议适配和协议转发，协议转发也就是基本的路由信息转发，本入门案例将演示一个 Spring Cloud Gateway 的基本路由转发功能，也就是通过 Spring Cloud Gateway 的 Path 路由断言工厂实现 url 直接转发。

1. 创建 Maven 工程

新建 Maven 工程 ch17-1-gateway，入门案例工程如 ch17-1-gateway 所示，工程结构如图 17-3 所示。

2. 配置依赖

如 ch17-1-gateway 工程所示，将 Spring Cloud Gateway 的 Maven 依赖添加到 pom.xml 文件中，主要核心依赖如代码清单 17-1 所示。

代码清单17-1 ch17-1/ch17-1-gateway/pom.xml

```xml
<dependency>
    <groupId>org.springframework.boot</groupId>
    <artifactId>spring-boot-starter-actuator</artifactId>
</dependency>
<!--Spring Cloud Gateway的Starter-->
<dependency>
    <groupId>org.springframework.cloud</groupId>
    <artifactId>spring-cloud-starter-gateway</artifactId>
</dependency>
```

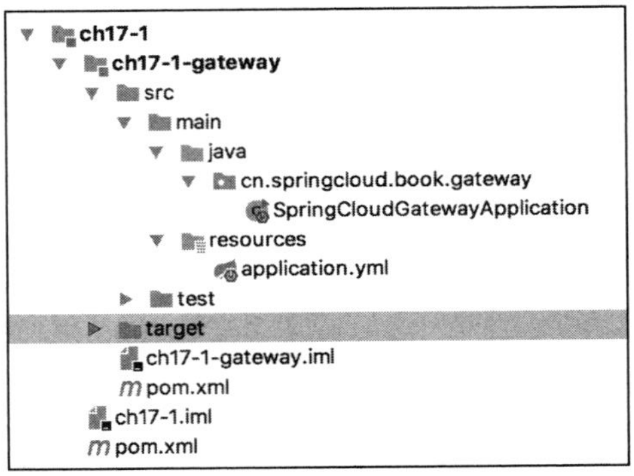

图 17-3 入门案例工程结构

3. 编写主入口程序

Spring Cloud Gateway Server 的主程序入口代码，如代码清单 17-2 所示：

代码清单17-2 ch17-1/ch17-1-gateway/src/main/java/cn/springcloud/book/gateway/SpringCloudGatewayApplication.java

```java
@SpringBootApplication
public class SpringCloudGatewayApplication {

    @Bean
    public RouteLocator customRouteLocator(RouteLocatorBuilder builder) {
        return builder.routes()
                //basic proxy
                .route(r ->r.path("/jd")
                        .uri("http://jd.com:80/").id("jd_route")
                ).build();
    }

    public static void main(String[] args) {
        SpringApplication.run(SpringCloudGatewayApplication.class, args);
    }
}
```

Spring Cloud Gateway 支持两种方式去配置路由信息，上述代码通过 Java 流式 API 自定义 RouteLocator 的方式定义 Spring Cloud Gateway 的路由信息。也可以通过如下 yml 文件的方式配置路由。

```yml
spring:
  cloud:
    gateway:
      routes: #当访问http://localhost:8080/jd直接转发到京东商城首页
        - id: jd_route
```

```yaml
      uri: http://jd.com:80/
      predicates:
      - Path=/jd
```

4. 配置 application.yml 文件

在 application.yml 文件中配置 Spring Cloud Gateway Server 的应用信息和日志信息。如代码清单 17-3 所示。

代码清单17-3 ch17-1/ch17-1-1-gateway/src/main/resources/application.yml

```yaml
server:
    port: 8080
spring:
    application:
        name: spring-cloud-gateway
logging: ## Spring Cloud Gateway的日志配置
    level:
        org.springframework.cloud.gateway: TRACE
        org.springframework.http.server.reactive: DEBUG
        org.springframework.web.reactive: DEBUG
        reactor.ipc.netty: DEBUG
```

为了演示 Gateway 两种配置方式都可以正常工作，同样创建 ch17-1-2-gateway 工程，不通过自定义 RouteLocator 方式配置路由，而是配置如代码清单 17-4 对应的路由信息。

代码清单17-4 ch17-1/ch17-1-2-gateway/src/main/resources/application.yml

```yaml
spring:
    cloud:
        gateway:
            routes: #当访问http://localhost:8080/baidu,直接转发到https://www.baidu.com/
            - id: baidu_route
              uri: http://baidu.com:80/
              predicates:
              - Path=/baidu
```

5. 开启端点

Spring Cloud Gateway 提供了一个 gateway actuator，该 EndPiont 提供了关于 Filter 及 routes 的信息查询以及指定 route 信息更新的 Rest API 接口。可以在 application.yml 中配置开启，配置如代码清单 17-5 所示。

代码清单17-5 ch17-1/ch17-1-1-gateway/src/main/resources/application.yml

```yaml
management:
    endpoints:
        web:
            exposure:
                include: '*'
    security:
        enabled: false
```

打开浏览器访问 http://localhost:8080/actuator/gateway/routes，可以看到返回所有的路由信

息如下所示：

```
[{
    "route_id": "jd_route",
    "route_object": {
        "predicate": "org.springframework.cloud.gateway.support.ServerWebExchange-
            Utils$$Lambda$283/1159234226@706859c0"
    },
    "order": 0
}]
```

Spring Cloud Gateway 其他的端点信息，可以通过查看 Gateway 的源码 org.springframework.cloud.gateway.actuate.GatewayControllerEndpoint.java 来学习，在这里不做过多阐述。

6. 案例测试

分别启动 ch17-1-1-gateway 和 ch17-1-2-gateway 两个工程。
- 打开浏览器访问 http://localhost:8080/jd，将结果转发到 https://www.jd.com/ 对应的首页。
- 打开浏览器访问 http://localhost:8081/baidu，将结果转发到 https://www.baidu.com/ 对应的首页。

17.4 Spring Cloud Gateway 的路由断言

通过上面的入门案例配置路由转发，我们快速学习 Spring Cloud Gateway 并入门。Spring Cloud Gateway 的路由匹配的功能是以 Spring WebFlux 中的 Handler Mapping 为基础实现的。Spring Cloud Gateway 也是由许多的路由断言工厂组成的。当 Http Request 请求进入 Spring Cloud Gateway 的时候，网关中的路由断言工厂会根据配置的路由规则，对 Http Request 请求进行断言匹配。匹配成功则进行下一步处理，否则断言失败直接返回错误信息。下面我们介绍一下 Spring Cloud Gateway 中经常使用的路由断言工厂。

17.4.1 After 路由断言工厂

After Route Predicate Factory 中会取一个 UTC 时间格式的时间参数，当请求进来的当前时间在配置的 UTC 时间之后，则会成功匹配，否则不能成功匹配。示例工程代码如 ch17-2/ch17-2-1-gateway 所示。

1）在主程序入口中通过 Java 代码的方式，将 After 路由断言的配置，配置到路由里面，如代码清单 17-6 所示。

代码清单17-6 ch17-2-1-gateway/src/main/java/cn/springcloud/book/gateway/SCGatewayApplication.java

```
@SpringBootApplication
public class SCGatewayApplication {
    @Bean
    public RouteLocator customRouteLocator(RouteLocatorBuilder builder) {
        //生成比当前时间早一个小时的UTC时间
        ZonedDateTime minusTime = LocalDateTime.now().minusHours(1).atZone
```

```
            (ZoneId.systemDefault());
        return builder.routes()
            .route("after_route", r -> r.after(minusTime)
                .uri("http://baidu.com"))
            .build();
    }
    public static void main(String[] args) {
        SpringApplication.run(SC1721GatewayApplication.class, args);
    }
}
```

也可以在 application.yml 文件中配置如下的 After 路由断言信息。

```
spring:
  cloud:
    gateway:
      routes:
      - id: after_route
        uri: http://baidu.com
        predicates:
        - After=2018-07-21T22:30:15.854+08:00[Asia/Shanghai]
```

> **说明** 其中的 -After=2018-07-21T22:30:15.854+08:00[Asia/Shanghai] 的 UTC 时间可以采用下面的 java 代码生成。

```
String minTime=ZonedDateTime.now().minusHours(1).format(DateTimeFormatter.ISO_ZONED_DATE_TIME);
System.out.println(minTime);
```

2）启动应用程序 ch17-2-1-gateway，然后访问 http://localhost:8081/ 成功转发到 https://www.baidu.com/ 对应的百度首页。能够成功转发的原因是请求的当前时间在配置的 UTC 时间之后，否则不能成功转发。

读者可以自行尝试通过下面的代码生成当前时间一个小时后的 UTC 时间：

```
String maxTime=ZonedDateTime.now().plusHours(1).format(DateTimeFormatter.ISO_ZONED_DATE_TIME);
System.out.println(maxTime);
```

将生成的 UTC 时间配置到 application.yml 中，然后重新启动应用，访问测试访问 http://localhost:8081/ 测试结果如图 17-4 所示。

图 17-4　After 路由断言失败

17.4.2 Before 路由断言工厂

Before 路由断言工厂会取一个 UTC 时间格式的时间参数，当请求进来的当前时间在路由断言工厂之前会成功匹配，否则不能成功匹配。示例工程代码如：ch17-2/ch17-2-2-gateway 所示。

1）自定义 RouteLocator 的路由 Java 代码配置如代码清单 17-7 所示。

代码清单17-7 ch17-2/ch17-2-2-gateway/src/main/java/cn/springcloud/book/gateway/SpringCloudGatewayApplication.java

```java
@Bean
public RouteLocator customRouteLocator(RouteLocatorBuilder builder) {
    ZonedDateTime datetime = LocalDateTime.now().plusDays(1).atZone(ZoneId.system-
        Default());
    return builder.routes()
            .route("before_route", r -> r.before(datetime)
                    .uri("http://baidu.com"))

            .build();
}
```

等价的 application.yml 文件的路由配置信息如下所示：

```yaml
spring:
  cloud:
    gateway:
      routes:
      - id: before_route
        uri: http://baidu.com
        predicates:
        - Before=2022-03-13T00:54:30.877+08:00[Asia/Shanghai]
```

2）启动应用 ch17-2-2-gateway，访问 http://localhost:8080/ 会转发到 https://www.baidu.com 对应的网站首页。

17.4.3 Between 路由断言工厂

Between 路由断言工厂会取一个 UTC 时间格式的时间参数，当请求进来的当前时间在配置的 UTC 时间工厂之间会成功匹配，否则不能成功匹配。示例工程代码如 ch17-2/ch17-2-3-gateway 所示。

1）自定义 RouteLocator 的路由配置代码，如代码清单 17-8 所示。

代码清单17-8 ch17-2/ch17-2-3-gateway/src/main/java/cn/springcloud/book/gateway/SCGatewayApplication.java

```java
@Bean
public RouteLocator customRouteLocator(RouteLocatorBuilder builder) {
    ZonedDateTime datetime1 = LocalDateTime.now().minusDays(1).atZone(ZoneId.
        systemDefault());
    ZonedDateTime datetime2 = LocalDateTime.now().plusDays(1).atZone(ZoneId.
        system-Default());
    return builder.routes()
```

```
            .route("between_route", r -> r.between(datetime1,datetime2))
            .uri("http://baidu.com")).build();
}
```

等价的 application.yml 中的路由配置如下所示:
```
spring:
  cloud:
    gateway:
      routes:
      - id: between_route
        uri: http://baidu.com
        predicates:
        - name: Between
          args:
            datetime1: 2018-03-15T00:02:48.513+08:00[Asia/Shanghai]
            datetime2: 2018-03-15T02:02:48.516+08:00[Asia/Shanghai]
```

2）启动应用 ch17-2-3-gateway，访问 http://localhost:8080/ 会转发到 https://www.baidu.com 对应的网站首页。

17.4.4　Cookie 路由断言工厂

Cookie 路由断言工厂会取两个参数——cookie 名称对应的 key 和 value。当请求中携带的 cookie 和 Cookied 断言工厂中配置的 cookie 一致，则路由匹配成功，否则匹配不成功。示例工程代码如 ch17-2/ch17-2-4-gateway 所示。

1）自定义 RouteLocator 的路由配置如代码清单 17-9 所示。

代码清单17-9　ch17-2/ch17-2-4-gateway/src/main/java/cn/springcloud/book/gateway/SpringCloudGatewayApplication.java

```java
@Bean
public RouteLocator customRouteLocator(RouteLocatorBuilder builder) {
    return builder.routes()
            .route("cookie_route", r -> r.cookie("chocolate", "ch.p")
            .uri("http://localhost:8071/test/cookie"))
            .build();
}
```

等价的 application.yml 文件的路由配置信息如下所示:
```
spring:
  cloud:
    gateway:
      routes:
        - id: cookie_route
          uri: http://localhost:8071/test/cookie
          predicates:
          - Cookie=chocolate, ch.p
```

 其中 http://localhost:8071/test/cookie 对应的源服务工程为 ch17-2-service。

2）先后启动 ch17-2-4-gateway 和 ch17-2-service 工程，使用 postman 设置请求结果如图 17-5 所示。

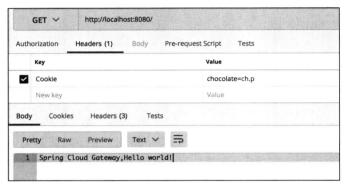

图 17-5　Cookie 路由断言处理结果

3）当设置为 Cookie 为 chocolate=ch.p，可以成功将请求转发到 ch17-2-service 的 TestController 中，通过 debug 调试，可以看到 Cookie 就是请求转发携带而来的 Cookie，如图 17-6 所示。

图 17-6　Debug 调试查看携带的 cookie 信息

4）当设置 chocolate=xujin 时，Cookie 断言工厂没有匹配到路由，返回 404 提示信息，如图 17-7 所示。

17.4.5　Header 路由断言工厂

Header 路由断言工厂用于根据配置的路由 header 信息进行断言匹配路由，匹配成功进行转发，否则不进行转发。示例工程代码如 ch17-2/ch17-2-5-gateway 所示。

1）自定义 RouteLocator 的路由配置如代码清单所示 17-10。

第 17 章　Spring Cloud Gateway 上篇　◆　409

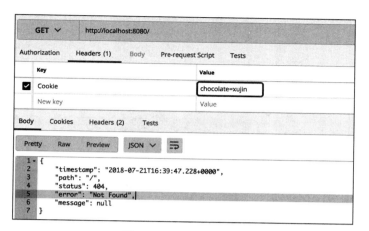

图 17-7　404 错误信息

代码清单17-10　ch17-2/ch17-2-5-gateway/src/main/java/cn/springcloud/book/gateway/SpringCloudGatewayApplication.java

```
@Bean
public RouteLocator customRouteLocator(RouteLocatorBuilder builder) {
    return builder.routes()
            .route("header_route", r -> r.header("X-Request-Id", "xujin")
                    .uri("http://localhost:8071/test/head"))
            .build();
}
```

等价的 application.yml 文件的路由配置信息如下所示：

```
spring:
  cloud:
    gateway:
      routes:
      - id: header_route
        uri: http://localhost:8071/test/head
        predicates:
        - Header=X-Request-Id, xujin
```

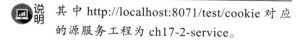

说明　其中 http://localhost:8071/test/cookie 对应的源服务工程为 ch17-2-service。

2）启动应用 ch17-2-service 和 ch17-2-5-gateway 测试。

打开 postman 访问 http://localhost:8080 设置正确的 header，测试结果如图 17-8 所示，设置错误的 Header 测试结果如图 17-9 所示。

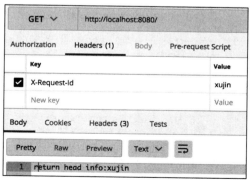

图 17-8　设置正确 Header 返回正常结果

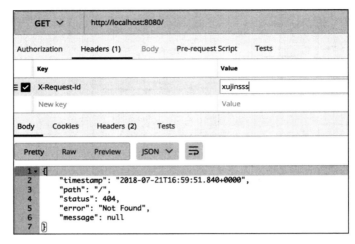

图 17-9　设置错误的 Header 返回 404

17.4.6　Host 路由断言工厂

Host 路由断言工厂根据配置的 Host，对请求中的 Host 进行断言处理，断言成功则进行路由转发，否则不转发。案例工程如 ch17-2-6-gateway 所示。

1）自定义 RouteLocator 的路由配置如代码清单 17-11 所示。

代码清单17-11　ch17-2/ch17-2-6-gateway/src/main/java/cn/springcloud/book/gateway/ ScGatewayApplication.java

```
@Bean
public RouteLocator customRouteLocator(RouteLocatorBuilder builder) {
    return builder.routes()
            .route("host_route", r -> r.host("**.baidu.com:8080")
                    .uri("http://jd.com"))
            .build();
}
```

等价的 application.yml 文件的路由配置信息如下所示：

```
spring:
  cloud:
    gateway:
      routes:
      - id: host_route
        uri: http://jd.com
        predicates:
        - Host=**.baidu.com:8080
```

> 说明　配置主机名的时候，如果 gateway 的端口为 80 则把 80 端口省略，如果 Gateway 有端口，如上所示需要配置 Host=**.baidu.com:8080 需要加上对应的端口。

2）通过 SwitchHosts 设置本地的二级域名 vip.baidu.com 来解析到本地 ip 127.0.0.1，如图 17-10

所示。

打开浏览器访问 http://vip.baidu.com:8080/，将会转发到网址 https://www.jd.com/ 对应的网站首页。

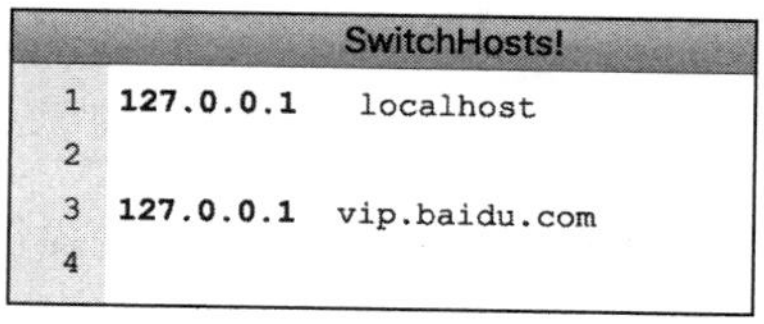

17.4.7 Method 路由断言工厂

Method 路由断言工厂会根据路由信息配置的 method 对请求方法是 Get 或者 Post 等进行断言匹配，匹配成功则进行转发，否则处理失败。案例工程代码如 ch17-2-7-gateway 所示。

图 17-10 通过 SwitchHosts 配置本地二级域名解析

1）自定义 RouteLocator 的路由配置，如代码清单 17-12 所示。

代码清单17-12 ch17-2/ch17-2-7-gateway/src/main/java/cn/springcloud/book/gateway/SpringCloudGatewayApplication.java

```
@Bean
public RouteLocator customRouteLocator(RouteLocatorBuilder builder) {
    return builder.routes()
            .route("method_route", r -> r.method("GET")
                    .uri("http://jd.com"))
            .build();
}
```

等价的 application.yml 文件的路由配置信息如下所示：

```
spring:
  cloud:
    gateway:
      routes:
      - id: method_route
        uri: http://jd.com
        predicates:
        - Method=GET
```

2）启用应用 ch17-2-7-gateway，访问 http://localhost:8080，由于是 Get 方法请求，因此可以成功转发到 https://www.jd.com/ 对应的首页。

17.4.8 Query 路由断言工厂

Query 路由断言工厂会从请求中获取两个参数，将请求中参数和 Query 断言路由中的配置进行匹配，比如 http://localhost:8080/?foo=baz 中的 foo=baz 和下面的 r.query("foo","baz") 配置一致则转发成功，否则转发失败。案例可以参考 ch17-2-8-gateway 中的代码。

1）使用 Java 代码自定义 RouteLocator 的配置，如代码清单 17-13 所示。

代码清单17-13 ch17-2/ch17-2-8-gateway/src/main/java/cn/springcloud/book/gateway/SpringCloudGatewayApplication.java

```
@Bean
public RouteLocator customRouteLocator(RouteLocatorBuilder builder) {
    return builder.routes()
```

```
            .route("query_route", r -> r.query("foo","baz")
                    .uri("http://baidu.com"))
                .build();
}
```

Spring Cloud Gateway 也提供属性文件的配置方式，application.yml 中的配置如下所示：

```
spring:
    cloud:
        gateway:
            routes:
            - id: query_route
                uri: http://baidu.com
                predicates:
                - Query=foo, baz
```

2）启动 ch17-2-8-gateway 应用，访问 http://localhost:8080/?foo=baz 成功转发到对应 https://www.baidu.com/?foo=baz 对应的百度首页。

17.4.9　RemoteAddr 路由断言工厂

RemoteAddr 路由断言工厂配置一个 IPv4 或 IPv6 网段的字符串或者 IP。当请求 IP 地址在网段之内或者和配置的 IP 相同，则表示匹配成功，成功转发，否则不能转发。例如 192.1680.1/16 是一个网段，其中 192.1680.1 是 IP 地址，16 是子网掩码，当然也可以直接配置一个 IP。案例代码可以参考 ch17-2/ch17-2-9-gateway 工程。

1）使用 Java 流式 API 的路由配置，如代码清单 17-14 所示。

代码清单17-14　ch17-2/ch17-2-9-gateway/src/main/java/cn/springcloud/book/gateway/SpringCloudGatewayApplication.java

```
@Bean
public RouteLocator customRouteLocator(RouteLocatorBuilder builder) {
return builder.routes()
        .route("remoteaddr_route", r -> r.remoteAddr("127.0.0.1"))
            .uri("http://baidu.com"))
            .build();
}
```

在 application.yml 中的等价的路由示例配置如下所示：

```
spring:
    cloud:
        gateway:
            routes:
            - id: remoteaddr_route
                uri: http://baidu.com
                predicates:
                - RemoteAddr=127.0.0.1
```

2）启动 ch17-2-9-gateway 中的主程序，访问 http://127.0.0.1:8080/，结果转发到 https://www.baidu.com/ 对应的首页。

17.5 Spring Cloud Gateway 的内置 Filter

Spring Cloud Gatewat 中内置很多的路由过滤工厂，当然可以自己根据实际应用场景需要定制的自己的路由过滤器工厂。路由过滤器允许以某种方式修改请求进来的 http 请求或返回的 http 响应。路由过滤器主要作用于需要处理的特定路由。Spring Cloud Gateway 提供了很多种类的过滤器工厂，过滤器的实现类将近二十多个。总得来说，可以分为七类：Header、Parameter、Path、Status、Redirect 跳转、Hytrix 熔断和 RateLimiter。下面介绍一下 Spring Cloud Gateway 中的 Filter 工厂。

17.5.1 AddRequestHeader 过滤器工厂

AddRequestHeader 过滤器工厂用于对匹配上的请求加上 header，案例如 ch17-3-1-gateway 所示。

1）使用 Java 流式 API 配置路由的配置，如代码清单 17-15 所示。

代码清单17-15　ch17-3/ch17-3-1-gateway/src/main/java/cn/springcloud/book/gateway/SpringCloudGatewayApplication.java

```
@Bean
public RouteLocator testRouteLocator(RouteLocatorBuilder builder) {
    return builder.routes()
            .route("add_request_header_route", r ->
                    r.path("/test").filters(f -> f.addRequestHeader("X-Request-
                        Acme", "ValueB"))
                            .uri("http://localhost:8071/test/head"))
            .build();
}
```

2）分别启动 ch17-3-1-gateway 和 ch17-3-service 应用。其中 ch17-3-service 用于验证是否把 header 添加到源服务 ch17-3-service，并将其返回。访问 http://localhost:8080/test 返回已经添加的 header 信息，如图 17-11 所示。

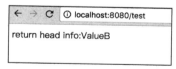

图 17-11　过滤器添加 Header 的返回结果

17.5.2 AddRequestParameter 过滤器

AddRequestParameter 过滤器作用是对匹配上的请求路由添加请求参数，案例工程如 ch17-3-2-gateway 所示。

1）使用 Java 代码配置 AddRequestParameter 过滤工厂配置，如代码清单 17-16 所示。

代码清单17-16　ch17-3/ch17-3-2-gateway/src/main/java/cn/springcloud/book/gateway/SCGatewayApplication.java

```
@Bean
public RouteLocator testRouteLocator(RouteLocatorBuilder builder) {
return builder.routes()
        .route("add_request_parameter_route", r ->
            r.path("/addRequestParameter").filters(f -> f.addRequestParameter
```

```
                ("example", "ValueB"))
            .uri("http://localhost:8071/test/addRequestParameter"))
            .build();
}
```

2）分别启动 ch17-3-2-gateway 和 ch17-3-service 应用。其中 ch17-3-service 用于将添加的请求参数返回。打开浏览器访问：http://localhost:8080/addRequestParameter，返回结果如图 17-12 所示。

图 17-12　Filter 添加请求参数的返回结果

17.5.3　RewritePath 过滤器

Spring Cloud Gateway 可以使用 RewritePath 替换 Zuul 的 StripPrefix 功能，而且功能更强大。在 Zuul 中使用如下的配置：

```
zuul:
    routes:
        demo:
            sensitiveHeaders: Access-Control-Allow-Origin,Access-Control-Allow-Methods
            path: /demo/**
            stripPrefix: true
            url: http://demo.com/
```

说明：这里的 stripPrefix 默认为 true，也就是所有 /demo/xxxx 的请求转发给 http://demo.com/xxxx，去除掉 demo 前缀。

Spring Cloud Gateway 实现类似的功能，使用的是 RewritePath 过滤器工厂。案例工程如 ch17-3-3-gateway 所示：

1）使用 Java 代码配置 RewritePath 过滤器工厂的代码，如代码清单 17-17 所示。

代码清单17-17　ch17-3/ch17-3-3-gateway/src/main/java/cn/springcloud/book/gateway/SCGatewayApplication.java

```
@Bean
public RouteLocator testRouteLocator(RouteLocatorBuilder builder) {
return builder.routes()
        .route("rewritepath_route", r ->
        r.path("/foo/**").filters(f -> f.rewritePath("/foo/(?<segment>.*)","/$\\
            {segment}"))
        .uri("http://www.baidu.com"))
        .build();
}
```

2）启动 ch17-3-3-gateway 应用，访问 http://localhost:8080/foo/cache/sethelp/help.html，路由转发结果如图 17-13 所示。

 说明　这里相当于把 foo 前缀去掉，直接访问 https://www.baidu.com/cache/sethelp/help.html，如图 17-14 所示。

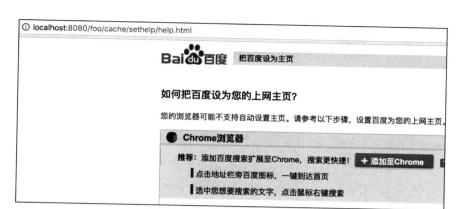

图 17-13　RewritePath 转发结果

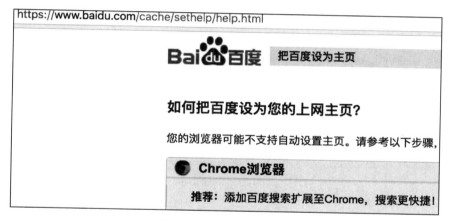

图 17-14　直接转发结果

17.5.4　AddResponseHeader 过滤器

AddResponseHeader 过滤器工厂的作用是对从网关返回的响应添加 Header。案例工程如 ch17-3-4-gateway 所示。

1）使用 Java 代码配置 AddResponseHeader 过滤器工厂的示例代码，如代码清单 17-18 所示。

代码清单17-18　ch17-3/ch17-3-4-gateway/src/main/java/cn/springcloud/book/gateway/SCGatewayApplication.java

```
@Bean
public RouteLocator testRouteLocator(RouteLocatorBuilder builder) {
return builder.routes()
        .route("add_request_header_route", r ->
        r.path("/test").filters(f -> f.addResponseHeader("X-Response-Foo", "Bar"))
        .uri("http://www.baidu.com"))
        .build();
}
```

2）启动工程 ch17-3-4-gateway，打开浏览器并开启控制台切换到 Network，访问 http://localhost:8080/test，添加的 Response Header 如图 17-15 所示。

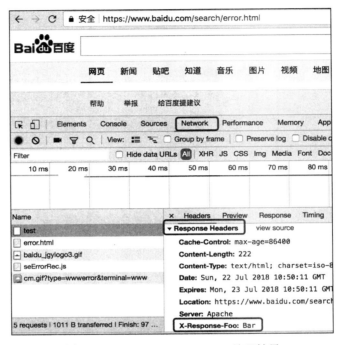

图 17-15　AddResponseHeader 处理结果

17.5.5　StripPrefix 过滤器

StripPrefixGatewayFilterFactory 是一个对针对请求 url 前缀进行处理的 filter 工厂，用于去除前缀。而 PrefixPathGatewayFilterFactory 是用于增加前缀。StripPrefixGatewayFilterFactory 的案例工程如 ch17-3-5-gateway 所示。

1）去除前缀的路由配置如代码清单 17-19 所示。

代码清单17-19　ch17-3/ch17-3-5-gateway/src/main/resources/application.yml

```yml
spring:
  cloud:
    gateway:
      routes:
      - id: baidu_route
        uri: http://www.baidu.com
        predicates:
        - Path=/baidu/test/**
        filters:
        - StripPrefix=2
```

2）启动 ch17-3-5-gateway 应用，访问 http://localhost:8080/baidu/test，可以成功去除前

缀 /baidu/test/，转到 https://www.baidu.com 对应的百度首页。

17.5.6 Retry 过滤器

网关作为所有请求流量的入口，网关对路由进行协议适配和协议转发处理的过程中，如果出现异常或网络抖动，为了保证后端服务请求的高可用，一般处理方式会对网络请求进行重试。下面通过案例来讲解怎么使用 Spring Cloud Gateway 中的重试 Filter 进行重试。案例代码如 ch17-3-6-gateway 所示。

1）新建 maven 工程 ch17-3-6-gateway，主要 Maven 依赖如代码清单 17-20 所示。

代码清单17-20　ch17-3/ch17-3-6-gateway/pom.xml

```
<dependencies>
    <dependency>
        <groupId>org.springframework.cloud</groupId>
        <artifactId>spring-cloud-starter-gateway</artifactId>
    </dependency>
</dependencies>
```

2）通过 Java 流式 API 的方式配置路由，如代码清单所示 17-21 所示。

代码清单17-21　ch17-3/ch17-3-6-gateway/src/main/java/cn/springcloud/book/gateway/CH1736GatewayApplication.java

```
@Bean
public RouteLocator retryRouteLocator(RouteLocatorBuilder builder) {
    return builder.routes()
            .route("retry_route", r -> r.path("/test/retry")
                    .filters(f ->f.retry(config -> config.setRetries(2).setStatuses
                            (HttpStatus.INTERNAL_SERVER_ERROR)))
                    .uri("http://localhost:8071/retry?key=abc&count=2"))
            .build();
}
```

说明　config.setRetries(2).setStatuses(HttpStatus.INTERNAL_SERVER_ERROR) 表示设置重试次数为两次，当代理服务调用失败时设置返回的状态码为 500，即服务器内部错误。

3）编写代理服务模拟请求处理异常。

在案例工程 ch17-3-service 中添加后端请求重试对应的服务，代码如代码清单所示 17-22 所示。

代码清单17-22　ch17-3/ch17-3-service/src/main/java/cn/springcloud/book/eureka/controller/TestController.java

```
ConcurrentHashMap<String, AtomicInteger> map = new ConcurrentHashMap<>();
@GetMapping("/retry")
public String testRetryByException(@RequestParam("key") String key, @RequestParam
        (name = "count") int count) {
    AtomicInteger num = map.computeIfAbsent(key, s -> new AtomicInteger());
    //对请求或重试次数计数
    int i = num.incrementAndGet();
```

```
        log.warn("重试次数: "+i);
        //计数i小于重试次数2抛出异常,让Spring Cloud Gateway进行重试
        if (i < count) {
            throw new RuntimeException("Deal with failure, please try again!");
        }
        //当重试两次时候,清空计数,返回重试两次成功
        map.clear();
        return "重试"+count+"次成功! ";
    }
```

说明 当 ch17-3-service 启动完毕之后,对 Spring Cloud Gateway 提供的重试请求服务地址为 http://localhost:8071/retry?key=abc&count=2,其中的 key=abc 主要用于作为 Concurrent-HashMap 的 key 统计计数,count 为 2 表示 Spring Cloud Gateway 的重试次数为 2 次。

4)启动测试。

按照顺序依次启动 ch17-3-service 和 ch17-3-6-gateway 应用,当应用启动完毕之后,打开浏览器访问网关提供的访问 URL:http://localhost:8080/test/retry,只需要请求 1 次即可,浏览器返回提示信息重试 2 次成功!此时在 ch17-3-service 工程的控台打印日志抛出异常并进行 2 次重试,如图 17-16 所示。

```
[nio-8071-exec-1] o.a.c.c.C.[Tomcat].[localhost].[/]        : Initializing Spring FrameworkServlet 'dispatcherServlet'
[nio-8071-exec-1] o.s.web.servlet.DispatcherServlet         : FrameworkServlet 'dispatcherServlet': initialization started
[nio-8071-exec-1] o.s.web.servlet.DispatcherServlet         : FrameworkServlet 'dispatcherServlet': initialization completed in 13 ms
[nio-8071-exec-1] c.s.b.eureka.controller.TestController    : 重试次数: 1
[nio-8071-exec-1] o.a.c.c.C.[.[.[/].[dispatcherServlet]     : Servlet.service() for servlet [dispatcherServlet] in context with path [] thr
failure, please try again!
    oller.TestController.testRetryByException(TestController.java:59) ~[classes!/:na] <14 internal calls>
ervice(HttpServlet.java:635) ~[tomcat-embed-core-8.5.28.jar:8.5.28] <1 internal calls>
ervice(HttpServlet.java:742) ~[tomcat-embed-core-8.5.28.jar:8.5.28]
tionFilterChain.internalDoFilter(ApplicationFilterChain.java:231) ~[tomcat-embed-core-8.5.28.jar:8.5.28]
```

图 17-16 重试 2 次打印日志信息

17.5.7 Hystrix 过滤器

在前面的章节介绍了什么是熔断器 Hystrix 以及怎么使用,Hystrix 可以提供熔断、自我保护、服务降级和快速失败等功能。Spring Cloud Gateway 对 Hystrix 进行集成提供路由层面的服务熔断和降级,最简单的使用场景是当通过 Spring Cloud Gateway 调用后端服务,后端服务一直出现异常、服务不可用的状态。此时为了提高用户体验,就需要对服务降级,返回友好的提示信息,在保护网关自身可用的同时保护后端服务高可用。下面通过案例介绍怎么使用 Spring Cloud Gateway 中内置的 Hytrix 过滤器,下面介绍一下主要的实现步骤,案例如源码工程 ch17-3-7-gateway 所示。

1. 创建 Maven 工程 ch17-3-7-gateway

由于使用了 Hystrix,所以除了配置 Spring Cloud Gateway 的核心 Maven 依赖之外,还需要配置 Hystrix 的依赖到 Maven 工程中,代码如代码清单 17-23 所示。

代码清单17-23   ch17-3/ch17-3-7-gateway/pom.xml

```
<dependency>
```

```xml
        <groupId>org.springframework.cloud</groupId>
        <artifactId>spring-cloud-starter-gateway</artifactId>
</dependency>
<dependency>
        <groupId>org.springframework.cloud</groupId>
        <artifactId>spring-cloud-starter-netflix-hystrix</artifactId>
</dependency>
```

2. 在 application.yml 中配置 Hystrix Filter 的路由配置，如代码清单 17-24 所示。

代码清单17-24　　ch17-3/ch17-3-7-gateway/src/main/resources/application.yml

```yaml
spring:
  cloud:
    gateway:
      routes:
      - id: prefix_route
        uri: http://localhost:8701/test/Hystrix?isSleep=true
        predicates:
        - Path=/test/Hystrix
        filters:
        - name: Hystrix # Hystrix Filter的名称
          args: # Hystrix配置参数
            name: fallbackcmd #HystrixCommand的名字
            fallbackUri: forward:/fallback #fallback对应的uri

#Hystrix的fallbackcmd的时间
hystrix.command.fallbackcmd.execution.isolation.thread.timeoutInMilliseconds: 5000
```

 http://localhost:8701/test/Hystrix?isSleep=true 是后端服务需要路由熔断处理对应的 URL。

3. 创建 fallbackUri

在 maven 工程 ch17-3-7-gateway 中创建 FallbackController 作为对应的 fallbackUri，代码如代码清单 17-25 所示。

代码清单17-25　　ch17-3/ch17-3-7-gateway/src/main/java/cn/springcloud/book/gateway/FallbackController.java

```java
@RestController
public class FallbackController {
    @GetMapping("/fallback")
    public String fallback() {
        return "Spring Cloud Gateway Fallback! ";
    }
}
```

4. 创建需要 Fallback 的后端服务

在 maven 工程 ch17-3-service 中编写需要 Fallback 的后端服务，代码如代码清单 17-26 所示。

代码清单17-26　　ch17-3/ch17-3-service/src/main/java/cn/springcloud/book/eureka/controller/TestController.java

```java
@GetMapping("/test/Hystrix/")
```

```
public String index(@RequestParam("isSleep") boolean isSleep) throws Interr-
   uptedException {
   log.info("issleep is " + isSleep);
   //isSleep为true开始睡眠,睡眠时间大于Gateway中的fallback设置的时间
   if (isSleep) {
           TimeUnit.MINUTES.sleep(10);
   }
   return "No Sleep";
}
```

5. 启动应用测试

按照顺序依次启动应用 ch17-3-service 和 ch17-3-7-gateway 启动完毕之后打开浏览器访问 http://localhost:8080/test/Hystrix,由于后端服务睡眠并且睡眠时间大于 Gateway 中的 fallback 设置的时间,因此 Spring Cloud Gateway 直接 fallback 到配置的 URI,如图 17-17 所示。

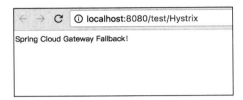

图 17-17　路由熔断处理结果

17.6　本章小结

本章首先介绍了什么是 Spring Cloud Gateway 以及工作原理,然后通过案例驱动的方式介绍了 Spring Cloud Gateway 中常用的路由断言工厂和常用的内置 Filter,其中重点介绍了重试 Filter 和 Hystrix Filter。

第 18 章 Spring Cloud Gateway 下篇

在第 17 章介绍了什么是 Spring Cloud Gateway 以及 Spring Cloud Gateway 的工作原理，还介绍了常见的路由断言工厂和 Filter 工厂，本章将在前面章节的基础上针对 Spring Cloud 的 Finchley 版本，对 Spring Cloud Gateway 进行深度实战剖析。

18.1 Gateway 基于服务发现的路由规则

18.1.1 Gateway 的服务发现路由概述

在前面介绍 Zuul 的章节中，我们知道 Spring Cloud 对 Zuul 进行封装处理之后，当通过 Zuul 访问后端微服务时，基于服务发现的默认路由规则是 :http://zuul_host:zuul_port/ 微服务在 Eureka 上的 serviceId/**，Spring Cloud Gateway 在设计的时候考虑从 Zuul 迁移到 Gateway 的兼容性和迁移成本等，Gateway 基于服务发现的路由规则和 Zuul 的设计类似，但是也有很大差别。但是 Spring Cloud Gateway 基于服务发现的路由规则，在不同注册中心下其差异如下：

- ❑ 如果把 Gateway 注册到 Eureka 上，通过网关转发服务调用，访问网关的 URL 是 http://Gateway_HOST:Gateway_PORT/ 大写的 serviceId/*，其中服务名默认必须是大写，否则会抛 404 错误，如果服务名要用小写访问，可以在属性配置文件里面加 spring.cloud.gateway.discovery.locator.lowerCaseServiceId=true 配置解决。
- ❑ 如果把 Gateway 注册到 Zookeeper 上，通过网关转发服务调用，服务名默认小写，因此不需要做任何处理。
- ❑ 如果把 Gateway 注册到 Consul 上，通过网关转发服务调用，服务名默认小写，也不需要做人为修改。

18.1.2 服务发现的路由规则案例

前面小节介绍了 Spring Cloud Gateway 基于服务发现的路由规则，下面用 Eureka 作为注册中心来剖析服务发现的路由规则。案例工程模块规划如表 18-1 所示，案例的聚合工程如 ch18-1 所示。

表 18-1 案例工程模块规划

模　　块	端口	说　　明
ch18-1	N/A	案例聚合父 pom 工程
ch18-1-consumer	8000	服务消费者
ch18-1-eureka	8761	Eureka 注册中心
ch18-1-gateway	9000	基于 Spring Cloud Gateway 的网关 Server
ch18-1-provider	8001	服务提供者

说明：由于 ch18-1-eureka 为 Eureka Server 注册中心，比较简单，前面章节已做详细介绍，在本案例中，将不再阐述。

1. 创建 Spring Cloud Gateway Server

1）创建 Maven 工程 ch18-1-gateway，Maven 的主要依赖如代码清单 18-1 所示。

代码清单18-1　ch18-1/ch18-1-gateway/pom.xml

```xml
<dependencies>
    <!-- Spring Cloud Gateway的依赖-->
    <dependency>
        <groupId>org.springframework.cloud</groupId>
        <artifactId>spring-cloud-starter-gateway</artifactId>
    </dependency>

    <!-- 由于需要把gateway注册到Eureka上，所以需要引入eureka-client-->
    <dependency>
        <groupId>org.springframework.cloud</groupId>
        <artifactId>spring-cloud-starter-netflix-eureka-client</artifactId>
    </dependency>
</dependencies>
```

2）创建 application.yml 文件，应用配置如代码清单 18-2 所示。

代码清单18-2　ch18-1/ch18-1-gateway/src/main/resources/application.yml

```
spring:
    application:
        name:sc-gateway-server
    cloud:
        gateway:
            discovery:
                locator:
                    enabled:true
                    lowerCaseServiceId:true
```

```yaml
server:
    port:9000  #网关服务监听9000端口
eureka:
    client:
        service-url:#指定注册中心的地址,以便使用服务发现功能
            defaultZone: http://localhost:8761/eureka/
logging:
    level:  #调整相关包的log级别,以便排查问题
        org.springframework.cloud.gateway:debug
```

主要配置说明:

- spring.cloud.gateway.discovery.locator.enabled:是否与服务发现组件进行结合,通过 serviceId 转发到具体的服务实例。默认为 false,若为 true 便开启基于服务发现的路由规则。
- spring.cloud.gateway.discovery.locator.lowerCaseServiceId=true:当注册中心为 Eureka 时,设置为 true 表示开启用小写的 serviceId 进行基于服务路由的转发。

3)创建 Spring Cloud Gateway 的主入口程序,如代码清单 18-3 所示。

代码清单18-3 ch18-1/ch18-1-gateway/src/main/java/cn/springcloud/book/GatewayServer-Application.java

```java
@SpringBootApplication
public class GatewayServerApplication {

    public static void main(String[] args) {
        SpringApplication.run(GatewayServerApplication.class, args);
    }

}
```

2. 创建服务提供者

创建服务提供者工程 ch18-1-provider,详细代码如工程源码所示。服务提供者提供简单的服务,如代码清单 18-4 所示。

代码清单18-4 ch18-1/ch18-1-provider/src/main/java/cn/springcloud/book/controller/HelloController.java

```java
@RestController
public class HelloController {

    @GetMapping("/hello")
    public String hello(@RequestParam String name) {
        return "Hello, " + name + "!";
    }
}
```

3. 创建服务消费者

创建服务消费者工程 ch18-1-consumer,服务消费者工程通过编写 Feign 的方式完成服务

消费，服务消费调用的主要代码如代码清单 18-5 所示。

代码清单18-5 ch18-1/ch18-1-consumer/src/main/java/cn/springcloud/book/controller/HelloController.java

```java
@RequestMapping("/hello")
@RestController
public class HelloController {

    @Autowired
    HelloFeignService helloRemote;

    @GetMapping("/{name}")
    public String index(@PathVariable("name") String name) {
        return helloRemote.hello(name) + "\n" + new Date().toString();
    }
}
```

4. 启动应用测试

依次启动 ch18-1-eureka、ch18-1-provider、ch18-1-consumer、ch18-1-gateway 应用。打开浏览器访问网关应用 ch18-1-gateway，当访问 http://localhost:9000/sc-consumer/hello/zhangsan 的时候，可以成功进行转发调用服务提供者，如图 18-1 所示。

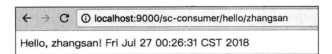

图 18-1　网关转发请求结果

从请求的控制台日志打印信息中，可以看出 Spring Cloud Gateway 自动地为服务消费者创建了类似下面的路由信息，然后自动转发调用服务提供者。

```
routes:
 - id:CompositeDiscoveryClient_SC-CONSUMER
     uri:lb://SC-CONSUMER
     order:0
     predicates:
        - Path=/SC-CONSUMER/**
     filters:
        - RewritePath=/SC-CONSUMER/(?<segment>.*),/$\{segment}
```

当 application.yml 中的 spring.cloud.gateway.discovery.locator.lowerCaseServiceId 为 false 或者不进行配置时，使用 Eureka 作为服务注册中心，访问 http://localhost:9000/sc-consumer/hello/zhangsan，会出现如下错误消息：

```
Failed to handle request [GET http://localhost:9000/sc-consumer/hello/zhangsan]: 
    Response status 404
```

报错的原因是当使用 Eureka 作为服务注册中心时，Spring Cloud Gateway 基于服务发现的默认路由规则是 http://Gateway_HOST:Gateway_PORT/ 大写的 serviceId/**，此时如果要通过网关进行正常的服务转发,需要把对应的 ServiceId 变为大写,然后进行访问,如图 18-2 所示。

图 18-2　ServiceId 为大写的转发结果

18.2　Gateway Filter 和 Global Filter

上面小节介绍了 Spring Cloud Gateway 基于服务发现默认的路由规则。本节将介绍 Spring Cloud Gateway 中 GatewayFilter 和 GlobalFilter 的区别和联系,以及如何自定义并使用 Gateway Filter 和 Global Filter。

18.2.1　Gateway Filter 和 Global Filter 概述

Spring Cloud Gateway 中的 Filter 从接口实现上分为两种:一种是 Gateway Filter,另外一种是 Global Filter。那要怎么理解这两种 Filter 的设计呢?通过和 Gateway 的核心设计者进行设计交流,我对这两者有了更深刻的了解。下面我们介绍一下 Gateway Filter 和 Global Filter 之间的区别和联系。

1. Gateway Filter 概述

Gateway Filter 是从 Web Filter 中复制过来的,相当于一个 Filter 过滤器,可以对访问的 URL 过滤,进行横切处理(切面处理),应用场景包括超时、安全等。

2. Global Filter 概述

Spring Cloud Gateway 定义了 Global Filter 的接口,让我们可以自定义实现自己的 Global Filter。Global Filter 是一个全局的 Filter,作用于所有路由。

3. Gateway Filter 和 Global Filter 的区别

从路由的作用范围来看,Global filter 会被应用到所有的路由上,而 Gateway filter 则应用到单个路由或者一个分组的路由上。从源码设计来看,Gateway Filter 和 Global Filter 两个接口中定义的方法一样都是 Mono filter(),唯一的区别就是 GatewayFilter 继承了 ShortcutConfigurable,而 GlobalFilter 没有任何继承。

18.2.2　自定义 Gateway Filter 案例

下面我们通过自定义 Gateway Filter 对路由转发的耗时进行统计,来学习如何自定义 Gateway Filter。

1. 创建自定义的 Gateway Filter

创建自定义的 Gateway Filter,如代码清单 18-6 所示。

代码清单18-6 ch18-2/ch18-2-gateway/src/main/java/cn/springcloud/book/filter/
CustomGatewayFilter.java

```java
/**
 * 统计某个或者某种路由的处理时长
 * @author xujin
 */
public class CustomGatewayFilter implements GatewayFilter, Ordered {
    private static final Log log = LogFactory.getLog(GatewayFilter.class);
    private static final String COUNT_Start_TIME = "countStartTime";

    @Override
    public Mono<Void> filter(ServerWebExchange exchange, GatewayFilterChain chain) {
        exchange.getAttributes().put(COUNT_Start_TIME, System.currentTimeMillis());
        return chain.filter(exchange).then(
            Mono.fromRunnable(() -> {
                Long startTime = exchange.getAttribute(COUNT_Start_TIME);
                Long endTime=(System.currentTimeMillis() - startTime);
                if (startTime != null) {
                    log.info(exchange.getRequest().getURI().getRawPath() + ": " +
                        endTime + "ms");
                }
            })
        );
    }

    @Override
    public int getOrder() {
        return Ordered.LOWEST_PRECEDENCE;
    }
}
```

2. 将 Gateway Filter 配置到路由上

由于 Gateway Filter 是作用于路由上，需要使用 Java 的流式 API 绑定 Gateway Filter 和路由，如代码清单 18-7 所示。

代码清单18-7： ch18-2/ch18-2-gateway/src/main/java/cn/springcloud/book/GatewayServer-
Application.java

```java
@Bean
public RouteLocator customerRouteLocator(RouteLocatorBuilder builder) {
    return builder.routes()
        .route(r -> r.path("/test")
            .filters(f -> f.filter(new CustomGatewayFilter()))
            .uri("http://localhost:8001/customFilter?name=xujin")
            .order(0)
            .id("custom_filter")
        ).build();
}
```

3. 启动测试

分别启动 ch18-2-gateway、ch18-2-provider 应用，打开浏览器访问 urlhttp://localhost:9000/

test?authToken=token，debug 断点调试代码，可以查看网关对请求的处理及耗时统计情况，如图 18-3 所示。

```
public class CustomGatewayFilter implements GatewayFilter, Ordered {
    private static final Log log = LogFactory.getLog(GatewayFilter.class);
    private static final String COUNT_Start_TIME = "countStartTime";

    @Override
    public Mono<Void> filter(ServerWebExchange exchange, GatewayFilterChain chain) { exchan
        exchange.getAttributes().put(COUNT_Start_TIME, System.currentTimeMillis());
        return chain.filter(exchange).then(
            Mono.fromRunnable(() -> {
                Long startTime = exchange.getAttribute(COUNT_Start_TIME);   startTime: 153
                Long endTime=(System.currentTimeMillis() - startTime);   endTime: 6986
                if (startTime != null) {  startTime: 1532695064275
                    log.info( exchange.getRequest().getURI().getRawPath() + ": " + endT
                })
            })
```

图 18-3 Debug 代码调试

当把 debug 断点调试放行之后，可以看到控制台打印请求的处理耗时如下所示：

[ctor-http-nio-5] o.s.cloud.gateway.filter.GatewayFilter   : /test: 6986ms

18.2.3 自定义 Global Filter 案例

下面通过简单定义一个名为 AuthSignatureFilter 的全局过滤器，对请求到网关的 URL 进行权限校验，判断请求的 URL 是否是合法请求。全局过滤器处理的逻辑是通过从 Gateway 的上下文 ServerWebExchange 对象中获取 authToken 对应的值进行判 Null 处理，读者可以根据需求定制开发更多复杂的校验逻辑。

1. 创建自定义的 Global Filter

创建代码如代码清单 18-8 所示：

代码清单18-8 ch18-2/ch18-2-gateway/src/main/java/cn/springcloud/book/filter/AuthSignatureFilter.java

```java
@Component
public class AuthSignatureFilter implements GlobalFilter, Ordered {

    @Override
    public Mono<Void> filter(ServerWebExchange exchange, GatewayFilterChain chain) {
        String token = exchange.getRequest().getQueryParams().getFirst("authToken");
        if (null==token || token.isEmpty()) {
            //当请求不携带Token或者token为空时，直接设置请求状态码为401，返回
            exchange.getResponse().setStatusCode(HttpStatus.UNAUTHORIZED);
            return exchange.getResponse().setComplete();
        }
        return chain.filter(exchange);
    }

    @Override
    public int getOrder() {
        return -400;
    }
}
```

2. 启动相关应用进行测试

分别启动 ch18-2-gateway、ch18-2-provider 应用，打开浏览器访问 URL http://localhost:9000/test，由于请求的 URL 中不携带 Token，请求直接返回 401 错误，如图 18-4 所示。

图 18-4 不带 Token 访问处理结果

当访问带 Token 的 URL（http://localhost:9000/test?authToken=token）时，能够显示正常的调用结果，如图 18-5 所示。

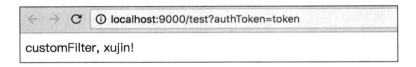

图 18-5 带 Token 的正常处理结果

18.3 Spring Cloud Gateway 实战

18.3.1 Spring Cloud Gateway 权重路由

WeightRoutePredicateFactory 是一个路由断言工厂，本节将使用 Spring Cloud Gateway 中的 WeightRoutePredicateFactory 对 URL 进行权重路由。

1. 权重路由的使用场景

在开发、测试的时候，或者线上发布、线上服务多版本控制的时候，需要对服务进行权重路由。最常见的使用场景就是一个服务有两个版本：旧版本 V1，新版本 V2。在线上灰度的时候，需要通过网关动态实时推送路由权重信息。比如 95% 的流量走服务 V1 版本，5% 的流量走服务 V2 版本。如图 18-6 所示。

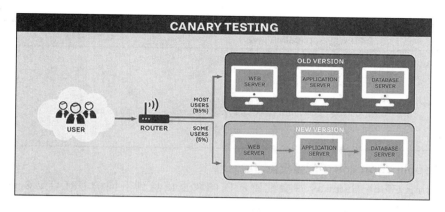

图 18-6　金丝雀灰度测试

2. 权重路由案例

Spring Cloud Gateway 会根据权重路由规则，针对特定的服务，把 95% 的请求流量分发给服务的 V1 版本，把剩余 5% 的流量分发给服务的 V2 版本，进行权重路由。

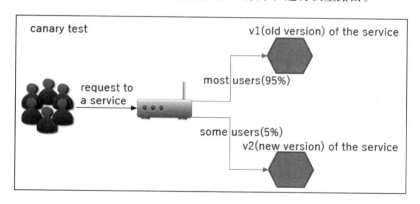

图 18-7　权重路由案例

1. 新建 Maven 工程 ch18-3-gateway

Spring Cloud Gateway 主要的 Maven 依赖如代码清单 18-9 所示。

代码清单18-9　ch18-3/ch18-3-gateway/pom.xml

```
<dependency>
    <groupId>org.springframework.cloud</groupId>
    <artifactId>spring-cloud-starter-gateway</artifactId>
</dependency>
```

2. 编写网关应用启动的入口主程序

Spring Cloud Gateway 应用的入口主程序代码，如代码清单 18-10 所示。

代码清单18-10　ch18-3/ch18-3-gateway/src/main/java/cn/springcloud/book/gateway/
GatewayApplication.java

```java
@SpringBootApplication
public class GatewayApplication {

    public static void main(String[] args) {
        SpringApplication.run(GatewayApplication.class, args);
    }
}
```

3. 在 application.yml 中配置权重路由信息

在 Spring Cloud Gateway 中会配置不同的权重信息到不同的 URL 上，Spring Cloud Gateway 会根据配置的路由权重信息，将请求分发到不同的源服务组上，权重配置信息如代码清单 18-11 所示。

代码清单18-11　ch18-3/ch18-3-gateway/src/main/resources/application.yml

```yaml
spring:
  cloud:
    gateway:
      routes:
      - id: service1_v1
        uri: http://localhost:8081/v1
        predicates:
        - Path=/test
        - Weight=service1, 95
      - id: service1_v2
        uri: http://localhost:8081/v2
        predicates:
        - Path=/test
        - Weight=service1, 5
```

说明：其中的 Weight=service1，95 和 Weight=service1，5 就是路由的权重信息。

4. 新建 Maven 工程 ch18-3-provider

ch18-3-provider 工程是一个 Spring Web 工程，主要提供两个版本的服务接口，用于接收 Spring Cloud Gateway 按权重路由转发的请求并返回结果。主要代码如代码清单 18-12 所示。

代码清单18-12　ch18-3/ch18-3-provider/src/main/java/cn/springcloud/book/service/
ServiceController.java

```java
@RestController
public class ServiceController {
    @GetMapping(value = "/v1")
    public String v1() {
        return "v1";
    }
    @GetMapping(value = "/v2")
    public String v2() {
        return "v2";
    }
}
```

5. 启动测试

分别启动 ch18-3-gateway、ch18-3-provider，访问 http://localhost:8080/test，多访问几次，会发现按权重信息返回的请求结果如图 18-8 和图 18-9 所示。

图 18-8 权重路由到 v1 版本

图 18-9 权重路由到 v2 版本

18.3.2　Spring Cloud Gateway 中 Https 的使用技巧

安全且需要大型互联网应用的生产环境基本是全站 Https。常规的做法是通过 Nginx 来配置 SSL 证书。如果使用 Spring Cloud Gateway 作为 API 网关，统管所有 API 请求的入口和出口，此时 Spring Cloud Gateway 就需要支持 Https。由于 Spring Cloud Gateway 是基于 Spring Boot 2.0 构建的，所以只需要将生成的 https 证书放到 Spring Cloud Gateway 应用的类路径下面即可。Spring Cloud Gateway 中使用 Https 的聚合工程说明，如表 18-2 所示。

表 18-2　Gateway 中使用 Https 聚合案例说明

模　　块	端口	说　　明
ch18-4	N/A	父模块聚合工程
ch18-4-eureka	8761	Eureka Server 注册中心，参考源码工程，本案例中不做过多介绍
ch18-4-gateway-https	8080	带有 Https 证书的网关 Server，使用 Https 协议访问
ch18-4-service-a	8071	后端代理服务 a，使用 Http 协议，用于负载均衡
ch18-4-service-b	8072	后端代理服务 b，使用 Http 协议，用于负载均衡

 说明　由于工程 ch18-4-eureka 和 ch18-4-service 相对比较简单，此处不做过多赘述，请读者参考源码工程学习。

1. 创建 Maven 模块工程 ch18-4-gateway-https

1）配置的 Gateway 的核心 Maven 依赖，如代码清单 18-13 所示。

代码清单18-13　ch18-4/ch18-4-gateway-https/pom.xml

```xml
<dependencies>
    <dependency>
        <groupId>org.springframework.cloud</groupId>
        <artifactId>spring-cloud-starter-gateway</artifactId>
    </dependency>
    <dependency>
        <groupId>org.springframework.cloud</groupId>
        <artifactId>spring-cloud-starter-netflix-eureka-client</artifactId>
    </dependency>
</dependencies>
```

2）创建 application.yml 并配置应用信息。在 application.yml 中配置使用 Https，具体配置信息如代码清单 18-14 所示。

代码清单18-14　ch18-4/ch18-4-gateway-https/src/main/resources/application.yml

```yaml
server:
ssl:
    key-alias: spring
    enabled: true
    key-password: spring
    key-store: classpath:selfsigned.jks
    key-store-type: JKS
    key-store-provider: SUN
    key-store-password: spring
```

关于如何生成 SSL 证书在这里就不再赘述，本案例提供一个已经生成好的 https 证书，放在应用的类路径下，如图 18-10 所示。

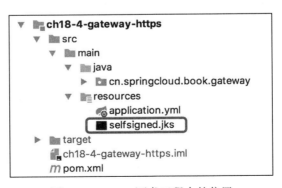

图 18-10　https 证书工程中的位置

2. 创建代理服务工程

创建 Maven 工程 ch18-4-service-a 和 ch18-4-service-b，用于接收 Spring Cloud Gateway 的转发请求，如代码清单 18-15 所示。

代码清单18-15　ch18-4/ch18-4-service-a/src/main/java/cn/springcloud/book/eureka/controller/TestController.java

```
@RestController
public class TestController {

    @GetMapping("/test")
    public String prefixpath(){
        return "https to Http";
    }
}
```

3. 启动应用测试

分别启动 ch18-4-eureka、ch18-4-service-a、ch18-4-service-b、ch18-4-gateway-https 应用进行测试。访问 https://localhost:8080/service-provider/test，浏览器报错，如图 18-11 所示。

图 18-11　Gateway 集成 https 访问报错

Gateway 的控制台报错消息如下：

```
io.netty.handler.codec.DecoderException: io.netty.handler.ssl.NotSslRecordException:
    not an SSL/TLS record: 485454502f312e3120343030200d0a5472616e736665722d456e
    636f64696e673a206368756e6b65640d0a446174653a204672692c203237204a756c2032303
    1382031343a32303a333920474d540d0a436f6e6e656374696f6e3a20636c6f73650d0a0d0a
    300d0a0d0a
    at io.netty.handler.codec.ByteToMessageDecoder.callDecode(ByteToMessageDecoder.
        java:459) ~[netty-codec-4.1.23.Final.jar:4.1.23.Final]
```

如图 18-11 所示，这个错误出现的原因是通过 Spring Cloud Gateway 请求进来的协议是 Https，而后端被代理的服务是 http 协议的请求，所以 Gateway 用 Https 请求转发调用 Http 协议的服务就会出现 "not an SSL/TLS record" 的错误。本质上是一个 Spring Cloud Gateway 将 Https 请求转发调用 Http 服务的问题。

由于服务的拆分，在微服务的应用集群中会存在很多服务提供者和服务消费者，而这些服务提供者和服务消费者基本都是部署在企业内网中，没必要全部加 Https 进行调用。因此 Spring Cloud Gateway 对外的请求是 Https，对后端代理服务的请求是 Http。通过 Debug 调试源码分析，LoadBalancerClientFilter#filter() 方法如下：

```
URI uri = exchange.getRequest().getURI();
String overrideScheme = null;
if (schemePrefix != null) {
```

```
            overrideScheme = url.getScheme();
    }

    URI requestUrl = loadBalancer.reconstructURI(new DelegatingServiceInstance(inst
        ance, overrideScheme), uri);

    log.trace("LoadBalancerClientFilter url chosen: " + requestUrl);
    exchange.getAttributes().put(GATEWAY_REQUEST_URL_ATTR, requestUrl);
```

从上面的代码可以看出，loadBalancer 对 Http 请求进行封装，如果从 Spring Cloud Gateway 进来的请求是 Https，它就用 Https 封装，如果是 Http 就用 Http 封装，而且没有预留任何扩展修改的接口，只能通过自定义 Global Filter 的方式对其修改。下面介绍两种修改方法，在实践中任选其中一种即可。

第一种修改方法：在 LoadBalancerClientFilter 执行之前将 Https 修改为 Http，如代码清单 18-16 所示。

代码清单18-16　ch18-4/ch18-4-gateway-https/src/main/java/cn/springcloud/book/gateway/filter/HttpsToHttpFilter.java

```java
/**
 * 在LoadBalancerClientFilter执行之前将Https修改为Http
 * https://github.com/spring-cloud/spring-cloud-gateway/issues/378
 */
@Component
public class HttpsToHttpFilter implements GlobalFilter, Ordered {

    private static final int HTTPS_TO_HTTP_FILTER_ORDER = 10099;

    @Override
    public Mono<Void> filter(ServerWebExchange exchange, GatewayFilterChain chain) {
        URI originalUri = exchange.getRequest().getURI();
        ServerHttpRequest request = exchange.getRequest();
        ServerHttpRequest.Builder mutate = request.mutate();
        String forwardedUri = request.getURI().toString();
        if (forwardedUri != null && forwardedUri.startsWith("https")) {
            try {
                URI mutatedUri = new URI("http",
                    originalUri.getUserInfo(),
                    originalUri.getHost(),
                    originalUri.getPort(),
                    originalUri.getPath(),
                    originalUri.getQuery(),
                    originalUri.getFragment());
                mutate.uri(mutatedUri);
            } catch (Exception e) {
                throw new IllegalStateException(e.getMessage(), e);
            }
        }
        ServerHttpRequest build = mutate.build();
        return chain.filter(exchange.mutate().request(build).build());
    }

    /**
```

```java
     * 由于LoadBalancerClientFilter的order是10100,
     * 要在LoadBalancerClientFilter执行之前将Https修改为Http,需要设置
     * order为10099
     * @return
     */
    @Override
    public int getOrder() {
        return HTTPS_TO_HTTP_FILTER_ORDER;
    }
}
```

第二种修改方法：在 LoadBalancerClientFilter 执行之后将 Https 修改为 http，拷贝 RibbonUtils 中的 upgradeConnection 方法自定义全局过滤器，如代码清单 18-17 所示。

代码清单18-17 ch18-4/ch18-4-gateway-https/src/main/java/cn/springcloud/book/gateway/filter/HttpSchemeFilter.java

```java
/**
 * 在LoadBalancerClientFilter执行之后将Https修改为http
 */
@Component
public class HttpSchemeFilter implements GlobalFilter, Ordered {

    private static final int HTTPS_TO_HTTP_FILTER_ORDER = 10101;
    @Override
    public Mono<Void> filter(ServerWebExchange exchange, GatewayFilterChain chain) {
        Object uriObj = exchange.getAttributes().get(GATEWAY_REQUEST_URL_ATTR);
        if (uriObj != null) {
            URI uri = (URI) uriObj;
            uri = this.upgradeConnection(uri, "http");
            exchange.getAttributes().put(GATEWAY_REQUEST_URL_ATTR, uri);
        }
        return chain.filter(exchange);
    }

    private URI upgradeConnection(URI uri, String scheme) {
        UriComponentsBuilder uriComponentsBuilder = UriComponentsBuilder.
            fromUri(uri).scheme(scheme);
        if (uri.getRawQuery() != null) {
            // When building the URI, UriComponentsBuilder verify the allowed
            //     characters and does not
            // support the '+' so we replace it for its equivalent '%20'.
            // See issue https://jira.spring.io/browse/SPR-10172
            uriComponentsBuilder.replaceQuery(uri.getRawQuery().replace("+", "%20"));
        }
        return uriComponentsBuilder.build(true).toUri();
    }

    /**
     * 由于LoadBalancerClientFilter的order是10100,
     * 所以设置HttpSchemeFilter的order是10101,
     * 在LoadBalancerClientFilter之后将https修改为http
     * @return
     */
    @Override
```

```
    public int getOrder() {
        return HTTPS_TO_HTTP_FILTER_ORDER;
    }
}
```

 说明　关于 Spring Cloud Gateway 中 Https 转 Http 的问题处理思路，可以参考扩展阅读下面两个项目的 issue，地址为 https://github.com/spring-cloud/spring-cloud-gateway/issues/160，https://github.com/spring-cloud/spring-cloud-gateway/issues/378。

重启应用 ch18-4-gateway-https，访问 https://localhost:8080/service-provider/test，会显示返回正确的结果，如图 18-12 所示。

图 18-12　Gateway 将 https 转 http 调用成功

18.3.3　Spring Cloud Gateway 集成 Swagger

Swagger 是一个可视化 API 测试工具，可以和应用完美融合。通过声明接口注解的方式，可以方便快捷地获取 API 调试界面进行测试。在前文中介绍了 Zuul 和 Swagger 的整合，由于 Spring Cloud Finchley 版是基于 Spring Boot 2.0 的，而 Spring Cloud Gateway 的底层是 WebFlux，且经验证，WebFlux 和 Swagger 不兼容。如果按照 Zuul 集成 Swagger 的方式，应用启动的时候会报错。下面通过案例介绍一下 Spring Cloud Gateway 如何集成 Swagger。聚合模块工程说明如表 18-3 所示。

表 18-3　聚合模块工程说明

模　　块	端口	说　　明
ch18-5	N/A	父模块聚合工程
ch18-5-eureka	8761	Eureka Server 注册中心，本案例中不做过多介绍
ch18-5-gateway	8081	带有 Https 证书的网关 Server，使用 Https 协议访问
ch18-5-service	8055	后端代理服务，提供加法计算服务

1. 创建 maven 模块工程 ch18-5-gateway

（1）将 Maven 核心依赖配置到 pom.xml 中

配置代码如代码清单 18-18 所示。

代码清单18-18　ch18-5/ch18-5-gateway/pom.xml

```xml
<dependencies>
    <dependency>
        <groupId>org.springframework.cloud</groupId>
        <artifactId>spring-cloud-starter-gateway</artifactId>
```

```xml
        </dependency>
        <dependency>
            <groupId>org.springframework.cloud</groupId>
            <artifactId>spring-cloud-starter-netflix-eureka-client</artifactId>
        </dependency>
        <dependency>
            <groupId>org.springframework.cloud</groupId>
            <artifactId>spring-cloud-starter-netflix-hystrix</artifactId>
        </dependency>
        <dependency>
            <groupId>io.springfox</groupId>
            <artifactId>springfox-swagger-ui</artifactId>
            <version>2.9.2</version>
        </dependency>
        <dependency>
            <groupId>io.springfox</groupId>
            <artifactId>springfox-swagger2</artifactId>
            <version>2.9.2</version>
        </dependency>
</dependencies>
```

（2）编写 SwaggerProvider 类

因为 Swagger 暂不支持 webflux 项目，所以不能在 Gateway 配置 SwaggerConfig，需要编写 GatewaySwaggerProvider 实现 SwaggerResourcesProvider 接口，用于获取 SwaggerResources，如代码清单 18-19 所示。

代码清单18-19 ch18-5/ch18-5-gateway/src/main/java/cn/springcloud/book/config/GatewaySwaggerProvider.java

```java
/**
 * @Primary注解的实例优先于其他实例被注入
 */
@Component
@Primary
public class GatewaySwaggerProvider implements SwaggerResourcesProvider {
    public static final String API_URI = "/v2/api-docs";
    private final RouteLocator routeLocator;
    private final GatewayProperties gatewayProperties;

    public GatewaySwaggerProvider(RouteLocator routeLocator, GatewayProperties
        gatewayProperties) {
        this.routeLocator = routeLocator;
        this.gatewayProperties = gatewayProperties;
    }

    @Override
    public List<SwaggerResource> get() {
        List<SwaggerResource> resources = new ArrayList<>();
        List<String> routes = new ArrayList<>();
        //取出Spring Cloud Gateway中的route
```

```
        routeLocator.getRoutes().subscribe(route -> routes.add(route.getId()));
        //结合application.yml中的路由配置，只获取有效的route节点
        gatewayProperties.getRoutes().stream().filter(routeDefinition -> routes.
            contains(routeDefinition.getId()))
            .forEach(routeDefinition -> routeDefinition.getPredicates().stream()
                .filter(predicateDefinition ->
                    ("Path").equalsIgnoreCase(predicateDefinition.getName()))
                .forEach(predicateDefinition ->
                    resources.add(swaggerResource(routeDefinition.getId(),
                        predicateDefinition.getArgs().get(NameUtils.GENERATED_NAME_
                    PREFIX + "0")
                        .replace("/**", API_URI)))));
        return resources;
    }

    private SwaggerResource swaggerResource(String name, String location) {
        SwaggerResource swaggerResource = new SwaggerResource();
        swaggerResource.setName(name);
        swaggerResource.setLocation(location);
        swaggerResource.setSwaggerVersion("2.0");
        return swaggerResource;
    }
}
```

（3）创建 Swagger-Resource 端点

因为没有在 Spring Cloud Gateway 中配置 SwaggerConfig，但是运行 Swagger-UI 的时候需要依赖一些接口，所以需要建立相应的 Swagger-Resource 端点，如代码清单 18-20 所示。

代码清单18-20　ch18-5/ch18-5-gateway/src/main/java/cn/springcloud/book/handler/SwaggerHandler.java

```
@RestController
@RequestMapping("/swagger-resources")
public class SwaggerHandler {
    @Autowired(required = false)
    private SecurityConfiguration securityConfiguration;
    @Autowired(required = false)
    private UiConfiguration uiConfiguration;
    private final SwaggerResourcesProvider swaggerResources;

    @Autowired
    public SwaggerHandler(SwaggerResourcesProvider swaggerResources) {
        this.swaggerResources = swaggerResources;
    }
    @GetMapping("/configuration/security")
    public Mono<ResponseEntity<SecurityConfiguration>> securityConfiguration() {
        return Mono.just(new ResponseEntity<>(
            Optional.ofNullable(securityConfiguration).orElse(SecurityConfigura
                tionBuilder.builder().build()), HttpStatus.OK));
    }

    @GetMapping("/configuration/ui")
    public Mono<ResponseEntity<UiConfiguration>> uiConfiguration() {
        return Mono.just(new ResponseEntity<>(
            Optional.ofNullable(uiConfiguration).orElse(UiConfigurationBuilder.
```

```
            builder().build()), HttpStatus.OK));
    }

    @GetMapping("")
    public Mono<ResponseEntity> swaggerResources() {
        return Mono.just((new ResponseEntity<>(swaggerResources.get(), HttpStatus.OK)));
    }
}
```

(4)创建 SwaggerHeaderFilter

在路由规则为 admin/test/{a}/{b} 时,Swagger 界面上会显示为 test/{a}/{b},缺少了 admin 这个路由节点。通过 Debug 断点调试发现,Swagger 会根据 X-Forwarded-Prefix 这个 Header 来获取 BasePath。因此需要将它添加到接口路径与 Host 之间才能正常工作。但是 Gateway 在做转发的时候并没有将这个 Header 添加到 Request 上,从而导致接口调试出现 404 错误。为了解决该问题,需要在 Gateway 中编写一个过滤器来添加这个 header。下面将会编写 GwSwaggerHeaderFilter 并继承 AbstractGatewayFilterFactory,如代码清单 18-21 所示。

代码清单18-21 ch18-5/ch18-5-gateway/src/main/java/cn/springcloud/book/filter/GwSwaggerHeaderFilter.java

```
@Component
public class GwSwaggerHeaderFilter extends AbstractGatewayFilterFactory {
    private static final String HEADER_NAME = "X-Forwarded-Prefix";
    @Override
    public GatewayFilter apply(Object config) {
        return (exchange, chain) -> {
            ServerHttpRequest request = exchange.getRequest();
            String path = request.getURI().getPath();
            if (!StringUtils.endsWithIgnoreCase(path, GatewaySwaggerProvider.API_URI)) {
                return chain.filter(exchange);
            }
            String basePath = path.substring(0, path.lastIndexOf(GatewaySwagger
                    Provider.API_URI));
            ServerHttpRequest newRequest = request.mutate().header(HEADER_NAME,
                    basePath).build();
            ServerWebExchange newExchange = exchange.mutate().request(newRequest).
                    build();
            return chain.filter(newExchange);
        };
    }
}
```

(5)配置 Gateway 的路由信息

配置代码如代码清单 18-22 所示。

代码清单18-22 ch18-5/ch18-5-gateway/src/main/resources/application.yml

```
spring:
  application:
    name: sc-gateway
  cloud:
```

```yaml
        gateway:
          locator:
            enabled: true
          routes:
          - id: sc-service
            uri: lb://sc-service
            predicates:
            - Path=/admin/**
            filters:
            - SwaggerHeaderFilter
            - StripPrefix=1
```

2. 创建 maven 模块工程 ch18-5-service

1）配置主要的 Maven 依赖，如代码清单 18-23 所示。

代码清单18-23　ch18-5/ch18-5-service/pom.xml

```xml
<dependencies>
    <dependency>
        <groupId>org.springframework.boot</groupId>
        <artifactId>spring-boot-starter-web</artifactId>
    </dependency>
    <dependency>
        <groupId>io.springfox</groupId>
        <artifactId>springfox-swagger2</artifactId>
        <version>2.9.2</version>
    </dependency>
    <dependency>
        <groupId>org.springframework.cloud</groupId>
        <artifactId>spring-cloud-starter-netflix-eureka-client</artifactId>
    </dependency>

    <dependency>
        <groupId>org.springframework.cloud</groupId>
        <artifactId>spring-cloud-starter-netflix-hystrix</artifactId>
    </dependency>
</dependencies>
```

2）配置 SwaggerConfig，如代码清单 18-24 所示。

代码清单18-24　ch18-5/ch18-5-service/src/main/java/cn/springcloud/book/config/SwaggerConfig.java

```java
@Configuration
@EnableSwagger2
public class SwaggerConfig {
    @Bean
    public Docket createRestApi() {

        return new Docket(DocumentationType.SWAGGER_2)
            .apiInfo(apiInfo())
            .select()
            .apis(RequestHandlerSelectors.withMethodAnnotation(ApiOperation.class))
            .paths(PathSelectors.any())
            .build();
    }
```

```
private ApiInfo apiInfo() {
    return new ApiInfoBuilder()
            .title("Swagger API")
            .description("test")
            .termsOfServiceUrl("")
            .contact(new Contact("Spring Cloud China", "http://springcloud.cn", ""))
            .version("2.0")
            .build();
}
```

3. 启动应用测试

依次按照顺序启动 ch18-5-eureka、ch18-5-service、ch18-5-gateway，启动完毕之后，打开浏览器访问 http://localhost:8081/swagger-ui.html，如图 18-13 所示。

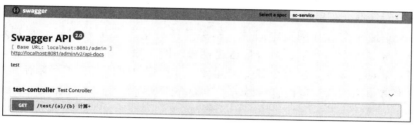

图 18-13　swagger-ui 主页面

打开对应的 URL，输入测试数据，如图 18-14 所示。

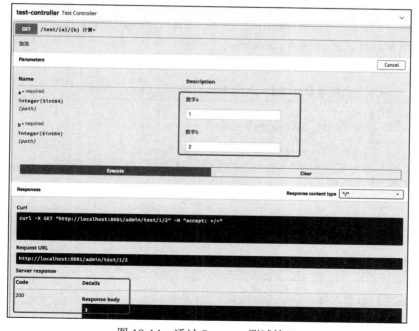

图 18-14　通过 Swagger 测试接口

18.3.4　Spring Cloud Gateway 限流

1. Spring Cloud Gateway 限流概述

在开发高并发系统时可以用三把利器来保护系统：缓存、降级和限流。缓存的目的是提升系统访问速度和增大系统处理的容量，是抗高并发流量的"银弹"；而降级是当服务出现问题或者影响到核心流程时，需要暂时将其屏蔽掉，待高峰过去之后或者问题解决后再打开；而有些场景并不能用缓存和降级来解决，比如稀缺资源（秒杀、抢购）、写服务（如评论、下单）、频繁的复杂查询，因此需要有一种手段来限制这些场景的并发/请求量，即限流。

限流的目的是通过对并发访问/请求进行限速或者对一个时间窗口内的请求进行限速来保护系统，一旦达到限制速率则可以拒绝服务（定向到错误页或友好的展示页）、排队或等待（比如秒杀、评论、下单等场景）、降级（返回兜底数据或默认数据）。

一般的中间件都会有单机限流框架，支持两种限流模式：控制速率和控制并发。Spring Cloud Gateway 是一个 API 网关中间件，网关是所有请求流量的入口。特别是像天猫双十一、双十二等高并发场景下，当流量迅速剧增，网关除了要保护自身之外，还要限流保护后端应用。常见的限流算法有令牌桶和漏桶，计数器也可以进行粗暴限流实现。对限流算法，这里不再阐述，读者可以自行学习。下面介绍如何在 Spring Cloud Gateway 中实现限流。

2. 自定义过滤器实现限流

在第 17 章介绍了很多过滤器，在 Spring Cloud Gateway 中实现限流比较简单，只需要编写一个过滤器就可以。Guava 中的 RateLimiter、Bucket4j、RateLimitJ 都是基于令牌桶算法实现的限流工具。下面介绍使用 Bucket4j 在 Spring Cloud Gateway 中实现限流。

（1）创建 Maven 工程 ch18-6-1-gateway

配置 Gateway 的主要 pom 依赖，如代码清单 18-25 所示。

代码清单18-25　ch18-6/ch18-6-1-gateway/pom.xml

```xml
<dependencies>
    <!-- Spring Cloud Gateway的依赖-->
    <dependency>
        <groupId>org.springframework.cloud</groupId>
        <artifactId>spring-cloud-starter-gateway</artifactId>
    </dependency>
    <!-- Bucket4j限流依赖-->
    <dependency>
        <groupId>com.github.vladimir-bukhtoyarov</groupId>
        <artifactId>bucket4j-core</artifactId>
        <version>4.0.0</version>
    </dependency>
</dependencies>
```

（2）自定义过滤器对特定资源进行限流

编写 GatewayRateLimitFilterByIp 并实现 GatewayFilter、Ordered 接口，如代码清单 18-26 所示。

代码清单18-26 ch18-6/ch18-6-1-gateway/src/main/java/cn/springcloud/book/filter/
GatewayRateLimitFilterByIp.java

```java
/**
 * 自定义过滤器进行限流
 * @author xujin
 */
public class GatewayRateLimitFilterByIp implements GatewayFilter, Ordered {

    private final Logger log = LoggerFactory.getLogger(GatewayRateLimitFilterByIp.class);

    /**
     * 单机网关限流用一个ConcurrentHashMap来存储 bucket,
     * 如果是分布式集群限流的话,可以采用 Redis等分布式解决方案
     */
    private static final Map<String, Bucket> LOCAL_CACHE = new ConcurrentHashMap<>();

    /**
     * 桶的最大容量,即能装载 Token 的最大数量
     */
    int capacity;
    /**
     * 每次 Token 补充量
     */
    int refillTokens;
    /**
     *补充 Token 的时间间隔
     */
    Duration refillDuration;

    private Bucket createNewBucket() {
        Refill refill = Refill.of(refillTokens, refillDuration);
        Bandwidth limit = Bandwidth.classic(capacity, refill);
        return Bucket4j.builder().addLimit(limit).build();
    }

    @Override
    public Mono<Void> filter(ServerWebExchange exchange, GatewayFilterChain chain) {
        String ip = exchange.getRequest().getRemoteAddress().getAddress().
            getHostAddress();
        Bucket bucket = LOCAL_CACHE.computeIfAbsent(ip, k -> createNewBucket());
        log.debug("IP:{} ,令牌桶可用的Token数量:{} " ,ip,bucket.getAvailableTokens());
        if (bucket.tryConsume(1)) {
            return chain.filter(exchange);
        } else {
            //当可用的令牌书为0时,进行限流返回429状态码
            exchange.getResponse().setStatusCode(HttpStatus.TOO_MANY_REQUESTS);
            return exchange.getResponse().setComplete();
        }
    }

    @Override
    public int getOrder() {
        return -1000;
    }
}
```

```
//此处省略Get和Set方法等
}
```

（3）通过 Java 流式 API 的方式配置路由规则

配置作用于某个具体路由的限流 Filter，如代码清单 18-27 所示。

代码清单18-27 ch18-6/ch18-6-1-gateway/src/main/java/cn/springcloud/book/filter/GatewayRateLimitFilterByIp.java

```
@Bean
public RouteLocator customerRouteLocator(RouteLocatorBuilder builder) {
    return builder.routes()
        .route(r -> r.path("/test/rateLimit")
            .filters(f -> f.filter(new GatewayRateLimitFilterByIp(10,1,Durati
                on.ofSeconds(1))))
            .uri("http://localhost:8000/hello/rateLimit")
            .id("rateLimit_route")
        ).build();
}
```

说明：.uri（"http://localhost:8000/hello/rateLimit"）对应的是代理后端服务，比较简单，这里不再赘述，读者可自行参考源码工程 ch18-6-provider 学习。

（4）启动测试

打开浏览器访问 http://localhost:8080/test/rateLimit，可以看到返回结果，如图 18-15 所示。

图 18-15　限流之前的正常请求结果

加快速度多次访问 http://localhost:8080/test/rateLimit，可以看到控制台打印的可用令牌数逐渐减少，如图 18-16 所示。

```
c.s.b.filter.RateLimitByIpGatewayFilter : IP:0:0:0:0:0:0:0:1 ,令牌通可用的Token数量:9
c.s.b.filter.RateLimitByIpGatewayFilter : IP:0:0:0:0:0:0:0:1 ,令牌通可用的Token数量:8
c.s.b.filter.RateLimitByIpGatewayFilter : IP:0:0:0:0:0:0:0:1 ,令牌通可用的Token数量:7
c.s.b.filter.RateLimitByIpGatewayFilter : IP:0:0:0:0:0:0:0:1 ,令牌通可用的Token数量:7
c.s.b.filter.RateLimitByIpGatewayFilter : IP:0:0:0:0:0:0:0:1 ,令牌通可用的Token数量:6
c.s.b.filter.RateLimitByIpGatewayFilter : IP:0:0:0:0:0:0:0:1 ,令牌通可用的Token数量:6
c.s.b.filter.RateLimitByIpGatewayFilter : IP:0:0:0:0:0:0:0:1 ,令牌通可用的Token数量:5
c.s.b.filter.RateLimitByIpGatewayFilter : IP:0:0:0:0:0:0:0:1 ,令牌通可用的Token数量:5
c.s.b.filter.RateLimitByIpGatewayFilter : IP:0:0:0:0:0:0:0:1 ,令牌通可用的Token数量:4
c.s.b.filter.RateLimitByIpGatewayFilter : IP:0:0:0:0:0:0:0:1 ,令牌通可用的Token数量:4
c.s.b.filter.RateLimitByIpGatewayFilter : IP:0:0:0:0:0:0:0:1 ,令牌通可用的Token数量:3
c.s.b.filter.RateLimitByIpGatewayFilter : IP:0:0:0:0:0:0:0:1 ,令牌通可用的Token数量:3
c.s.b.filter.RateLimitByIpGatewayFilter : IP:0:0:0:0:0:0:0:1 ,令牌通可用的Token数量:2
c.s.b.filter.RateLimitByIpGatewayFilter : IP:0:0:0:0:0:0:0:1 ,令牌通可用的Token数量:2
c.s.b.filter.RateLimitByIpGatewayFilter : IP:0:0:0:0:0:0:0:1 ,令牌通可用的Token数量:1
c.s.b.filter.RateLimitByIpGatewayFilter : IP:0:0:0:0:0:0:0:1 ,令牌通可用的Token数量:1
c.s.b.filter.RateLimitByIpGatewayFilter : IP:0:0:0:0:0:0:0:1 ,令牌通可用的Token数量:0
c.s.b.filter.RateLimitByIpGatewayFilter : IP:0:0:0:0:0:0:0:1 ,令牌通可用的Token数量:0
```

图 18-16　可用令牌数减少

当可用的令牌数量为 0 时，Spring Cloud Gateway 中自定义的限流过滤器开始拒绝请求，直接返回 429 状态码（因为请求太多，限流返回 429 状态码），浏览器访问结果如图 18-17 所示。

图 18-17　网关限流拒绝请求

3. Gateway 内置过滤器工厂限流

在前面的小节是通过自定义限流过滤器实现限流的功能，其实 Spring Cloud Gateway 内置了一个名为 RequestRateLimiterGatewayFilterFactory 的过滤器工厂，可以用来直接限流。下面通过案例工程 ch18-6-2-gateway 来进行讲解说明。打开 RequestRateLimiter 过滤器工厂所在的 jar，如图 18-18 所示，可以看到使用名为 request_rate_limiter.lua 的 lua 脚本实现限流。

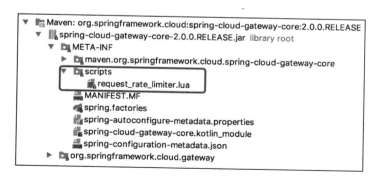

图 18-18　限流 Lua 脚本

（1）新建 Maven 工程 ch18-6-2-gateway

由于 RequestRateLimiterGatewayFilterFactory 的实现依赖 Redis，所以需要引入 spring-boot-starter-data-redis-reactive 相关的依赖，如代码清单 18-28 所示。

代码清单18-28　ch18-6/ch18-6-2-gateway/pom.xml

```xml
<dependencies>
    <dependency>
        <groupId>org.springframework.cloud</groupId>
        <artifactId>spring-cloud-starter-gateway</artifactId>
    </dependency>
    <dependency>
        <groupId>org.springframework.boot</groupId>
        <artifactId>spring-boot-starter-data-redis-reactive</artifactId>
    </dependency>
</dependencies>
```

（2）Gateway 相关限流配置

在 application.yml 中配置 Gateway 相关的限流配置，如代码清单 18-29 所示。

代码清单18-29　ch18-6/ch18-6-2-gateway/src/main/resources/application.yml

```yaml
spring:
  application:
      name: ch18-6-1-gateway
  redis:
        host: localhost
        port: 6379
  cloud:
    gateway:
      routes:
        - id: rateLimit_route
          uri: http://localhost:8000/hello/rateLimit
          order: 0
          predicates:
            - Path=/test/rateLimit
          filters:
              #filter名称必须是RequestRateLimiter
            - name: RequestRateLimiter
              args:
                  #使用SpEL按名称引用bean
                  key-resolver: "#{@remoteAddrKeyResolver}"
                  #允许用户每秒处理多少个请求
                  redis-rate-limiter.replenishRate: 1
                  #令牌桶的容量，允许在一秒钟内完成的最大请求数
                  redis-rate-limiter.burstCapacity: 5
```

　redis.host=localhost 和 redis.port=6379 分别配置了 Redis 的主机和端口，注意 Redis 需要读者自行安装，关于 Redis 的安装不做过多扩展。

（3）编写 key-resolver 对应的 remoteAddrKeyResolver

具体代码如代码清单 18-30 所示。

代码清单18-30　ch18-6/ch18-6-2-gateway/src/main/java/cn/springcloud/book/
RemoteAddrKeyResolver.java

```java
public class RemoteAddrKeyResolver implements KeyResolver {

    public static final String BEAN_NAME = "remoteAddrKeyResolver";
    @Override
```

```
    public Mono<String> resolve(ServerWebExchange exchange) {
        return Mono.just(exchange.getRequest().getRemoteAddress().getAddress().
            getHostAddress());
    }
}
```

把 Key 对应的解析器配置加载到 Spring 容器中，如代码清单 18-31 所示。

代码清单18-31 ch18-6/ch18-6-2-gateway/src/main/java/cn/springcloud/book/
GatewayApplication.java

```
@SpringBootApplication
public class GatewayApplication {
    @Bean(name = RemoteAddrKeyResolver.BEAN_NAME)
    public RemoteAddrKeyResolver remoteAddrKeyResolver() {
        return new RemoteAddrKeyResolver();
    }
    public static void main(String[] args) {
        SpringApplication.run(GatewayApplication.class, args);
    }
}
```

（4）启动 ch18-6-2-gateway 应用测试

启动 Redis，当 ch18-6-2-gateway 启动完毕之后，访问 http://localhost:8081/test/rateLimit，可以正常返回结果，如图 18-19 所示。

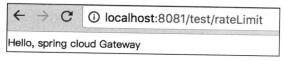

图 18-19　返回正常结果

当快速多次请求访问时，浏览器返回 429 状态码，如图 18-20 所示。

图 18-20　网关限流拒绝请求

4. 基于 CPU 的使用率进行限流

在实际项目应用中对网关进行限流时,需要参考的因素比较多。可能会根据网络请求连接数、请求流量、CPU 使用率等流控。可以通过 Spring Boot Actuator 提供的 Metrics 获取当前 CPU 的使用情况,当 CPU 使用率高于某个阈值就开启限流,否则不开启限流。

在 Actuator 1.x 里通过 SystemPublicMetrics 来获取 CPU 的使用情况,但是在 Actuator 2.x 里只能通过 MetricsEndpoint 来获取。详细案例代码如 ch18-6-3-gateway 所示。下面介绍主要的实现步骤。

1)自定义过滤器 Filter 并开启基于 CPU 使用情况的限流,代码如代码清单 18-32 所示。

代码清单18-32 ch18-6/ch18-6-3-gateway/src/main/java/cn/springcloud/book/filter/GatewayRateLimitFilterByCpu.java

```java
@Component
public class GatewayRateLimitFilterByCpu implements GatewayFilter, Ordered {

    private final Logger log = LoggerFactory.getLogger(GatewayRateLimitFilterByCpu.class);

    @Autowired
    private MetricsEndpoint metricsEndpoint;

    private static final String METRIC_NAME = "system.cpu.usage";

    private static final double MAX_USAGE = 0.50D;

    @Override
    public Mono<Void> filter(ServerWebExchange exchange, GatewayFilterChain chain) {
        //获取网关所在机器的CPU使用情况
        Double systemCpuUsage = metricsEndpoint.metric(METRIC_NAME, null)
            .getMeasurements()
            .stream()
            .filter(Objects::nonNull)
            .findFirst()
            .map(MetricsEndpoint.Sample::getValue)
            .filter(Double::isFinite)
            .orElse(0.0D);

        boolean isOpenRateLimit = systemCpuUsage >MAX_USAGE;
        log.debug("system.cpu.usage: {}, isOpenRateLimit:{} ",systemCpuUsage ,
            isOpenRateLimit);
        if (isOpenRateLimit) {
            //当CPU的使用超过设置的最大阈值开启限流
            exchange.getResponse().setStatusCode(HttpStatus.TOO_MANY_REQUESTS);
            return exchange.getResponse().setComplete();
        } else {
            return chain.filter(exchange);
        }
    }

    @Override
    public int getOrder() {
        return 0;
    }

}
```

2）通过 Java 流式 API 的方式配置过滤器作用于某个路由，如代码清单 18-33 所示。

代码清单18-33　ch18-6/ch18-6-3-gateway/src/main/java/cn/springcloud/book/GatewayApplication.java

```java
@SpringBootApplication
public class GatewayApplication {

    @Autowired
    private GatewayRateLimitFilterByCpu gatewayRateLimitFilterByCpu;

    @Bean
    public RouteLocator customerRouteLocator(RouteLocatorBuilder builder) {
        return builder.routes()
                .route(r -> r.path("/test/rateLimit")
                        .filters(f -> f.filter(gatewayRateLimitFilterByCpu))
                        .uri("http://localhost:8000/hello/rateLimit")
                        .id("rateLimit_route")
                ).build();
    }
    public static void main(String[] args) {
        SpringApplication.run(GatewayApplication.class, args);
    }
}
```

3）启动测试。

a）打开 postman 访问 http://localhost:8082/actuator/metrics/system.cpu.usage，CPU 的使用情况如图 18-21 所示。

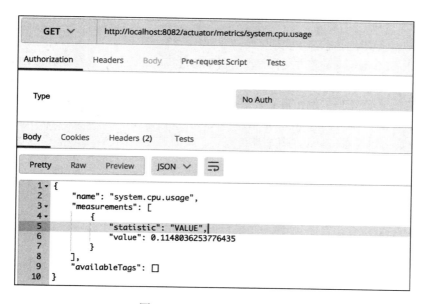

图 18-21　CPU 的使用情况

b）打开浏览器访问 http://localhost:8082/test/rateLimit，可以正常返回 "Hello，spring cloud Gateway"，控制台的日志打印信息如图 18-22 所示。

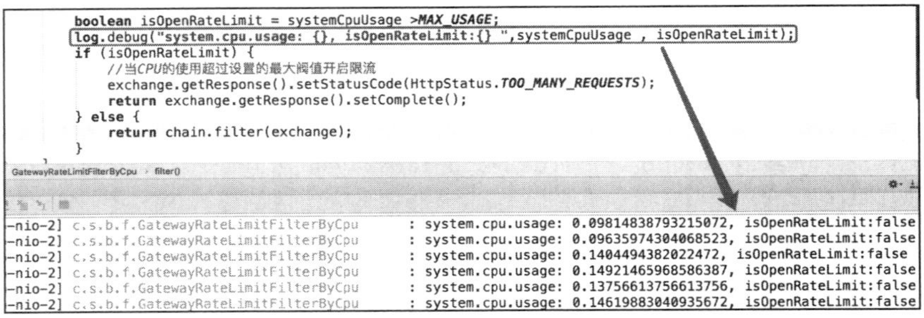

图 18-22　控制台打印 CPU 限流信息

18.3.5　Spring Cloud Gateway 的动态路由

网关中有两个重要的概念，那就是路由配置和路由规则。路由配置是指配置某请求路径路由到指定的目的地址。而路由规则是指匹配到路由配置之后，再根据路由规则进行转发处理。Spring Cloud Gateway 作为所有请求流量的入口，在实际生产环境中为了保证高可靠和高可用，以及尽量避免重启，需要实现 Spring Cloud Gateway 动态路由配置。前面章节介绍了 Spring Cloud Gateway 提供的配置路由规则的两种方法，但都是在 Spring Cloud Gateway 启动时将路由配置和规则加载到内存里，无法做到不重启网关就可以动态地对应路由的配置和规则进行增加、修改和删除。

1. Gateway 简单的动态路由实现

Spring Cloud Gateway 的官方文档并没有讲如何进行动态配置，查看 Spring Cloud Gateway 的源码，发现在 org.springframework.cloud.gateway.actuate.GatewayControllerEndpoint 类中提供了动态配置的 Rest 接口，但是需要开启 Gateway 的端点，而且其提供的功能不是很强大。通过参考与 GatewayControllerEndpoint 相关的代码，可以自己编码实际动态路由配置。下面通过案例来讲解如何实现 Gateway 的动态路由配置。案例工程如 ch18-7-gateway 所示。

（1）新建 Maven 工程 ch18-7-gateway

配置主要的核心依赖，如代码清单 18-34 所示。

代码清单18-34　ch18-7/ch18-7-gateway/pom.xml

```xml
<dependencies>
    <dependency>
        <groupId>org.springframework.cloud</groupId>
        <artifactId>spring-cloud-starter-gateway</artifactId>
    </dependency>
    <dependency>
        <groupId>org.springframework.boot</groupId>
        <artifactId>spring-boot-starter-webflux</artifactId>
```

```xml
        </dependency>
        <dependency>
            <groupId>org.springframework.boot</groupId>
            <artifactId>spring-boot-starter-actuator</artifactId>
        </dependency>
</dependencies>
```

(2)定义数据传输模型

根据 Spring Cloud Gateway 的路由模型定义数据传输模型,分别创建 GatewayRouteDefinition.java、GatewayPredicateDefinition.java 和 GatewayFilterDefinition.java 这三个类。

1)创建路由定义模型如代码清单 18-35 所示。

代码清单18-35 ch18-7/ch18-7-gateway/src/main/java/cn/springcloud/book/gateway/model/ GatewayRouteDefinition.java

```java
public class GatewayRouteDefinition {

    //路由的id
    private String id;

    //路由断言集合配置
    private List<GatewayPredicateDefinition> predicates = new ArrayList<>();

    //路由过滤器集合配置
    private List<GatewayFilterDefinition> filters = new ArrayList<>();

    //路由规则转发的目标uri
    private String uri;

    // 路由执行的顺序
    private int order = 0;
    //此处省略get和set方法
}
```

2)创建过滤器定义模型,代码如代码清单 18-36 所示。

代码清单18-36 ch18-7/ch18-7-gateway/src/main/java/cn/springcloud/book/gateway/model/ GatewayFilterDefinition.java

```java
public class GatewayFilterDefinition {

    //Filter Name
    private String name;

    //对应的路由规则
    private Map<String, String> args = new LinkedHashMap<>();

    //此处省略Get和Set方法
}
```

3)路由断言定义模型,代码如代码清单 18-37 所示。

代码清单18-37　ch18-7/ch18-7-gateway/src/main/java/cn/springcloud/book/gateway/model/
　　　　　　　GatewayPredicateDefinition.java

```java
public class GatewayPredicateDefinition {
    //断言对应的Name
    private String name;

    //配置的断言规则
    private Map<String, String> args = new LinkedHashMap<>();

    //此处省略Get和Set方法
}
```

（3）编写动态路由实现类

编写 DynamicRouteServiceImpl 并实现 ApplicationEventPublisherAware 接口，代码如代码清单18-38所示。

代码清单18-38　ch18-37/ch18-7-gateway/src/main/java/cn/springcloud/book/gateway/route/
　　　　　　　DynamicRouteServiceImpl.java

```java
@Service
public class DynamicRouteServiceImpl implements ApplicationEventPublisherAware {

    @Autowired
    private RouteDefinitionWriter routeDefinitionWriter;

    private ApplicationEventPublisher publisher;

    //增加路由
    public String add(RouteDefinition definition) {
        routeDefinitionWriter.save(Mono.just(definition)).subscribe();
        this.publisher.publishEvent(new RefreshRoutesEvent(this));
        return "success";
    }

    //更新路由
    public String update(RouteDefinition definition) {
        try {
            this.routeDefinitionWriter.delete(Mono.just(definition.getId()));
        } catch (Exception e) {
            return "update fail,not find route  routeId: "+definition.getId();
        }
        try {
            routeDefinitionWriter.save(Mono.just(definition)).subscribe();
            this.publisher.publishEvent(new RefreshRoutesEvent(this));
            return "success";
        } catch (Exception e) {
            return "update route  fail";
        }
    }

    //删除路由
    public Mono<ResponseEntity<Object>> delete(String id) {
        return this.routeDefinitionWriter.delete(Mono.just(id))
```

```
                .then(Mono.defer(() -> Mono.just(ResponseEntity.ok().build())))
                .onErrorResume(t -> t instanceof NotFoundException, t -> Mono.
                    just(ResponseEntity.notFound().build()));
    }

    @Override
    public void setApplicationEventPublisher(ApplicationEventPublisher
        applicationEventPublisher) {
        this.publisher = applicationEventPublisher;
    }

}
```

（4）编写 Rest 接口

编写 RouteController 类的提供 Rest 接口，用于动态路由配置，如代码清单 18-39 所示。

代码清单18-39　ch18-7/ch18-7-gateway/src/main/java/cn/springcloud/book/gateway/controller/RouteController.java

```
@RestController
@RequestMapping("/route")
public class RouteController {

    @Autowired
    private DynamicRouteServiceImpl dynamicRouteService;
    //增加路由
    @PostMapping("/add")
    public String add(@RequestBody GatewayRouteDefinition gwdefinition) {
        try {
            RouteDefinition definition = assembleRouteDefinition(gwdefinition);
            return this.dynamicRouteService.add(definition);
        } catch (Exception e) {
            e.printStackTrace();
        }
        return "succss";
    }

    //删除路由
    @DeleteMapping("/routes/{id}")
    public Mono<ResponseEntity<Object>> delete(@PathVariable String id) {
        return this.dynamicRouteService.delete(id);
    }

    //更新路由
    @PostMapping("/update")
    public String update(@RequestBody GatewayRouteDefinition gwdefinition) {
        RouteDefinition definition = assembleRouteDefinition(gwdefinition);
        return this.dynamicRouteService.update(definition);
    }
```

（5）配置 application.yml 文件

在 application.yml 文件配置应用的配置信息，并开启 Spring Cloud Gateway 对外提供的端点 Rest 接口，如代码清单 18-40 所示。

代码清单18-40　　ch18-7/ch18-7-gateway/src/main/resources/application.yml

```yaml
# 配置输出日志
logging:
    level:
        org.springframework.cloud.gateway: TRACE
        org.springframework.http.server.reactive: DEBUG
        org.springframework.web.reactive: DEBUG
        reactor.ipc.netty: DEBUG

#开启端点
management:
    endpoints:
        web:
            exposure:
                include: '*'
    security:
        enabled: false
```

（6）启动 ch18-7-gateway 应用测试

1）启动 ch18-7-gateway 应用之后，由于开启了端点，首先打开浏览器访问端点 URL：http://localhost:8080/actuator/gateway/routes，查看路由信息返回为空，如图 18-23 所示。

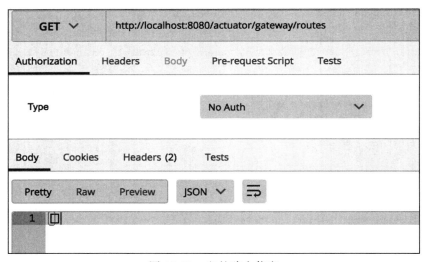

图 18-23　空的路由信息

2）打开 postman，访问 http://localhost:8080/route/add，发起 Post 请求，如图 18-24 所示，返回 success 说明向 Gateway 增加路由配置成功。

然后再打开 postman 访问 http://localhost:8080/actuator/gateway/routes，查看路由信息，返回结果如图 18-25 所示，可以看到已经添加的路由配置。

3）打开浏览器访问 http://localhost:8080/jd，可以正常转发 https://www.jd.com/ 对应的京东商城首页。

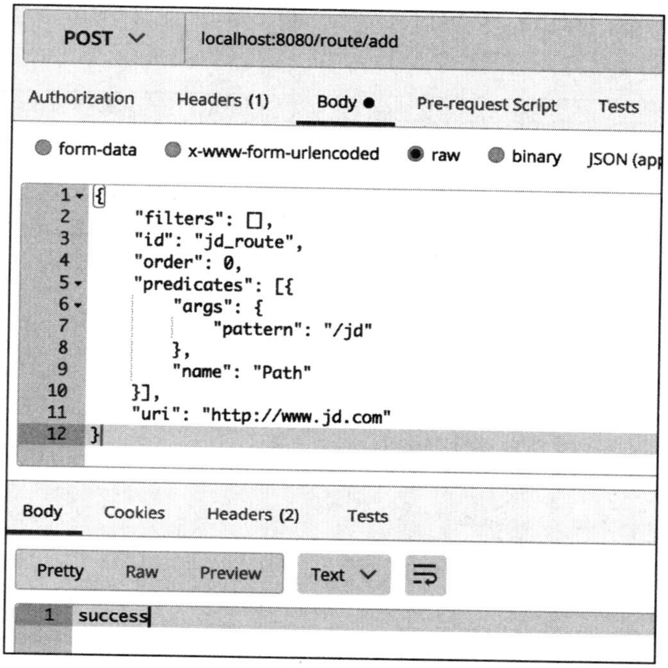

图 18-24　动态添加路由成功

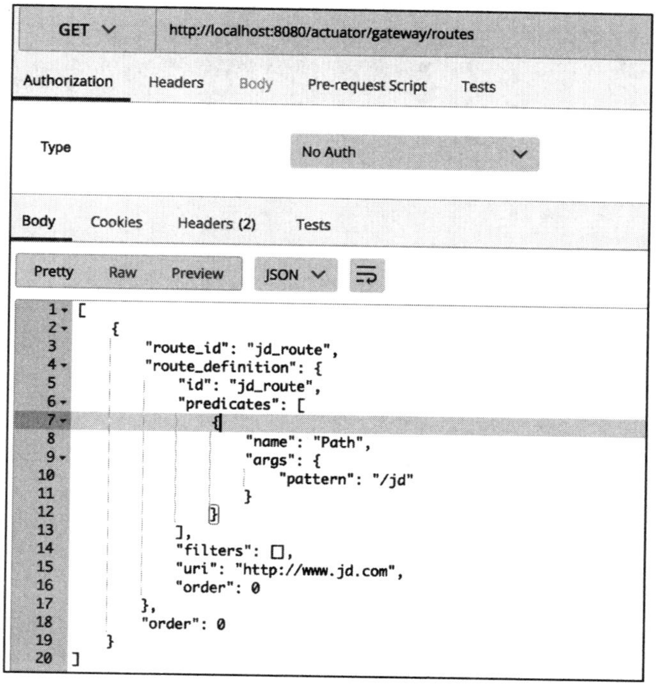

图 18-25　路由端点返回结果

4）通过访问 http://localhost:8080/route/update，对 id 为 jd_route 的路由更新配置，如图 18-26 所示。

图 18-26　更新路由配置

然后再访问路由端点 URL，发现路由配置已经被更新了，如图 18-27 所示。

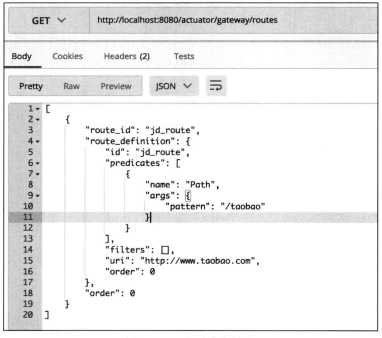

图 18-27　查看路由端点

然后通过浏览器访问 http://localhost:8080/taobao，可以成功转发到淘宝网。

5）访问 http://localhost:8080/route/delete/jd_route，其中的 id 为路由对应的 id，删除路由结果，如图 18-28 所示。

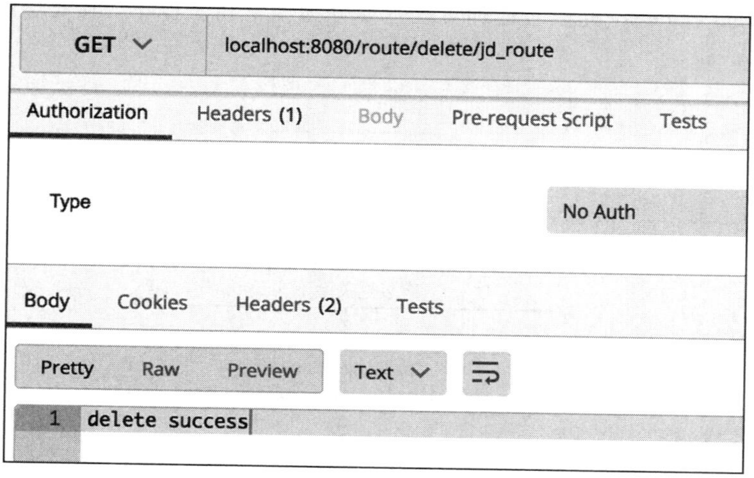

图 18-28　删除路由成功

2. Gateway 集群下的动态路由

前文通过 ch18-7-gateway 简单地实现了单机 Gateway 的动态路由，单机 Gateway 中的 Route 信息保存在当前实例的内存中，无法做到整个 Gateway 集群的动态路由修改。通过分析 Spring Cloud Gateway 源码可以发现，默认的 RouteDefinitionWriter 实现类是 InMemoryRouteDefinitionRepository，如图 18-29 所示。

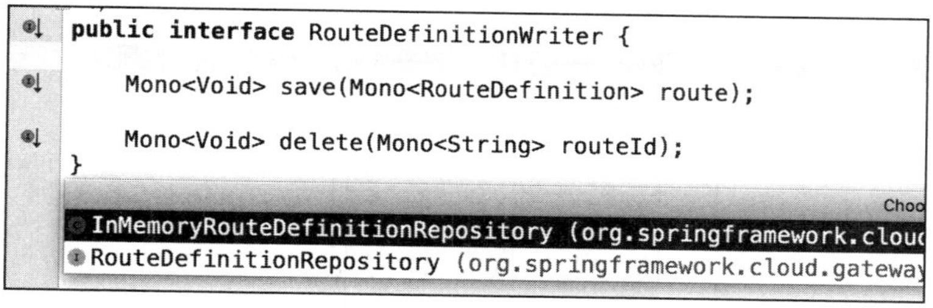

图 18-29　RouteDefinitionWriter 实现类

RouteDefinitionRepository 继承了 RouteDefinitionWriter，是 Spring Cloud Gateway 官方预留的接口，从而可以通过下面两种方式来实现集群下的动态路由修改：RouteDefinitionWriter 接口和 RouteDefinitionRepository 接口。在这里推荐实现 RouteDefinitionRepository 这个接口，从数据库或者从配置中心获取路由进行动态配置，可以参考上面单机版的动态路由实现，在这里不做过多阐述。

18.4 Spring Cloud Gateway 源码篇

18.4.1 Spring Cloud Gateway 的处理流程

Spring Cloud Gateway 的核心处理流程主要分为如下几个步骤，其处理流程如图 18-30 所示。

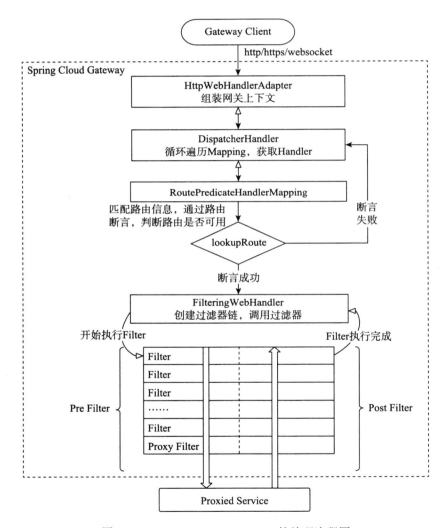

图 18-30　Spring Cloud Gateway 的处理流程图

- ❑ HttpWebHandlerAdapter：构建组装网关请求的上下文。
- ❑ DispatcherHandler：所有请求的分发处理器，负责分发请求到对应的处理器。

- RoutePredicateHandlerMapping：路由断言处理映射器，用于路由的查找，以及找到路由后返回对应的 WebHandler，DispatcherHandler 会依次遍历 HandlerMapping 集合进行处理。
- FilteringWebHandler：创建过滤器链，使用 Filter 链表处理请求。
RoutePredicateHandlerMapping 找到路由后返回对应的 FilteringWebHandler 后对请求进行处理，FilteringWebHandler 负责组装 Filter 链表并调用 Filter 处理请求。

18.4.2　Gateway 中 ServerWebExchange 构建分析

不管是 Zuul、Spring Cloud Gateway，还是基于 Netty 的自研网关，都会对请求进来的 Request，或者返回的 Response 进行包装，转换提取为网关运行的上下文，而在 Spring Cloud Gateway 中网关的上下文为 ServerWebExchange。当使用 Spring Cloud Gateway 进行开发时，难免会遇到卡壳或无法解决的问题。因此了解 Spring Cloud Gateway 执行的生命周期相当重要，特别是了解从哪个请求入口开始处理。下面从 Spring Cloud Gateway 的请求入口开始分析 Gateway 是如何组装构建网管的上下文的。

1. 入口 HttpServerRequest 和 HttpServerResponse 转换

Spring Cloud Gateway 的请求入口，在 ReactorHttpHandlerAdapter#apply 方法中，代码截图如图 18-31 所示。

```java
@Override
public Mono<Void> apply(HttpServerRequest request, HttpServerResponse response) {
    NettyDataBufferFactory bufferFactory = new NettyDataBufferFactory(response.alloc());
    ServerHttpRequest adaptedRequest;
    ServerHttpResponse adaptedResponse;
    try {
        adaptedRequest = new ReactorServerHttpRequest(request, bufferFactory);
        adaptedResponse = new ReactorServerHttpResponse(response, bufferFactory);
    }
    catch (URISyntaxException ex) {
        if (logger.isWarnEnabled()) {
            logger.warn("Invalid URL for incoming request: " + ex.getMessage());
        }
        response.status(HttpResponseStatus.BAD_REQUEST);
        return Mono.empty();
    }

    if (adaptedRequest.getMethod() == HttpMethod.HEAD) {
        adaptedResponse = new HttpHeadResponseDecorator(adaptedResponse);
    }

    return this.httpHandler.handle(adaptedRequest, adaptedResponse)
            .doOnError(ex -> logger.error("Handling completed with error", ex))
            .doOnSuccess(aVoid -> logger.debug("Handling completed with success"));
}
```

图 18-31　请求包装转换代码

代码来源于 spring-web-5.0.5.RELEASE.jar，此方法为 Spring Cloud Gateway 的请求入口方法，该方法的作用就是把接收到的 HttpServerRequest 或者最终需要返回的 HttpServerResponse 包装转换为 ReactorServerHttpRequest 和 ReactorServerHttpResponse。

2. 构造 Spring Cloud gateway 的上下文 ServerWebExchange

在 org.springframework.web.server.adapter.HttpWebHandlerAdapter 中，具体代码如下所示：

```
@Override
public Mono<Void> handle(ServerHttpRequest request, ServerHttpResponse response)
{
    ServerWebExchange exchange = createExchange(request, response);
    return getDelegate().handle(exchange)
        .onErrorResume(ex -> handleFailure(request, response, ex))
        .then(Mono.defer(response::setComplete));
}
```

其中的 createExchange() 方法将 ServerHttpRequest 与 ServerHttpResponse 构建成网关上下文 ServerWebExchange。这里 org.springframework.web.server.handler.WebHandlerDecorator.getDelegate() 通过委托的方式获取一系列需要处理的 WebHandler。

18.4.3 DispatcherHandler 源码分析

DispatcherHandler 实现了 WebHandler 和 ApplicationContextAware 接口，用于 HTTP 请求处理器/控制器的中央分发处理器，把请求分发给已经注册的处理程序处理。

```
@Override
public Mono<Void> handle(ServerWebExchange exchange) {
    if (logger.isDebugEnabled()) {
        ServerHttpRequest request = exchange.getRequest();
        logger.debug("Processing " + request.getMethodValue() + " request for ["
            + request.getURI() + "]");
    }
    //校验handlerMapping集合是否为空
    if (this.handlerMappings == null) {
        return Mono.error(HANDLER_NOT_FOUND_EXCEPTION);
    }
    //依次遍历handlerMapping集合进行请求处理
    return Flux.fromIterable(this.handlerMappings)
        .concatMap(mapping ->
        //通过mapping获取mapping对应的handler
        mapping.getHandler(exchange))
        .next()
        .switchIfEmpty(Mono.error(HANDLER_NOT_FOUND_EXCEPTION))
        .flatMap(handler ->
        //调用handler处理
        invokeHandler(exchange, handler))
        .flatMap(result -> handleResult(exchange, result));
}

private Mono<HandlerResult> invokeHandler(ServerWebExchange exchange, Object handler) {
    if (this.handlerAdapters != null) {
        for (HandlerAdapter handlerAdapter : this.handlerAdapters) {
            //判断当前handlerAdapter与handler是否匹配
            if (handlerAdapter.supports(handler)) {
                return handlerAdapter.handle(exchange, handler);
            }
        }
    }
```

```
        }
        return Mono.error(new IllegalStateException("No HandlerAdapter: " + handler));
}
```

DispatcherHandler 的 handler 执行顺序如下：

1）校验 handlerMapping。

2）遍历 Mapping 获取 Mapping 对应的 handler（此处会找到 gateway 对应的 RoutePredicate-HandlerMapping），并通过 RoutePredicateHandlerMapping 获取 handler（FilteringWebHandler）。

3）通过 handler 对应的 HandlerAdapter 对 handler 进行调用（Gateway 使用的 SimpleHandler-Adapter），即 FilteringWebHandler 与 SimpleHandlerAdapter 对应。

18.4.4 RoutePredicateHandlerMapping 源码分析

RoutePredicateHandlerMapping 的执行顺序是先通过路由定位器获取全部路由，然后通过路由断言过滤掉不匹配的路由信息，找到路由信息之后将路由信息设置到上下文中，最后返回 Gateway 固定的 webhandler，即 FilteringWebHandler。代码如下所示，更多内容请查看源码文件 RoutePredicateHandlerMapping.java。

```
public class RoutePredicateHandlerMapping extends AbstractHandlerMapping {
    private final FilteringWebHandler webHandler;
    private final RouteLocator routeLocator;

    public RoutePredicateHandlerMapping(FilteringWebHandler webHandler,
        RouteLocator routeLocator) {
        this.webHandler = webHandler;
        this.routeLocator = routeLocator;
        setOrder(1);
    }

    @Override
    protected Mono<?> getHandlerInternal(ServerWebExchange exchange) {
        //设置mapping到上下文环境
        exchange.getAttributes().put(GATEWAY_HANDLER_MAPPER_ATTR, getClass().
            getSimpleName());

        //查找路由
        return lookupRoute(exchange)
            // .log("route-predicate-handler-mapping", Level.FINER) //name this
            .flatMap((Function<Route, Mono<?>>) r -> {
                exchange.getAttributes().remove(GATEWAY_PREDICATE_ROUTE_ATTR);
                if (logger.isDebugEnabled()) {
                    logger.debug("Mapping [" + getExchangeDesc(exchange) + "] to " + r);
                }

                //将找到的路由信息设置到上下文环境中
                exchange.getAttributes().put(GATEWAY_ROUTE_ATTR, r);
                //返回mapping对应的WebHandler即FilteringWebHandler
                return Mono.just(webHandler);
            }).switchIfEmpty(Mono.empty().then(Mono.fromRunnable(() -> {
                //当前未找到路由时返回空，并移除GATEWAY_PREDICATE_ROUTE_ATTR
```

```java
            exchange.getAttributes().remove(GATEWAY_PREDICATE_ROUTE_ATTR);
            if (logger.isTraceEnabled()) {
                logger.trace("No RouteDefinition found for [" +
                    getExchangeDesc(exchange) + "]");
            }
    })));
}

protected Mono<Route> lookupRoute(ServerWebExchange exchange) {
    //通过路由定位器获取路由信息
    return this.routeLocator.getRoutes()
        .filter(route -> {
            // add the current route we are testing
            exchange.getAttributes().put(GATEWAY_PREDICATE_ROUTE_ATTR,
                route.getId());
            //返回通过谓语过滤的路由信息
            return route.getPredicate().test(exchange);
        })
        // .defaultIfEmpty() put a static Route not found
        // or .switchIfEmpty()
        // .switchIfEmpty(Mono.<Route>empty().log("noroute"))
        .next()
        //TODO: error handling
        .map(route -> {
            if (logger.isDebugEnabled()) {
                logger.debug("Route matched: " + route.getId());
            }
            //校验路由，目前空实现
            validateRoute(route, exchange);
            return route;
        });
}
```

因为 Spring Cloud Gateway 的 GatewayWebfluxEndpoint 提供 HTTP API，不需要经过网关，而是通过 RequestMappingHandlerMapping 进行请求匹配处理，并且它的执行顺序需要在 RoutePredicateHandlerMapping 之前执行，所以设置 RoutePredicateHandlerMapping 的 order 为 0。

18.4.5　FilteringWebHandler 源码分析

FilteringWebHandler 的执行顺序是先构建一个包含全局过滤器的集合，获取上下文中的路由信息，然后将路由里的过滤器添加到集合中，对过滤器集合进行排序。通过过滤器集合组装 Filter 链并调用 Gateway 默认的 Filter 链，最后通过过滤器来处理请求并转发到代理服务中。核心代码如下所示，其源码文件为 org.springframework.cloud.gateway.handler.FilteringWebHandler.java。

```java
public class FilteringWebHandler implements WebHandler {
    protected static final Log logger = LogFactory.getLog(FilteringWebHandler.class);

    /**
```

```java
     * 全局过滤器
     */
    private final List<GatewayFilter> globalFilters;

    public FilteringWebHandler(List<GlobalFilter> globalFilters) {
        this.globalFilters = loadFilters(globalFilters);
    }

    //包装加载全局的过滤器，将全局过滤器包装成GatewayFilter
    private static List<GatewayFilter> loadFilters(List<GlobalFilter> filters) {
        return filters.stream()
                .map(filter -> {
                    //将所有的全局过滤器包装成网关过滤器
                    GatewayFilterAdapter gatewayFilter = new GatewayFilterAdapter(filter);
                    //判断全局过滤器是否实现了可排序接口
                    if (filter instanceof Ordered) {
                        int order = ((Ordered) filter).getOrder();
                        //包装成可排序的网关过滤器
                        return new OrderedGatewayFilter(gatewayFilter, order);
                    }
                    return gatewayFilter;
                }).collect(Collectors.toList());
    }


    @Override
    public Mono<Void> handle(ServerWebExchange exchange) {
        //获取请求上下文设置的路由实例
        Route route = exchange.getRequiredAttribute(GATEWAY_ROUTE_ATTR);
        //获取路由定义下的网关过滤器集合
        List<GatewayFilter> gatewayFilters = route.getFilters();

        //组合全局的过滤器与路由配置的过滤器
        List<GatewayFilter> combined = new ArrayList<>(this.globalFilters);
        //将路由配置的过滤器添加到集合尾部
        combined.addAll(gatewayFilters);
        //对过滤器进行排序
        //TODO: needed or cached?
        AnnotationAwareOrderComparator.sort(combined);

        logger.debug("Sorted gatewayFilterFactories: "+ combined);
        //创建过滤器链表，对其进行链式调用
        return new DefaultGatewayFilterChain(combined).filter(exchange);
    }
}
```

18.4.6 执行 Filter 源码分析

1. 执行进入 Filter 链

在 org.springframework.cloud.gateway.handler.FilteringWebHandler#handle 方法中，具体代码如下所示：

```java
@Override
```

```java
public Mono<Void> handle(ServerWebExchange exchange) {
    Route route = exchange.getRequiredAttribute(GATEWAY_ROUTE_ATTR);
    List<GatewayFilter> gatewayFilters = route.getFilters();

    List<GatewayFilter> combined = new ArrayList<>(this.globalFilters);
    combined.addAll(gatewayFilters);
    //TODO: needed or cached?
    AnnotationAwareOrderComparator.sort(combined);

    logger.debug("Sorted gatewayFilterFactories: "+ combined);

    return new DefaultGatewayFilterChain(combined).filter(exchange);
}
```

上述代码中，通过 new DefaultGatewayFilterChain(combined).filter(exchange) 创建 Gateway，默认为过滤器链。

2. 执行 Filter 链

通过点击 DefaultGatewayFilterChain，进入执行 Filter 链的代码逻辑，主要代码如下所示：

```java
@Override
public Mono<Void> filter(ServerWebExchange exchange) {
    return Mono.defer(() -> {
        if (this.index < filters.size()) {
            GatewayFilter filter = filters.get(this.index);
            DefaultGatewayFilterChain chain = new DefaultGatewayFilterChain(this,
                this.index + 1);
            return filter.filter(exchange, chain);
        } else {
            return Mono.empty(); // complete
        }
    });
}
```

3. Gateway Filter 委托给 Gloable Filter 执行

org.springframework.cloud.gateway.handler.FilteringWebHandler 中的如下代码所示，将 Gateway Filter 委托给 Gloable Filter 执行：

```java
private static class GatewayFilterAdapter implements GatewayFilter {

    private final GlobalFilter delegate;

    public GatewayFilterAdapter(GlobalFilter delegate) {
        this.delegate = delegate;
    }

    @Override
    public Mono<Void> filter(ServerWebExchange exchange, GatewayFilterChain chain) {
        // Gateway Filter委托为Gloable Filter执行
        return this.delegate.filter(exchange, chain);
    }
    @Override
    public String toString() {
```

```
            final StringBuilder sb = new StringBuilder("GatewayFilterAdapter{");
            sb.append("delegate=").append(delegate);
            sb.append('}');
            return sb.toString();
        }
    }
```

18.5 本章小结

本章首先通过案例讲解了 Spring Cloud Gateway 基于服务发现的默认路由规则，以及什么是 Global Filter 和 Gateway Filter 并实现自定义运用。然后讲解了 Spring Cloud Gateway 的实战技巧，比如权重路由、集成 Swagger、限流以及 Gateway 中 HTTPS 的使用。最后通过源码分析 Spring Cloud Gateway 的请求处理流程，从而帮助开发者快速学习源码，加深对原理的理解，从而帮读者更好地解决学习开发中遇到的问题。

第 19 章

Spring Cloud 与 gRPC 上篇

在 Spring Cloud 构建的微服务系统中，大多数开发者使用官方提供的服务调用组件 Feign 来进行内部服务通信，这种声明式的 HTTP 客户端使用起来极为简单、优雅、方便。然而，在使用 Feign 消费服务时，相对于 Dubbo、gRPC 等 RPC 框架来说，性能显得十分低。那么在 Spring Cloud 构建的微服务系统中，能否使用其他更高效的 RPC 框架呢？答案是肯定的，本章和下一章将为大家讲解如何在 Spring Cloud 项目中，将 Google 开源的 gRPC 整合运用到实际项目中。

19.1 Spring Cloud 为什么需要 gRPC

微服务架构的风格，就是将单一程序开发成一个微服务，每个微服务运行在自己的进程中，并使用轻量级机制通信，通常是 HTTP RESTFUL API。这些服务是围绕业务能力来划分、构建的，并通过完全自动化的机制来独立部署。这些服务可以使用不同的编程语言，以及不同数据存储技术，以保证最低限度的集中式管理。

按照业务划分的微服务单元要独立部署，并运行在各自的进程中。微服务单元之间的通信方式一般倾向于使用 HTTP 这种简单的通信机制，更多的时候使用 Rest API。接受请求、处理业务逻辑、返回数据的 HTTP 模式非常高效，并且这种通信机制与平台和语言无关。例如用 Java 写的服务可以消费用 Go 语言写的服务，用 Go 写的服务又可以消费用 Ruby 写的服务。不同的服务采用不同的语言实现，采用不同的平台来部署，它们之间使用 HTTP 进行通信，如图 19-1 所示。

Spring Cloud 通过 REST API 的方式完成服

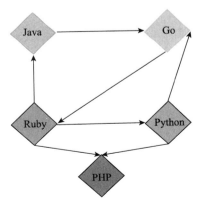

图 19-1 不同语言构建的微服务可以相互调用

务之间的调用，使用 HTTP 协议传输 JSON 序列化后的数据，这种请求－响应的方式非常便捷、通用且高效。但是通常情况下，HTTP 不会开启 KeepAlive 功能，即为短连接，每次请求都需要建立 TCP 连接，这使得其耗时非常低效。对外提供 RESTAPI 是可以理解的，但内部服务之间进行服务调用时若也采用 HTTP 就会显得性能低下，Spring Cloud 默认使用 Feign 进行内部的服务调用，而 Feign 底层使用 HTTP 协议进行服务之间的调用。

目前，Spring Cloud 官方并没有提供除了 RESTAPI 之外的服务调用方案，在业内，许多公司自研了 RPC 框架，自研的目的就是要 RESTAPI 低效问题。现有的更高效的内部服务调用方案是 GRPC。

首先，对比一下 gRPC 和 HTTP/JSON 客户端的传输效率（来自 gRPC 官网的资料）。本次测试为基于 Java 语言的客户端的基准测试，在相同配置的机器上发送 50KB 的消息，HTTP/JSON 客户端和创建了 9 个通道的 gRPC 在吞吐量方面的数据如图 19-2 所示。

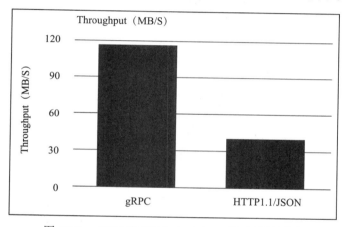

图 19-2　gRPC 和 HTTP 1.1/JSON 吞吐量的测试

由上图可知，gRPC 的吞吐量是 HTTP/JSON 客户端的 3 倍，不仅如此，在 gRPC 只使用了 1/4 的 CPU 资源，而 HTTP/JSON 几乎用了全部。两者在单个 CPU 上的吞吐量对比，如图 19-3 所示。

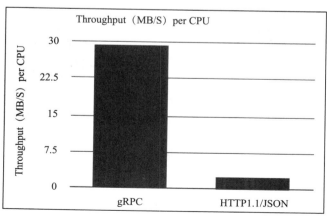

图 19-3　gRPC 和 HTTP 1.1/JSON 在单个 CPU 下的吞吐量的测试

你一定想知道，gRPC 相比于 HTTP/JSON 客户端为何在性能方面表现得如此优异，二者差距为何如此之大吧？主要有三方面的原因：一是 gRPC 采用了 Proto Buffer 作为序列化工具，这比采用 JSON 方式进行序列化性能提高了不少；二是 gRPC 采用了 HTTP2 协议，进行了头部信息压缩，对连接进行了复用，减少了 TCP 的连接次数；三是 gRPC 采用了 Netty 作为 IO 处理框架，提高了性能。

下面就对 gRPC 进行详细介绍。

19.2　gRPC 简介

gRPC 是谷歌开源的一个高性能的、通用的 RPC 框架。gRPC 和其他 RPC 一样，客户端应用程序可以直接调用远程服务的方法，就好像调用本地方法一样。RPC 框架隐藏了底层的实现细节，包括序列化（XML、JSON、二进制）、数据传输（TCP、HTTP、UDP）、反序列化等，开发人员只需要关注业务本身，而不需要关注 RPC 的技术细节，这不仅提高了开发效率，从另一维度上看，这也使创建分布式系统变得更简单。

与其他 RPC 框架一样，gRPC 也遵循定义服务（类似于定义接口）的思想。gRPC 客户端通过定义方法名、方法参数和返回类型来声明一个可以远程调用的接口方法。在 gRPC 服务端，实现 gRPC 客户端定义的接口方法并运行一个 gRPC 服务器来处理 gRPC 客户端调用。需要注意的是，gRPC 客户端和 gRPC 服务端共用一个接口方法。

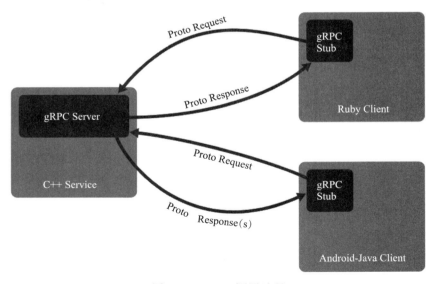

图 19-4　gRPC 调用过程

gRPC 客户端和服务端可以在各种环境中运行和相互调用。它具有运行环境无关性和开发语言无关性。gRPC 通过第三方的开源扩展组件，几乎支持绝大部分的开发语言，包括 C++、Java、Python、Go、C#、Node.js、PHP 等主流开发语言。如图 19-4 所示，用 Ruby 和

Android-Java 写的 gRPC 客户端可以通过 Proto 请求来获取用 C++ 写的 gRPC 服务端的 Proto 响应。这种与语言无关的 RPC 框架，更适合开发团队来构建分布式应用，尤其是微服务系统，开发者不需要使用团队固定的开发语言或者陈旧的技术，他们可以自由选择最适合自己的语言和技术，这大大提高了开发效率。

19.3 gRPC 的一些核心概念

19.3.1 服务定义

和其他 RPC 框架一样，gRPC 遵循服务定义（Service definition）的思想。在默认的情况下，gRPC 使用 protocol buffers 作为接口定义语言（IDL），这种接口语言描述了服务接口和负载消息的结构。关于 protocol buffers，会在后面小节详细介绍。定义一个 HelloService 的代码如下：

```
service HelloService {
    rpc SayHello (HelloRequest) returns (HelloResponse);
}

message HelloRequest {
    string greeting = 1;
}
message HelloResponse {
    string reply = 1;
}
```

gRPC 可以定义 4 种服务方法：

1）Unary RPCS（一元 RPC）：客户端向服务端发送单个请求并获取单个响应，就像调用本地的的方法函数一样。比如下面的代码。

```
rpc SayHello(HelloRequest) returns (HelloResponse){
}
```

2）Server streaming RPCS（服务端流式 RPC）：客户端向服务端发送一个请求，并且客户端获取一个流以读取消息序列。客户端持续从流中读取消息，直至消息读取完毕。

```
rpc LotsOfReplies(HelloRequest) returns (stream HelloResponse){
}
```

3）Client streaming RPCS（客户端流式 RPC）：客户端使用服务端提供的流写入消息序列，并发送给服务端。一旦完成写入消息，客户端会等待服务端读取这些消息并给出响应。

```
rpc LotsOfGreetings(stream HelloRequest) returns (HelloResponse) {
}
```

4）Bidirectional streaming RPCS（双向流式 RPC）：客户端和服务端双方使用读写流发送消息序列。这两个流独立运行，因此客户端和服务器可以按任何顺序读取和写入。例如，服务器可以等待在写入其响应之前接收所有客户端消息，或者可以交替读取消息然后再写入消息，或读写其他组合。每个流中的消息顺序都会保留。

```
rpc BidiHello(stream HelloRequest) returns (stream HelloResponse){
}
```

19.3.2 使用 API

从 .proto 文件中的服务定义开始，gRPC 提供了生成客户端和服务端代码的 protocol buffer 编译器插件。gRPC 用户通常在客户端生成这些 API、在服务端实现相应的 API。

- 在服务端实现声明的服务方法，并运行 gRPC 服务来处理客户端调用。服务端接收客户端传入的请求，执行服务方法并返回服务响应。在接收请求和返回响应之间有编码、解码的过程。
- 在客户端，客户端有一个称为 stub 的本地对象，它实现了与服务相同的方法。然后，客户端可以在 stub 本地对象上调用这些方法，并将该调用的参数包装在适当的 protocol buffer 消息类型中。然后客户端将请求消息发送到服务器，如果是同步调用，需要等待服务端的响应。

19.3.3 同步 vs 异步

同步 RPC 调用会阻塞线程直到客户端收到从服务端返回的响应。通常来讲，网络本质上是异步的，在许多情况下，能够启动 RPC 而不阻塞当前线程是非常有用的。gRPC 支持同步调用和异步调用。

19.4 RPC 的生命周期

现在让我们进一步了解，当 gRPC 客户端调用了一个 gRPC 服务端的方法后到底会发生什么。

1. 一元 RPC（Unary RPC）

Unary RPC（一元 RPC），是 gRPC 中最简单的 RPC，也是应用最广泛的 RPC，客户端发送一个简单的请求，并获取服务端的一个响应。

- 一旦客户端调用 RPC 方法，服务端会接到客户端传输过来的元数据、方法名和过期时间等信息。
- 服务端可以选择发送自己的初始元数据（必须在任何响应之前发送），或者不发送这些元数据，等待客户端发送真正的请求消息。
- 一旦服务端获得了客户端的请求消息，服务端会根据客户端传过来的消息，执行具体的方法业务逻辑，并包装到响应里面。如果业务逻辑执行成功，响应、状态信息（状态码以及其他的状态信息）和可选的其他的元数据信息一起返回给客户端。
- 如果状态码为 OK，则客户端将获得响应，一个完整的 RPC 就完成了。

2. 服务端流式 RPC

服务端流式 RPC 与 Unary RPC 类似，但服务端获取客户端的请求消息后发送的是响应流。

在服务端发送完所有的流数据之后,会在响应里面加上状态信息(状态码及其他状态信息)和可选的其他的元数据信息,将之一起返回给客户端。客户端收到服务端的所有响应之后,一个完整的服务端流式 RPC 就完成了。

3. 客户端流式 RPC

客户端流式 RPC 与 Unary RPC 类似,只不过客户端向服务端发送的是请求流而不是单个请求消息。服务端通常但不一定在收到所有客户端的请求(包括流以及其状态详细信息和可选的尾随元数据)后才发回单个响应,也可以边收到客户端的流消息,边给出响应。

4. 双向流式 RPC

在双向流式 RPC 中,客户端可以多次调用服务端的方法,并发送与客户端相关的流数据,服务端可以发送流数据到客户端,也可以等到更多的流数据到达后再发送流数据。客户端和服务端可以按照任何顺序去读取和写入流数据,它们之间完全独立。

调用方法的客户端接收客户端元数据,方法名称和截止日期的服务器再次发起调用。服务端可以再次选择发送其初始元数据,或者等待客户端开始发送请求。例如,服务端可能会等收到所有客户端的消息后再编写响应,或者服务器和客户端可以"ping-pong":服务器获取请求并发回响应,然后客户端发送另一个基于响应的请求。

5. 截止时间/超时

gRPC 允许客户端指定它们愿意等待多久再完成 RPC 的调用,直到 RPC 以 DEADLINE_EXCEEDED 错误终止。服务端可以查询特定的 RPC 是否超时,或者剩余多少时间来完成 RPC。

6. RPC 终止

在 gRPC 中,客户端和服务端都会对 RPC 调用是否成功进行各自独立的本地比较和判断,它们的结论有可能不一致。这意味着,有可能出现在服务端成功完成 RPC("我已发送所有响应!")但在客户端失败("响应在我的截止时间后到达!")的情况。服务端可以决定在客户端发送的所有请求都到达服务端之前终止。

7. 取消 RPC

客户端或服务端可以随时取消 RPC。取消操作会立即终止 RPC,之后不再进行任何数据传输。取消不是"撤销",进行取消操作之前所做的工作不会被回滚。

8. 元数据

元数据是指关于特定 RPC 调用的信息(例如认证细节),以键值对列表的形式出现,其中键是字符串,值通常也是字符串(也可以是二进制数据)。元数据对 gRPC 本身是不透明的,它允许客户端提供与服务端调用相关的信息,反之亦然。对元数据的访问取决于具体开发语言。

9. 管道

gRPC 管道(Channels)提供了与指定主机和端口上的 gRPC 服务端的连接,这个连接在创建客户端时生成，客户端和服务端就可以通过管道来进行数据传输了。客户端可以指定管道的相关参数来修改 gRPC 的默认行为，例如打开或者关闭消息压缩。管道具有状态，包括已连接和空闲的状态。gRPC 如何处理、关闭管道取决于具体开发语言，例如有些语言允许查询管道的状态，而有些语言没有提供这样的方法。

19.5 gRPC 依赖于 Protocol Buffers

19.5.1 Protocol Buffers 的特点

Protocol Buffers（简称 Protobuf），是一种高效、轻便、易用的结构化数据存储格式，它与平台无关、与语言无关、可扩展性强，广泛用于通信协议和数据存储等领域。

数据存储格式使用最多的是 JSON、XML，它们有轻便、易读、平台无关性、语言无关性等特点，所以广泛使用在各种场景中。既然有了 JSON、XML，为什么还需要 Protocol Buffers 呢？原因如下：一是 Protocol Buffers 具有 JSON、XML 格式的所有优点，且易于扩展；二 Protocol Buffers 是解析速度非常快，与 XML 对比，解析速度快了 20~100 倍，比 JSON 快了 3~5 倍；三是，序列化数据非常紧凑、简洁，与 XML 对比，采用 Protocol Buffers 序列化的数据大概是采用 XML 序列化数据体积的 10%~30%。

19.5.2 使用 Protocol Buffers 的 Maven 插件

在本小节中，将通过实际案例来讲解如何使用 Protocol Buffers。使用 Protocol Buffers 并非我们想象的那么简单，但也不是一件非常困难的事情，它不仅需要掌握额外的接口定义语言（IDL），还需要 Protocol Buffers 编译组件的支持。在本案例中，将采用实际项目中使用最广泛、最方便的 Maven 插件的方式来讲解如何使用 Protocol Buffers。Maven 的 Protocol Buffers 插件屏蔽了手动安装 Protocol Buffers 编译组件的过程，插件会自动安装编译所需的组件。

新建一个 Maven 工程，在工程的 pom 文件中加上 protobuf-java 的依赖包，具体代码如代码清单 19-1 所示。

代码清单19-1　/ch19-1/pom.xml

```xml
<dependency>
    <groupId>com.google.protobuf</groupId>
    <artifactId>protobuf-java</artifactId>
    <version>3.5.1</version>
</dependency>
```

然后安装 Protocol Buffers 的 Maven 插件，具体安装可以参考 https://www.xolstice.org/protobuf-maven-plugin 中的介绍。在工程的 pom 文件中加上 Protocol Buffers 的 Maven 插件，具体代码如代码清单 19-2 所示。

代码清单19-2　/ch19-1/pom.xml

```xml
<build>
<extensions>
        <extension>
            <groupId>kr.motd.maven</groupId>
            <artifactId>os-maven-plugin</artifactId>
            <version>1.5.0.Final</version>
        </extension>
    </extensions>
    <plugins>
        <plugin>
            <groupId>org.xolstice.maven.plugins</groupId>
            <artifactId>protobuf-maven-plugin</artifactId>
            <version>0.5.1</version>
            <configuration>
  <protoSourceRoot>${project.basedir}/src/main/proto</protoSourceRoot>
                <!--默认值-->
<!--<outputDirectory>${project.build.directory}/generated-sources/protobuf/
    java</outputDirectory>-->
<outputDirectory>${project.build.directory}/generated-sources/protobuf/
    java</outputDirectory>
                <!--设置是否在生成Java文件之前清空outputDirectory的文件，默认
                    值为true，设置为false时也会覆盖同名文件-->
                <clearOutputDirectory>true</clearOutputDirectory>
                <protocArtifact>com.google.protobuf:protoc:3.5.1:exe:${os.
                    detected.classifier}</protocArtifact>
                <pluginId>grpc-java</pluginId>
                <pluginArtifact>io.grpc:protoc-gen-grpc-java:${grpc.
                    version}:exe:${os.detected.classifier}</pluginArtifact>
            </configuration>
            <executions>
                <execution>
                    <goals>
                        <goal>compile</goal>
                        <goal>compile-custom</goal>
                    </goals>
                </execution>
            </executions>
        </plugin>
    </plugins>
</build>
```

其中，os-maven-plugin 为当前操作系统和体系结构自动生成分类器，Google 的 protobuf 团队使用相同的插件为 protoc 组件生成分类器。所以强烈建议使用这个扩展依赖。

利用 protobuf-maven-plugin 的 configuration 标签下可以配置各种属性，比如 protoSourceRoot 标签为配置 Proto 的源 IDL 文件，默认地址为 ${project.basedir}/src/main/proto；outputDirectory 标签为配置编译后输出的文件地址，默认地址为 ${project.build.directory}/generated-sources/protobuf/java；protocArtifact 标签为配置 Protocol Buffers 的编译器地址，它是用 artifact、groupId、vsersion 来确定唯一性的。关于 protobuf-maven-plugin 的更多的配置信息，请在网页 https://www.xolstice.org/protobuf-maven-plugin/compile-mojo.html 中查看。

在工程的 /src/main 目录下新建一个 proto 的目录用于存放 Protocol Buffers 的 IDl 文件，这些文件是以 .proto 后缀结尾的。为了能够在开发工具 IDEA 中高亮显示 Protocol Buffers IDl 文件，需要安装一个 IDEA 的 Protobuf Support 插件。打开 IDEA 的 settings，在 plugins 下搜索 Protobuf Support，点击安装即可，安装完成后应重启 IDEA。

在 /src/main/proto 目录下新建一个文件 person.proto 如代码清单 19-3 所示。

代码清单19-3　ch19-1/src/main/proto/person.proto

```
syntax = "proto3";
option java_package = "cn.springcloud.proto";
option java_outer_classname = "PersonModel";
message Person {
    int32 id = 1;
    string name = 2;
    string email = 3;
}
```

在上面的代码中，syntax 指定了 Protocol Buffers 的版本，当前我们使用的是 proto3，另外还有一个 proto2 版本。option java_package 指定了生成类的包名，option java_outer_classname 指定了生成类的名称。定义一个消息（message）Person，相当于一个 Java 实体。在这个消息中有三个变量，分别为 id、name、email。类型 int32 对应 Java 的 Interger 类型，更多 Proto Buffer 的变量类型可参考官方文档。

打开终端，进入工程的根目录，即工程的 pom 文件，执行 mvn protobuf:compile 命令，就会在工程的 /target/genenrrated-sources/java 目录下生成 cn.springcloud.protod 包，在包下生成一个 PersonModel 类。在该类中代码很多，都是与 Protocol Buffers 相关的类和方法，具体在此不展开讲解。另外，也可以借助 IDEA 工具来执行 mvn protobuf:compile 命令，点击 Plugins 下的 protobuf 的 protobuf:compile 按钮即可，具体如图 19-5 所示。

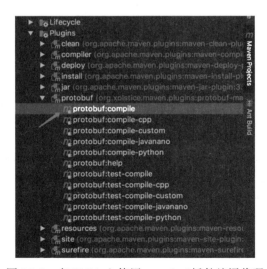

图 19-5　在 IDEA 上使用 protobuf 插件编译代码

生成 PersonModel 类后，再来演示如何使用它来做序列化和反序列化，如代码清单 19-4 所示。

代码清单19-4　ch19-1/src/main/java/cn/springcloud/PersonUseCase.java

```java
public class PersonUseCase {

    public static void main(String[] args) {
        PersonModel.Person forezp= PersonModel.Person.newBuilder()
                .setId(1)
                .setName("forezp")
                .setEmail("miles02@163.com").build();


        for(byte b : forezp.toByteArray()){
            System.out.print(b);
        }

        System.out.println("\n" + "bytes长度" + forezp.toByteString().size());
        System.out.println("===== forezp Byte 结束 =====");

        System.out.println("===== forezp 反序列化生成对象开始 =====");
        PersonModel.Person forezpCopy = null;
        try {
            forezpCopy = PersonModel.Person.parseFrom(forezp.toByteArray());
        } catch (InvalidProtocolBufferException e) {
            e.printStackTrace();
        }
        System.out.print(forezpCopy.toString());
        System.out.println("===== forezp 反序列化生成对象结束 =====");
    }
}
```

运行 PersonUseCase 的 main 方法，输出如下：

```
811861021111141011221122615109105108101115485064495451469911109
bytes长度27
===== forezp Byte 结束 =====
===== forezp 反序列化生成对象开始 =====
id: 1
name: "forezp"
email: "miles02@163.com"
===== forezp 反序列化生成对象结束 =====
```

在上面的案例中，演示了使用 Proto Buffer 来序列化和反序列化一个 Java 对象的过程，这跟使用 JSON 进行序列化和反序列化类似。采用 Proto Buffer 序列化的数据可读性很差，但是数据体积很小，所以能够大大提高传输效率。

19.5.3　Proto Buffer 语法介绍

1）定义一个复杂类型的消息，在该消息中，定义一个 Java 的 List 集合类的变量，需要在变量前使用 repeated 关键字，如下所示：

```
message ComplexObject {
```

```
        repeated string sons = 4; // List列表
        repeated Person persons= 6; // 复杂的对象List
}
```

2）定义一个 Map 类型，这和 Java 语法一样，直接使用 map 关键字，如下所示：

```
message ComplexObject {
    map<string, Person > map = 8;
}
```

3）定义枚举类，使用 enum 关键字，代码如下：

```
message ComplexObject {
    ...
    Gender gender = 5;
    ...
}
enum Gender {
    MAN = 0;
    WOMAN = 1;
}
```

Proto Buffer IDL 语法简单明了，能够满足大多数的需求，如果想深入了解 Proto Buffer 语法，可以参考官方文档，文档地址为 https://developers.google.com/protocol-buffers/docs/proto3。

19.6　gRPC 基于 HTTP2

HTTP2 即超文本传输协议版本 2，HTTP2 通过对头部字段进行压缩，在客户端和服务端建立连接之后允许双向传输数据来提高传输效率、减少延迟。HTTP2 还允许在客户端与服务器未建立连接的情况下由服务端提送消息到客户端。

下面介绍 HTTP2 的一些概念和特性。

1．二进制流

HTTP2 是一个二进制协议，基于二进制的 HTTP2 对帧（Frame）的使用变得更为简单。帧是数据传输的最小单位。每个帧都属于一个特定的流（Stream）。一个请求或者响应可能由多个帧组成。HTTP2 的请求和响应以帧为单位进行了的划分，所有的帧都在流上进行传输，帧可以不按照顺序发送，然后在接收端重新按照头部字段 Stream Identifier 进行排序。帧具有如下的公告字段：Type、Length、Flags、Stream Identifier、Frame Payload。

2．多路复用

客户端和服务端建立的连接可以包含多个并发的流，流被客户端和服务端共享，可以让客户端和服务端同时发送消息（请求或者响应），也可以单方使用发送消息，也可以被任意一方关闭。流的多路复用，提高了连接的利用率。

3. 头压缩

HTTP 是一种无状态的协议，即服务器不需要保存 HTTP 的所有细节，发一次请求产生一次响应，不需要保留之前请求的元数据。HTTP2 也遵循这个范式。客户端只有通过发送 cookies 才能使服务器记住客户端的一些状态，cookies 需要包含的信息可能比较复杂，并且需要每次数据发送都要携带 cookies，这会严重拖累 HTTP 请求的速度，所以头压缩是非常有必要的。

4. 服务器推送

服务器推送是 HTTP2 具有的一个强大的功能，这个功能被称为缓存推送。服务器推送可以让服务器在客户端不发送请求的情况下，主动将资源推送给客户端，并让客户端缓存起来，从而当需要请求该资源时，直接从缓存中取。比如，浏览器请求某网页时，客户端发送请求 HTML，这时服务器不仅把客户端请求的 HTML 返回给客户端，还把 HTML 所需的 CSS 文件推送给客户端。服务器推送，需要客户端主动开启，开启之后，客户端可以随时选择是否终止该推送服务。如果客户端不需要服务器推送，可以通过发送一个 RST_STREAM 帧来终止。

gRPC 基于 HTTP2 协议，所以 gRPC 继承了 HTTP2 的优点，带来诸如双向流、流控、头部压缩、单 TCP 连接上的多复用请求等特性。从实现和特性来看，gRPC 目前更多考虑的是移动端和服务端的通信，gRPC 高效地实现了 HTTP2 的连接多路复用、Body 和 Header 的压缩。这些功能给移动端带来了非常多的好处，比如节省了流量、响应速度更快、节约了 CPU 资源。正因为 gRPC 有这么多的好处，它在服务端使用非常广泛。

19.7 gRPC 基于 Netty 进行 IO 处理

Netty 是由 JBOSS 开源的一个异步事件驱动的 Java 网络应用框架，用于快速开发可维护的、高性能的协议服务器和客户端。Netty 是一款 NIO 客户端、服务器框架，可以快速且轻松地开发网络应用程序，比如协议服务器、客户端应用等。它极大地简化了 TCP 和 UDP 的 Socket 服务器等网络编程过程。

gRPC 的 Java 版本用 Netty 作为 NIO 框架。gRPC 的服务端（Server）和客户端（Client）均使用 Netty Channel 作为数据传输通道，使用 Proto Buffer 作为数据序列化和反序列化的工具，每个请求被封装成符合 HTTP2 规范的数据流。客户端连接上服务端的 Channel 之后，仍保持长连接，这样就做到了连接复用，而不需要每次请求都需要客户端重新连接上服务端，从而大大提高了数据交互的效率。

为什么使用 Netty 能够提高数据传输性能？这要从线程模型角度来分析。在早期的网络编程框架中，使用的是阻塞 IO 来处理网络请求，伪代码如下：

```
ServeSocket serverSocket = new ServeSocket(8080);
Socket socket = serverSocket.accept();
BufferReader in = .... ;
String request ;
```

```
while((request = in.readLine()) != null){
    //TODO
}
```

在上面的代码中，只能同时处理一个连接，如果需要处理多个客户端的连接，必须创建多个 ServeSocket，并且需要重新创建一个线程去处理该连接的请求，该线程会处于阻塞状态。在大多数情况下，这些线程是处于空闲状态，浪费了大量的内存资源。再者，线程对于服务器来说是一种有限资源，大量的客户端连接意味着需要创建大量的线程，这是不合理且不可能的。图 19-6 所示为使用阻塞 IO 处理多个请求的示意。

在 Java NIO 框架中，使用选择器消除了 Java IO 的一些弊端。Selector 是关键，它使用了事件通知 API 确定哪些线程是可以进行读写操作的。这样单个线程就可处理多个并发的连接，如图 19-7 所示。Netty 是 Java NIO 的一个封装，能够快速帮助使用者搭建一个功能强大的网络处理框架。

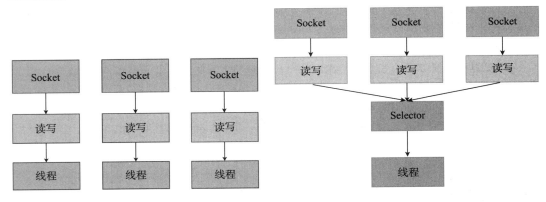

图 19-6　阻塞线程模型　　　　　　　图 19-7　NIO 线程模型

gRPC 的 Java 版本采用 Netty 作为 IO 处理框架，其继承了 Netty 的优点，所以 gRPC 在 IO 处理上也是非常高效的。

19.8　gRPC 案例实战

在此案例中，包含三个 Module 工程：grpc-simple-server 工程作为 gRPC 的服务端；grpc-simple-client 工程作为 gRPC 的客户端；grpc-lib-1 工程作为存放 gRPC 的 .proto 文件的 libary 工程。

首先在主 Maven 工程的父 pom 文件中引入 gRPC 工程所必须的依赖，包括 grpc-netty、grpc-protobuf 和 grpc-stub，如代码清单 19-5 所示。

代码清单19-5　ch19-2/pom.xml

```xml
<dependencies>
    <dependency>
        <groupId>io.grpc</groupId>
        <artifactId>grpc-netty</artifactId>
```

```xml
        <version>${grpc.version}</version>
    </dependency>
    <dependency>
        <groupId>io.grpc</groupId>
        <artifactId>grpc-protobuf</artifactId>
        <version>${grpc.version}</version>
    </dependency>
    <dependency>
        <groupId>io.grpc</groupId>
        <artifactId>grpc-stub</artifactId>
        <version>${grpc.version}</version>
    </dependency>
</dependencies>
```

作为存放 .proto 文件的 lib 工程，在 grpc-lib-1 工程中，需要根据 .proto 文件生成 Java 文件，以供另外两个工程使用。grpc-lib-1 的 pom 工程需要加上 os-maven-plugin 和 protobuf-maven-plugin 这两个插件的依赖，这在本章之前已经介绍过，在此不再重复。在 /src/main/proto 文件夹下，新建一个 HiService.proto 文件，代码如代码清单 19-6 所示。

代码清单19-6　ch19-2/grpc-lib/src/main/proto/HiService.proto

```
syntax = "proto3";
package cn.springcloud.grpc;
option java_multiple_files = true;
message HelloRequest {
    string name = 1;
    int32 age = 2;
    repeated string hobbies = 3;
    map<string, string> tags = 4;
}
message HelloResponse {
    string greeting = 1;
}
service HelloService {
    rpc hello(HelloRequest) returns (HelloResponse);
}
```

HiService.proto 文件中创建了 HelloRequest 和 HelloResponse 两个实体，并创建了 HelloService 服务，该服务有一个方法 hello，该方法是一个 RPC 方法，需要传入 HelloRequest 参数，并返回 HelloResponse 响应。

创建完工程之后，执行 mvn clean install 命令，protobuf-maven-plugin 插件会根据 .proto 文件生成对应的 Java 文件、Maven 插件并将该 grpc-lib-1 打成 Jar 包，然后安装到本地的 Maven 仓库中，以供另外两个工程使用。

在 grpc-server 工程中，其 pom 文件需要继承主 Maven 工程的 pom 文件，另外需要加上 grpc-lib-1 依赖。

在工程里新建一个 HelloService 类，该类继承自 HelloServiceGrpc.HelloServiceImplBase 类，其中 HelloServiceGrpc 为 grpc-lib-1 中的 HiService.proto 文件生成的类，需要 grpc-lib-1 编译完之后才能使用，不然 IDEA 会报错。HelloService 复写了父类的 hello 方法，在

该方法中打印了从客户端发送过来的 HelloRequest 对象，并根据 HelloRequest 对象生成一个 HelloResponse，其中 responseObserver.onNext(response) 方法向流中写入 HelloResponse 对象，responseObserver.onCompleted() 向流中写入结束。具体代码如代码清单 19-7 所示。

代码清单19-7 ch19-2/grpc-simple-server/src/main/java/cn/springcloud/grpcsimpleserver/servcie/HelloService.java

```java
public class HelloService extends HelloServiceGrpc.HelloServiceImplBase {
    @Override
    public void hello(cn.springcloud.grpc.HelloRequest request,
    io.grpc.stub.StreamObserver<cn.springcloud.grpc.HelloResponse> responseObserver) {
        System.out.println( request );
        String greeting = "Hi " + request.getName() + " you are " + request.getAge()
            + " years old" +
            " your hoby is " + (request.getHobbiesList()) + " your tags " + request.
                getTagsMap();
        HelloResponse response = HelloResponse.newBuilder().setGreeting(
            greeting ).build();
        responseObserver.onNext( response );
        responseObserver.onCompleted();
    }
}
```

创建 MyGrpcServer 类，该类用 main 函数作为 gRPCServer 的函数，gRPCServer 的监听端口为 8082，并将 HelloService 注册到该 gRPCServer 中，然后启动 gRPCServer。启动成功后 gRPCServer 处于等待终止状态，具体代码如代码清单 19-8 所示。

代码清单19-8 ch19-2/grpc-simple-server/src/main/java/cn/springcloud/grpcsimpleserver/MyGrpcServer.java

```java
public class MyGrpcServer {
    static public void main(String[] args) throws IOException, InterruptedException {
        Server server = ServerBuilder.forPort( 8082 )
            .addService( new HelloService() )
            .build();
        System.out.println( "Starting server..." );
        server.start();
        System.out.println( "Server started!" );
        server.awaitTermination();
    }
}
```

在 grpc-simple-client 工程中，pom 文件需要继承主 Maven 工程的 pom 文件，另外需要加上 grpc-lib-1 依赖。在 gRPCClient 端，先创建一个连接 gRPCServer 的 Channel，其中 usePlaintext() 表明用纯文本创建连接，在默认的情况下，会使用 TLS 安全连接机制。然后根据该 Channel 创建一个阻塞的 Stub。使用该 Stub 向 gRPCServer 发送一条 HelloRequest 消息，并阻塞线程直到接到 gRPCServer 发回的 HelloResponse 响应，在控制台上打印该响应，最后关闭该 Channel。具体如代码清单 19-9 所示。

代码清单19-9 ch19-2/grpc-simple-client/src/main/java/cn/springcloud/grpcsimpleclient/
MyGrpcClient.java

```java
public class MyGrpcClient {
    public static void main(String[] args) throws InterruptedException {
        ManagedChannel channel = ManagedChannelBuilder.forAddress("localhost",
            8082)
                .usePlaintext()
                .build();
        HelloServiceGrpc.HelloServiceBlockingStub stub =
                HelloServiceGrpc.newBlockingStub(channel);
        HelloResponse helloResponse = stub.hello(
            HelloRequest.newBuilder()
                    .setName("forezp")
                    .setAge(17)
                    .addHobbies("football").putTags( "how?","wonderful" )
                    .build());
        System.out.println(helloResponse);
        channel.shutdown();
    }
}
```

运行 MyGrpcServer 的 main 函数，启动 gRPC 服务端。运行 MyGrpcClient 的 main 函数，启动 gRPC 客户端，并和服务端创建连接，发送 HelloRequest 消息，并收到 HelloResponse 的响应。响应如下：

greeting: "Hi forezp you are 17 years old your hoby is [football] your tags {how?=wonderful}"

可见 MyGrpcClient 客户端通过 gRPC 向 MyGrpcServer 服务端发送了一条消息。服务端接到客户端的消息后，经过逻辑处理，给客户端发送了响应。

19.9 本章小结

在本章中，首先讲述了为什么 Spring Cloud 需要 gRPC；然后讲解了 gRPC 的一些概念；接着讲解了 gRPC 基于 Proto Buffer，对 Proto Buffer 进行了详细介绍，以及 gRPC 基于 HTTP2、Netty 的案例的讲解。通过对本章的学习，相信大家对 gRPC 有了全面的了解，下章将讲述如何在 Spring Cloud 项目中使用 gRPC。

第 20 章

gRPC 在 Spring Cloud 与 gRPC 下篇

在 19 章，介绍了 gRPC 的基本情况、原理、性能为何如此优异，也阐述了 Spring Cloud 需要 gRPC 的理由。本章介绍如何在 Spring Cloud 项目中使用 gRPC。

目前国内外有许多大型公司在使用 gRPC，比如谷歌、Netflix、CoreOs、Ngnix、京东、迅雷等公司。到目前为止，Spring Cloud 官方并不支持 gRPC，但有非常优秀的第三方开源项目对其做出了很好的支持，比如开源项目 grpc-spring-boot-starter，项目地址为 https://github.com/yidongnan/grpc-spring-boot-starter。该项目作者目前就职于腾讯，有非常丰富的微服务实战经验。该项目也是 Spring Cloud 中国社区推荐的一个 gRPC 项目，已经有公司将该项目应用到生产环境中，有很好的可靠性和稳定性。

20.1 gRPC Spring Boot Starter 介绍

该项目是 java-grpc 结合 Spring Boot、Spring Cloud 使用的一个开源项目，既可以结合 Spring Boot 项目单独使用，也可以结合 Spring Cloud 在微服务架构下使用。该项目具有如下特点或者特性。

- ❑ 基于 Spring Boot 的自动配置，实现了起步依赖，只需要写简单的配置和相应的注解，就可以启动 gRPC Server。
- ❑ 支持在 Spring Cloud 项目中使用，支持主流的 Spring Cloud Eureka、Spring Cloud Consul 的服务注册发现组件。
- ❑ 支持分布式链路追踪，结合 Spring Cloud Sleuth，可以展示服务调用链。
- ❑ 支持对 gRPC Server 的健康检查。

20.2 gRPC Spring Boot Starter 架构设计

在该框架中，涉及两个角色，一个是 gRPC Server，即服务提供者，另一个是 gRPC Client，即服务消费者。在 Java 项目中 gRPC 进行远程调度，是一件非常简单容易的事，在上一章已经有专门案例介绍过了。但 gRPC 如何与 Spring Cloud 结合使用，需要自己去写一些代码定制，gRPC Spring Boot Starter 框架就做了这样的代码定制。其实现原理如下：

1）gRPC Server 和 gRPC Client 都为一个 Discovery Client，需要向服务注册中心（Eureka 或者 Consule）注册。另外 gRPC Server 是一个 Netty Server，需要在配置文件中配置 Netty Server 的启动端口，如果设置为 0，则随机分配一个端口。

2）当 gRPC Server 向注册中心注册时，会携带 gRPC Server 的 Netty Server 的端口信息。这样做的目的在于，在获取服务注册列表的时候，能够知道其他服务的 Netty Server 的端口信息，从而建立连接，进行通信，这是十分关键的一点。

3）自定义的 @GrpcService 注解包含 @Servcie 注解，所以在 gRPC Server 启动时，Spring IoC 容器初始化注入被 @GrpcService 注解修饰类的 Bean，并会给这些 Bean 设置 ServerInterceptor（包括全局的 ServerInterceptor 和单个 Bean 自定义的 ServerInterceptor），这些 ServerInterceptor 是打印请求日志和实现链路追踪的关键。这些 ServerInterceptor 类继承了 BindableService，最后将这些 Bean 包装在 ServerServiceDefinition 类中。

说明：

1）gRPC Server 在创建 NettyServerBuilder 时，设置它的端口为在配置文件配置的端口，并将上一步中的 ServerServiceDefinition 类的信息注册到 NettyServerBuilder 中，供 gRPC Client 调用。

2）在 gRPC Client 端，只需要创建一个可以连接 gRPC Server 的 Netty Channel，就可以进行对 gRPC Server 的远程调用。gRPC Client 端的 Channel 的创建是在创建 Bean 的后置处理过程中完成的。首先需要一个被 @Service 修饰的服务类，在该类定义一个被自定义注解 @GrpcClient 修饰的 Channel 变量。当 @Service 注解修饰的类创建完之后，会取出被 @GrpcClient 注解的 Bean 进行处理，处理的过程是先取出 @GrpcClient 注解的值，该值是要调用 gRPC Server 的服务名，根据服务名，去获取服务注册列表的中的 gRPC Server 的 host 和 gRPC Server 的 Netty Server 端口（如果不结合 Spring Cloud 使用，则是取配置文件配置的 host 和端口），如果有多个实例则采用负载的策略选取。创建 Channel 完成后，通过反射设置到 Bean 的成员变量中。

20.3 gRPC Spring Boot Starter 源码分析

gRPC Spring Boot Starter 框架设计得十分精妙，用不多代码实现了 gRPC 与 Spring Cloud 结合的功能。为了加深读者对该框架的理解，我觉得十分有必要对该框架做一个源码解析。另外该框架可以作为一个定制 Spring Cloud 开发的范例。

20.3.1 gRPC Server Spring Boot Starter 源码解析

在源码 grpc-server-spring-boot-autoconfigure 工程中的 /src/main/resources/META_INF 目录下有一个 spring.factories 文件，该文件在 Spring Boot 启动时会自动加载，该文件一共配置了三个类，分别为 GrpcMetedataEurekaConfiguration、GrpcMetedataConsulConfiguration 和 GrpcServerAutoConfiguration。其中 GrpcMetedataEurekaConfiguration 是在 Spring Cloud 环境下，当

服务注册中心为 Eureka 时,将 Netty Server 的端口信息设置到服务注册的元数据 Map 中,具体代码如下:

```
@Configuration
@EnableConfigurationProperties
@ConditionalOnBean(EurekaInstanceConfig.class)
public class GrpcMetedataEurekaConfiguration {

    @Autowired
    private EurekaInstanceConfig instance;
    @Autowired
    private GrpcServerProperties grpcProperties;

    @PostConstruct
    public void init() {
        this.instance.getMetadataMap().put("gRPC", String.valueOf(grpcProperties.
            getPort()));
    }
}
```

GrpcMetedataConsulConfiguration 是当服务注册中心为 Consul 时将 Netty Server 的端口信息设置到服务注册的元数据 Map 中,与 GrpcMetedataEurekaConfiguration 相同。GrpcServer-AutoConfiguration 类做了很多默认的配置,其中注入了一个 AnnotationGrpcServiceDiscoverer 的 Bean,该 Bean 的作用就是找到有被 @GrpcService 注解修饰的类的 Bean,代码如下:

```
@ConditionalOnMissingBean
@Bean
public AnnotationGrpcServiceDiscoverer defaultGrpcServiceFinder() {
    return new AnnotationGrpcServiceDiscoverer();
}
```

自定义的注解 @GrpcService,修饰类或者接口,并且包含了 @Service 注解,也就是 @GrpcService 注解具有 @Service 注解的功能,被 @GrpcService 修饰的类会生成 Bean,并注入 Spring IoC 中。代码如下:

```
@Target(ElementType.TYPE)
@Retention(RetentionPolicy.RUNTIME)
@Documented
@Service
public @interface GrpcService {

    Class<?> value();

    Class<? extends ServerInterceptor>[] interceptors() default {};
}
```

在 AnnotationGrpcServiceDiscoverer 类中有个核心方法为 findGrpcServices(),该方法根据 @GrpcService 注解来获取所有被 @GrpcService 修饰的所有 BindableService 类型的 Bean,在获取所有的 ServerInterceptor 类型的 Bean 之后,将 ServerInterceptor 类型的 Bean 设置到 BindableService 类型的 Bean 中,最后包成一个 GrpcServiceDefinition,具体代码如下:

```
public Collection<GrpcServiceDefinition> findGrpcServices() {
```

```
    Collection<String> beanNames = findGrpcServiceBeanNames();
    List<GrpcServiceDefinition> definitions = Lists.newArrayListWithCapacit
        y(beanNames.size());
    GlobalServerInterceptorRegistry globalServerInterceptorRegistry =
        applicationContext.getBean(GlobalServerInterceptorRegistry.class);
    List<ServerInterceptor> globalInterceptorList = globalServerInterceptorRegistry.
        getServerInterceptors();
    for (String beanName : beanNames) {
        BindableService bindableService = this.applicationContext.
            getBean(beanName, BindableService.class);
        ServerServiceDefinition serviceDefinition = bindableService.
            bindService();
        GrpcService grpcServiceAnnotation = applicationContext.
            findAnnotationOnBean(beanName, GrpcService.class);
        serviceDefinition = bindInterceptors(serviceDefinition,
            grpcServiceAnnotation, globalInterceptorList);
        definitions.add(new GrpcServiceDefinition(beanName, bindableService.
            getClass(), serviceDefinition));
        log.debug("Found gRPC service: " + serviceDefinition.
            getServiceDescriptor().getName() + ", bean: " + beanName + ", class:
            " + bindableService.getClass().getName());
    }
    return definitions;
}
```

在上面的代码中有个关键的点是对每一个 BindableService 的 Bean 设置 ServerInterceptor，这个 ServerInterceptor 是可以自定义的，具体怎么自定义可以参考后面的实战小节。其中比较关键的一点是注入了一个 TraceServerInterceptor 类的 Bean，该类是在 gRPC 中实现链路追踪的关键，读者可以自行阅读 TraceServerInterceptor 类的源码。TraceServerAutoConfiguration 的配置类代码如下：

```
@Configuration
    @ConditionalOnProperty(value = "spring.sleuth.scheduled.enabled", matchIf-
        Missing = true)
    @AutoConfigureAfter({TraceAutoConfiguration.class})
    @ConditionalOnClass(value={Tracing.class, GrpcTracing.class})
    protected static class TraceServerAutoConfiguration {
        @Bean
        Public GrpcTracing grpcTracing(Tracing tracing){retur GrpcTracing.
            create(tracing);
        }
        @Bean
        public GlobalServerInterceptorConfigurerAdapter globalTraceServerInterc
            eptorConfigurerAdapter(final GrpcTacing grpcTracing) {
            return new GlobalServerInterceptorConfigurerAdapter() {
                @Override
                public void addServerInterceptors(GlobalServerInterceptorRegist
                    ry registry) {
                    registry.addServerInterceptors(grpcTracing.newServer
                        Interceptor());
                }
            };
        }
    }
```

在 GrpcServerAutoConfiguration 配置类注入了一个 NettyGrpcServerFactory，该 Netty-GrpcServerFactory 创建了 NettyServerBuilder，并将之前包装好的 GrpcServiceDefinition 注入 NettyServerBuilder 中，作为服务提供者，供服务消费者远程调用。代码如下：

```java
@Override
    public Server createServer() {
        NettyServerBuilder builder = NettyServerBuilder.forAddress(
            new InetSocketAddress(InetAddresses.forString(getAddress()),
            getPort()));

            builder.addService(healthStatusManager.getHealthService());

        for (GrpcServiceDefinition service : this.services) {
            String serviceName = service.getDefinition().getServiceDescriptor().
                getName();
            log.info("Registered gRPC service: " + serviceName + ", bean: " +
                service.getBeanName() + ", class: " + service.getBeanClazz().
                getName());
            builder.addService(service.getDefinition());
            healthStatusManager.setStatus(serviceName, HealthCheckResponse.
                ServingStatus.SERVING);
        }

        if (this.properties.getSecurity().getEnabled()) {
            File certificateChain = new File(this.properties.getSecurity().
                getCertificateChainPath());
            File certificate = new File(this.properties.getSecurity().
                getCertificatePath());
            builder.useTransportSecurity(certificateChain, certificate);
        }
        if(properties.getMaxMessageSize() > 0) {
            builder.maxInboundMessageSize(properties.getMaxMessageSize());
        }

        return builder.build();
```

最后 NettyGrpcServerFactory 被设置到 GrpcServerLifecycle 类的 Bean 中，由 GrpcServer-Lifecycle 实现了 SmartLifecycle 接口。实现了 SmartLifecycle 接口的 Bean，当 Spring 容器加载所有 Bean 并完成初始化之后，会回调实现 SmartLifecycle 接口的类中的 start() 方法，跟踪代码可知，最终 Netty Server 在 start() 方法里启动，并等待中断命令。

20.3.2　gRPC Client Spring Boot Starter 源码解析

gRPC Client Spring Boot Starter 实现的功能是服务的调用，关于如何调用服务，请看本章的案例实战小节。在工程的 /src/main/resouces/META_INF 目录下有一个 spring.factories 文件，在该文件中配置了 Spring IoC 初始化的配置类 GrpcClientAutoConfiguration，而 Grpc-ClientAutoConfiguration 类中又配置了一些 Bean，列举如下：

- ❑ GrpcChannelsProperties 类型的 Bean，从配置文件 application 中读取 GrpcChannel 相关的配置，存储在这个 Bean 中。

- GlobalClientInterceptorRegistry 类型的 Bean 为全局 gRPC 拦截器。
- LoadBalancer.Factory 类型的 Bean，是客户端负载均衡处理 Bean。
- DiscoveryClientChannelFactory 类型的 Bean，它是根据负载均衡的 Bean、拦截器的 Bean、服务发现客户端的 Bean，以及相关配置来创建 Channel 的工厂类的 Bean。
- GrpcClientBeanPostProcessor 类型的 Bean，它是用来后置处理带有被 @GrpcClient 注解修饰的变量的 Bean 的。

其中，比较重要的一个类是 GrpcClientBeanPostProcessor。当某个 Bean 的变量被 @GrpcClient 注解修饰后，该 Bean 会被 GrpcClientBeanPostProcessor 后置处理。GrpcClientBeanPost-Processor 类实现了 BeanPostProcessor 接口，在 postProcessBeforeInitialization() 方法上，找出 Bean 中的变量（包括这个 Bean 父类的变量）被 @GrpcClient 修饰的 Bean，将这些 Bean 存储在 Map 中，等待下一步处理，代码如下：

```java
@Override
public Object postProcessBeforeInitialization(Object bean, String beanName)
    throws BeansException {
    Class clazz = bean.getClass();
    do {
        for (Field field : clazz.getDeclaredFields()) {
            if (field.isAnnotationPresent(GrpcClient.class)) {
                if (!beansToProcess.containsKey(beanName)) {
                    beansToProcess.put(beanName, new ArrayList<Class>());
                }
                beansToProcess.get(beanName).add(clazz);
            }
        }
        clazz = clazz.getSuperclass();
    } while (clazz != null);
    return bean;
}
```

其中 @GrpcClient 注解包含了 value（）方法，该值是服务提供者的 name，即 gRPC Server 的 name，另外 interceptors（）方法是配置拦截器的。代码如下：

```java
@Target({ElementType.FIELD})
@Retention(RetentionPolicy.RUNTIME)
@Documented
@Inherited
public @interface GrpcClient {
    String value();
    Class<? extends ClientInterceptor>[] interceptors() default {};
}
```

在 GrpcClientBeanPostProcessor 的 postProcessAfterInitialization() 方法中，如果所处理的 Bean 在上一步的 Map 中，则进行处理。先取出修饰变量的 @GrpcClient 注解，再取出注解的 value 值，即服务调用者的 name，以及服务调用配置的拦截器，将 value 及服务提供者的服务名和拦截器交给 channelFactory 去创建一个 Channel 对象，最后通过反射将该 channel 对象设置给该变量。也就是说被 @GrpcClient 修饰的变量，在这里被设置成为一个可被使用的 Channel 对象。代码如下：

```java
public Object postProcessAfterInitialization(Object bean, String beanName)
    throws BeansException {
    if (beansToProcess.containsKey(beanName)) {
        Object target = getTargetBean(bean);
        for (Class clazz : beansToProcess.get(beanName)) {
            for (Field field : clazz.getDeclaredFields()) {
                GrpcClient annotation = AnnotationUtils.getAnnotation(field,
                    GrpcClient.class);
                if (null != annotation) {

                    List<ClientInterceptor> list = Lists.newArrayList();
                    for (Class<? extends ClientInterceptor> clientInterceptorClass
                        : annotation.interceptors()) {
                        ClientInterceptor clientInterceptor;
                        if (beanFactory.getBeanNamesForType(ClientInterceptor.
                            class).length > 0) {
                            clientInterceptor = beanFactory.getBean(clientInter
                                ceptorClass);
                        } else {
                            try {
                                clientInterceptor = clientInterceptorClass.
                                    newInstance();
                            } catch (Exception e) {
                                throw new BeanCreationException("Failed to
                                    create interceptor instance", e);
                            }
                        }
                        list.add(clientInterceptor);
                    }

                    Channel channel = channelFactory.createChannel(annotation.
                        value(), list);
                    ReflectionUtils.makeAccessible(field);
                    ReflectionUtils.setField(field, target, channel);
                }
            }
        }
    }
    return bean;
}
```

创建 Channel 对象是在 DiscoveryClientChannelFactory 类中进行的,它是根据被调用的服务提供者从 GrpcChannelProperties 中获取相应的配置,作负载均衡的策略,以及一些其他的配置,并为这个 Channel 设置拦截器。具体代码如下:

```java
public Channel createChannel(String name, List<ClientInterceptor> interceptors) {
    GrpcChannelProperties channelProperties = properties.getChannel(name);
    NettyChannelBuilder builder = NettyChannelBuilder.forTarget(name)
            .loadBalancerFactory(loadBalancerFactory)
            .nameResolverFactory(new DiscoveryClientResolverFactory(client,
                this))
            .usePlaintext(properties.getChannel(name).isPlaintext());
    if (channelProperties.isEnableKeepAlive()) {
        builder.keepAliveWithoutCalls(channelProperties.isKeepAlive-
            WithoutCalls())
```

```
            .keepAliveTime(channelProperties.getKeepAliveTime(),
                TimeUnit.SECONDS)
            .keepAliveTimeout(channelProperties.getKeepAliveTimeout(),
                TimeUnit.SECONDS);
    }
    if(channelProperties.getMaxInboundMessageSize() > 0) {
        builder.maxInboundMessageSize(channelProperties.getMaxInboundMessageSize());
    }
    Channel channel = builder.build();

    List<ClientInterceptor> globalInterceptorList = globalClientInterceptorRegistry.
        getClientInterceptors();
    Set<ClientInterceptor> interceptorSet = Sets.newHashSet();
    if (globalInterceptorList != null && !globalInterceptorList.isEmpty()) {
        interceptorSet.addAll(globalInterceptorList);
    }
    if (interceptors != null && !interceptors.isEmpty()) {
        interceptorSet.addAll(interceptors);
    }
    return ClientInterceptors.intercept(channel, Lists.newArrayList(interceptorSet));
}
```

被 @GrpcClient 注解修饰的变量，在 Spring 启动之后，是一个连接了 gRPC Server 的 Channel 对象，该对象具有向 gRPC Server 发送消息和接收 gRPC Server 响应的能力。

20.4 案例实战

在本案例中，将使用上述的 gRPC 框架，即 gRPC Spring Boot Starter 框架进行远程服务的调用，取代了 Spring Cloud 原生支持的 HTTP 客户端，比如 FeignClient、RestTemplate。在本案例中采用的 Maven 为多个 Module 的形式，一共分为四个工程，如表 20-1 所示。

表 20-1 案例工程清单

工　　程	作　　用	备　　注
cloud-eureka-server	Eureka 注册中心	它作为 Eureka Server
grpc-lib	存放 gRPC 的 IDL 文件	它需要被 cloud-grpc-server 和 cloud-grpc-client 引用
cloud-grpc-server	gRPC 的服务端，即服务提供者	它同时还是 Eureka Client、Zipkin Client
cloud-grpc-client	gRPC 的客户端，即服务消费者	它同时还是 Eureka Client、Zipkin Client

需要说明的是，gRPC Spring Boot Starter 不仅支持 Eureka，还支持 Consul，本案例使用的是 Eureka。

20.4.1 注册中心

cloud-eureka-server 工程为使用 Eureka 组件实现的服务注册中心，在这里不做过多的描述，具体代码可以参照源码，也可以参照前面第 2 章。在本案例中 Eureka Server 的端口为 8761。

20.4.2 链路追踪服务端

在 gRPC Spring Boot Starter 框架中集成了 Spring Cloud Slueth 作为链路追踪工具。在最新的 Spring Cloud 版本中，Spring Cloud Slueth 是以 Jar 包的形式启动的，官方默认不允许修改其 Jar 包，启动方式如下：

```
curl -sSL https://zipkin.io/quickstart.sh | bash -s
java -jar zipkin.jar
```

也可以到网址 https://dl.bintray.com/openzipkin/maven/io/zipkin/java/zipkin-server/ 下载具体版本的 Jar，然后以 java -jar 命令启动下载的 Jar 包即可。zipkin-server 默认启动的端口为 9411。

20.4.3 gRPC 的 lib 工程

grpc-lib 工程作为 gRPC 的 IDL 文件存放工程，承担了 libary 的角色，不仅需要为 cloud-grpc-server 和 cloud-grpc-client 工程提供公共依赖，还需要编译 IDL 文件供 cloud-grpc-server 和 cloud-grpc-client 工程调用。在工程的 pom 文件需要引入以下依赖，如代码清单 20-1 所示。

代码清单20-1　ch20/grpc-lib/pom.xml

```xml
<dependency>
    <groupId>io.grpc</groupId>
    <artifactId>grpc-netty</artifactId>
</dependency>
<dependency>
    <groupId>io.grpc</groupId>
    <artifactId>grpc-protobuf</artifactId>
</dependency>
<dependency>
    <groupId>io.grpc</groupId>
    <artifactId>grpc-stub</artifactId>
</dependency>
<dependency>
    <groupId>io.netty</groupId>
    <artifactId>netty-common</artifactId>
</dependency>
```

除了上面 4 个必须的依赖外，还需要在 pom 文件中引入编译 Proto Buffer 的 IDL 文件的两个 Maven 插件，分别为 os-maven-plugin 和 protobuf-maven-plugin，具体可参考上一章。

在工程的 /src/main 文件下创建一个 proto 的文件夹，用来存放定义的 IDL 文件。在 proto 文件夹下，新建一个 helloword.proto 文件，如代码清单 20-2 所示。

代码清单20-2　ch20/grpc-lib/src/main/proto/helloworld.proto

```
syntax = "proto3";

option java_multiple_files = true;
```

```
option java_package = "cn.springcloud.grpc.lib";
option java_outer_classname = "HelloWorldProto";
service Simple {
        rpc SayHello (HelloRequest) returns (HelloReply) {
    }
}
message HelloRequest {
    string name = 1;
}
message HelloReply {
    string message = 1;
}
```

在上述文件中，采用的 Proto Buffer 的语言版本为 3；IDL 文件编译生成的 Java 文件的包名为 "cn.springcloud.grpc.lib"，文件名为 HelloWorldProto；定义了一个请求的消息为 HelloRequest，该消息包含一个变量为 name；定义了一个响应消息为 HelloReply，该消息包含一个变量为 message；最后定义了一个服务，方法为 SayHello，参数为 HelloRequest，返回 HelloReply。

然后，在 grpc-lib 的工程目录下，执行 mvn clean insatll 命令，执行成功后，会在 /target/generated-sources/protobuf 包下面生成具体的 java 类，包括 SimpleGrpc、HelloReply、HelloRequest 等。

20.4.4 gRPC 服务端

cloud-grpc-server 作为 gRPC 服务端，为 gRPC 客户端提供服务功能。当它作为 Eureka Client 时，需要在工程的 pom 文件引入 Eureka 的起步依赖 spring-cloud-starter-eureka；作为 Zipkin Client 时，需要引入 Zipkin 的起步依赖 spring-cloud-starter-zipkin；作为 gRPC Server 时，需要引入起步依赖 grpc-server-spring-boot-starter，引入 grpc-lib 工程。具体依赖版本见工程源码，如代码清单 20-3 所示。

代码清单20-3　ch20/cloud-grpc-server/pom.xml

```xml
<dependency>
        <groupId>org.springframework.cloud</groupId>
        <artifactId>spring-cloud-starter-netflix-eureka-client</artifactId>
</dependency>
<dependency>
        <groupId>org.springframework.boot</groupId>
        <artifactId>spring-boot-starter-web</artifactId>
</dependency>
<dependency>
        <groupId>org.springframework.boot</groupId>
        <artifactId>spring-boot-starter-actuator</artifactId>
</dependency>
<dependency>
        <groupId>org.springframework.cloud</groupId>
        <artifactId>spring-cloud-starter-zipkin</artifactId>
</dependency>
<dependency>
```

```xml
        <groupId>cn.springcloud</groupId>
        <artifactId>grpc-lib</artifactId>
        <version>0.0.1-SNAPSHOT</version>
</dependency>
<dependency>
        <groupId>net.devh</groupId>
            <artifactId>grpc-server-spring-boot-starter</artifactId>
        </dependency>
        <dependency>
            <groupId>io.zipkin.brave</groupId>
            <artifactId>brave-instrumentation-grpc</artifactId>
            <version>${brave.instrumentation.grpc}</version>
</dependency>
```

在工程的配置文件 application.yml 做相关的工程配置，其中应用名为 cloud-grpc-server。Spring Boot 启动端口为 8081，其中 grpc.server.port 配置的是 gRPC 的 netty 端口，会随机分配端口；spring.sleuth.sampler.probability 配置的是链路数据上传给 Zipkin Server 的比例，在本案例中是 100% 上传，具体如代码清单 20-4 所示。

代码清单20-4 \ch20\cloud-grpc-server\src\main\resources\application.yml

```yaml
spring:
    application:
        name: cloud-grpc-server
    sleuth:
        sampler:
            probability: 1
server:
    port: 0
grpc:
    server:
        port: 8081
eureka:
    instance:
        prefer-ip-address:true
        instanceId: ${spring.application.name}:${vcap.application.instance_
            id:${spring.application.instance_id:${random.value}}}
    client:
        register-with-eureka: true
        fetch-registry: true
        service-url:
            defaultZone: http://localhost:8761/eureka/
```

在程序的启动类 CloudGrpcServerApplication 加 @SpringBootApplication 注解、@EnableEurekaClient 注解和 @EnableDiscoveryClient 注解，如代码清单 20-5 所示。

代码清单20-5 ch20\cloud-grpc-server\CloudGrpcServerApplication.java

```java
@SpringBootApplication
@EnableEurekaClient
@EnableDiscoveryClient
public class CloudGrpcServerApplication {
```

```java
    public static void main(String[] args) {
        SpringApplication.run( CloudGrpcServerApplication.class, args );
    }
}
```

写一个 LogGrpcInterceptor，该类实现了 gRPC 的 ClientInterceptor，用来拦截请求，并输出请求的具体方法的日志，然后释放请求，如代码清单 20-6 所示。

代码清单20-6　ch20\cloud-grpc-server\ClientInterceptor.java

```java
public class LogGrpcInterceptor implements ClientInterceptor {
    private static final Logger log = LoggerFactory.getLogger(LogGrpcInterceptor.class);
    @Override
    public <ReqT, RespT> ClientCall<ReqT, RespT> interceptCall(MethodDescriptor
        <ReqT, RespT> method, CallOptions callOptions, Channel next) {
        log.info(method.getFullMethodName());
        return next.newCall(method, callOptions);
    }
}
```

创建一个 GlobalClientInterceptorConfiguration 类，加上 @Configuration 注解（表明该类为 Spring 的配置类），加上 @Order 注解（该注解配置了 Bean 的注入优先级，默认优先级最低）。然后注入一个 GlobalClientInterceptorConfigurerAdapter 的 Bean，在 Bean 中注册一个 LogGrpcInterceptor，如代码清单 20-7 所示。

代码清单20-7　ch20\cloud-grpc-server\GlobalClientInterceptorConfiguration.java

```java
@Order
@Configuration
public class GlobalClientInterceptorConfiguration {

    @Bean
    public GlobalClientInterceptorConfigurerAdapter globalInterceptorConfigurer
        Adapter() {
        return new GlobalClientInterceptorConfigurerAdapter() {

            @Override
            public void addClientInterceptors(GlobalClientInterceptorRegistry
                registry) {
                registry.addClientInterceptors(new LogGrpcInterceptor());
            }
        };
    }
}
```

写一个 gRPC 的服务类 GrpcServerService，该类继承了 SimpleGrpc.SimpleImplBase（在 grpc-lib 工程中根据 helloworld.proto 生成的 Java 类），在该类上加 @GrpcService(SimpleGrpc.class) 注解，注册 SimpleGrpc 类为一个 GrpcService。然后在 sayHello 方法中，获取 HelloRequest 的 name，创建一个 HelloReply，并返回，最后结束响应，如代码清单 20-8 所示。

代码清单20-8 ch20\cloud-grpc-server\src\main\java\cn\springcloud\cloudgrpcserver\GrpcServerService.java

```java
@GrpcService(SimpleGrpc.class)
public class GrpcServerService extends SimpleGrpc.SimpleImplBase {

    @Override
    public void sayHello(HelloRequest req, StreamObserver<HelloReply> responseObserver) {
        HelloReply reply = HelloReply.newBuilder().setMessage("Hello =============> " + req.getName()).build();
        responseObserver.onNext(reply);
        responseObserver.onCompleted();
    }
}
```

至此，gRPC 服务端实现了向 Eureka Server 注册，并作为 Zipkin Client，还提供了 gRPC 服务的全部功能。

20.4.5 gRPC 客户端

cloud-grpc-client 作为 gRPC 的客户端，调用了 cloud-grpc-server 提供的 SimpleGrpc 的服务。除了需要在工程的 pom 文件引入 spring-cloud-starter-eureka、spring-cloud-starter-zipkin、grpc-lib、spring-boot-starter-actuator 依赖外，还需要引入 grpc-client-spring-boot-starter 依赖，具体依赖版本见工程源码，如代码清单 20-9 所示。

代码清单20-9 ch20\cloud-grpc-client\pom.xml

```xml
<dependency>
    <groupId>net.devh</groupId>
    <artifactId>grpc-client-spring-boot-starter</artifactId>
</dependency>
```

在工程的配置文件 application.yml 中配置程序的端口为 8080，服务名为 cloud-grpc-client，以及 zipkin 与服务注册的一些配置信息，如代码清单 20-10 所示。

代码清单20-10 ch20\ cloud-grpc-client \src\main\resources\application.yml

```yml
server:
    port: 8080
spring:
    application:
        name: cloud-grpc-client
    sleuth:
        sampler:
            probability: 1
eureka:
instance:
    prefer-ip-address: true
    status-page-url-path: /actuator/info
```

```yaml
    health-check-url-path: /actuator/health
nstanceId: ${spring.application.name}:${vcap.application.instance_
    id:${spring.application.instance_id:${random.value}}}
client:
    register-with-eureka: true
    fetch-registry: true
    service-url:
        defaultZone: http://localhost:8761/eureka/
```

在工程的启动类 CloudGrpcClientApplication 中加上相关的注解，同 cloud-grpc-server 的启动类，另外需要配置 LogGrpcInterceptor 类，打印请求日志，具体实现同 cloud-grpc-server 工程。

写一个服务类 GrpcClientService，在类上加 @Service 注解，将该类的 Bean 注入 Spring 容器中。并使用 @GrpcClient 注入一个 Channel 变量，该 Channel 为客户端和服务名为 cloud-grpc-server 的 gRPC 服务端创建了一个 Channel 连接。使用该 Channel 创建一个阻塞的 Stub，使用该 Stub 向 gRPC 服务端发送一条消息，并阻塞线程，直至接收到服务响应，具体如代码清单 20-11 所示。

代码清单20-11　ch20\cloud-grpc-client\GrpcClientService.java

```java
@Service
public class GrpcClientService {

    @GrpcClient("cloud-grpc-server")
    private Channel serverChannel;

    public String sendMessage(String name) {
        SimpleGrpc.SimpleBlockingStub stub = SimpleGrpc.newBlocking
            Stub(serverChannel);
        HelloReply response = stub.sayHello(HelloRequest.newBuilder().
            setName(name).build());
        return response.getMessage();
    }
}
```

写一个 GrpcClientController 类，加上 @RestController 注解，表明该类为 RestController。写一个 Rest API，在 API 方法中调用 GrpcClientService 的 sendMessage 方法，即调用 gRPC 客户端向 gRPC 服务端发送消息，获取 gRPC 服务端的响应后返回给 API 请求，如代码清单 20-12 所示。

代码清单20-12　ch20\cloud-grpc-client\GrpcClientController.java

```java
@RestController
public class GrpcClientController {

    @Autowired
    private GrpcClientService grpcClientService;

    @RequestMapping("/")
    public String printMessage(@RequestParam(defaultValue = "Spring Cloud")
```

```
            String name) {
        return grpcClientService.sendMessage(name);
    }
}
```

依次启动 cloud-eureka-server、cloud-zipkin-server、cloud-grpc-server 和 cloud-grpc-client 工程。在浏览器上访问 cloud-grpc-client 工程提供的 API http://localhost:8080/，浏览器显示如下：

```
Hello ==============> Spring Cloud
```

该字符串为 cloud-grpc-server 返回给 cloud-grpc-client 的消息，可见 cloud-grpc-client 通过 gRPC 消费了 cloud-grpc-server 提供的服务。

打开浏览器 http://localhost:9411/，进入 Zipkin Server 的展示页面，该页面可以看到服务的调用链，如图 20-1 所示。

图 20-1　Zipkin Server 展示界面

单击一个 trace，可以看见具体的链路情况，首先通过 HTTP 调用了 cloud-grpc-client，由 cloud-grpc-client 调用 cloud-grpc-server，然后各个阶段将数据返回，可以查看具体的消耗时间，如图 20-2 所示，总耗时为 66.743ms。

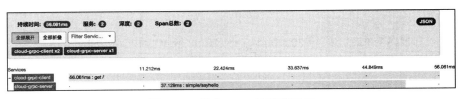

图 20-2　单个 trace 的链路展示

单击依赖分析，可以查看服务之间的链路关系，在本案例中 cloud-grpc-client 依赖于 cloud-grpc-server 工程，如图 20-3 所示。

图 20-3　服务的依赖情况

20.5　本章小结

本章全面介绍了 gRPC Spring Boot Starter 这个项目。首先介绍了它的架构设计和实现原理，紧接着讲述了它的源码分析。最后，以案例实战作为结束。通过上一章和本章的学习，相信读者对 gRPC 有一个较全面的认识，如果有任何问题，可以和我联系。

第 21 章

Spring Cloud 版本控制与灰度发布

软件系统的版本更新与发布上线，是几乎所有 IT 公司都会面临的一个问题，也是一个难点。如何保证系统的平滑上线？如何在上线过程中兼顾迭代功能的准确性？如何保证上线过程中不影响用户使用系统功能，让用户毫无感知？如何契合当今流行的 DevOps 理念？这一系列问题无不考验着广大 IT 工程师，让我们带着这些问题来阅读。本章将会就各种主流发布更新方式，以及 Spring Cloud 微服务体系下的系统版本迭代过程，进行一次全面的讲解。

21.1 背景

如果你经历过一次完整的系统迭代上线过程，对于其中遇到的一些问题一定不会陌生。在传统的应用当中，我们可能会首先备份老版本包，接着才会停掉当前应用服务并且上传新包，最后重启应用服务。姑且不论这样的发布方式有多繁琐，中间应用宕调的那一段时间是十分致命的，这在流量较大的互联网公司是无法忍受的，因为无法找到一个完全没有用户请求的时间点来做这些事情。有一些技术团队稍微高明一些，在应用的上层准备一台 Nginx，当需要更新应用时，先在冗余机器或者当前机器使用不同端口启动新版本的应用，然后使用 Nginx 把流量切过来，这样虽然免去了应用的重启过程，把问题抛给 Nginx，但是也有一些问题，比如需要两倍的硬件资源，以及 Nginx 的重启耗时。总而言之，传统应用要做到用户无感知的系统迭代与发布上线，依靠固化的思维是存在很大问题的，为了解决这些问题，业界衍生出一些发布管理方式，或在应用启动的同时做到瞬时切换，或在线测试新版本应用，功能验证完毕达到发布标准之后再切换上线，技术团队只需结合自身应用场景来选取合适的发布管理方式，解决系统上线遇到的种种问题即可。

由于 Spring Cloud 具有微服务体系完善的应用组件，其应用系统完全拆分为服务的粒度，对于发布管理方式有天然的实现优势。在第 8 章已经对基于 Zuul 的灰度发布方式进行了简单的陈述与实践，其中应用到的原理无非就是注册中心的 metadata 搭配特定的负载均衡机制，然后整合成一个 Filter 注册到 Zuul 的上下文环境，这样请求就可以带上版本标签经过路由访问到相应的服务版本，详细内容请翻阅 8.5 节。

21.2 常见发布方式

本书在 8.5 节简单地介绍过灰度发布的概念，鉴于本章的主题，在引入灰度发布概念之前，我们先来谈一谈目前主流的发布方式，在有对比的情况下，各种方式的异同以及优缺点就会慢慢浮现。

21.2.1 蓝绿发布

蓝绿发布（Blue & Green Deployment）是一种可以"零停服"的应用发布方式，可以保证应用平滑切换上线。其原理十分简单，即用蓝与绿来区分两套版本不一样的环境，一般用绿环境来表示旧版本，此时全部的流量由该环境承担，蓝环境则是一套独立的新版本应用环境，在部署之后如果测试通过，则将流量由绿切换到蓝，完成一次几乎无感知的上线过程，其过程如图 21-1、图 21-2 所示。

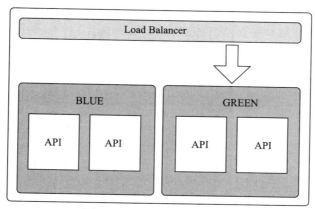

图 21-1　蓝绿发布第一阶段

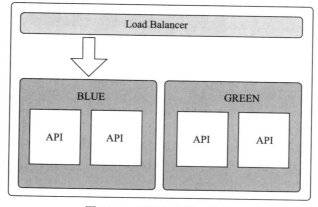

图 21-2　蓝绿发布第二阶段

21.2.2 滚动发布

滚动发布（Rolling Update Deployment）一般是将集群里的少量节点进行更新上线，然后再更新其他节点，每次只更新少量节点，直到全部更新完毕，整个系统变为新版本，如图 21-3、图 21-4 所示。

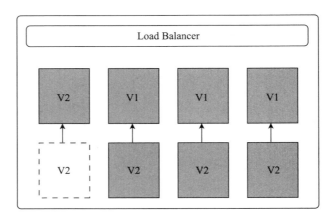

图 21-3　滚动发布第一阶段

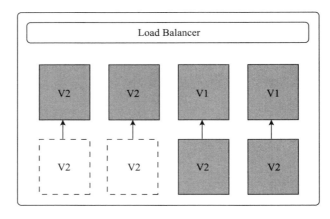

图 21-4　滚动发布第二阶段

21.2.3 灰度发布

在本书 8.5.1 节中已经有对它的初步描述，灰度发布（Gray Deployment）是一个统称，它是目前应用面比较广的一种发布方式，它的思想是在线上环境中部署一个新版本应用，然后引入一小部分流量进入其中，如果没有发现问题就切换上线，这种方式对于测试应用在线上环境的实际表现情况极其有效，我们在灰度阶段就可以发现问题，及时调整，并且用户无感知。目前灰度发布可以分为两种，一种是 A/B 测试，另一种是金丝雀部署。

1. A/B 测试（A/B Testing）

A 版本是线上稳定版本，B 版本是迭代版本，如果一下子切到 B 环境，可能用户会难以适应，所以先部署一个 B 环境，分一部分流量过来，收集用户反馈然后逐步改进 B 版本，直到用户可以接受完全用 B 版本替换 A 版本的程度，如图 21-5 所示。

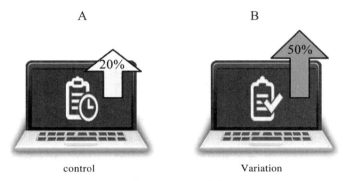

图 21-5　A/B 测试

可能读者会觉得蓝绿发布与 A/B 测试是一个概念，其实不然，蓝绿发布关注的是新版本的发布，而 A/B 测试关注的是一个测试的过程，两者还是有本质的区别的。

2. 金丝雀部署（Canary Deployment）

背景出自"矿井中的金丝雀"：17 世纪，英国矿井工人发现，金丝雀对瓦斯这种气体十分敏感。哪怕空气中只有极其微量的瓦斯，金丝雀也会停止歌唱；当瓦斯含量超过一定限度时，虽然鲁钝的人类毫无察觉，金丝雀却早已毒发身亡。当时在采矿设备相对简陋的条件下，工人们每次下井都会带上一只金丝雀作为瓦斯检测指标，以便及时发现危险状况并紧急撤离。

后来的软件工程师借鉴这个典故来升级应用，在黑白之间平滑发布过渡，一般是在应用集群中部署一台实例作为"金丝雀"，引入一小部分流量，收集问题，及时调整，待达到上线标准再替换集群中其他实例，如图 21-6 所示。

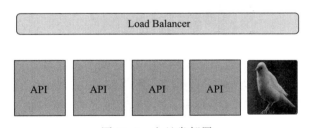

图 21-6　金丝雀部署

21.2.4　对比

对于上述几种常见发布方式，可以结合自身应用的场景与特性来选取适宜的方式，现对几种发布方式的优劣进行总结，如表 21-1 所示。

表 21-1　常见发布方式优劣总结

发布方式	优　势	劣　势
蓝绿发布	1）升级与回滚速度较快； 2）无须停服，风险较小	1）需要考虑好升级切换与回滚的边界； 2）资源消耗是线上应用的两倍，因为需要做冗余； 3）两个版本的业务一致性问题
滚动发布	用户感知较小	1）发布过程较长； 2）对发布工具要求较高
A/B 测试	用户感知较小	搭建复杂度较高
金丝雀部署	用户体验影响小，灰度发布过程出现问题只影响少量用户	发布自动化程度不够，发布期间可引发服务中断

在"开发 – 测试 – 上线"阶段，对于用户与系统服务商，不难看出，对整个流程皆要兼顾的是灰度发布，这也就是目前灰度发布较为流行的原因，本章的实战内容就基于灰度发布来展开。

21.3　版本控制与灰度发布实战

目前 Spring Cloud 技术栈的版本与灰度实战项目已有不少，但由于缺少支持与活跃度，一直止步不前。截至目前，笔者所见过的相关项目中，维护度最高、功能最丰富，且有实际落地案例的项目，来自于 Spring Cloud 中国社区核心成员任浩军，感谢其贡献项目回馈社区，项目地址为 https://github.com/Nepxion/Discovery。

Nepxion Discovery 是一款对 Spring Cloud 的服务注册发现的增强中间件，其功能包括多版本灰度发布，黑 / 白名单的 IP 地址过滤，限制注册等，支持 Eureka、Consul 和 Zookeeper，支持 Spring Cloud Gateway、Zuul 和微服务的灰度发布，支持用户自定义和编程灰度路由策略，支持 Nacos 和 Redis 为远程配置中心，支持 Spring Cloud Edgware 版和 Finchley 版。现有的 Spring Cloud 微服务可以很方便地引入该插件，代码零侵入。使用者只需要关注如下事项：

- 引入相关 Plugin Starter 依赖到 pom.xml。
- 必须为微服务定义一个版本号（version），必须为微服务自定义一个便于为微服务归类的 Key，例如组名（group）或者应用名（application）。两者定义在 application.properties 或者 yaml 的 metadata 里，便于远程配置中心推送和灰度界面分析。
- 使用者只需要关注相关规则推送，可以采用如下任一方式：
  - 通过远程配置中心推送规则。
  - 通过控制台界面推送规则。
  - 通过客户端工具（例如 Postman）推送。

本节我们只关注该项目在版本控制与灰度发布的部分，鉴于篇幅原因，本次实战讲解完整的基于图形界面的版本灰度切换过程，注册中心为 Eureka，路由组件为 Zuul。

21.3.1　Discovery 项目

由于图形界面是一个 swing 项目，所以为了方便考虑，这里采用本地部署的方式，后期如需使用此项目，可以自行打包。部署完成后，运行 discovery-console-desktop 项目下的 ConsoleLauncher 主类，如图 21-7 所示。

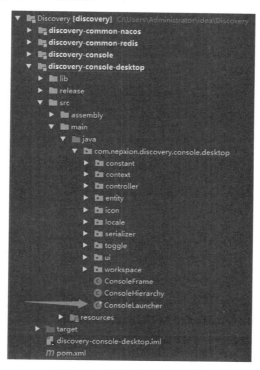

图 21-7　图形界面启动主类

运行完成后即可弹出一个图形界面窗口，这便是进行版本灰度发布的主界面，如图 21-8 所示。

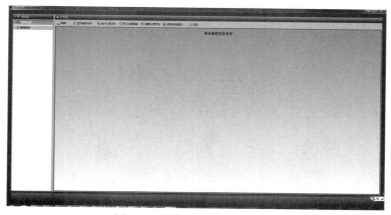

图 21-8　版本灰度发布主界面

21.3.2 实战案例

实战案例一共有 4 个工程，注册中心 Eureka、路由组件 Zuul、Discovery 独立控制台 console，以及源服务，具体信息如表 21-2 所示。

表 21-2 版本灰度实战案例工程描述

项目名称	描　　述
ch21-1-eureka-server	注册中心
ch21-1-zuul-server	路由网关
ch21-1-discovery-console	Discovery 独立控制台，注意 图形界面项目需配置此项目地址与端口号
ch21-1-original-service	内置 7 个启动主类，7 套配置，用于启动 A1、A2、B1、B2、C1、C2、C3 服务实例

由于代码量较大，这里就只讲解核心部分，首先我们需要了解案例调用拓扑关系，如图 21-9 所示。

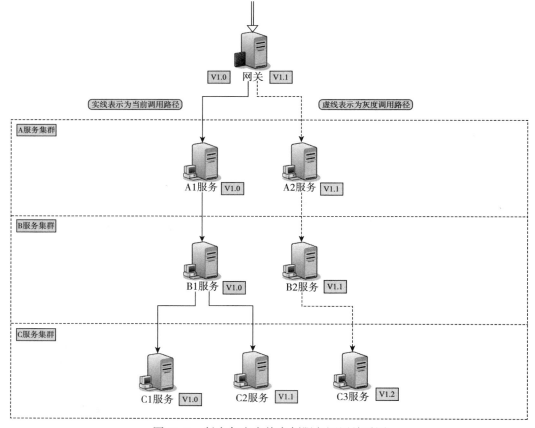

图 21-9 版本灰度实战案例服务调用关系图

图 21-9 中实线表示正常的调用路径，虚线表示我们将要使用图形界面发布的灰度调用路径。

在 Discovery 项目中，我们可以使用 xml 文件或者 json 文件来描述版本之间的调用关系，例如 A1 服务 V1.0 版本调用 B1 服务 V1.0 版本，用 xml 描述为：

```xml
<?xml version="1.0" encoding="UTF-8"?>
<rule>
    <discovery>
        <version>
            <service consumer-service-name="A1" provider-service-name="B1"
                consumer-version-value="1.0" provider-version-value="1.0"/>
        </version>
    </discovery>
</rule>
```

详细案例代码见本书源码部分 ch21-1。

21.3.3 实战测试

1）分别启动 ch21-1-eureka-server、ch21-1-zuul-server、ch21-1-discovery-console、ch21-1-original-service（original-service 有 7 个启动主类）。

2）在图形界面单击"显示服务拓扑"，弹出框内选择"example-service-group"，如图 21-10 所示。

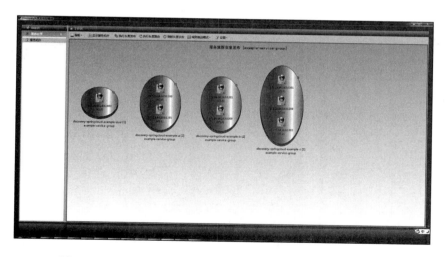

图 21-10　图形界面 example-service-group 服务 group 显示图

3）查看当前调用路径，右键单击 Zuul 服务图标，选择"执行灰度路由"，在弹出的界面上方添加"服务列表"下拉框中的服务，单击右上角的"执行路由"，结果如图 21-11 所示。

4）使用 postman 测试，如图 21-12、图 21-13 所示，结果完全符合图 21-11 的服务调用关系。

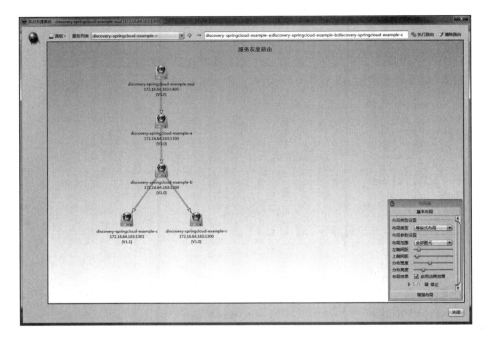

图 21-11　原始执行路径

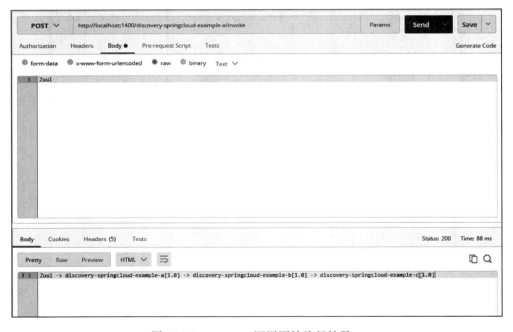

图 21-12　postman 调用原始路径结果一

图 21-13　postman 调用原始路径结果二

5）将调用版本 V1.0 切换为 V1.1，也就是将图 21-9 的实线切换到虚线，这时的操作需要右键单击 Zuul 服务，在弹出框中选择"执行灰度发布"，在弹出框的文本框填写 1.1，表示切换到 V1.1 版本，填好之后单击"更新灰度版本"，弹出框显示 OK 即表示切换成功，此时回到主页面，Zuul 图标会闪烁。本次操作界面如图 21-14 所示。

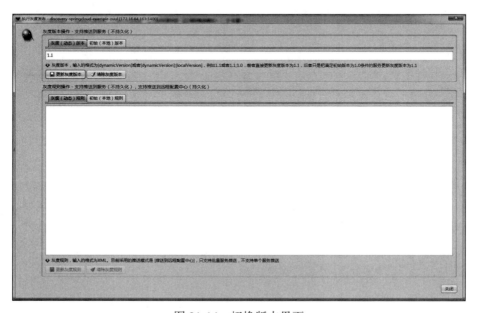

图 21-14　切换版本界面

再次查看 Zuul 服务中的路由关系，重复第 3 步操作，如图 21-15 所示。

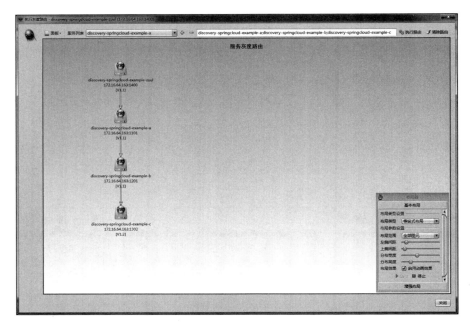

图 21-15　执行灰度路由后的调用关系

6）使用 postman 调用，结果如图 21-16 所示，完全符合图 21-15 中的调用关系。

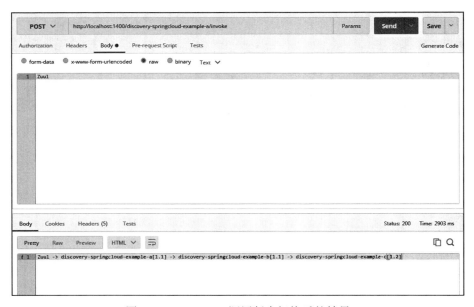

图 21-16　postman 调用版本切换后的结果

21.4 本章小结

本章就市面上常见的发布方式做了一些讲解，借此引出灰度发布，说明了灰度发布流行的原因以及它的实现方式，并且讲解了一个开源版本控制与灰度发布项目，基于图形化界面实现灰度发布的过程。

第 22 章

Spring Cloud 容器化

容器化技术的出现标准化了服务的基础设施，统一了应用的打包分发、部署及操作系统相关类库等，解决了测试及生产部署时环境差异的问题，更方便分析排查问题。对运维来说，由于镜像的不可变性，更容易进行服务部署升级及回滚。另外利用诸如 Kubernetes 之类的容器管理平台，更容易实现一键部署、扩容、缩容等操作，非常方便，更能将微服务架构、DevOps、不可变基础设施的思想落地下来。本章就来讲述一下如何将 Spring Cloud 使用 Docker 容器化，然后使用 Kubernetes 进行管理。

22.1 Java 服务 Docker 化

22.1.1 基础镜像选择

操作系统层面，可以选择传统的 Centos、Ubuntu 或者轻量级的 Alpine。像 Ubuntu 16.04 版本的镜像大小约为 113M，压缩后大约 43M；Centos 7 版本的镜像大小约为 199M，压缩后大约 73M；而 Alpine 3.7 版本镜像大小约为 4.15M，压缩后约为 2M。

关于基础镜像的选择，一个是考虑镜像大小，一个是只提供最小的依赖包。关于第二点不同的服务应用依赖包是不同的，这里不再展开，只从镜像大小角度考虑的话，Alpine 是首选，镜像小，远程推拉镜像的速度快，更为方便，这里采用 Alpine 镜像作为基础镜像。

如果从 Docker 镜像分层缓存的机制来考虑，如果选择了比较大的基础镜像，dockerfile 编写时适当分层，然后集中在几台镜像打包机上处理镜像打包及上传，这样可以充分利用打包机镜像分层缓存的机制，减少上传镜像的耗时。但是对于分布式服务的 Docker 部署，目标服务实例部署的机器比较多而且是随机的，就没办法利用这个机制，减少镜像下载速度。

22.1.2　Dockerfile 编写

选择 Alpine 有个麻烦的地方就是 Alpine 采用的是 musl libc 的 C 标准库，而 Oracle 或 OpenJDK 提供的版本则主要是以 glibc 为主，虽然 OpenJDK 在一些早期版本会放出使用 musl libc 编译好的版本，不过在正式发布的时候，并没有单独的 musl libc 编译版本可以下载，需要自己单独编译，稍微有些不便。因此我们考虑在 Alpine 上加上 glibc，然后添加 glibc 的 JDK 编译版本作为基础镜像。

1. Alpine+glibc

这里我们选择 Alpine3.7 版本，glibc 采用 Sgerrand 开源的 glibc 安装包（https://github.com/sgerrand/alpine-pkg-glibc），版本为 2.27-r0。其 Dockerfile 如代码清单 22-1 所示：

代码清单22-1　ch22-1\ch22-1-2-dockerfile\alpine-glibc\Dockerfile

```
FROM alpine:3.7
MAINTAINER caibosi <caibosi@139.com>
RUN apk add --no-cache ca-certificates curl openssl binutils xz tzdata \
    && ln -sf /usr/share/zoneinfo/Asia/Shanghai /etc/localtime \
    && echo "Asia/Shanghai" > /etc/timezone \
    && GLIBC_VER="2.27-r0" \
    && ALPINE_GLIBC_REPO="https://github.com/sgerrand/alpine-pkg-glibc/releases/
       download" \
    && curl -Ls ${ALPINE_GLIBC_REPO}/${GLIBC_VER}/glibc-${GLIBC_VER}.apk > /
       tmp/${GLIBC_VER}.apk \
    && apk add --allow-untrusted /tmp/${GLIBC_VER}.apk \
    && curl -Ls https://www.archlinux.org/packages/core/x86_64/gcc-libs/download
       > /tmp/gcc-libs.tar.xz \
    && mkdir /tmp/gcc \
    && tar -xf /tmp/gcc-libs.tar.xz -C /tmp/gcc \
    && mv /tmp/gcc/usr/lib/libgcc* /tmp/gcc/usr/lib/libstdc++* /usr/glibc-
       compat/lib \
    && strip /usr/glibc-compat/lib/libgcc_s.so.* /usr/glibc-compat/lib/
       libstdc++.so* \
    && curl -Ls https://www.archlinux.org/packages/core/x86_64/zlib/download >
       /tmp/libz.tar.xz \
    && mkdir /tmp/libz \
    && tar -xf /tmp/libz.tar.xz -C /tmp/libz \
    && mv /tmp/libz/usr/lib/libz.so* /usr/glibc-compat/lib \
    && apk del binutils \
    && rm -rf /tmp/${GLIBC_VER}.apk /tmp/gcc /tmp/gcc-libs.tar.xz /tmp/libz /
       tmp/libz.tar.xz /var/cache/apk/*
```

这里有几点需要注意：

- 由于 Docker 镜像采用的是分层机制，因此安全类库或软件的命令最好在同一行命令中，减少分层，以降低最后镜像的大小。
- 命令中间安装了类库或软件包，需要在同一行命令中删除 apk 的 cache，这样才能有效删除 apk，减少镜像大小。
- 这里安装了 openssl、curl、xz、tzdata，同时把 timezone 改为了 Asia/Shanghai。
- 这里笔者已经创建好一个版本，上传到了阿里云的镜像仓库上，有兴趣的读者可以直接 pull 这个镜像：

```
docker pull registry.cn-hangzhou.aliyuncs.com/springcloud-cn/alpine-3.7:glibc-2.27-r0
```

2. Alpine+glibc+JDK8

对于 JDK 版本的选择，有 Oracle 的 Hotspot JDK，也有 OpenJDK。对于 Oracle 的 JDK，个人使用及非商业使用是免费的，而对于商业使用来说，需进行企业订阅，在 2019 年 1 月之后才能继续获得 Java SE 8 更新。Oracle 已经建议选择不订阅或不继续订阅的公司在订阅结束之前，把 JDK 版本迁移到 OpenJDK，以确保相关应用程序不受影响。

这里我们对 JDK 8 版本采用 Oracle 的 server-jre-8u172 版本，而对于 JDK9、10 及 11 版本采取 OpenJDK 来构建。

Oracle 的 JDK 8 版本的 Dockerfile 如代码清单 22-2 所示：

代码清单22-2　ch22-1\ch22-1-2-dockerfile\java8 \Dockerfile

```
FROM registry.cn-hangzhou.aliyuncs.com/springcloud-cn/alpine-3.7:glibc-2.27-r0
MAINTAINER caibosi <caibosi@139.com>
ADD server-jre-8u172-linux-x64.tar.gz /opt/
RUN chmod +x /opt/jdk1.8.0_172
ENV JAVA_HOME=/opt/jdk1.8.0_172
ENV PATH="$JAVA_HOME/bin:${PATH}"
```

为了方便使用，可以直接使用笔者构建好的镜像，使用如下命令拉取：

```
docker pull registry.cn-hangzhou.aliyuncs.com/springcloud-cn/java:8u172-jre-alpine
```

验证如下：

```
docker run --rm -it registry.cn-hangzhou.aliyuncs.com/springcloud-cn/java:8u172-jre-alpine java -version
java version "1.8.0_172"
Java(TM) SE Runtime Environment (build 1.8.0_172-b11)
Java HotSpot(TM) 64-Bit Server VM (build 25.172-b11, mixed mode)
```

3. Alpine+glibc+OpenJDK9

OpenJDK 9 的 Dockerfile 如代码清单 22-3 所示：

代码清单22-3　ch22-1\ch22-1-2-dockerfile\java9 \Dockerfile

```
FROM registry.cn-hangzhou.aliyuncs.com/springcloud-cn/alpine-3.7:glibc-2.27-r0
MAINTAINER caibosi <caibosi@139.com>
ADD openjdk-9u181_linux-x64_bin.tar.gz /opt/
RUN chmod +x /opt/jdk-9
ENV JAVA_HOME=/opt/jdk-9
ENV PATH="$JAVA_HOME/bin:${PATH}"
```

为了方便使用，可以直接使用笔者构建好的镜像，使用如下命令拉取：

```
docker pull registry.cn-hangzhou.aliyuncs.com/springcloud-cn/java:openjdk-9u181-alpine
```

验证如下：

```
docker run --rm -it registry.cn-hangzhou.aliyuncs.com/springcloud-cn/java:openjdk-
    9u181-alpine java -version
openjdk version "9"
OpenJDK Runtime Environment (build 9+181)
OpenJDK 64-Bit Server VM (build 9+181, mixed mode)
```

4. Alpine+glibc+OpenJDK10

OpenJDK10 的 Dockerfile 如代码清单 22-4 所示：

代码清单22-4　ch22-1\ch22-1-2-dockerfile\java10\Dockerfile

```
FROM registry.cn-hangzhou.aliyuncs.com/springcloud-cn/alpine-3.7:glibc-2.27-r0
MAINTAINER caibosi <caibosi@139.com>
ADD openjdk-10.0.1_linux-x64_bin.tar.gz /opt/
RUN chmod +x /opt/jdk-10.0.1
ENV JAVA_HOME=/opt/jdk-10.0.1
ENV PATH="$JAVA_HOME/bin:${PATH}"
```

为了方便使用，可以直接使用笔者构建好的镜像，使用如下命令拉取：

```
docker pull registry.cn-hangzhou.aliyuncs.com/springcloud-cn/java:openjdk-
    10.0.1-alpine
```

验证如下：

```
docker run --rm -it registry.cn-hangzhou.aliyuncs.com/springcloud-cn/java:openjdk-
    10.0.1-alpine java -version
openjdk version "10.0.1" 2018-04-17
OpenJDK Runtime Environment (build 10.0.1+10)
OpenJDK 64-Bit Server VM (build 10.0.1+10, mixed mode)
```

5. Alpine+glibc+OpenJDK11

OpenJDK11 的 Dockerfile 如代码清单 22-5 所示：

代码清单22-5　ch22-1\ch22-1-2-dockerfile\java11\Dockerfile

```
FROM registry.cn-hangzhou.aliyuncs.com/springcloud-cn/alpine-3.7:glibc-2.27-r0
MAINTAINER caibosi <caibosi@139.com>
ADD openjdk-11-ea_19_linux-x64_bin.tar.gz /opt/
RUN chmod +x /opt/jdk-11
ENV JAVA_HOME=/opt/jdk-11
ENV PATH="$JAVA_HOME/bin:${PATH}"
```

为了方便使用，可以直接使用笔者构建好的镜像，使用如下命令拉取：

```
docker pull registry.cn-hangzhou.aliyuncs.com/springcloud-cn/java:openjdk-11-ea19-
    alpine
```

验证如下：

```
docker run --rm -it registry.cn-hangzhou.aliyuncs.com/springcloud-cn/java:openjdk-
    11-ea19-alpine java -version
openjdk version "11-ea" 2018-09-25
OpenJDK Runtime Environment 18.9 (build 11-ea+19)
OpenJDK 64-Bit Server VM 18.9 (build 11-ea+19, mixed mode)
```

目前使用的是 ea19 版本，等到本书出版的时候，OpenJDK 11 版本的正式版应该发布了，可以直接使用正式版。

22.1.3 镜像构建插件

这里我们以 maven 构建为例，选用的是 com.spotify 的插件 dockerfile-maven-plugin，其 maven 的 pom 配置如代码清单 22-6 所示：

代码清单22-6　ch22-1\ch22-1-3-jdk8-docker\pom.xml

```xml
    <properties>
<project.build.sourceEncoding>UTF-8</project.build.sourceEncoding>
<project.reporting.outputEncoding>UTF-8</project.reporting.outputEncoding>
        <java.version>1.8</java.version>
        <dockerfile.maven.version>1.4.3</dockerfile.maven.version>
<docker.image.prefix>registry.cn-hangzhou.aliyuncs.com/springcloud-cn</docker.image.prefix>
    </properties>
    <dependencies>
        <dependency>
            <groupId>org.springframework.boot</groupId>
            <artifactId>spring-boot-starter-actuator</artifactId>
        </dependency>
        <dependency>
            <groupId>org.springframework.boot</groupId>
            <artifactId>spring-boot-starter-webflux</artifactId>
        </dependency>
    </dependencies>
    <build>
        <plugins>
            <plugin>
                <groupId>org.springframework.boot</groupId>
                <artifactId>spring-boot-maven-plugin</artifactId>
            </plugin>
            <plugin>
                <groupId>com.spotify</groupId>
                <artifactId>dockerfile-maven-plugin</artifactId>
                <version>${dockerfile.maven.version}</version>
                <executions>
                    <execution>
                        <id>default</id>
                        <goals>
                            <goal>build</goal>
                            <goal>push</goal>
                        </goals>
                    </execution>
                </executions>
                <configuration>
<repository>${docker.image.prefix}/${project.artifactId}</repository>
                    <tag>${project.version}</tag>
                    <buildArgs>
<JAR_FILE>${project.build.finalName}.jar</JAR_FILE>
                    </buildArgs>
                </configuration>
```

```
            </plugin>
        </plugins>
    </build>
```

这里我们使用 spring-boot-maven-plugin 的 1.4.3 版本，另外设置的镜像前缀为 registry.cn-hangzhou.aliyuncs.com/springcloud-cn，tag 为 ${project.version}，repository 为 registry.cn-hangzhou.aliyuncs.com/springcloud-cn/${project.artifactId}，另外这里还传递了一个 Docker 的 buildArg 为 JAR_FILE，其值为 ${project.build.finalName}.jar。

对应的 Dockerfile 如代码清单 22-7 所示：

代码清单22-7　ch22-1\ch22-1-3-jdk8-docker\Dockerfile

```
FROM registry.cn-hangzhou.aliyuncs.com/springcloud-cn/java:8u172-jre-alpine
ARG JAR_FILE
ENV PROFILE default
ADD target/${JAR_FILE} /opt/app.jar
EXPOSE 8080
ENTRYPOINT java ${JAVA_OPTS} -Djava.security.egd=file:/dev/./urandom -Duser.
    timezone=Asia/Shanghai -Dfile.encoding=UTF-8 -Dspring.profiles.
    active=${PROFILE} -jar /opt/app.jar
```

1. 打包构建

具体代码如下：

```
mvn clean package -Dmaven.test.skip=true
// ……
[INFO] Step 1/6 : FROM registry.cn-hangzhou.aliyuncs.com/springcloud-cn/java:8u172-
    jre-alpine
[INFO]
[INFO] Pulling from springcloud-cn/java
[INFO] Digest: sha256:fa30aaf978dee3d146ee21085c786422fabca0ab273dfca59749f8a2b23ef9c7
[INFO] Status: Image is up to date for registry.cn-hangzhou.aliyuncs.com/springcloud-
    cn/java:8u172-jre-alpine
[INFO]  ---> cbf57828b6e0
[INFO] Step 2/6 : ARG JAR_FILE
[INFO]
[INFO]  ---> Using cache
[INFO]  ---> b6aee85766ff
[INFO] Step 3/6 : ENV PROFILE default
[INFO]
[INFO]  ---> Using cache
[INFO]  ---> 4992cc067714
[INFO] Step 4/6 : ADD target/${JAR_FILE} /opt/app.jar
[INFO]
[INFO]  ---> d52678ff6e42
[INFO] Step 5/6 : EXPOSE 8080
[INFO]
[INFO]  ---> Running in 37d8da7677e3
[INFO] Removing intermediate container 37d8da7677e3
[INFO]  ---> bfd221fbf91c
[INFO] Step 6/6 : ENTRYPOINT java ${JAVA_OPTS} -Djava.security.egd=file:/dev/./urandom
    -Duser.timezone=Asia/Shanghai -Dfile.encoding=UTF-8 -Dspring.profiles.
```

```
        active=${PROFILE} -jar /opt/app.jar
[INFO]
[INFO] ---> Running in fa64ff8be86b
[INFO] Removing intermediate container fa64ff8be86b
[INFO]  ---> c25687ea630a
[INFO] Successfully built c25687ea630a
[INFO] Successfully tagged registry.cn-hangzhou.aliyuncs.com/springcloud-cn/
    ch22-1-3-jdk8-docker:0.0.1-SNAPSHOT
[INFO]
[INFO] Detected build of image with id c25687ea630a
[INFO] Building jar: /Users/caibosi/workspace/sccode/ch22-1/ch22-1-3-jdk8-
    docker/target/ch22-1-3-jdk8-docker-0.0.1-SNAPSHOT-docker-info.jar
[INFO] Successfully built registry.cn-hangzhou.aliyuncs.com/springcloud-cn/
    ch22-1-3-jdk8-docker:0.0.1-SNAPSHOT
```

2. Push 镜像

具体代码如下：

```
mvn dockerfile:push -Ddockerfile.username=xxx -Ddockerfile.password=xxx
[INFO] Scanning for projects...
[INFO]
[INFO] ------------------------------------------------------------------------
[INFO] Building ch22-1-3-jdk8-docker 0.0.1-SNAPSHOT
[INFO] ------------------------------------------------------------------------
[INFO]
[INFO] --- dockerfile-maven-plugin:1.4.3:push (default-cli) @ ch22-1-3-jdk8-docker ---
[INFO] The push refers to repository [registry.cn-hangzhou.aliyuncs.com/springcloud-
    cn/ch22-1-3-jdk8-docker]
```

这里使用 -Ddockerfile.username 来指定 docker registry 的账户名，使用 -Ddockerfile.password 来指定 docker registry 的密码。

3. 运行镜像

具体代码如下：

```
docker run -p 8080:8080 --rm \
-e JAVA_OPTS='-server -Xmx1g -Xms1g -XX:MetaspaceSize=64m -verbose:gc -verbose:sizes
    -XX:+UseG1GC  -XX:MaxGCPauseMillis=50  -XX:+UnlockDiagnosticVMOptions
    -XX:+HeapDumpOnOutOfMemoryError  -XX:HeapDumpPath=/  -XX:+PrintGCDetails
    -XX:+PrintGCTimeStamps -XX:+PrintGCDateStamps -XX:+PrintTenuringDistribution
    -Xloggc:/opt/gc.log  -XX:+UseGCLogFileRotation  -XX:NumberOfGCLogFiles=5
    -XX:GCLogFileSize=20M -Djava.io.tmpdir=/tmp' \
-e PROFILE='default' \
registry.cn-hangzhou.aliyuncs.com/springcloud-cn/ch22-1-3-jdk8-docker:0.0.1-
    SNAPSHOT
```

22.1.4　JDK8+ 的 Docker 支持

1. JDK8&9

Java 8u131 及以上版本开始支持了 Docker 的 cpu 和 memory 限制。

对于 cpu 限制，即如果没有显式指定 -XX：ParalllelGCThreads 或者 -XX：CICompilerCount，

那么 JVM 使用 Docker 的 cpu 限制。如果 Docker 有指定 cpu limit，jvm 参数也有指定 -XX：ParalllelGCThreads 或者 -XX：CICompilerCount，那么以指定的参数为准。

对于 memory 限制，需要加上 -XX：+UnlockExperimentalVMOptions -XX：+UseCGroupMemoryLimitForHeap 才能使得 Xmx 感知 Docker 的 memory limit。

2. JDK10

JDK10 版本废弃了 UseCGroupMemoryLimitForHeap，同时新引入了一个 ActiveProcessorCount 可以用来强制指定 cpu 的个数。

3. JDK11

JDK11 正式移除 UseCGroupMemoryLimitForHeap，同时新引入 UseContainerSupport 配置，默认为 ture，即默认支持 Docker 的 cpu 及 memory 限制，也可以设置为 false 禁用容器支持。

22.1.5　JDK9+ 镜像优化

JDK9 及以上的版本与之前的版本有一个比较大的变动，就是 JDK9 及以上的版本支持了模块系统 JPMS，同时 JDK 自身也模块化。里头的 Modular Run-Time Images 功能特性以及 jlink 工具对于镜像的优化非常有帮助，可根据所需模块来精简 JDK。本节就来演示一下如何对使用 JDK9 及以上版本的 Docker 镜像进行精简。

1. jlink

jlink 工具可以用来将已有的 JDK 按所需模块进行优化，并重新组装成一个自定义的 runtime image。其基本语法如下：

```
jlink [options] --module-path modulepath --add-modules module [,module...]
```

其中 module-path 参数用于指定需要 jlink 的 JDK 的 jmods 路径，options 的部分参数如下：
- add-mobules，用来指定所需要的模块名称，比如 java.xml。
- compress，用来指定压缩级别，0 为不压缩，1 为常量字符串共享，2 为 zip 压缩。
- no-hreader-files，表示排除掉 header 文件。
- output，指定输出精简后的 JDK 的文件夹路径。

这里，我们的实例工程为 ch22\ch22-1-4-jdk10-docker，其 Dockerfile 如代码请求 22-8 所示：

代码清单22-8　ch22-1\ch22-1-4-jdk10-docker\Dockerfile

```
FROM registry.cn-hangzhou.aliyuncs.com/springcloud-cn/java:openjdk-10.0.1-
    alpine as packager

## jlink
RUN /opt/jdk-10.0.1/bin/jlink \
    --module-path /opt/jdk-10.0.1/jmods \
    --verbose \
    --add-modules java.base,java.logging,java.xml,jdk.unsupported,java.sql,java.
```

```
            desktop,java.management,java.naming,java.instrument,jdk.jstatd,jdk.
            jcmd,jdk.management \
    --compress 2 \
    --no-header-files \
    --output /opt/jdk-10-jlinked

# copy jdk after jlink
FROM registry.cn-hangzhou.aliyuncs.com/springcloud-cn/alpine-3.7:glibc-2.27-r0
COPY --from=packager /opt/jdk-10-jlinked /opt/jdk-10.0.1
ENV JAVA_HOME=/opt/jdk-10.0.1
ENV PATH=$JAVA_HOME/bin:$PATH

## add jar
ARG JAR_FILE
ENV PROFILE default
ADD target/${JAR_FILE} /opt/app.jar
EXPOSE 8080
ENTRYPOINT java ${JAVA_OPTS} -Djava.security.egd=file:/dev/./urandom
    -Duser.timezone=Asia/Shanghai -Dfile.encoding=UTF-8 -Dspring.profiles.
    active=${PROFILE} -jar /opt/app.jar
```

这里我们指定了依赖 java.base、java.logging、java.xml、jdk.unsupported、java.sql、java.desktop、java.management、java.naming、java.instrument、jdk.jstatd、jdk.jcmd、jdk.management 这几个模块。

2. 构建

这里我们使用 maven 打包，记得用 JDK10 来编译：

```
mvn clean package
```

构建镜像如下：

```
docker build --build-arg JAR_FILE=ch22-1-4-jdk10-docker-0.0.1-SNAPSHOT.jar \
-t registry.cn-hangzhou.aliyuncs.com/springcloud-cn/ch22-1-4-jdk10-docker:0.0.1-
    SNAPSHOT .
```

查看下镜像大小：

```
docker images | grep jdk10
registry.cn-hangzhou.aliyuncs.com/springcloud-cn/ch22-1-4-jdk10-docker
    0.0.1-SNAPSHOT 32c741362c3e 34 minutes ago 91.09 MB
```

可以看到，精简后的 JDK 包括 app.jar，在 100M 以内。

3. 运行

使用如下命令运行：

```
docker run -p 8080:8080 --rm \
-e JAVA_OPTS='-server -XX:+UseG1GC -XX:MaxGCPauseMillis=50 -XX:+Unlock-
    DiagnosticVMOptions -XX:+UnlockExperimentalVMOptions -XX:+UseCGroupMemor
    yLimitForHeap -XX:ActiveProcessorCount=1 -XX:+HeapDumpOnOutOfMemoryError
    -XX:HeapDumpPath=/ -Xlog:age*,gc*=info:file=gc-%p-%t.log:time,tid,tags:filecou
    nt=5,filesize=10m -Djava.io.tmpdir=/tmp' \
-e PROFILE='default' \
```

```
registry.cn-hangzhou.aliyuncs.com/springcloud-cn/ch22-1-4-jdk10-docker:0.0.1-
SNAPSHOT
```

然后登录进去，查看精简后的 JDK 大小：

```
docker exec -it a68ec905a170 sh
du -sh /opt/jdk-10.0.1/
53.6M /opt/jdk-10.0.1/
du -sh /opt/app.jar
17.5M /opt/app.jar
```

可以看到精简后的 JDK 大小为 53.6M，实例 app.jar 为 17.5M，共 71.1M。而对比下安装包 jdk-10.0.1_linux-x64_bin 为 556M，serverjre-10.0.1_linux-x64_bin 为 164M，可以发现 jlink 功能非常强大，JDK 精简了不少。

22.2　Spring Cloud 组件的 Docker 化

本节主要简单介绍下 Spring Cloud 的一些组件的 Docker 化。主要分为 config-server、eureka-server、gateway、turbine、spring admin 这几个组件的 Docker 化。

22.2.1　Docker 化配置

这里我们基于 dockerfile-maven-plugin 来构建，该 plugin 版本为 1.4.3，公共 maven 的配置文件如代码清单 22-9 所示：

代码清单22-9　ch22-2\pom.xml

```xml
<build>
    <plugins>
        <plugin>
            <groupId>org.springframework.boot</groupId>
            <artifactId>spring-boot-maven-plugin</artifactId>
        </plugin>
        <plugin>
            <groupId>com.spotify</groupId>
            <artifactId>dockerfile-maven-plugin</artifactId>
            <version>${dockerfile.maven.version}</version>
            <executions>
                <execution>
                    <id>default</id>
                    <goals>
                        <goal>build</goal>
                        <goal>push</goal>
                    </goals>
                </execution>
            </executions>
            <configuration>
                <repository>${docker.image.prefix}/${project.artifactId}</repository>
                <tag>${project.version}</tag>
                <buildArgs>
```

```xml
<JAR_FILE>${project.build.finalName}.jar</JAR_FILE>
            </buildArgs>
        </configuration>
    </plugin>
  </plugins>
</build>
```

dockerfile 的模板基本一致,这里都放到每个工程的根目录下面,跟 pom.xml 文件一个目录,如下:

```
FROM registry.cn-hangzhou.aliyuncs.com/springcloud-cn/java:8u172-jre-alpine
ARG JAR_FILE
ENV PROFILE default
ADD target/${JAR_FILE} /opt/app.jar
EXPOSE 8080
ENTRYPOINT java ${JAVA_OPTS} -Djava.security.egd=file:/dev/./urandom -Duser.
    timezone=Asia/Shanghai -Dfile.encoding=UTF-8 -Dspring.profiles.
    active=${PROFILE} -jar /opt/app.jar
```

- 注意这里的 JAR_FILE 是构建镜像的参数,必须在 spring-boot-maven-plugin 的配置中指定,这里指定为 ${project.build.finalName}.jar。
- 端口号,这里统一默认为 8080,最后映射出来的端口号可以根据每个应用进行不同的映射。
- 这里统一指定了时区为 Asia/Shanghai。
- 暴露了 JAVA_OPTS 环境变量,允许不同应用去指定 jvm 参数。
- 暴露了 PROFILE 环境变量,允许不同应用通过环境变量去启动不同 profile。

22.2.2　config-server 的 Docker 化

1. 初始化配置

这里以 jdbc 的存储形式来演示,创建表的 SQL 如代码清单 22-10 所示:

代码清单22-10　ch22-2\ch22-2-1-config-server\src\main\resources\db\schema.sql

```sql
CREATE TABLE IF NOT EXISTS PROPERTIES (
    KEY         VARCHAR(2048),
    VALUE       VARCHAR(4096),
    APPLICATION VARCHAR(128),
    PROFILE     VARCHAR(128),
    LABEL       VARCHAR(128),
    PRIMARY KEY (`KEY`, `APPLICATION`, `PROFILE`, `LABEL`)
);
```

初始化各个服务配置项的 SQL 部分,如代码清单 22-11 所示:

代码清单22-11　ch22-2\ch22-2-1-config-server\src\main\resources\db\data.sql

```sql
INSERT INTO PROPERTIES (APPLICATION, PROFILE, LABEL, KEY, VALUE)
VALUES ('demo', 'default', 'master', 'app.greet.name', 'Demo');
-- eureka server
```

```sql
INSERT INTO PROPERTIES (APPLICATION, PROFILE, LABEL, KEY, VALUE)
VALUES ('eureka-server', 'default', 'master', 'server.port', '${SERVER_PORT:8761}');
INSERT INTO PROPERTIES (APPLICATION, PROFILE, LABEL, KEY, VALUE)
VALUES ('eureka-server', 'default', 'master', 'eureka.instance.preferIpAddress', 'true');

-- gateway
INSERT INTO PROPERTIES (APPLICATION, PROFILE, LABEL, KEY, VALUE)
VALUES ('gateway', 'default', 'master', 'server.port', '${SERVER_PORT:8000}');

-- turbine
INSERT INTO PROPERTIES (APPLICATION, PROFILE, LABEL, KEY, VALUE)
VALUES ('turbine', 'default', 'master', 'turbine.appConfig', 'gateway,biz-service');

-- spring-admin
INSERT INTO PROPERTIES (APPLICATION, PROFILE, LABEL, KEY, VALUE)
VALUES ('spring-admin', 'default', 'master', 'server.port', '${SERVER_PORT:8002}');

-- biz-service
INSERT INTO PROPERTIES (APPLICATION, PROFILE, LABEL, KEY, VALUE)
VALUES ('biz-service', 'default', 'master', 'server.port', '${SERVER_PORT:8003}');
```

2. application 配置

application.yml 文件如代码清单 22-12 所示:

代码清单22-12　ch22-2/ch22-2-1-config-server/src/main/resources/application.yml

```yaml
spring:
  application:
    name: config-server
  profiles:
    active: jdbc
  h2:
    console:
      enabled: true
  cloud:
    config:
      label: master
      server:
        bootstrap: true
        jdbc: true
  datasource:
    url: jdbc:h2:mem:testdb
    driver-class-name: org.h2.Driver
    username: sa
    password:
    continue-on-error: false
    schema: classpath:db/schema.sql
    data: classpath:db/data.sql
    initialization-mode: always
    type: org.apache.tomcat.jdbc.pool.DataSource
management:
  endpoints:
    web:
```

```yaml
    exposure:
      include: '*'
```

其他配置可以具体详见代码工程 ch22-2\ch22-2-1\config-server。

3. Build & Run

打包构建镜像及启动的代码如下：

```
mvn clean package
docker run -p 8888:8080 --rm \
-e JAVA_OPTS='-server -Xmx1g' \
-e PROFILE='jdbc' \
-e SERVER_PORT=8080 \
registry.cn-hangzhou.aliyuncs.com/springcloud-cn/ch22-2-1-config-server:0.0.1-
    SNAPSHOT
```

笔者是在 mac 的 docker-machine 上执行的，可以使用如下命令来确定 docker machine 的 ip 地址：

```
docker-machine ip default
192.168.99.100
```

这里的 default 是你创建的 docker machine 的名称，默认为 default。

4. 验证

访问 eureka-server 的配置来验证下：

```
curl -i http://192.168.99.100:8888/eureka-server/default/master
HTTP/1.1 200
Content-Type: application/json;charset=UTF-8
Transfer-Encoding: chunked
Date: Mon, 02 Jul 2018 07:31:48 GMT

{"name":"eureka-server","profiles":["default"],"label":"master","versi
    on":null,"state":null,"propertySources":[{"name":"eureka-server-
    default","source":{"server.port":"${SERVER_PORT:8761}","eureka.
    instance.preferIpAddress":"true","eureka.instance.lease-renewal-
    interval-in-seconds":"10","eureka.instance.lease-expiration-
    duration-in-seconds":"30","eureka.client.registerWithEureka":"true
    ","eureka.client.fetchRegistry":"true","eureka.client.serviceUrl.
    defaultZone":"http://${EUREKA_SERVER_HOST}:${EUREKA_SERVER_PORT}/
    eureka/","eureka.server.waitTimeInMsWhenSyncEmpty":"0","eureka.server.en
    ableSelfPreservation":"false","eureka.server.eviction-interval-timer-in-
    ms":"10000"}}]}eka-server/default/master
```

22.2.3　eureka-server 的 Docker 化

1. bootstrap 配置

这里我们是以两个 Eureka Server 为例，其 bootstrap.yml 文件配置如代码清单 22-13 所示：

代码清单22-13　ch22-2\ch22-2-2-eureka-server\src\main\resources\bootstrap.yml

```yml
spring:
  application:
    name: eureka-server
  cloud:
    config:
      fail-fast: true
      label: master
      uri: http://${CONFIG_SERVER_HOST}:${CONFIG_SERVER_PORT}
management:
  endpoints:
    web:
      exposure:
        include: '*'
```

2. application 配置

访问 http://192.168.99.100:8888/eureka-server/default/master，查看其在 config-server 配置的属性如下：

```
{
    "name": "eureka-server",
    "profiles": [
        "default"
    ],
    "label": "master",
    "version": null,
    "state": null,
    "propertySources": [
        {
            "name": "eureka-server-default",
            "source": {
                "server.port": "${SERVER_PORT:8761}",
                "eureka.instance.preferIpAddress": "true",
                "eureka.instance.lease-renewal-interval-in-seconds": "10",
                "eureka.instance.lease-expiration-duration-in-seconds": "30",
                "eureka.client.registerWithEureka": "true",
                "eureka.client.fetchRegistry": "true",
                "eureka.client.serviceUrl.defaultZone": "http://${EUREKA_SERVER_HOST}:${EUREKA_SERVER_PORT}/eureka/",
                "eureka.server.waitTimeInMsWhenSyncEmpty": "0",
                "eureka.server.enableSelfPreservation": "false",
                "eureka.server.eviction-interval-timer-in-ms": "10000"
            }
        }
    ]
}
```

可以看到这里通过 EUREKA_SERVER_HOST 以及 EUREKA_SERVER_PORT 的环境变量来指定 Eureka Server 的信息。

3. 构建和启动

打包构建镜像及启动的代码如下：

```
mvn clean package
docker run -p 8761:8080 --rm \
-e JAVA_OPTS='-server -Xmx1g' \
-e PROFILE='default' \
-e SERVER_PORT=8080 \
-e CONFIG_SERVER_HOST=192.168.99.100 \
-e CONFIG_SERVER_PORT=8888 \
-e EUREKA_SERVER_HOST=192.168.99.100 \
-e EUREKA_SERVER_PORT=8762 \
registry.cn-hangzhou.aliyuncs.com/springcloud-cn/ch22-2-2-eureka-server:0.0.1-
    SNAPSHOT


docker run -p 8762:8080 --rm \
-e JAVA_OPTS='-server -Xmx1g' \
-e PROFILE='default' \
-e SERVER_PORT=8080 \
-e CONFIG_SERVER_HOST=192.168.99.100 \
-e CONFIG_SERVER_PORT=8888 \
-e EUREKA_SERVER_HOST=192.168.99.100 \
-e EUREKA_SERVER_PORT=8761 \
registry.cn-hangzhou.aliyuncs.com/springcloud-cn/ch22-2-2-eureka-server:0.0.1-
    SNAPSHOT
```

值得注意的是，这里我们需要给出接入 config-client 的指定 config-server 的地址及端口，通过环境变量设置。对于两个 Eureka Server，分别通过环境变量指定对方为注册及复制的节点。

4. 验证

启动之后，分别访问 192.168.199.100:8761 以及 192.168.99.100:8762，观察是否启动正常。

22.2.4　gateway 的 Docker 化

1. bootstrap 配置

这里的 gateway 以 netflix-zuul 为例，其 bootstrap.yml 配置如代码清单 22-14 所示：

代码清单22-14　ch22-2\ch22-2-3-gateway\src\main\resources\bootstrap.yml

```
spring:
  application:
    name: gateway
  cloud:
    config:
      fail-fast: true
      label: master
      uri: http://${CONFIG_SERVER_HOST}:${CONFIG_SERVER_PORT}
management:
  endpoints:
    web:
      exposure:
        include: '*'
```

2. application 配置

访问 http://192.168.99.100:8888/gateway/default/master，查看其在 config-server 配置的属性如下：

```
{
    "name": "gateway",
    "profiles": [
        "default"
    ],
    "label": "master",
    "version": null,
    "state": null,
    "propertySources": [
        {
            "name": "gateway-default",
            "source": {
                "server.port": "${SERVER_PORT:8000}",
                "eureka.instance.preferIpAddress": "true",
                "eureka.instance.lease-renewal-interval-in-seconds": "10",
                "eureka.instance.lease-expiration-duration-in-seconds": "30",
                "eureka.client.registerWithEureka": "true",
                "eureka.client.fetchRegistry": "true",
                "eureka.client.serviceUrl.defaultZone": "http://${EUREKA_SERVER1_HOST}:${EUREKA_SERVER1_PORT}/eureka/,http://${EUREKA_SERVER2_HOST}:${EUREKA_SERVER2_PORT}/eureka/"
            }
        }
    ]
}
```

3. 构建 & 启动

打包构建镜像及启动的代码如下：

```
mvn clean package
docker run -p 8000:8080 --rm \
-e JAVA_OPTS='-server -Xmx1g' \
-e PROFILE='default' \
-e SERVER_PORT=8080 \
-e CONFIG_SERVER_HOST=192.168.99.100 \
-e CONFIG_SERVER_PORT=8888 \
-e EUREKA_SERVER1_HOST=192.168.99.100 \
-e EUREKA_SERVER1_PORT=8761 \
-e EUREKA_SERVER2_HOST=192.168.99.100 \
-e EUREKA_SERVER2_PORT=8762 \
registry.cn-hangzhou.aliyuncs.com/springcloud-cn/ch22-2-3-gateway:0.0.1-SNAPSHOT
```

注意，对于接入 Eureka Client 的应用，因为这里是两个 Eureka Server，所以配置文件使用 EUREKA_SERVER1_HOST、EUREKA_SERVER1_PORT、EUREKA_SERVER2_HOST、EUREKA_SERVER2_PORT 来指定它们的地址及端口。

4. 验证

```
curl -i 192.168.99.100:8000/actuator
```

```
HTTP/1.1 200
Content-Type: application/vnd.spring-boot.actuator.v2+json;charset=UTF-8
Transfer-Encoding: chunked
Date: Mon, 02 Jul 2018 08:39:50 GMT

{"_links":{"self":{"href":"http://192.168.99.100:8000/actuator","templated":
false},"archaius":{"href":"http://192.168.99.100:8000/actuator/archaius
","templated":false},"auditevents":{"href":"http://192.168.99.100:8000/
actuator/auditevents","templated":false},"beans":{"href":"ht
tp://192.168.99.100:8000/actuator/beans","templated":false},"health":{
"href":"http://192.168.99.100:8000/actuator/health","templated":false}
,"conditions":{"href":"http://192.168.99.100:8000/actuator/conditions"
,"templated":false},"configprops":{"href":"http://192.168.99.100:8000/
actuator/configprops","templated":false},"env-toMatch":{"href":"ht
tp://192.168.99.100:8000/actuator/env/{toMatch}","templated":true},"env":{
"href":"http://192.168.99.100:8000/actuator/env","templated":false},"info"
:{"href":"http://192.168.99.100:8000/actuator/info","templated":false},"lo
ggers":{"href":"http://192.168.99.100:8000/actuator/loggers","templated":f
alse},"loggers-name":{"href":"http://192.168.99.100:8000/actuator/loggers/
{name}","templated":true},"heapdump":{"href":"http://192.168.99.100:8000/
actuator/heapdump","templated":false},"threaddump":{"href":"ht
tp://192.168.99.100:8000/actuator/threaddump","templated":false},"metrics"
:{"href":"http://192.168.99.100:8000/actuator/metrics","templated":false},
"metrics-requiredMetricName":{"href":"http://192.168.99.100:8000/actuator/
metrics/{requiredMetricName}","templated":true},"scheduledtasks":{"href":"h
ttp://192.168.99.100:8000/actuator/scheduledtasks","templated":false},"htt
ptrace":{"href":"http://192.168.99.100:8000/actuator/httptrace","templated
":false},"mappings":{"href":"http://192.168.99.100:8000/actuator/mappings"
,"templated":false},"refresh":{"href":"http://192.168.99.100:8000/actuator/
refresh","templated":false},"features":{"href":"http://192.168.99.100:8000/
actuator/features","templated":false},"service-registry":{"href":
"http://192.168.99.100:8000/actuator/service-registry","templated":false},
"routes-format":{"href":"http://192.168.99.100:8000/actuator/routes/{format
}","templated":true},"routes":{"href":"http://192.168.99.100:8000/actuator/
routes","templated":false},"filters":{"href":"http://192.168.99.100:8000/
actuator/filters","templated":false},"hystrix.stream":{"href":
"http://192.168.99.100:8000/actuator/hystrix.stream","templated":false}}}
```

22.2.5　turbine 的 Docker 化

1. bootstrap 配置

turbine 用来聚合各个应用实例的 Hystrix 数据，其 bootstrap.yml 配置如代码清单 22-15 所示：

代码清单22-15　ch22-2\ch22-2-4-turbine \src\main\resources\bootstrap.yml

```yaml
spring:
  application:
    name: turbine
  cloud:
    config:
      fail-fast: true
      label: master
```

```yaml
            uri: http://${CONFIG_SERVER_HOST}:${CONFIG_SERVER_PORT}
management:
  endpoints:
    web:
      exposure:
        include: '*'
```

2. application 配置

访问 http://192.168.99.100:8888/turbine/default/master，查看其在 config-server 配置的属性如下：

```
{
    "name": "turbine",
    "profiles": [
        "default"
    ],
    "label": "master",
    "version": null,
    "state": null,
    "propertySources": [
        {
            "name": "turbine-default",
            "source": {
                "turbine.appConfig": "gateway,biz-service",
                "server.port": "${SERVER_PORT:8001}",
                "eureka.instance.preferIpAddress": "true",
                "eureka.instance.lease-renewal-interval-in-seconds": "10",
                "eureka.instance.lease-expiration-duration-in-seconds": "30",
                "eureka.client.registerWithEureka": "true",
                "eureka.client.fetchRegistry": "true",
                "eureka.client.serviceUrl.defaultZone":  "http://${EUREKA_SERVER1_
                    HOST}:${EUREKA_SERVER1_PORT}/eureka/,http://${EUREKA_SERVER2_
                    HOST}:${EUREKA_SERVER2_PORT}/eureka/"
            }
        }
    ]
}
```

3. 构建 & 启动

打包构建镜像及启动的代码如下：

```
mvn clean package
docker run -p 8001:8080 --rm \
-e JAVA_OPTS='-server -Xmx1g' \
-e PROFILE='default' \
-e SERVER_PORT=8080 \
-e CONFIG_SERVER_HOST=192.168.99.100 \
-e CONFIG_SERVER_PORT=8888 \
-e EUREKA_SERVER1_HOST=192.168.99.100 \
-e EUREKA_SERVER1_PORT=8761 \
-e EUREKA_SERVER2_HOST=192.168.99.100 \
-e EUREKA_SERVER2_PORT=8762 \
registry.cn-hangzhou.aliyuncs.com/springcloud-cn/ch22-2-4-turbine:0.0.1-SNAPSHOT
```

4. 验证

通过访问 http://192.168.99.100:8001/hystrix 来查看 turbine 是否启动起来，正常的话，其界面如图 22-1 所示。

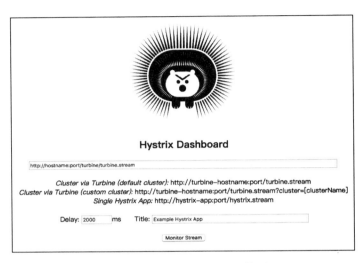

图 22-1　hystrix dashboard 首页

22.2.6　Spring Admin 的 Docker 化

Spring Admin 是 codecentric 开源的一个管理 Spring Cloud 应用实例的组件，其 GitHub 地址为 https://github.com/codecentric/spring-boot-admin。这里我们用这个组件来监控 Spring Cloud 的各个应用实例。

1. bootstrap 配置

其 bootstrap.yml 配置如代码清单 22-16 所示：

代码清单22-16　ch22-2\ch22-2-5-spring-admin \src\main\resources\bootstrap.yml

```yaml
spring:
  application:
    name: spring-admin
  cloud:
    config:
      fail-fast: true
      label: master
      uri: http://${CONFIG_SERVER_HOST}:${CONFIG_SERVER_PORT}
management:
  endpoints:
    web:
      exposure:
        include: '*'
```

2. application 配置

访问 http://192.168.99.100:8888/spring-admin/default/master，查看其在 config-server 配置的属性如下：

```
{
    "name": "spring-admin",
    "profiles": [
        "default"
    ],
    "label": "master",
    "version": null,
    "state": null,
    "propertySources": [
        {
            "name": "spring-admin-default",
            "source": {
                "server.port": "${SERVER_PORT:8002}",
                "eureka.instance.preferIpAddress": "true",
                "eureka.instance.lease-renewal-interval-in-seconds": "10",
                "eureka.instance.lease-expiration-duration-in-seconds": "30",
                "eureka.client.registerWithEureka": "true",
                "eureka.client.fetchRegistry": "true",
                "eureka.client.serviceUrl.defaultZone": "http://${EUREKA_SERVER1_HOST}:${EUREKA_SERVER1_PORT}/eureka/,http://${EUREKA_SERVER2_HOST}:${EUREKA_SERVER2_PORT}/eureka/"
            }
        }
    ]
}
```

3. 构建 & 启动

打包构建镜像及启动代码如下：

```
mvn clean package
docker run -p 8002:8080 --rm \
-e JAVA_OPTS='-server -Xmx1g' \
-e PROFILE='default' \
-e SERVER_PORT=8080 \
-e CONFIG_SERVER_HOST=192.168.99.100 \
-e CONFIG_SERVER_PORT=8888 \
-e EUREKA_SERVER1_HOST=192.168.99.100 \
-e EUREKA_SERVER1_PORT=8761 \
-e EUREKA_SERVER2_HOST=192.168.99.100 \
-e EUREKA_SERVER2_PORT=8762 \
registry.cn-hangzhou.aliyuncs.com/springcloud-cn/ch22-2-5-spring-admin:0.0.1-SNAPSHOT
```

4. 验证

访问 http://192.168.99.100:8002/#/wallboard，其界面如图 22-2 所示。

访问 http://192.168.99.100:8002/#/applications，其界面如图 22-3 所示。

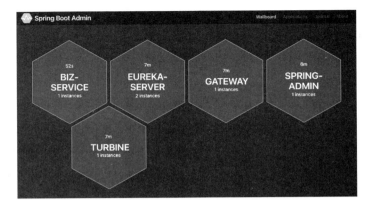

图 22-2　spring boot admin 的 Wallboard 功能界面

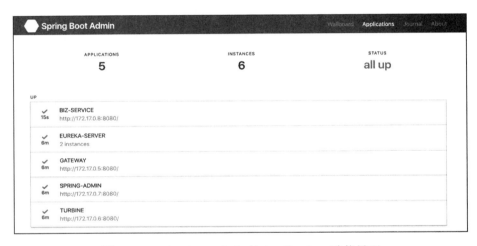

图 22-3　spring boot admin 的 Applications 功能界面

22.2.7　biz-service 的 Docker 化

biz-service 是这里的一个业务应用例子。

1. bootstrap 配置

其 bootstrap.yml 配置如代码清单 22-17 所示：

代码清单22-17　ch22-2\ch22-2-6-biz-service \src\main\resources\bootstrap.yml

```
spring:
  application:
    name: spring-admin
  cloud:
    config:
      fail-fast: true
      label: master
```

```yaml
      uri: http://${CONFIG_SERVER_HOST}:${CONFIG_SERVER_PORT}
management:
  endpoints:
    web:
      exposure:
        include: '*'
```

2. application 配置

访问 http://192.168.99.100:8888/biz-service/default/master，查看其在 config-server 配置的属性如下：

```
{
    "name": "biz-service",
    "profiles": [
        "default"
    ],
    "label": "master",
    "version": null,
    "state": null,
    "propertySources": [
        {
            "name": "biz-service-default",
            "source": {
                "server.port": "${SERVER_PORT:8003}",
                "eureka.instance.preferIpAddress": "true",
                "eureka.instance.lease-renewal-interval-in-seconds": "10",
                "eureka.instance.lease-expiration-duration-in-seconds": "30",
                "eureka.client.registerWithEureka": "true",
                "eureka.client.fetchRegistry": "true",
                "eureka.client.serviceUrl.defaultZone": "http://${EUREKA_SERVER1_HOST}:${EUREKA_SERVER1_PORT}/eureka/,http://${EUREKA_SERVER2_HOST}:${EUREKA_SERVER2_PORT}/eureka/"
            }
        }
    ]
}
```

3. 构建 & 启动

打包构建镜像及启动如下：

```
mvn clean package
docker run -p 8003:8080 --rm \
-e JAVA_OPTS='-server -Xmx1g' \
-e PROFILE='default' \
-e SERVER_PORT=8080 \
-e CONFIG_SERVER_HOST=192.168.99.100 \
-e CONFIG_SERVER_PORT=8888 \
-e EUREKA_SERVER1_HOST=192.168.99.100 \
-e EUREKA_SERVER1_PORT=8761 \
-e EUREKA_SERVER2_HOST=192.168.99.100 \
-e EUREKA_SERVER2_PORT=8762 \
registry.cn-hangzhou.aliyuncs.com/springcloud-cn/ch22-2-6-biz-service:0.0.1-SNAPSHOT
```

4. 验证

访问 http://192.168.99.100:8003/actuator/health，返回如下：

```
{
    "status": "UP"
}
```

22.2.8　网卡选择

在 Docker 中使用 Spring Cloud Discovery Client 组件，可以通过配置 inetutils 相关属性来忽略或者优先选择指定的网卡，这样注册到服务中心的时候就会以指定网卡的 ip 地址来注册，主要有如下几个参数：

- spring.cloud.inetutils.useOnlySiteLocalInterfaces：设置为 true，则强制使用本地地址，默认为 false。
- spring.cloud.inetutils.ignoredInterfaces：以数组形式赋值，用于指定要忽略的网卡，支持正则表达式，比如 docker0、veth.*。
- spring.cloud.inetutils.preferredNetworks：用于指定优先使用的 ip 地址，比如 192.168.99.100，也可以指定 ip 段，比如 10.0。也支持正则表达。

22.2.9　小结

本节列举了 config-server、eureka-server、gateway、turbine、spring admin、biz-service 这几个工程来演示 Spring Cloud 组件及业务服务的 Docker 化配置及其运行。但仅仅是 Docker 化而已，并没有管理起来，下一节我们会将本节的应用实例部署到 kubernetes 上进行管理。

22.3　使用 Kubernetes 管理

22.3.1　概述

kubernetes 是 Google 开源的容器集群管理系统，支持 Docker 及 Rocket 容器技术，目前已经是云原生应用的基础设施。其基本组件如图 22-4 所示。

在使用 Kubernetes 管理之前，需要了解如下几个基本概念：

1）Namespace：Kubernetes 使用 namespace 来划分为多个虚拟集群，可以供不同的团队或业务群组来使用，进行资源等的隔离。

可以使用如下命令来创建一个新的 namespace：

```
kubectl create namespace springcloud-cn
```

2）Pod：Pod 是 kubernetes 的最小调度单元，可以看作应用服务的逻辑宿主机，通常可以这样理解——部署几个实例，就是部署几个 Pod。

3）Replication Controller：Pod 的控制器，用于监控指定条件 Pod 的实例的状态，并维持其健康个数为指定的副本数量。其主要功能包括确保 Pod 实例数量、确保 Pod 健康、弹性伸缩、滚动升级。

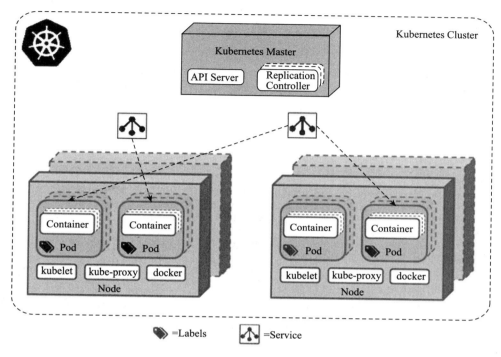

图 22-4 kubernetes 组件图

4）Service：为一组提供相同服务的 Pod 提供对外访问，具备一定 Pod 的服务发现、路由的能力。

5）Label：用来关联 Pod 及 Service，每个 Pod 可以定义 Label，然后 Service 可以通过 Label 去选择服务哪些 Pod。

6）Deployment：Replication Controller 的升级版，对 Deployment 进行了更细展示及控制，如可以查看升级的详细进度及状态，对每次 Deployment 进行版本管理，方便回滚，也可以对升级行为进行暂停及继续。

22.3.2 本地安装 Kubernetes

为了方便大家学习，kubernetes 提供了 Minikube 版本，可以在本地进行安装，官方文档在 https://kubernetes.io/docs/tasks/tools/install-minikube/，这里笔者以在 Mac 上安装为例，简述一下安装过程，而 Linux 或 Windows 系统可以参照官方文档下载对应的软件即可。

1. 安装 VisualBox

VisualBox 的安装比较简单，到官网 https://www.virtualbox.org/wiki/Downloads 选择对应的操作系统的版本安装即可。

2. 安装 kubectl

kubectl 是一个可以操作本地或远程 Kubernetes 集群的命令行工具，不同操作系统的安装，可以参考官方文档 https://kubernetes.io/docs/tasks/tools/install-kubectl/，这里使用命令行在 Mac 上下载最新稳定版 v1.11.0，代码如下：

```
curl -LO https://storage.googleapis.com/kubernetes-release/release/v1.11.0/bin/
    darwin/amd64/kubectl
```

然后赋予执行权限放到 /usr/local/bin/ 目录下：

```
chmod +x kubectl && sudo mv kubectl /usr/local/bin/
```

最后验证一下：

```
kubectl version
Client Version: version.Info{Major:"1", Minor:"11", GitVersion:"v1.11.0", Git
    Commit:"91e7b4fd31fcd3d5f436da26c980becec37ceefe", GitTreeState:"clean",
    BuildDate:"2018-06-27T20:17:28Z", GoVersion:"go1.10.2", Compiler:"gc",
    Platform:"darwin/amd64"}
The connection to the server localhost:8080 was refused - did you specify the
    right host or port?
```

3. 安装 Minikube

首先到官网地址（https://github.com/kubernetes/minikube/releases）下载对应操作系统的版本，这里笔者下载的是 v0.25.2 版本的 minikube-darwin-amd64。

然后赋予执行权限安装如下：

```
chmod +x minikube-darwin-amd64 && sudo mv minikube-darwin-amd64 /usr/local/bin/
    minikube
```

验证一下：

```
minikube version
minikube version: v0.25.2
```

4. 启动 Minikube

使用如下命令启动 Minikube，第一次启动的时候会下载安装 Minikube ISO，代码如下：

```
minikube start
Starting local Kubernetes v1.9.4 cluster...
Starting VM...
Downloading Minikube ISO
    142.22 MB / 142.22 MB [============================================] 100.00% 0s
Getting VM IP address...
Moving files into cluster...
Downloading localkube binary
    163.02 MB / 163.02 MB [============================================] 100.00% 0s
    0 B / 65 B [--------------------------------------------------------]   0.00%
    65 B / 65 B [==========================================================] 100.00% 0sSetting
        up certs...
Connecting to cluster...
Setting up kubeconfig...
Starting cluster components...
```

```
Kubectl is now configured to use the cluster.
Loading cached images from config file.
```

第一次启动的时候会下载安装相关组件。中间安装过程可能比较长，可以使用如下命令查看是否有异常：

```
minikube logs
```

发现如下报错信息：

```
Jul 03 05:47:13 minikube localkube[3129]: E0703 05:47:13.293734 3129 remote_runtime.
    go:92] RunPodSandbox from runtime service failed: rpc error: code = Unknown
    desc = failed pulling image "gcr.io/google_containers/pause-amd64:3.0": Error
    response from daemon: Get https://gcr.io/v2/: net/http: request canceled
    while waiting for connection (Client.Timeout exceeded while awaiting headers)
Jul 03 05:47:13 minikube localkube[3129]: E0703 05:47:13.293888 3129 kuberuntime_
    sandbox.go:54] CreatePodSandbox for pod "kube-addon-manager-minikube_kube-
    system(c4c3188325a93a2d7fb1714e1abf1259)" failed: rpc error: code = Unknown
    desc = failed pulling image "gcr.io/google_containers/pause-amd64:3.0": Error
    response from daemon: Get https://gcr.io/v2/: net/http: request canceled
    while waiting for connection (Client.Timeout exceeded while awaiting headers)
```

原因是拉取不到镜像 gcr.io/google_containers/pause-amd64:3.0，这里我们从阿里云拉取下来，再改下 tag：

```
minikube ssh
docker pull registry.cn-hangzhou.aliyuncs.com/google-containers/pause-amd64:3.0
docker tag registry.cn-hangzhou.aliyuncs.com/google-containers/pause-amd64:3.0
    gcr.io/google_containers/pause-amd64:3.0
```

其他镜像也会有相关问题，这里一并重新拉取下：

```
docker pull registry.cn-hangzhou.aliyuncs.com/google_containers/kube-addon-
    manager:v6.5
docker tag registry.cn-hangzhou.aliyuncs.com/google_containers/kube-addon-
    manager:v6.5 gcr.io/google-containers/kube-addon-manager:v6.5
docker pull registry.cn-hangzhou.aliyuncs.com/google_containers/k8s-dns-kube-
    dns-amd64:1.14.5
docker tag registry.cn-hangzhou.aliyuncs.com/google_containers/k8s-dns-kube-
    dns-amd64:1.14.5 k8s.gcr.io/k8s-dns-kube-dns-amd64:1.14.5
docker pull registry.cn-hangzhou.aliyuncs.com/google_containers/kubernetes-
    dashboard-amd64:v1.8.1
docker tag registry.cn-hangzhou.aliyuncs.com/google_containers/kubernetes-
    dashboard-amd64:v1.8.1 k8s.gcr.io/kubernetes-dashboard-amd64:v1.8.1
docker pull registry.cn-hangzhou.aliyuncs.com/google_containers/storage-
    provisioner:v1.8.1
docker tag registry.cn-hangzhou.aliyuncs.com/google_containers/storage-
    provisioner:v1.8.1 gcr.io/k8s-minikube/storage-provisioner:v1.8.1
docker pull registry.cn-hangzhou.aliyuncs.com/google_containers/k8s-dns-sidecar-
    amd64:1.14.5
docker tag registry.cn-hangzhou.aliyuncs.com/google_containers/k8s-dns-sidecar-
    amd64:1.14.5 k8s.gcr.io/k8s-dns-sidecar-amd64:1.14.5
docker pull registry.cn-hangzhou.aliyuncs.com/google_containers/k8s-dns-dnsmasq-
    nanny-amd64:1.14.5
docker tag registry.cn-hangzhou.aliyuncs.com/google_containers/k8s-dns-dnsmasq-
    nanny-amd64:1.14.5 k8s.gcr.io/k8s-dns-dnsmasq-nanny-amd64:1.14.5
```

然后查询下 kube-system 系统的 Pod：

```
kubectl get pods -n kube-system
NAME                                        READY    STATUS     RESTARTS    AGE
kube-addon-manager-minikube                 1/1      Running    2           54m
kube-dns-54cccfbdf8-vxmrh                   3/3      Running    6           34m
kubernetes-dashboard-77d8b98585-sdz2r       1/1      Running    0           34m
storage-provisioner                         1/1      Running    0           34m
```

可以看到组件正常。

打开 dashboard：

```
minikube dashboard
Opening kubernetes dashboard in default browser...
```

可以看到如图 22-5 所示的界面。

图 22-5　Kubernetes Dashboard 界面

恭喜你，你已经成功在本地部署好了单机版的 Kubernetes。

22.3.3　部署到 Kubernetes

1. 创建 namespace

这里我们创建 springcloud-cn 的 namespace，命令如下：

```
kubectl create namespace springcloud-cn
namespace/springcloud-cn created
```

查询所有的 namespace：

```
kubectl get namespaces
NAME              STATUS    AGE
default           Active    1h
kube-public       Active    1h
kube-system       Active    1h
springcloud-cn    Active    2s
```

2. 创建 replication controller

config-server 的 replication controller 定义如代码清单 22-18 所示：

代码清单22-18　ch22-3\ch22-3-1\k8s\config-server-rc.yml

```yaml
apiVersion: v1
kind: ReplicationController
metadata:
    name: config-server
    namespace: springcloud-cn
    labels:
        app: config-server
spec:
    replicas: 1
    template:
        metadata:
            labels:
                app: config-server
        spec:
            containers:
            - name: config-server
                image: registry.cn-hangzhou.aliyuncs.com/springcloud-cn/ch22-2-
                    1-config-server:0.0.1-SNAPSHOT
                imagePullPolicy: Always
                resources:
                    requests:
                        cpu: 100m
                        memory: 256Mi
                    limits:
                        cpu: 1000m
                        memory: 2Gi
                env:
                    - name: PROFILE
                        value: "jdbc"
                    - name: SERVER_PORT
                        value: "8080"
                    - name: JAVA_OPTS
                        value: "
-server \
-XX:+PrintGCDetails \
-XX:+PrintTenuringDistribution \
-XX:+PrintGCTimeStamps \
-XX:+HeapDumpOnOutOfMemoryError \
-XX:HeapDumpPath=/ \
-Xloggc:/gc.log \
-XX:+UseGCLogFileRotation \
-XX:NumberOfGCLogFiles=5 \
-XX:GCLogFileSize=10M"
                ports:
                    - name: http
                        containerPort: 8080
```

这里我们重点来讲解了一下 replication controller 的定义：

❑ metadata：定义了元数据，这里定义了 name、namespace，以及 label。

- spec.replicas：用于定义所需的副本数量。
- spec.template：用于定义扩容时创建 Pod 使用的模板。
- resources.requests 及 limits 部分：用来定义请求的 cpu 及内存资源大小及其上限，这里 cpu 为 100m，可以大致理解为 0.1 个 cpu，memory 为 256Mi，表示 256M 的内存。
- env 部分：用于定义启动 Docker 实例时的环境变量，这里我们定义了 PROFILE、SERVER_PORT 以及 JAVA_OPTS 这几个环境变量。
- ports 部分：定义了容器里头的端口名称及端口号。

对于 eureka-server 的 replication controller 定义，这里我们分为两个 peer，分别是 eureka-server-1 及 eureka-server2，它们的 replication controller 定义如代码清单 22-19 所示：

代码清单22-19　ch22-3\ch22-3-1\k8s\eureka-server-1-rc.yml

```yaml
apiVersion: v1
kind: ReplicationController
metadata:
    name: eureka-server-1
    namespace: springcloud-cn
    labels:
        app: eureka-server
        peer: "1"
spec:
    replicas: 1
    template:
        metadata:
            labels:
                app: eureka-server
                peer: "1"
        spec:
            containers:
            - name: eureka-server
              image: registry.cn-hangzhou.aliyuncs.com/springcloud-cn/ch22-2-2-eureka-server:0.0.1-SNAPSHOT
              imagePullPolicy: Always
              resources:
                  requests:
                      cpu: 100m
                      memory: 256Mi
                  limits:
                      cpu: 500m
                      memory: 2Gi
              env:
              - name: PROFILE
                value: "default"
              - name: SERVER_PORT
                value: "8080"
              - name: CONFIG_SERVER_HOST
                value: "192.168.99.101"
              - name: CONFIG_SERVER_PORT
                value: "8888"
              - name: EUREKA_SERVER_HOST
                value: "192.168.99.101"
```

```
              - name: EUREKA_SERVER_PORT
                value: "8762"
              - name: JAVA_OPTS
                value: "
-server \
-XX:+PrintGCDetails \
-XX:+PrintTenuringDistribution \
-XX:+PrintGCTimeStamps \
-XX:+HeapDumpOnOutOfMemoryError \
-XX:HeapDumpPath=/ \
-Xloggc:/gc.log \
-XX:+UseGCLogFileRotation \
-XX:NumberOfGCLogFiles=5 \
-XX:GCLogFileSize=10M"
            ports:
              - name: http
                containerPort: 8080
```

注意我们在这里的 label 中添加了一个 peer 便签，用于标识是哪个 peer。对于 CONFIG_SERVER 相关的环境变量，这里由于我们之前定义的环境变量名称跟 Kubernetes 给 config-server 自动创建的名称不一样（CONFIG_SERVER_SERVICE_HOST 及 CONFIG_SERVER_SERVICE_PORT），所以我们使用 Service 映射出来的 config-server 的地址及端口。另外 peer1 配置的是 peer2 的 eureka 地址及端口（同样这里我们采用的是 Service 映射出来的 eureka 地址及端口）。peer2 的 replication 定义文件如代码清单 22-20 所示：

代码清单22-20   ch22-3\ch22-3-1\k8s\eureka-server-2-rc.yml

```
apiVersion: v1
kind: ReplicationController
metadata:
    name: eureka-server-2
    namespace: springcloud-cn
    labels:
        app: eureka-server
        peer: "2"
spec:
    replicas: 1
    template:
        metadata:
            labels:
                app: eureka-server
                peer: "2"
        spec:
            containers:
            - name: eureka-server
                image: registry.cn-hangzhou.aliyuncs.com/springcloud-cn/ch22-2-
                    2-eureka-server:0.0.1-SNAPSHOT
                imagePullPolicy: Always
                resources:
                    requests:
                        cpu: 100m
                        memory: 256Mi
                    limits:
```

```
                            cpu: 500m
                            memory: 2Gi
                  env:
                  - name: PROFILE
                      value: "default"
                  - name: SERVER_PORT
                      value: "8080"
                  - name: CONFIG_SERVER_HOST
                      value: "192.168.99.101"
                  - name: CONFIG_SERVER_PORT
                      value: "8888"
                  - name: EUREKA_SERVER_HOST
                      value: "192.168.99.101"
                  - name: EUREKA_SERVER_PORT
                      value: "8761"
                  - name: JAVA_OPTS
                      value: "
-server \
-XX:+PrintGCDetails \
-XX:+PrintTenuringDistribution \
-XX:+PrintGCTimeStamps \
-XX:+HeapDumpOnOutOfMemoryError \
-XX:HeapDumpPath=/ \
-Xloggc:/gc.log \
-XX:+UseGCLogFileRotation \
-XX:NumberOfGCLogFiles=5 \
-XX:GCLogFileSize=10M"
                  ports:
                  - name: http
                      containerPort: 8080
```

跟 peer1 不同的是，它的 label 的 peer 值为 2，另外 eureka 地址为 peer1 的地址。

Gateway 作为 config client 及 eureka client，需要配置它们的地址，其 replication controller 的定义如代码清单 22-21 所示：

代码清单22-21　ch22-3\ch22-3-1\k8s\gateway-rc.yml

```
apiVersion: v1
kind: ReplicationController
metadata:
    name: gateway
    namespace: springcloud-cn
    labels:
        app: gateway
spec:
    replicas: 1
    template:
        metadata:
            labels:
                app: gateway
        spec:
            containers:
            - name: gateway
                image: registry.cn-hangzhou.aliyuncs.com/springcloud-cn/ch22-2-
```

```yaml
          3-gateway:0.0.1-SNAPSHOT
        imagePullPolicy: Always
        resources:
            requests:
                cpu: 100m
                memory: 256Mi
            limits:
                cpu: 1000m
                memory: 2Gi
        env:
            - name: PROFILE
              value: "default"
            - name: SERVER_PORT
              value: "8080"
            - name: CONFIG_SERVER_HOST
              value: "192.168.99.101"
            - name: CONFIG_SERVER_PORT
              value: "8888"
            - name: EUREKA_SERVER1_HOST
              value: "192.168.99.101"
            - name: EUREKA_SERVER1_PORT
              value: "8761"
            - name: EUREKA_SERVER2_HOST
              value: "192.168.99.101"
            - name: EUREKA_SERVER2_PORT
              value: "8762"
            - name: JAVA_OPTS
              value: "
-server \
-XX:+PrintGCDetails \
-XX:+PrintTenuringDistribution \
-XX:+PrintGCTimeStamps \
-XX:+HeapDumpOnOutOfMemoryError \
-XX:HeapDumpPath=/ \
-Xloggc:/gc.log \
-XX:+UseGCLogFileRotation \
-XX:NumberOfGCLogFiles=5 \
-XX:GCLogFileSize=10M"
        ports:
            - name: http
              containerPort: 8080
```

注意，由于我们采用的是两个 eureka server，因此这里配置了两个 eureka peer 的地址。

turbine、spring-admin、biz-service 的 replication controller 的定义跟 gateway 类似，这里不再重复，具体可以见工程源码。

3. 创建 service

前面有提到了 service 可以对一组 pod 进行路由，以两个 eureka peer 为例，我们分别看下它们的 service 定义，peer1 的 service 文件定义如代码清单 22-22 所示：

代码清单22-22　ch22-3\ch22-3-1\k8s\eureka-server-1-svc.yml

```yaml
apiVersion: v1
kind: Service
metadata:
    name: eureka-server-1
    namespace: springcloud-cn
    labels:
        service: eureka-server-svc
spec:
    selector:
        app: eureka-server
        peer: "1"
    ports:
    - name: http
      port: 8761
      targetPort: 8080
    clusterIP: 10.96.0.3
    externalIPs:
    - 192.168.99.101
```

可以看到，这里的 selector 选择了 label 中 app 为 eureka-server，且 peer 值为 1 的 pod 来路由，其中 externalIPs 选择了 192.168.99.101 这个 node 来暴露，暴露的端口为 8761。peer2 的 service 文件定义如代码清单 22-23 所示：

代码清单22-23　ch22-3\ch22-3-1\k8s\eureka-server-2-svc.yml

```yaml
apiVersion: v1
kind: Service
metadata:
    name: eureka-server-2
    namespace: springcloud-cn
    labels:
        service: eureka-server-svc
spec:
    selector:
        app: eureka-server
        peer: "2"
    ports:
    - name: http
      port: 8762
      targetPort: 8080
    clusterIP: 10.96.0.4
    externalIPs:
    - 192.168.99.101
```

peer2 的 service 则选择的是 peer 值为 2 的 eureka server，然后 externalIPs 选择了 192.168.99.101 这个 node 来暴露，暴露的端口为 8762。

这里选择将 eureka 暴露出来，就可以让外部通过 node 的 ip 地址及映射的端口来访问 eureka，因此生产环境必须加上鉴权才比较安全。实际应用过程中，如无必要，可尽量避免使用 service 来暴露，而是采用 kubernetes 自动创建的环境变量来进行引用。

4. 运行

使用如下命令，分别创建 replication controller 及 service：

```
kubectl create -f config-server-rc.yml
kubectl create -f config-server-svc.yml
kubectl create -f eureka-server-1-rc.yml
kubectl create -f eureka-server-1-svc.yml
kubectl create -f eureka-server-2-rc.yml
kubectl create -f eureka-server-2-svc.yml
kubectl create -f gateway-rc.yml
kubectl create -f gateway-svc.yml
kubectl create -f turbine-rc.yml
kubectl create -f turbine-svc.yml
kubectl create -f spring-admin-rc.yml
kubectl create -f spring-admin-svc.yml
kubectl create -f biz-service-rc.yml
```

执行完之后，使用 get pods 命令查询：

```
kubectl get pods -n springcloud-cn
NAME                        READY   STATUS    RESTARTS   AGE
biz-service-cxngx           1/1     Running   0          4m
config-server-r45vk         1/1     Running   0          4m
eureka-server-1-zp27x       1/1     Running   0          4m
eureka-server-2-9589x       1/1     Running   0          4m
gateway-zxklf               1/1     Running   0          4m
spring-admin-qzzpq          1/1     Running   0          4m
turbine-77pxf               1/1     Running   0          4m
```

通过 minikube dashboard 访问，选择 springcloud-cn 的 namespace，其界面如图 22-6 所示。

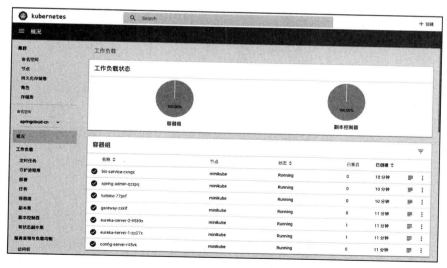

图 22-6　kubernetes 的 namespace 容器组视图

至此，我们就成功将上一节的工程项目通过 kubernetes 管理了起来，接下来我们会演示如何使用 kubernetes 进行扩容、缩容以及滚动升级。

22.3.4 一键伸缩

1. 准备工作

这里我们创建一个简单的 SpringBoot 工程，具体详见 ch22-3-2-example-service 工程。然后部署到 kubernetes 上 springcloud-cn 的命名空间，代码如下：

```
kubectl create -f example-service-rc.yml
kubectl get pods -n springcloud-cn
NAME READY STATUS RESTARTS AGE
example-service-bm27m 1/1 Running 0 1m
```

2. 扩容、缩容

kubernetes 有个 scale 命令，可以用来控制副本个数，既可用来扩容也可以用来缩容。命令实例如下：

```
kubectl scale --replicas=2 -f example-service-rc.yml
```

或者：

```
kubectl scale --replicas=2 rc/example-service -n springcloud-cn
```

由于之前 example-service 的 replicas 为 1，因此设定为 2，相当于扩容。

使用 get pods 命令显示如下：

```
kubectl get pods -n springcloud-cn
NAME                       READY   STATUS    RESTARTS   AGE
example-service-bm27m      1/1     Running   0          2m
example-service-z2h4l      1/1     Running   0          0s
```

再将 example-service 实例缩小为 1，即进行缩容操作：

```
kubectl scale --replicas=1 -f example-service-rc.yml
```

或者：

```
kubectl scale --replicas=1 rc/example-service -n springcloud-cn
```

使用 get pods 命令显示如下：

```
kubectl get pods -n springcloud-cn
NAME                       READY   STATUS        RESTARTS   AGE
example-service-bm27m      1/1     Running       0          3m
example-service-z2h4l      1/1     Terminating   0          44s
```

可以看到，kubernetes 帮忙终止了 1 个实例。

3. 自动伸缩

Kubernetes1.2 及以上版本提供了 HPA(Horizontal Pod Autoscaling) 功能特性，即可以根据 CPU 利用率等指标完成自动伸缩。这里我们需要组件来采集 cpu 利用率等指标，通常可以使用 heapster 或者轻量级的 Kubernetes Metrics Server。

这里我们使用 metrics-server 的 v0.2.1 版本，其下载地址为 https://github.com/kubernetes-incubator/metrics-server/releases/tag/v0.2.1。这里笔者也有下载一份，略做修改放到了工程里头

(主要将 metrics-server-deployment 的 imagePullPolicy 由 Always 改为 IfNotPresent，另外如果镜像拉取问题，也需要自行拉取改下 tag），详见 ch22-3/metrics-server-0.2.1。

然后使用如下命令创建 metrics-server：

```
kubectl create -f metrics-server-0.2.1
clusterrolebinding.rbac.authorization.k8s.io/metrics-server:system:auth-
    delegator created
rolebinding.rbac.authorization.k8s.io/metrics-server-auth-reader created
apiservice.apiregistration.k8s.io/v1beta1.metrics.k8s.io created
serviceaccount/metrics-server created
deployment.extensions/metrics-server created
service/metrics-server created
clusterrole.rbac.authorization.k8s.io/system:metrics-server created
clusterrolebinding.rbac.authorization.k8s.io/system:metrics-server created
```

执行如下命令，可以验证是否成功：

```
kubectl get --raw "/apis/metrics.k8s.io/v1beta1/nodes"
{"kind":"NodeMetricsList","apiVersion":"metrics.k8s.io/v1beta1","metadata":{"se
    lfLink":"/apis/metrics.k8s.io/v1beta1/nodes"},"items":[{"metadata":{"name":
    "minikube","selfLink":"/apis/metrics.k8s.io/v1beta1/nodes/minikube","creati
    onTimestamp":"2018-07-04T11:02:12Z"},"timestamp":"2018-07-04T11:02:00Z","wi
    ndow":"1m0s","usage":{"cpu":"184m","memory":"1250416Ki"}}]}
```

创建 hpa：

```
kubectl autoscale rc example-service --cpu-percent=10 --min=1 --max=3 -n
    springcloud-cn
```

然后查看：

```
kubectl get hpa -n springcloud-cn
NAME REFERENCE TARGETS MINPODS MAXPODS REPLICAS AGE
example-service ReplicationController/example-service 1%/10% 1 3 1 16m
```

首先我们暴露一下 example-service 的地址，其 service 文件如代码清单 22-24 所示：

代码清单22-24　ch22-3/ch22-3-2-example-service\example-service-svc.yml

```yaml
apiVersion: v1
kind: Service
metadata:
    name: example-service
    namespace: springcloud-cn
    labels:
        service: example-service
spec:
    selector:
        app: example-service
    ports:
    - name: http
        port: 8080
        targetPort: 8080
    clusterIP: 10.96.0.8
    externalIPs:
        - 192.168.99.101
```

对 example-service 进行压测，代码如下：

```
wrk -t8 -c200 -d30s http://192.168.99.101:8080/actuator/metrics
Running 30s test @ http://192.168.99.101:8080/actuator/metrics
  8 threads and 200 connections
  Thread Stats   Avg      Stdev     Max   +/- Stdev
    Latency   588.15ms  341.11ms   1.99s    71.03%
    Req/Sec    50.93     40.15    242.00    68.86%
  10061 requests in 30.11s, 6.46MB read
  Socket errors: connect 0, read 0, write 0, timeout 74
Requests/sec:    334.18
Transfer/sec:    219.60KB
```

压测之前的 hpa 状态如下：

```
kubectl get hpa -n springcloud-cn
NAME REFERENCE TARGETS MINPODS MAXPODS REPLICAS AGE
example-service ReplicationController/example-service 1%/10% 1 3 1 21m
```

压测一阵子之后查看如下：

```
kubectl get hpa -n springcloud-cn
NAME REFERENCE TARGETS MINPODS MAXPODS REPLICAS AGE
example-service ReplicationController/example-service 107%/10% 1 3 1 21m
```

可以看到 cpu 飙升，紧接着查看 example-service 的 pod 实例：

```
kubectl get pods -n springcloud-cn
NAME                      READY   STATUS    RESTARTS   AGE
example-service-62sz7     1/1     Running   0          2m
example-service-bm27m     1/1     Running   0          2h
example-service-fpqhj     1/1     Running   0          2m
```

可以看到自动扩容到了 3 个。查看 dashboard，如图 22-7 所示。

图 22-7　kubernetes 自动扩容实例

也可以看到 example 从 1 个扩容到了 3 个。
压测结束后，查看下 hpa，如下：

```
kubectl get hpa -n springcloud-cn
NAME REFERENCE TARGETS MINPODS MAXPODS REPLICAS AGE
example-service ReplicationController/example-service 2%/10% 1 3 3 24m
```

可以看到 cpu 已经下降下来了，replicas 字段为 3，表示已经扩容到了 3 个实例。

在等一阵子，查看 hpa，可以看到：

```
kubectl get hpa -n springcloud-cn
NAME REFERENCE TARGETS MINPODS MAXPODS REPLICAS AGE
example-service ReplicationController/example-service 1%/10% 1 3 1 32m
```

查看 pod 状态如下：

```
kubectl get pods -n springcloud-cn
NAME READY STATUS RESTARTS AGE
example-service-bm27m 1/1 Running 0 2h
```

由于 cpu 利用率已经降低到阈值，于是 hpa 自动缩容到 1 个副本。

查看 dashboard 的事件，如图 22-8 所示。

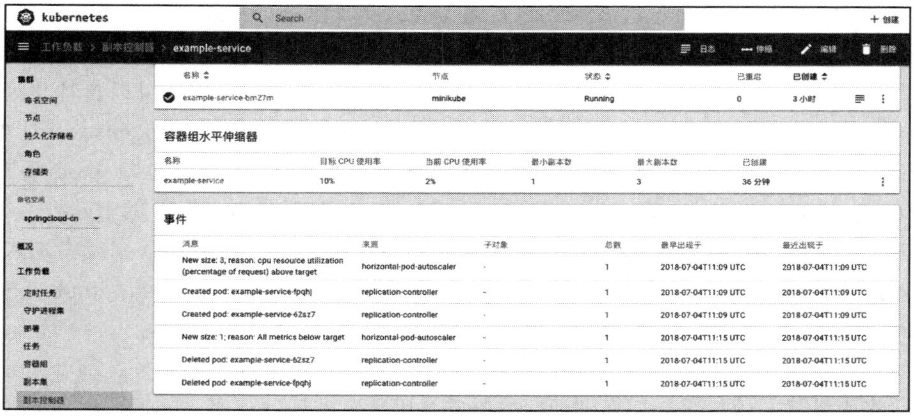

图 22-8　kubernetes 扩缩容事件列表

可见，图 22-8 中清晰描述了扩容、缩容产生的原因及时间点。

至此 hpa 基于 cpu 指标的自动扩容缩容实例演示完成。在最开始版本，hpa 只支持根据 cpu 指标来进行扩容缩容，kubernetes1.6 版本支持了基于多个 metric 的扩容缩容，而且还支持自定义 metric，更加灵活强大。

22.3.5　滚动升级

1. 部署模式概述

一般而言，系统部署模式可以分为如下几种：

- 金丝雀部署（canary deployments）/ 灰度发布。
- LB 部署。
- 蓝绿部署（blue green deployments）。

金丝雀部署的主要思路是将生产服务的一小部分实例更新为最新版本，然后提供给一小部分用户进行测试，同时对新版本的服务进行监控，观察是否有异常。如果发现异常，则进行快速回滚，将影响降低到最小。通常金丝雀部署时间不会持续太久，观察没问题，即可全量更新部署，因此生产上的服务新旧版本不会并存太多时间。

灰度发布的话，通常用来进行 A/B 测试，因此生产上会同时并存两个版本或更多的服务，按比例将流量对不同版本的服务进行路由，然后进行观测。金丝雀发布可以看作灰度发布的一个瞬时版。

基于 LB（Load Balancer）的部署方式比较常见，它有以下几个步骤：

1）从 LB 中摘除。
2）更新到新版本。
3）观察 / 自测是否有异常。
4）加入 LB。
5）观察生产流量过来之后有没有异常。

没有异常就结束，继续更新该服务的其他实例；有异常的话，就重新走下 1～5 步，只是第 2 步是回退到旧版本。

这种方式人工操作的话，非常机械，配合 /health 以及自动化检测，可以自动化处理。比如 Nginx 有个 dynamic upstream（https://github.com/cubicdaiya/ngx_dynamic_upstream）的模块，可以动态修改 upstream 而无须重启 nginx，另外还有 nginx_upstream_check_module（https://github.com/yaoweibin/nginx_upstream_check_module），相当于 Nginx 版本的 Hystrix。可以用来增强对 upstream 的 health 检查，然后自动摘除不健康的 upstream 的 server 或者加入恢复健康的 upstream 的 server，它的前身是 healthcheck_nginx_upstreams（https://github.com/cep21/healthcheck_nginx_upstreams）模块。

蓝绿部署要求在部署的时候，额外提供一套生产环境来部署新版的服务，部署完验证没问题，然后将流量从旧版的服务切换到新版的服务，用生产的流量观察一阵子，没问题的话，再销毁旧版的服务实例，如果发现有问题，立即将流量切换回旧版的服务。蓝绿部署属于"一刀切"的方式，这个比较适合服务实例不是太多的情况，毕竟新部署一套系统需要临时暂用额外的资源。

kubernetes 的 deployment 有个滚动升级的功能，也是属于蓝绿部署的范畴之类，只是更灵活一点，它可以根据生产上的资源情况，控制升级过程中额外所需占用的资源。比如先新拉起一个新版的服务实例，确认 ok 了，然后销毁一个旧版的服务实例。具体可以通 strategy.rollingUpdate.maxSurge 来控制滚动升级的时候，最大新拉起的服务个数，以及 strategy.rollingUpdate.maxUnavailable 来控制整个滚动升级期间最大的不可用服务个数，如果 maxUnavailable 为 0 就相当于蓝绿部署了。下面我们就来演示一下如何使用 deployment 进行滚动升级。

2. Deployment 定义

首先我们定义一下 example-service 的 deployment 文件，如代码清单 22-25 所示：

代码清单22-25　ch22-3\ch22-3-2-example-service\example-service-dm.yml

```yaml
apiVersion: extensions/v1beta1
kind: Deployment
metadata:
    name: example-service
    namespace: springcloud-cn
    labels:
        app: example-service
spec:
    replicas: 3
#   minReadySeconds: 60          #滚动升级时60s后认为该pod就绪
    strategy:
        rollingUpdate:    ##由于replicas为3,则整个升级,pod个数在2-4个之间
            maxSurge: 1           #滚动升级时会先启动1个pod
            maxUnavailable: 1 #滚动升级时允许的最大Unavailable的pod个数
    template:
        metadata:
            labels:
                app: example-service
        spec:
            terminationGracePeriodSeconds: 60 ##k8s将会给应用发送SIGTERM信号,可以用
                来正确、优雅地关闭应用,默认为30秒
            containers:
            - name: example-service
                image: registry.cn-hangzhou.aliyuncs.com/springcloud-cn/ch22-3-
                    2-example-service:0.0.1-SNAPSHOT
                imagePullPolicy: Always
                livenessProbe:  #kubernetes认为该pod是存活的,不存活则需要重启
                    httpGet:
                        path: /acutator/health
                        port: 8080
                        scheme: HTTP
                    initialDelaySeconds: 60 ## equals to the maximum startup time
                        of the application + couple of seconds
                    timeoutSeconds: 5
                    successThreshold: 1
                    failureThreshold: 5
                readinessProbe: #kubernetes认为该pod是启动成功的
                    httpGet:
                        path: /actuator/health
                        port: 8080
                        scheme: HTTP
                    initialDelaySeconds: 30 ## equals to minimum startup time of
                        the application
                    timeoutSeconds: 5
                    successThreshold: 1
                    failureThreshold: 5
                resources:
                    requests:
                        cpu: 50m
                        memory: 256Mi
                    limits:
                        cpu: 500m
                        memory: 512Mi
                env:
```

```yaml
        - name: PROFILE
          value: "default"
      ports:
        - name: http
          containerPort: 8080
```

然后删除前面部署的 example-service 的 replication controller，部署 example-service 的 deployment 如下：

```
kubectl delete rc example-service -n springcloud-cn
kubectl create -f example-service-dm.yml
```

查看 deployment 如下：

```
kubectl get deployment -n springcloud-cn
NAME             DESIRED   CURRENT   UP-TO-DATE   AVAILABLE   AGE
example-service  3         3         3            0           13s
```

3. 新版本升级

假设我们对 example-service 工程进行了升级，从 0.0.1-SNAPSHOT 升级到 0.0.1 版本，升级版的工程见 ch22-3/ch22-3-3-example-service，构建如下：

```
mvn clean package
```

得到新版本的镜像 registry.cn-hangzhou.aliyuncs.com/springcloud-cn/ch22-3-3-example-service：0.0.1。

下面我们使用 deployment 进行滚动升级：

```
kubectl set image deployment/example-service example-service=registry.cn-hangzhou.aliyuncs.com/springcloud-cn/ch22-3-3-example-service:0.0.1
    --namespace=springcloud-cn
```

这里通过 set image 更新 example-service 的 deployment 中名为 example-service 的容器的镜像为 registry.cn-hangzhou.aliyuncs.com/springcloud-cn/ch22-3-3-example-service:0.0.1。

查看升级状态如下：

```
kubectl describe deployment/example-service -n springcloud-cn
Name:                   example-service
Namespace:              springcloud-cn
CreationTimestamp:      Thu, 05 Jul 2018 09:45:45 +0800
Labels:                 app=example-service
Annotations:            deployment.kubernetes.io/revision=2
Selector:               app=example-service
Replicas:               3 desired | 3 updated | 3 total | 2 available | 1 unavailable
StrategyType:           RollingUpdate
MinReadySeconds:        0
RollingUpdateStrategy:  1 max unavailable, 1 max surge
Pod Template:
  Labels:  app=example-service
  Containers:
    example-service:
      Image:registry.cn-hangzhou.aliyuncs.com/springcloud-cn/ch22-3-3-
        example-service:0.0.1
```

```
            Port:           8080/TCP
            Host Port:      0/TCP
            Limits:
              cpu:          500m
              memory:       512Mi
            Requests:
              cpu:          50m
              memory:       256Mi
            Liveness:       http-get http://:8080/acutator/health delay=60s
              timeout=5s period=10s #success=1 #failure=5
            Readiness:      http-get http://:8080/actuator/health delay=30s
              timeout=5s period=10s #success=1 #failure=5
            Environment:
              PROFILE:      default
            Mounts:         <none>
      Volumes:              <none>
Conditions:
    Type            Status  Reason
    ----            ------  ------
    Available       True    MinimumReplicasAvailable
OldReplicaSets:     <none>
NewReplicaSet:      example-service-85f4ccc8c (3/3 replicas created)
Events:
    Type    Reason              Age         From                    Message
    ----    ------              ----        ----                    -------
    Normal  ScalingReplicaSet   29m         deployment-controller   Scaled up replica
      set example-service-769fc9b45c to 3
    Normal  ScalingReplicaSet   56s         deployment-controller   Scaled up replica
      set example-service-85f4ccc8c to 1
    Normal  ScalingRe plicaSet  56s         deployment-controller   Scaled down replica
      set example-service-769fc9b45c to 2
    Normal  ScalingReplicaSet   56s         deployment-controller   Scaled up replica
      set example-service-85f4ccc8c to 2
    Normal  ScalingReplicaSet   0s          deployment-controller   Scaled down replica
      set example-service-769fc9b45c to 1
    Normal  ScalingReplicaSet   0s          deployment-controller   Scaled up replica
      set example-service-85f4ccc8c to 3
    Normal  ScalingReplicaSet   <invalid>   deployment-controller   Scaled down replica
      set example-service-769fc9b45c to 0
```

查看升级历史如下：

```
kubectl rollout history deployments -n springcloud-cn
deployments "example-service"
REVISION    CHANGE-CAUSE
1           <none>
2           <none>
```

上面演示的仅仅是执行顺利的过程，如果中途需要暂停、继续升级或者回滚升级，kubectl rollout 都有相应的命令可以一键完成：

暂停升级：

```
kubectl rollout pause deployment/example-service --namespace=springcloud-cn
```

继续升级：

```
kubectl rollout resume deployment/example-service --namespace=springcloud-cn
```

回滚升级：

```
kubectl rollout undo deployment/example-service --namespace=springcloud-cn
```

本节给大家演示了如何使用 kubernetes 的 deployment 进行滚动升级，其核心在于 deployment 文件的定义，这里有几个参数需要注意一下：

- strategy，这里设置为 rollingUpdate。
- strategy.rollingUpdate.maxUnavailable 为 0 相当于蓝绿部署，可以结合 service 及 label 处理下路由。
- terminationGracePeriodSeconds 用于关闭应用实例时发送 SIGTERM 信号之后等待的时间，默认为 30 秒，如果应用实例对优雅关闭有要求，则需要考虑调整该值。
- livenessProbe 与 readinessProbe，这里笔者为了演示，都设置为了 /actuator/health 路径，其中 livenessProbe 用于判断 pod 是否存活或者启动成功，如果判断不通过则 kubernetes 会重新 kill 掉再重新启动，对于该值不同的业务应用，其启动时间各不相同，需要自己调整，不然就很容易出现因为应用启动时间慢，导致 livenessProbe 不通过而被频繁 kill 和 create。readinessProbe 是 kubernetes 用来判断 pod 是否可以接收流量的，对于分布式应用来说，Tomcat 或 Netty 等服务器启动起来之后，并不意味着可以立即接受请求，通常还需要做一些初始化动作，比如注册到服务中心、初始化一些业务资源等，因此业务应用可以根据需要自定义一个 endpoint 作为 readinessProbe，蚂蚁金服开源的 SOFABoot 就专门定制了这样一个 endpoint。

22.4　本章小结

本章讲解了 Java 应用的 Docker 化及其优化，然后展示了 Spring Cloud 相关组件的 Docker，紧接着演示了如何将各个组件部署到 kubernetes 上，让大家对 kubernetes 有个初步的了解。之后介绍了如何使用 kubernetes 进行一键扩容、缩容以及如何使用 hpa 进行自动伸缩，还有如何进行 deployment 的滚动升级，让大家领略一下 kubernetes 的强大。相信通过本章的介绍，大家能够对在 kubernetes 部署 Spring Cloud 微服务有个初步的认知。关于 Spring Cloud 部署到 kubernetes 上面，本章演示的是将 Spring Cloud 的组件 Docker 化，然后部署到 kubernetes 上，其实有些分布式组件可以直接利用 kubernetes 已有的功能特性进行部署，对此，可以详见 spring-cloud-incubator 里的 spring-cloud-kubernetes 工程（https://github.com/spring-cloud-incubator/spring-cloud-kubernetes），有兴趣的读者可以深入研究下。

第 23 章 Chapter 23

Dubbo 向 Spring Cloud 迁移

目前，很多公司有很多系统使用的是 Dubbo 架构。Dubbo 是一个非常优秀的 RPC 框架，有着优异的服务调用性能，但 Dubbo 并没有像 Spring Cloud 那样能够满足服务治理方方面面的需求。随着微服务的流行和普及，Spring Cloud 成为了微服务架构的不二选择。一些公司可能老服务使用 Dubbo，而新服务尝试使用 Spring Cloud，这样就涉及两类不同架构的服务如何相互调用的问题。

本章我们主要研究 SpringCloud 的服务如何调用既有的 Dubbo 服务的问题，这里我们提供两个解决方案：一个是将现有的 Dubbo 服务 RESTful 化，通过集成 Eureka 等纳入 Spring Cloud 的微服务中，这样二者共用 Spring Cloud 的服务注册与发现，通过 FeignClient 相互调用；另一个是扩展 FeignClient 的契约，通过 Dubbo 的 RPC 协议无缝调用现有 Dubbo 服务，同时将 Spring Cloud 的 RESTful API 转换为 Dubbo 服务提供方并注册到 Dubbo 服务中心，既有的 Dubbo 服务也可以无缝调用 SpringCloud 的服务。下面我们就来一一介绍。

23.1 将 Dubbo 服务纳入 Spring Cloud 体系中

这是一种比较简单且容易实现的方式，没有任何改造难点，只需要在原本的 Dubbo 系统增加一些代码就可以实现。这种改造的核心思想是：原本的 Dubbo 服务，内部继续使用 Dubbo 调用；Dubbo 服务与 Spring Cloud 服务系统之间使用 REST 进行服务消费，这需要将 Dubbo 服务暴露为 REST API，同时为了能使用 Spring Cloud 服务调用 Feign 客户端的 Ribbon 的服务负载均衡的能力，需要将 Dubbo 系统纳入 Spring Cloud 系统中。下面将为大家介绍如何实现改造。

23.1.1 将 Dubbo 项目改造成 Spring Boot 项目

在很多老的 Dubbo 系统中，还不是 Spring Boot 项目，而 Spring Cloud 项目是基于 Spring

Boot 的，所以首先需要将 Dubbo 系统改造成 Spring Boot 项目。首先在项目中的 pom 文件中加上 Spring Boot 的相关依赖，示例代码如下：

```xml
<dependency>
    <groupId>org.springframework.boot</groupId>
    <artifactId>spring-boot-starter-web</artifactId>
</dependency>
<dependencyManagement>
    <dependency>
        <groupId>org.springframework.boot</groupId>
        <artifactId>spring-boot-dependencies</artifactId>
        <version>${spring.boot.version}</version>
        <type>pom</type>
        <scope>import</scope>
    </dependency>
</dependencyManagement>
<build>
    <plugins>
        <plugin>
            <groupId>org.springframework.boot</groupId>
            <artifactId>spring-boot-maven-plugin</artifactId>
            <executions>
                <execution>
                    <goals>
                        <goal>repackage</goal>
                    </goals>
                </execution>
            </executions>
        </plugin>
    </plugins>
</build>
```

然后，替换原项目启动方式，采用 Spring Boot 的启动方式，并废弃原 web.xml 等配置文件。在启动类中通过注解加载 Spring 的 xml 配置。示例代码如下：

```java
@SpringBootApplication
@ImportResource("classpath:spring-root.xml")
public class ApplicationSupport {
    public static void main(String[] args) {
        SpringApplication.run(ApplicationSupport.class, args);
    }
}
```

在 resources 下新增 application.yml，对改造后的项目进行配置，比如将应用名设为 user，应用的 tomcat 端口设为 8080。示例代码如下：

```yaml
spring:
    application:
        name: user
server:
    port: 8080
```

23.1.2　集成 Spring Cloud 组件

将原有非 SpringBoot 的 Dubbo 工程升级为 SpringBoot 之后，接下来我们需要将 Dubbo

纳入 Spring Cloud 体系中。这样做有两个好处：一个是使原有的 Spring Cloud 项目，能通过 Feign 的 Ribbon 去负载均衡调用 Dubbo 提供的 REST API；另一个是改造后的 Dubbo 项目也能过通过 Feign 调用原有的 Spring Cloud 项目的 REST API。

如何将 Spring Boot 项目纳入 Spring Cloud 体系中呢？很简单，无非需要在 pom 文件中引入服务注册发现组件，比如 Eureka Client 的依赖；引入 Spring Cloud Config 的配置；在路由网关，比如 Zuul、Gateway，加上该服务的路由。由于在之前的章节这方面的知识已经讲得很透彻，所在此处略过，读者可自行查看前面的章节。将 Dubbo 项目纳入 Spring Cloud 体系中，这样项目就会同时向两个注册中心提供注册服务，如图 23-1 所示。此时的服务既具有 Dubbo 提供服务的能力，又提供了 REST API 服务的能力。

图 23-1　Dubbo 项目的服务提供者同时有 2 个注册中心

23.1.3　将 Dubbo 服务暴露为 RESTful API

将 Dubbo 项目纳入 Spring Cloud 体系之后，在 Dubbo 项目中引入 Feign 客户端，就可以消费服务了。那么如何将 Dubbo 服务暴露为 REST API 呢？笔者的建议是重写服务代码，将 Dubbo 服务提供者的服务再写一遍，暴露成 RESTful API 的形式。之所以这样做，是因为原本的 Dubbo 系统之间的服务调用还可以继续采用 Dubbo 进行调用，同时原本的 Spring Cloud 项目可以用 REST API 调用 Dubbo 的服务。这样虽然带来了一定的工作量，但是代码分层合理，其实工作量很少。再者，Spring Cloud 项目调用 Dubbo 的场景其实并不多。

为 Dubbo 项目增加 REST API 服务，下面用具体案例来介绍实现过程。首先为 Dubbo 项目的服务提供者添加 Feign 依赖。然后增加原 Dubbo 服务定义的 API 接口，支持 Feign 调用，以对外二方库的形式，对外提供服务。示例代码如下：

```
@FeignClient("provider")
public interface HelloService {
    @GetMapping("/hello")
    String hello();
}
```

增加上述 Feign 接口的实现类，在类上添加 @RestController 接口，对外提供 RESTful API，示例代码如下：

```
@RestController
public class HelloServiceImpl implements HelloService {
    @Override
    public String hello() {
        return "hello at " + System.currentTimeMillis();
```

		}
	}

经过这样的改造，Dubbo 服务中就增加了可以被 Spring Cloud 微服务消费的 RESTful API。Spring Cloud 项目的服务消费者只需要注入 Dubbo 项目对外二方库的 Feign Client 就可以消费该服务了。

23.2 将 Spring Cloud 服务 Dubbo 化

在上一节中，笔者已经很清晰地描述了如何将 Dubbo 服务升级为 Spring Cloud 架构，并在 Dubbo 项目中通过多写代码的形式暴露出 REST API，这样 Dubbo 和 Spring Cloud 项目就可以互相消费服务了。

在本节中，笔者将向大家演示 Spring Cloud 项目如何在用户无感知的情况下，不通过 REST API 的形式去消费服务，而是采用 Dubbo 的 RPC 去消费 Dubbo 的服务。同时原本的 Spring Cloud 项目不用修改任何代码，就能提供 Dubbo 类型的服务。可能读者觉得这是一件不可思议的事情，但事实是 Spring Cloud 中国社区已经实现该框架，并在 Github 上开源。项目地址 https://github.com/SpringCloud/spring-cloud-dubbo。

该框架的核心原理是替换 Spring Cloud 的 Feign 组件的底层服务调用契约，将原本交由 Http Client（Feign 目前支持三种 Htpp Client，分别为 HttpURLConnection、HttpClient 和 OKHttp）处理的请求交给 Dubbo RPC 处理；同时原本对外提供的 REST API 转为提供 Dubbo 服务。另外该框架对于原本的服务无任何侵入，只需要在 pom 文件中加一个依赖，并在配置文件 application 中做一些配置即可。

这里我们用一个案例来演示一下如何使用 spring-cloud-dubbo 组件。本案例主要分为四个工程：服务注册中心 eureka-server 工程、服务提供者对外发布的二方库 demo-dubbo-provider-api、服务提供方的实现工程 demo-dubbo-provider、服务消费者工程 demo-dubbo-consumer。下面我们来一一介绍。

23.2.1 服务注册中心

本案例使用 Eureka 作为服务注册中心，案例中的注册中心的代码实现跟之前章节的介绍的没有任何区别。首先需要在 pom 文件中加上 Eureka Server 的起步依赖 spring-cloud-starter-netflix-eureka-server；然后在配置文件 application.yml 中做相关的配置，其中服务端口为 8761；最后在程序的启动类加上 @EnableEurekaServer 注解，开启 EurekaServer 的功能，具体参考工程 ch23-2/eureka-server。

23.2.2 服务提供者

本案例中，服务提供者分为两个部分：一个部分为服务提供者对外提供的二方库，另一部分为作为对外提供服务的实现类，这二者分别为 demo-dubbo-provider-api 和 demo-dubbo-provider。

首先在 demo-dubbo-provider-api 工程的 pom 文件中引入相关依赖，具体如代码清单 23-1 所示。

代码清单 23-1　ch23-2/demo-dubbo-provider-api/pom.xml

```xml
<dependencies>
    <dependency>
        <groupId>org.springframework.cloud</groupId>
        <artifactId>spring-cloud-openfeign-core</artifactId>
    </dependency>
    <dependency>
        <groupId>org.springframework</groupId>
        <artifactId>spring-web</artifactId>
    </dependency>
</dependencies>
```

然后写一个 FeignClient，该 FeignClient 作为服务提供者的对外二方库，任何使用了服务提供者的对外二方库，都具备调用该服务提供者的服务能力，具体如代码清单 23-2 所示。

代码清单 23-2　ch23-2/demo-dubbo-provider-api/src/main/java/cn/springcloud/book/service/HelloService.java

```java
@FeignClient("provider")
public interface HelloService {
    @GetMapping("/hello")
    String hello();
}
```

这样服务提供者的对外二方库就写好了。而服务提供者为一个 Spring Boot 工程，这里需要引入 spring-cloud-dubbo-starter 和 demo-dubbo-provider-api 依赖，如代码清单 23-3 所示。

代码清单 23-3　ch23-2/demo-dubbo-provider/pom.xml

```xml
<dependencies>
    <dependency>
        <groupId>cn.springcloud.book</groupId>
        <artifactId>demo-dubbo-provider-api</artifactId>
        <version>0.0.1-SNAPSHOT</version>
    </dependency>
    <dependency>
        <groupId>cn.springcloud.dubbo</groupId>
        <artifactId>spring-cloud-dubbo-starter</artifactId>
    </dependency>
</dependencies>
```

然后需要在工程配置文件中给 spring-cloud-dubbo-starter 组件配置 dubbo 相关信息，具体如代码清单 23-4 所示。

代码清单 23-4　ch23-2/demo-dubbo-provider/src/main/resources/application.yml

```yaml
spring:
  application:
    name: demo-provider
```

```yaml
eureka:
    client:
        serviceUrl:
            defaultZone: http://localhost:8761/eureka/
dubbo:
    application:
        name: demo-provider
    registry:
        address: zookeeper://127.0.0.1:2181
    protocol:
        name: dubbo
        port: 20880
    scan:
        basePackages: cn.springcloud.book
```

接下来我们来实现 HelloService 接口，并提供 RESTful API，具体如代码清单 23-5 所示。

代码清单23-5 ch23-2/demo-dubbo-provider/src/main/java/cn/springcloud/book/service/HelloServiceImpl.java

```java
@RestController
public class HelloServiceImpl implements HelloService {
    @Override
    public String hello() {
        return "hello at " + System.currentTimeMillis();
    }
}
```

至此，服务提供者就可以提供服务了。需要说明的是，此时的用法和 Spring Cloud 正常写一个 Feign Client 并实现其接口并没有什么不同，但是因为引入了 spring-cloud-dubbo-starter 的 jar 包，所以其实此时服务提供者对外提供了 Dubbo 的服务。

23.2.3 服务消费者

这里引用 dubbo-spring-boot-starter 来作为一个 Dubbo 服务的消费者，另外还需要引用服务提供者的对外二方库，具体如代码清单 23-6 所示。

代码清单23-6 ch23-2/demo-dubbo-consumer/pom.xml

```xml
<dependencies>
    <dependency>
        <groupId>cn.springcloud.book</groupId>
        <artifactId>demo-dubbo-provider-api</artifactId>
        <version>0.0.1-SNAPSHOT</version>
    </dependency>
    <dependency>
        <groupId>com.alibaba.spring.boot</groupId>
        <artifactId>dubbo-spring-boot-starter</artifactId>
        <version>2.0.0</version>
    </dependency>
    <dependency>
        <groupId>com.101tec</groupId>
```

```xml
            <artifactId>zkclient</artifactId>
            <version>0.10</version>
            <exclusions>
                <exclusion>
                    <groupId>org.slf4j</groupId>
                    <artifactId>slf4j-log4j12</artifactId>
                </exclusion>
                <exclusion>
                    <groupId>log4j</groupId>
                    <artifactId>log4j</artifactId>
                </exclusion>
            </exclusions>
        </dependency>
        <dependency>
            <groupId>org.springframework.boot</groupId>
            <artifactId>spring-boot-starter-test</artifactId>
            <scope>test</scope>
        </dependency>
</dependencies>
```

引入相关依赖之后，需要在工程的配置文件 application.yml 中进行具体配置，如代码清单 23-7 所示。

代码清单23-7 ch23-2/demo-dubbo-consumer/src/main/resources/application.yml

```yaml
dubbo:
    application:
        name: demo-consumer
    registry:
        address: zookeeper://127.0.0.1:2181
    protocol:
        name: dubbo
    scan:
        basePackages: cn.springcloud.book.service
```

配置好配置文件之后，还需要在工程的启动文件 DemoDubboConsumerApplication 中加上 @EnableDubbo 注解，开启 Dubbo 相关配置的支持，具体如代码清单 23-8 所示。

代码清单23-8 ch23-2/demo-dubbo-consumer/src/main/java/cn/springcloud/book/DemoDubboConsumerApplication.java

```java
@SpringBootApplication
@EnableDubbo
public class DemoDubboConsumerApplication {

    public static void main(String[] args) {
        SpringApplication.run(DemoDubboConsumerApplication.class, args);
    }

}
```

接下来我们就可以通过 dubbo 的 Reference 注解来引用 HelloService 服务了，具体如代码清单 23-9 所示。

代码清单23-9　ch23-2/demo-dubbo-consumer/src/main/java/cn/springcloud/book/service/DemoConsumer.java

```java
@Component
public class DemoConsumer {

    @Reference
    private HelloService helloService;

    public String callHello(){
        return helloService.hello();
    }

}
```

最后我们通过一个 Junit 的 Test 来验证一下是否可以成功消费 HelloService 这个 Dubbo 服务，具体如代码清单 23-10 所示。

代码清单23-10　ch23-2/demo-dubbo-consumer/src/test/java/cn/springcloud/book/DemoDubboConsumerApplicationTests.java

```java
@RunWith(SpringRunner.class)
@SpringBootTest
public class DemoDubboConsumerApplicationTests {

    @Autowired
    DemoConsumer demoConsumer;

    @Test
    public void callhello() {
        String result = demoConsumer.callHello();
        Assert.assertTrue(result.startsWith("hello"));
    }

}
```

执行该测试方法，可以看到验证通过，部分日志如下：

```
2018-08-03 09:21:01.412  INFO 2338 --- [           main] c.a.d.r.zookeeper.ZookeeperRegistry      : [DUBBO] Register: consumer://10.2.240.134/cn.springcloud.book.service.HelloService?application=demo-consumer&category=consumers&check=false&dubbo=2.6.0&interface=cn.springcloud.book.service.HelloService&methods=hello&pid=2338&side=consumer&timestamp=1533259260880, dubbo version: 2.6.0, current host: 10.2.240.134
2018-08-03 09:21:01.459  INFO 2338 --- [           main] c.a.d.r.zookeeper.ZookeeperRegistry      : [DUBBO] Subscribe: consumer://10.2.240.134/cn.springcloud.book.service.HelloService?application=demo-consumer&category=providers,configurators,routers&dubbo=2.6.0&interface=cn.springcloud.book.service.HelloService&methods=hello&pid=2338&side=consumer&timestamp=1533259260880, dubbo version: 2.6.0, current host: 10.2.240.134
2018-08-03 09:21:01.502  INFO 2338 --- [           main] c.a.d.r.zookeeper.ZookeeperRegistry      : [DUBBO] Notify urls for subscribe url consumer://10.2.240.134/cn.springcloud.book.service.HelloService?application=demo-consumer&categor
```

```
        y=providers,configurators,routers&dubbo=2.6.0&interface=cn.springcloud.
        book.service.HelloService&methods=hello&pid=2338&side=consumer&timesta
        mp=1533259260880, urls: [dubbo://10.2.240.134:20880/cn.springcloud.book.
        service.HelloService?anyhost=true&application=demo-provider&dubbo=2.6.2&ge
        neric=false&interface=cn.springcloud.book.service.HelloService&methods=he
        llo&pid=1779&side=provider&timestamp=1533258363131, empty://10.2.240.134/
        cn.springcloud.book.service.HelloService?application=demo-consumer
        &category=configurators&dubbo=2.6.0&interface=cn.springcloud.book.
        service.HelloService&methods=hello&pid=2338&side=consumer&timesta
        mp=1533259260880, empty://10.2.240.134/cn.springcloud.book.service.
        HelloService?application=demo-consumer&category=routers&dubbo=2.6.0&inter
        face=cn.springcloud.book.service.HelloService&methods=hello&pid=2338&sid
        e=consumer&timestamp=1533259260880], dubbo version: 2.6.0, current host:
        10.2.240.134
2018-08-03 09:21:01.860   INFO 2338 --- [              main] c.a.d.remoting.
        transport.AbstractClient   : [DUBBO] Successed connect to server
        /10.2.240.134:20880 from NettyClient 10.2.240.134 using dubbo version 2.6.0,
        channel is NettyChannel [channel=[id: 0x7578e06a, /10.2.240.134:54049 =>
        /10.2.240.134:20880]], dubbo version: 2.6.0, current host: 10.2.240.134
2018-08-03 09:21:01.861   INFO 2338 --- [              main] c.a.d.remoting.
        transport.AbstractClient   : [DUBBO] Start NettyClient /10.2.240.134 connect
        to the server /10.2.240.134:20880, dubbo version: 2.6.0, current host:
        10.2.240.134
2018-08-03 09:21:02.059   INFO 2338 --- [              main] com.alibaba.dubbo.
        config.AbstractConfig   : [DUBBO] Refer dubbo service cn.springcloud.book.
        service.HelloService from url zookeeper://127.0.0.1:2181/com.alibaba.dubbo.
        registry.RegistryService?anyhost=true&application=demo-consumer&check=false
        &dubbo=2.6.0&generic=false&interface=cn.springcloud.book.service.HelloServi
        ce&methods=hello&pid=2338&register.ip=10.2.240.134&remote.timestamp=1533258
        363131&side=consumer&timestamp=1533259260880, dubbo version: 2.6.0, current
        host: 10.2.240.134
2018-08-03 09:21:02.168   INFO 2338 --- [              main] c.a.d.c.s.b.f.a.Refe
        renceBeanBuilder        : <dubbo:reference object="com.alibaba.dubbo.common.
        bytecode.proxy0@52b6319f" singleton="true" interface="cn.springcloud.book.
        service.HelloService" uniqueServiceName="cn.springcloud.book.service.
        HelloService" generic="false" filter="" listener="" id="cn.springcloud.
        book.service.HelloService" /> has been built.
```

23.2.4 Spring Cloud Dubbo 框架原理

通过上面的案例我们可以看到，在服务提供者端引入 spring-cloud-dubbo-starter 依赖后，就会提供 Dubbo 的 @Service 注解的服务。那么为什么加上 spring-cloud-dubbo-starter 就能提供 Dubbo 的 @Service 注解的服务呢？

在 SpringCloudDubboAutoConfiguration 类配置程序启动的 Bean 中，当配置文件中存在"dubbo"前缀的配置并且工程包中有 AbstractConfig 类的时候，会注入 FeignClientToDubboProviderBeanPostProcessor 类的 Bean。该 Bean 是一个扩展 BeanFactoryPostProcessor 的标准 SPI 类的 Bean。在服务启动时会进行包的扫描（该包在配置文件中配置），找到被 @FeignClient 注解修饰的组件，比如本案例中的 ProviderServiceImpl 组件。找到组件之后，检查该组件是否具有 Dubbo 的 @Service 注解，如果没有则默认加一个；如果有则取出 Dubbo 的 @Service 注解的值，然后构建一个具有 Dubbo 的 @Serice 注解功能的 Bean。这样只要项目中引入 springcloud-dubbo-starter 依赖，并在配置文件中做了 Dubbo 的相关配置，就会使得被 @FeignClient 修饰的接口的实现类具备 Dubbo 的 @Service 的功能。

在服务消费者端，在 SpringCloudDubboAutoConfiguration 类配置程序启动的 Bean 中，当配置文件中存在"dubbo"前缀的配置并且工程包中有 AbstractConfig 类的时候，会注入 DubboFeignBuilder 类的 Bean。该 Bean 实现了将 FeignClient 的底层调用契约由 Http 换成由 ReferenceBeanBuilder 构建的 Bean，即换成了 Dubbo 调用。

该框架极其精简强悍，有非常高的可阅读性，由此可见该框架作者具有极其深厚的功力，在此感谢该框架的作者。希望读者能够阅读该框架的源码，以了解该框架的实现原理和设计的精妙之处。

23.3 本章小结

本章介绍了两种 Spring Cloud 调用 Dubbo 服务的方案：第一种方案，将 Dubbo 服务注册到 Spring Cloud 注册中心，并提供 REST API，这样 Dubbo 项目和 Spring Cloud 服务就可以通过 REST API 相互调用；第二种方案，使用 Spring Cloud 中国社区提供的 spring-cloud-dubbo-starter 组件将 SpringCloud 的服务 Dubbo 化，该框架是在服务提供者端被 @FeignClient 修饰的 Bean 具备 Dubbo 的 @Service 注解的功能，在服务消费者端替换掉 FeignClient 的底层契约，使之前的由 HTTP 调用转换为 Dubbo 调用。对于第一种方案，存在将既有 Dubbo 服务升级为 RESTful 服务的工作量，但好处是都可以纳入 SpringCloud 微服务架构体系中；对于第二种方案，好处是不用改动既有的 Dubbo 服务，引入 spring-cloud-dubbo-starter 组件就可以将 SpringCloud 服务 Dubbo 化，非常简便，这也体现了 SpringCloud 服务组件的灵活性及可扩展性。

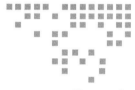

第 24 章

Spring Cloud 与分布式事务

使用 Spring Cloud 构建微服务应用绕不过的一个难点就是跨服务的业务操作的数据一致性问题，本章就来研究一下这个问题。本章首先概述下相关的基础理论，接着介绍一些开源的解决方案，最后选择几个案例结合 Spring Boot/Cloud 进行实战演练。相信通过阅读本章，大家能够对微服务的分布式事务要解决的问题、当前的解决方案及其落地有一个基本的认识。

24.1 概述

24.1.1 ACID

在单体架构的时代，业务系统通常是使用关系型数据库来做存储的，如果一个业务操作需要对同一个数据库进行多次操作，其数据的一致性由数据库的事务来保证，这依赖的就是关系型数据库的 ACID 理论。

ACID 是 Atomic、Consistency、Isolation、Durability 四个单词的缩写，它们的含义如下：

1) Atomic（原子性）：事务的原子性要求事务必须是一个原子的操作系列。一个事务中包含的各个操作在一次执行过程中，要么全部执行，要么全部不执行，任何一个操作失败，整个事务回滚；只有该事务中所有操作都执行成功，整个事务才提交。

2) Consistency（一致性）：事务的一致性要求事务提交之后，原有的约束及规则不被破坏，如果一个事务产生了违反已有约束及规则的数据，那么它们需要被回退到事务开始时的状态。这里的一致性，对数据库来说就是不违反相关约束，比如唯一约束、外键约束等；对于应用系统的业务层面来说，则是事务提交之后相关数据满足业务规则的要求。

3) Isolation（隔离性）：事务的隔离性要求并发中的事务是相互隔离的，即不同的事务操作相同的数据时，每个事务是独立相互不干扰的，具体而言有读未提交、读已提交、可重复读、串行化 4 种事务隔离级别。如果没有事务隔离的话，那么就可能会有更新丢失、脏读、不可重复读、幻读的异常。

4）Durability（持久性）：事务的持久性要求事务一旦提交，则它对数据的变更应该是持久的，即当事务提交数据持久化之后，如果系统崩溃，那么重启之后需要能够看到该事务修改之后的数据。

24.1.2 X/Open DTP 模型与 XA 接口

1. X/Open DTP 模型

如果一个业务操作涉及对多个数据源进行操作，那么使用原来单一数据库的事务（本地事务）来控制就会不能满足（全局事务）数据的一致性要求。为此 X/Open 组织定义了 DTP（Distributed Transaction Processing）模型（见图 24-1），来规范全局事务处理。

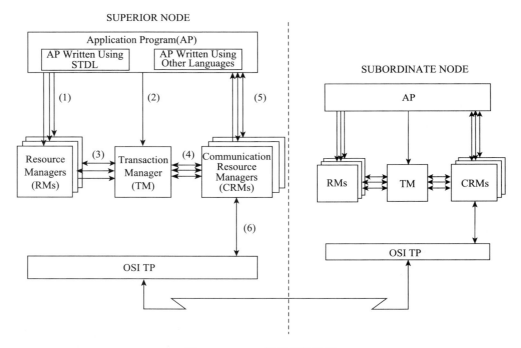

图 24-1 DTP 模型组件视图

如图 24-1 所示，它定义了几个组件，分别如下：

- AP（Application Program，应用程序）：即需要使用分布式事务的应用服务。
- RM（Resource Manager，资源管理器）：比如数据库或文件系统，提供对共享资源的访问，保障资源的 ACID 特性。
- TM（Transaction Manager，事务管理器）：主要是给事务分配唯一标识，负责事务的启动、提交及回滚，保障全局事务的原子性。

- CRMs（Communication Resource Managers，通信资源管理器）：负责控制分布式应用在 TM domain（一组使用同一个 TM 的实例集合）之内或跨 TM domain 之间的通信，该通信使用的是 OSI TP（Open Systems Interconnection Distributed Transaction Processing）服务。早期规范里并没有提出 CRM 组件的概念，这是在后面版本的规范中提出的。
- 通信协议：即由通信资源管理器支持，在分布式应用使用的通信协议。

2. OSI TP 与 2PC

ISO/IEC 10026-1:1998 Information technology -- Open Systems Interconnection -- Distributed Transaction Processing 标准（简称 OSI TP），提出了 2PC 协议（two-phase commit with presumed rollback protocol）。该协议具体如下：

在阶段一，事务管理器 TM 请求所有的资源管理器 RM 预提交各自的事务分支。资源管理器 RM 如果能够执行提交，则它会记录相关的事务日志；如果资源管理器 RM 不能提交事务，则返回失败，同时回滚已经处理的操作，然后释放该事务分支的资源。

在阶段二，事务管理器 TM 向所有的资源管理器 RM 发送提交事务或回滚事务分支的请求。如果第一阶段资源管理器 RM 返回的都是成功，则发送提交事务请求；只要有一个资源管理器 RM 返回失败，就发送回滚事务请求。在所有资源管理器对共享资源提交或回滚变更之后，反馈执行结果给事务管理器 TM，然后事务管理器 TM 释放占用的相关资源。

成功提交事务的过程如图 24-2 所示。

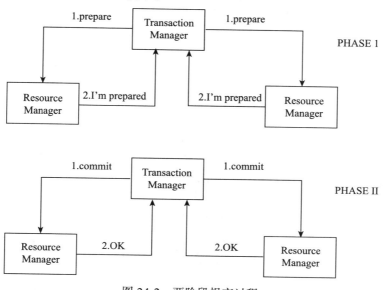

图 24-2 两阶段提交过程

采用两阶段提交协议，优点就是简单，但是它也有几个缺点，分别如下：
- 两阶段的过程都是同步阻塞的，需要等待各个参与者的响应，这会影响分布式事务的操作性能。

- 事务管理器 TM 在整个过程中负责协调管理，如果它自身发生故障，那么资源管理器 RM 就会阻塞状态，事务无法进行下去。
- 在数据一致性方面，如果在阶段二有部分资源管理器 RM 在收到提交请求后提交了事务，而部分资源管理器 RM 由于网络异常等未能收到提交事务请求，就会造成数据的不一致。

3. XA 接口与 JTA

XA 接口是 X/Open DTP 在 2PC 的基础上给事务管理器 TM 与资源管理器 RM 之间通信定义的接口。在 Java 领域，J2EE（Java Enterprise Edition）定义了 JTA（Java Transaction API）规范，具体见 JSR 907，它遵循 X/Open XA 接口，是高级版本的 API 规范。2000 年起草了第一版 J2EE，最新版是 2018 年 3 月份发布的 1.3 版本。另外还有 JTS（Java Transaction Service）规范，其定义了更底层的实现事务管理器 TM 的相关接口，其中包括支持高级别的 JTA 接口以及标准的 CORBA Object Transaction Service 到 Java 的映射。JTS 对于 Java 的 XA 事务规范，类似于 JMS（Java Message Service）对于 Java 的消息中间件规范。JTA 与 JTS 的关系如图 24-3 所示。

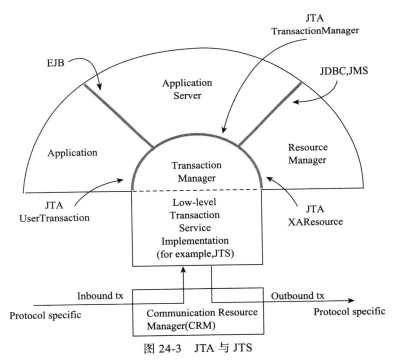

图 24-3　JTA 与 JTS

JTA 主要定义如下几个类：
- javax.transaction.UserTransaction。
- javax.transaction.TransactionManager。
- javax.transaction.xa.XAResource。

- javax.transaction.Synchronization。
- javax.transaction.xa.Xid。
- javax.transaction.TransactionSynchronizationRegistry。
- javax.transaction.Transactional。
- javax.transaction.Transactional.TxType。
- javax.transaction.TransactionScoped。

JDBC 与 XA 相关的几个类如下：
- javax.sql.XADataSource。
- javax.sql. XAConnection。

关于 JTA，在接下来的实战部分我们会使用 atomikos 的实现来进一步演示。

24.1.3 CAP 与 BASE 定理

ACID 与 CAP 它们中都有一个 C，不过其一致性的含义不太一样：ACID 的 C 主要关注的是数据在跨节点操作情况下的操作原子性，而 CAP 的 C 关注的是同一份数据在多个副本之间同步的一致性。

BASE 是 Basically Available（基本可用）、Soft state（软状态）、Eventually consistent（最终一致性）这四个单词的缩写。在单体架构往分布式架构迁移时，数据一致性要做到满足 ACID 特性往往比较困难，于是 BASE 定理就提出在一些非严格要求强一致的应用场景下面，应用可以根据情况采取适当的措施来达到最终一致性，其宗旨是通过牺牲强一致来获取可用性。

使用 XA 事务来实现分布式事务，其可用性就比较差，于是业界提出了 TCC 及 SAGA 等解决方案，其思想跟 BASE 定理类似，通过打破 ACID 特性中的一个或几个来提升分布式事务的可用性。

24.2 解决方案

24.2.1 Java 事务编程接口 JTA

满足 J2EE 规范的容器支持了 JTA 规范，在纯 J2SE 领域，Spring Boot 官方文档推荐了 atomikos、bitronix、narayana 这三个组件。这里分别对它们简单介绍一下：

- atomikos：希腊语的含义就是原子性，以这个词作为分布式事务组件的名称，用于暗指其可靠的事务管理能力。关于 atomikos 有开源的版本（TransactionsEssentials），也有增强的收费版（ExtremeTransactions）。TransactionsEssentials 最新版本为 2018 年 1 月发布的 4.0.6 版本，相应的 spring-boot 的组件为 spring-boot-starter-jta-atomikos。
- bitronix：BTM（Bitronix Transaction Manager）是一款简单但是完整实现了 JTA1.1 规范的事务管理器，目前该项目已经搁置，最新的版本为 2013 年 9 月份发布的 2.1.4 版本，相应的 spring-boot 的组件为 spring-boot-starter-jta-bitronix。

- arayana：它是 jboss 提供的一款分布式事务管理器，有基于 J2EE 容器的版本，也有 standalone 的版本，目前实现了 JTA1.2 版本的规范，最新版本为 2018 年 7 月发布的 5.9.0.Final 版本。

24.2.2 分布式事务 TCC 模式

1. 定义

2007 年 Amazon 的 Pat Helland 在 Conference on Innovative Database Research 上发表了一篇文章——《Life Beyond Distributed Transactions:An apostate's opinion》，首次提出了解决分布式事务一致性的 Tentative Operations、Confirmation 和 Cancellation 模型。10 年后的 2017 年 Pat Helland 在 acmqueue 上发表了一篇同名文章，对前一篇文章的内容进行了概括及更新。其基本定义如下：

- Tentative Operation：为了在多个实体之间达成一致，要求一个实体必须能够接受另一个实体对请求执行的不确定性，比如，在发出执行操作请求之后又发出取消该操作的请求。这类后续可能请求取消的操作，就称为 Tentative Operation。
- Confirmation：如果请求方认为 Tentative Operation 没问题，那么就可以发出 Confirmation 的执行请求，最终确定这个操作。
- Cancellation：如果请求方决定撤回 Tentative Operation，那么就发出 Cancellation 的执行请求，取消这个操作。

当一个实体统一执行 Tentative Operation 的时候，就意味着它接受了这种不确定性，允许另外的实体通过发出 Confirmation 或者 Cancellation 请求，来降低这种不确定性，最终完成这个 Tentative Operation。

Gregor Hohpe 在其维护的 www.enterpriseintegrationpatterns.com 网站中，将 Pat Helland 提出的 Tentative Operations、Confirmation 和 Cancellation 模型概括为 Tentative Operation 模式（http://www.enterpriseintegrationpatterns.com/patterns/conversation/TryConfirmCancel.html），然后提出与之类似的 TCC（Try-Confirm-Cancel）事务模型，其状态图如图 24-4 所示。

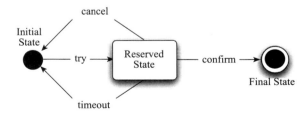

图 24-4　TCC 状态图

TCC 事务模型总共有 Initial、Reserved、Final 三个状态：

- Initial 状态：是最初始的状态，接到 Try 请求时变成 Reserved 状态。
- Reserved 状态：接收 Confirm 请求时变成 Final 状态，如果接收 Cancel 请求或者是等待超时则退回到 Initial 状态。

❏ Final 状态：TCC 事务成功状态。

2. TCC 与 2PC

与 2PC 不同，TCC 的模式把原来 2PC 的 prepare 操作从事务管理中剥离出来，规范为 try 操作，并且变为 Reserved 状态。同理也将撤销 prepare 的操作规范为 cancel 操作，并且变为 Initial 状态。通过这样的规范，使得 TCC 模式可以更好地适应 SOA 架构的分布式事务场景，并规范原来黑盒的操作，委托给各个业务场景去实现，将业务服务纳入整个分布式事务的处理当中。这样做的好处就是把原来的黑盒操作暴露出来，在出现不一致的场景（比如 timeout 的时候出现了 confirm 请求，或者 confirm 的时候出现失败）可以通过手工等方式介入。

3. TCC 与 ACID

TCC 模式不满足 Isolation 的特性，即 TCC 是采用预留资源的方式来做 try 以及 cancel 操作的，那么在达到 final 状态之前，其他事务如果读取处于中间状态的资源的时候，读到的是"脏"数据。

另外 TCC 要求 confirm、cancel 操作必须是幂等的，因为实际场景会因为网络等其他问题导致这些操作被重试，因此 TCC 实现的是最终一致性的分布式事务。

4. TCC 开源框架

在国内，蚂蚁金服的大部分业务系统采用了 TCC 的方式接入分布式事务（https://yq.aliyun.com/articles/609854），除此之外国内开源的分布式事务框架中也有大部分是基于 TCC 的，这里进行整理，如表 24-1 所示。

表 24-1 国内开源分布式事务框架

框架名称	支持 TCC	GitHub 地址	Start 数量
社区开源项目 dts		https://github.com/venusteam/dts	111
tcc-transaction	是	https://github.com/changmingxie/tcc-transaction	2206
Hmily	是	https://github.com/yu199195/hmily	1206
ByteTCC	是	https://github.com/liuyangming/ByteTCC	1141
myth		https://github.com/yu199195/myth	897
EasyTransaction	是	https://github.com/QNJR-GROUP/EasyTransaction	805
tx-lcn		https://github.com/codingapi/tx-lcn/	802

关于上述分布式事务框架的性能，海信 HICS 技术团队进行了专门的压测，结果发表在 springcloud 中国社区的主站上（http://springcloud.cn/view/374），有兴趣的读者可以前往深入研究。

24.2.3 分布式事务 SAGA 模式

1. 定义

1987 年普林斯顿大学的 Hector Garcaa-Molrna 与 Kenneth Salem 发表了一篇名为《SAGAS》的论文，提出了 SAGA 的事务模式。它涉及一个概念——LLT（Long Lived Transaction）。LLT 指的是持有数据库资源相对较长的长活事务。如果一个 LLT 可以被拆解为一序列可以跟其他事务错开执行（interleaved with）的子事务，而且里面的每个子事务保持自身 ACID 特性，要么可以成功提交，要么可以通过补偿事务（compensating transaction）来进行恢复，从而达到最终一致性，那么这个 LLT 就可称为 SAGA。

该论文同时还定义了两种恢复方式，一种是正向恢复 Forward Recovery，一种是逆向恢复 Backward Recovery。每一个内部事务 T 都有一个对应的补偿事务 C，补偿事务 C 用于逆向恢复已提交的事务 T，来使数据恢复到 T 事务执行之前的状态。逆向恢复的示例过程为 T1T2T3C3C2C1，即对每个已经提交的事务 T，按照后提交先补偿的顺序依次执行补偿事务 C。正向恢复则不同，它则是充分利用保存点 sp（save points），在 sp 的基础上不断重试，使得整个大事务不断往前执行下去，其过程示例为 T1（sp）T2（sp）T3（sp）T4。当然正向恢复并不适用于所有的场景。

虽然 SAGA 理论提出的比较早，最初提出来是用于替代数据库的 Long Lived Transaction。最近几年也被应用在分布式系统当中用于解决分布式事务的问题。2015 年 Caitie McCaffrey 及 Kyle Kingsbury 发布了一篇名为《Distributed Sagas》的文章，将 SAGAS 模式应用到分布式系统当中，提出了 Distributed Sagas 模式。该模式定义了 SEC（Saga Execution Coordinator）以及 TEC（Transaction Execution Coordinator）。整个 SEC 的执行逻辑如图 24-5 所示。

对于整个 SEC，启动的时候为 Saga start，中断的时候为 Saga abort，执行完成的时候为 Saga done。从图 24-5 可以看出，每次操作之前都会先 log 一下，另外还维护了一个变量 i，用来控制 SAGA 的子事务是否执行完或者补偿完，然后结束整个流程。

Distributed Saga 对正常请求及补偿请求有如下几点要求：

- 正常请求以及补偿请求都必须是幂等的。
- 补偿请求是 commutative 的，commutative 在这里的意思是：如果一个正常请求在补偿请求之后到达，那么正常请求不应该被执行。
- 正常请求可以被中断（触发补偿请求），但是补偿请求必须执行完成，没有中断操作。

除了 Caitie McCaffrey 及 Kyle Kingsbury 提出 Distributed Sagas 外，cloudfoundry.com 创始人、《POJOS IN ACTION》的作者 Chris Richardson 在 2017 年 11 月份的 QCONSF 会议中发表了名为《ACID Is So Yesterday：Maintaining Data Consistency with Sagas》的演讲，同时展示了使用 Eventuate Tram Saga framework 通过 SAGA 模式处理微服务中的分布式事务的方法。Chris Richardson 在 http://microservices.io/ 网站以及预计在 2018 年 9 月份正式出版的书籍《Microservices Patterns：With examples in Java》中将 SAGA 模式列为微服务的模式之一。Chris Richardson 还提出了 SAGA 的两种协调方式，分别是 Orchestration 和 Choreography。

Orchestration 通常翻译为编制，其字面意思为管弦乐编曲。管弦乐有一个指挥官来进行集中控制，这里引申为对分布式事务的调度及协调。Choreography 通常翻译为编排，其字面意思为舞蹈中的编舞，舞者一般跟着音乐节奏做出相应的动作及舞步，这里引申为根据外部事件等进行相应反应。在实现过程中，Orchestration 一般是通过 BPM 流程中心或类似的管理器来实现统一控制，而 Choreography 则一般是通过 Event Driven 的事件方式来进行触发执行其他服务的本地事务。

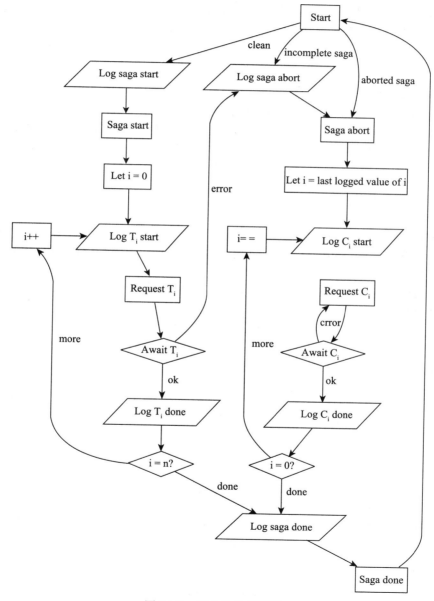

图 24-5　SEC 的执行逻辑

2. SAGA 与 2PC

与 2PC 相比，SAGA 与 TCC 通过牺牲 ACID 的部分特性来提升分布式事务的可用性。另外 SAGA 及 TCC 可以理解为是服务层面的事务模式，将分布式事务的控制由数据库事务层提升到服务层。

3. SAGA 与 TCC

SAGA 与 TCC 最显著的区别在于 TCC 采用的是预留资源的方式，其状态里有个 Reserved 状态；而 SAGA 则没有这个预留资源的动作，事务直接提交，然后采取的是补偿事务的方式来进行撤回。与 2PC 相比，2PC 采用事务 prepare 以及 rollback 动作，整个都在一个大事务中，而 SAGA 是将这些大事务拆分为一个个本地事务。

4. SAGA 与 ACID

SAGA 模式不满足 Isolation 的特性，因为其将 LLT 划分为一个个本地事务，一旦本地事务提交，在整个 SAGA 执行完毕之前，中间如果有其他事务也访问到了共享资源，则会读到"未完成"的事务的数据。与 TCC 类似，SAGA 不满足 C 特性，SAGA 实现的是最终一致性，但是拆分出来的一个个本地事务则满足 ACID 特性。

5. SAGA 开源框架

支持 SAGA 的有如下几个开源框架：

- Axon framework：是一款采用 Java 编写的、面向可伸缩及高性能的 CQRS 框架，该框架内置了 SagaRepository、SagaManager、SagaStore 等组件，对 SAGA 模式提供了原生的支持。
- Eventuate.io：eventuate-tram-sagas 是 Chris Richardson 开源的一款为使用 JDBC/JPA 的微服务准备的 SAGA 框架。
- Narayana LRA：Narayana 是 jboss 提供的一款分布式事务管理器，其中实现 SAGA 模式的事务管理器为 Narayana LRA，LRA 为 Long Running Action 的缩写。
- ServiceComb Saga：是华为于 2017 年 7 月份开源的一款基于 SAGA 模式的分布式事务框架，目前在 Apache 孵化中。

24.3 实战

24.3.1 Atomikos JTA

这里我们选用 spring-boot-starter-jta-atomikos 组件来展示如何使用 JTA 来保证跨多个数据源操作时数据的一致性。这里我们创建一个 springboot 的工程，演示如何对 dborder 及 dblog 这两个数据库的表的操作进行 JTA 事务管理。

1. maven 的 pom 文件配置

这里引入 spring-boot-starter-jta-atomikos 及 spring-boot-starter-data-jpa，数据库使用的是 h2，连接池使用的是 tomcat-jdbc，具体配置如代码清单 24-1 所示。

代码清单24-1　ch24-3/ch24-3-1-jta-atomikos/pom.xml

```xml
<dependencies>
    <dependency>
        <groupId>org.springframework.boot</groupId>
        <artifactId>spring-boot-starter-web</artifactId>
    </dependency>
    <dependency>
        <groupId>org.springframework.boot</groupId>
        <artifactId>spring-boot-starter-actuator</artifactId>
    </dependency>
    <dependency>
        <groupId>org.springframework.boot</groupId>
        <artifactId>spring-boot-starter-jta-atomikos</artifactId>
    </dependency>
    <dependency>
        <groupId>org.springframework.boot</groupId>
        <artifactId>spring-boot-starter-web</artifactId>
    </dependency>
    <dependency>
        <groupId>org.springframework.boot</groupId>
        <artifactId>spring-boot-starter-data-jpa</artifactId>
    </dependency>
    <dependency>
        <groupId>org.apache.tomcat</groupId>
        <artifactId>tomcat-jdbc</artifactId>
    </dependency>
    <dependency>
        <groupId>com.h2database</groupId>
        <artifactId>h2</artifactId>
        <scope>runtime</scope>
    </dependency>
    <dependency>
        <groupId>org.springframework.boot</groupId>
        <artifactId>spring-boot-starter-test</artifactId>
        <scope>test</scope>
    </dependency>
</dependencies>
```

2. application.yml 配置

springboot 提供了以 spring.jta 为前缀的配置选项，这里由于我们使用的是 atomikos 提供的 JTA，因此其数据源配置在 spring.jta.atomikos.datasource 下，具体如代码清单 24-2 所示。

代码清单24-2　ch24-3/ch24-3-1-jta-atomikos/src/main/resources/application.yml

```yaml
spring:
  application:
    name: atomikos-demo
  h2:
    console:
      enabled: true
  jta:
```

```yaml
        enabled: true
        atomikos:
            datasource:
                order:
                    xa-properties.url: jdbc:h2:mem:dborder
                    xa-properties.user: sa
                    xa-properties.password:
                    xa-data-source-class-name: org.h2.jdbcx.JdbcDataSource
                    unique-resource-name: order
                    max-pool-size: 10
                    min-pool-size: 1
                    max-lifetime: 10000
                    borrow-connection-timeout: 10000
                log:
                    xa-properties.url: jdbc:h2:mem:dblog
                    xa-properties.user: sa
                    xa-properties.password:
                    xa-data-source-class-name: org.h2.jdbcx.JdbcDataSource
                    unique-resource-name: log
                    max-pool-size: 10
                    min-pool-size: 1
                    max-lifetime: 10000
                    borrow-connection-timeout: 10000
management:
    endpoints:
        web:
            exposure:
                include: '*'
```

这里我们指定了两个数据源：一个是 order 数据源，使用的是 dborder 数据库；另外一个是 log 数据源，使用的是 dblog 数据库。xa-data-source-class-name 配置的是每个数据库支持 XA 事务的数据源，对于 h2 数据库就是 org.h2.jdbcx.JdbcDataSource。

3. atomikos jta 与 jpa 配置

这里我们使用 hibernate 作为 JPA 的实现，由于 atomikos 并没有提供对 hibernate 的 AbstractJtaPlatform 的实现，因此这里需要自己适配一下，具体如代码清单 24-3 所示。

代码清单24-3 ch24/ch24-jta-atomikos/src/main/java/cn/springcloud/book/config/AtomikosJtaPlatform.java

```java
public class AtomikosJtaPlatform extends AbstractJtaPlatform {

    private static TransactionManager transactionManager;

    private static UserTransaction userTransaction;

    public static void setTransactionManager(TransactionManager transactionManager) {
        AtomikosJtaPlatform.transactionManager = transactionManager;
    }

    public static void setUserTransaction(UserTransaction userTransaction) {
        AtomikosJtaPlatform.userTransaction = userTransaction;
    }
```

```
    @Override
    protected TransactionManager locateTransactionManager() {
        return transactionManager;
    }

    @Override
    protected UserTransaction locateUserTransaction() {
        return userTransaction;
    }
}
```

这个类主要是重写了 locateTransactionManager 及 locateUserTransaction 方法，分别返回 transactionManager 及 userTransaction。

由于这里使用了两个数据源，因此需要对这两个数据源的 JPA 进行配置。log 数据源的配置如代码清单 24-4 所示。

代码清单24-4 ch24/ch24-jta-atomikos/src/main/java/cn/springcloud/book/config/LogDatasourceConfig.java

```java
@Configuration
@EnableConfigurationProperties
@EnableJpaRepositories(basePackages = "cn.springcloud.book.dao.log",
    entityManagerFactoryRef = "logEntityManager",
        transactionManagerRef = "transactionManager")
public class LogDatasourceConfig {

    @Bean(name = "logDatasource")
    @Qualifier("logDatasource")
    @ConfigurationProperties(prefix="spring.jta.atomikos.datasource.log")
    public DataSource logDatasource() {
        return new AtomikosDataSourceBean();
    }

    @Bean(name = "logEntityManager")
    public LocalContainerEntityManagerFactoryBean logEntityManager(TransactionMana
        ger transactionManager, UserTransaction userTransaction) throws Throwable {
        AtomikosJtaPlatform.setTransactionManager(transactionManager);
        AtomikosJtaPlatform.setUserTransaction(userTransaction);

        HibernateJpaVendorAdapter jpaVendorAdapter = new HibernateJpaVendorAdapter();
        jpaVendorAdapter.setGenerateDdl(true);
        jpaVendorAdapter.setDatabase(Database.H2);
        jpaVendorAdapter.setDatabasePlatform("org.hibernate.dialect.H2Dialect");

        Map<String, Object> properties = new HashMap<String, Object>();
        properties.put("hibernate.transaction.jta.platform", AtomikosJtaPlatform.
            class.getName());
        properties.put("javax.persistence.transactionType", "JTA");
        properties.put("hibernate.hbm2ddl.auto","update");

        LocalContainerEntityManagerFactoryBean entityManager = new LocalContain
            erEntityManagerFactoryBean();
```

```
            entityManager.setJtaDataSource(logDatasource());
            entityManager.setJpaVendorAdapter(jpaVendorAdapter);
            entityManager.setPackagesToScan("cn.springcloud.book.domain.log");
            entityManager.setPersistenceUnitName("logPersistenceUnit");
            entityManager.setJpaPropertyMap(properties);
            return entityManager;
     }
}
```

这里根据 spring.jta.atomikos.datasource.log 前缀的配置创建了 AtomikosDataSourceBean，另外 logEntityManager 是 JPA 相关的配置，这里将 transactionManager 及 userTransaction 设置到 AtomikosJtaPlatform 中，同时指定 hibernate.transaction.jta.platform 为 AtomikosJtaPlatform.class.getName()，设置 javax.persistence.transactionType 为 JTA 类型。order 数据源的 JTA 及 JPA 配置跟 log 数据源配置类似，这里不再赘述，其配置如代码清单 24-5 所示。

代码清单24-5 ch24/ch24-jta-atomikos/src/main/java/cn/springcloud/book/config/OrderDatasourceConfig.java

```
@Configuration
@EnableConfigurationProperties
@EnableJpaRepositories(basePackages = "cn.springcloud.book.dao.order",
    entityManagerFactoryRef = "orderEntityManager",
        transactionManagerRef = "transactionManager")
public class OrderDatasourceConfig {

    @Bean(name = "orderDatasource")
    @Qualifier("orderDatasource")
    @ConfigurationProperties(prefix="spring.jta.atomikos.datasource.order")
    @Primary
    public DataSource orderDatasource() {
        return new AtomikosDataSourceBean();
    }

    @Bean(name = "orderEntityManager")
    @Primary
    public LocalContainerEntityManagerFactoryBean orderEntityManager(Transact
        ionManager transactionManager,UserTransaction userTransaction) throws
        Throwable {
        AtomikosJtaPlatform.setTransactionManager(transactionManager);
        AtomikosJtaPlatform.setUserTransaction(userTransaction);

        HibernateJpaVendorAdapter jpaVendorAdapter = new HibernateJpaVendorAdapter();
        jpaVendorAdapter.setGenerateDdl(true);
        jpaVendorAdapter.setDatabase(Database.H2);
        jpaVendorAdapter.setDatabasePlatform("org.hibernate.dialect.H2Dialect");

        Map<String, Object> properties = new HashMap<String, Object>();
        properties.put("hibernate.transaction.jta.platform", AtomikosJtaPlatform.
            class.getName());
        properties.put("javax.persistence.transactionType", "JTA");
        properties.put("hibernate.hbm2ddl.auto","update");

        LocalContainerEntityManagerFactoryBean entityManager = new LocalContain
```

```
            erEntityManagerFactoryBean();
        entityManager.setJtaDataSource(orderDatasource());
        entityManager.setJpaVendorAdapter(jpaVendorAdapter);
        entityManager.setPackagesToScan("cn.springcloud.book.domain.order");
        entityManager.setPersistenceUnitName("orderPersistenceUnit");
        entityManager.setJpaPropertyMap(properties);
        return entityManager;
    }
}
```

4. 使用 JTA

这里我们在 log 数据源中创建名为 EventLog 的实体对象，其定义如代码清单 24-6 所示。

代码清单 24-6　ch24/ch24-jta-atomikos/src/main/java/cn/springcloud/book/domain/log/EventLog.java

```
@Entity
@PersistenceUnit(unitName="logPersistUnit")
public class EventLog {
    @Id
    @GeneratedValue(strategy = GenerationType.AUTO)
    private Integer id;
    private String operation;
    private String operator;
    //…… getter, setter
}
```

其中，dao 为 EventLogDao。

在 order 数据库中我们创建了 UserOrder 实体对象，其定义如代码清单 24-7 所示。

代码清单 24-7　ch24/ch24-jta-atomikos/src/main/java/cn/springcloud/book/domain/order/UserOrder.java

```
@Entity
@PersistenceUnit(unitName="orderPersistUnit")
public class UserOrder {
    @Id
    @GeneratedValue(strategy = GenerationType.AUTO)
    private Integer id;
    private String userId;
    private String productCode;
    private Integer quantity;
    //…… getter、setter
}
```

其中，dao 为 EventLogDao。

接下来，我们就可以使用 Transactional 注解来进行 XA 事务管理，这里我们在 OrderService 类里定义了 newOrder 及 newOrderRollback 两个方法，用于演示 JTA 事务正常提交及回滚。newOrder 方法模拟的操作语义是在 order 数据库中新增 order 数据，同时在 log 数据库中新增 log 记录。OrderService 的定义如代码清单 24-8 所示。

代码清单24-8　ch24/ch24-jta-atomikos/src/main/java/cn/springcloud/book/service/OrderService.java

```java
@Component
public class OrderService {

    @Autowired
    UserOrderDao userOrderDao;

    @Autowired
    EventLogDao eventLogDao;

    @Transactional
    public void newOrder(String userId,String productCode,int quantity){
        UserOrder userOrder = new UserOrder();
        userOrder.setUserId(userId);
        userOrder.setProductCode(productCode);
        userOrder.setQuantity(quantity);
        userOrderDao.save(userOrder);

        EventLog eventLog = new EventLog();
        eventLog.setOperation("new order");
        eventLog.setOperator(userId);
        eventLogDao.save(eventLog);
    }

    @Transactional
    public void newOrderRollback(String userId,String productCode,int quantity){
        UserOrder userOrder = new UserOrder();
        userOrder.setUserId(userId);
        userOrder.setProductCode(productCode);
        userOrder.setQuantity(quantity);
        userOrderDao.save(userOrder);

        EventLog eventLog = new EventLog();
        eventLog.setOperation("new order");
        eventLog.setOperator(userId);
        eventLogDao.save(eventLog);

        throw new RuntimeException("test jta rollback");
    }
}
```

5. 验证

这里我们使用 SpringBootTest 来模拟对 orderService 的业务方法的调用，其定义如代码清单 24-9 所示。

代码清单24-9　ch24/ch24-jta-atomikos/src/test/java/cn/springcloud/book/Ch24-JtaAtomikosApplicationTests.java

```java
@RunWith(SpringRunner.class)
@SpringBootTest(webEnvironment=SpringBootTest.WebEnvironment.DEFINED_PORT)
public class Ch24JtaAtomikosApplicationTests {
```

```
    @Autowired
    OrderService orderService;

    @Test
    public void testJtaCommit() throws InterruptedException {
        try{
            orderService.newOrder("tom","0001",100);
        }catch (Exception e){
            e.printStackTrace();
        }
        TimeUnit.MINUTES.sleep(10);
    }

    @Test
    public void testJtaRollback() throws InterruptedException {
        try{
            orderService.newOrderRollback("tom","0001",100);
        }catch (Exception e){
            e.printStackTrace();
        }
        TimeUnit.MINUTES.sleep(10);
    }

}
```

执行 testJtaCommit 方法，然后打开 localhost:8080/h2-console，使用 h2 的 web 进行管理后台，在 JDBC URL 处输入 jdbc:h2:mem:dborder 以访问 dborder 数据库，并查看 dborder 的 userorder 表，这时就可以看到新增了一条记录；在 JDBC URL 处输入 jdbc：h2：mem：dblog 以访问 dblog 数据库，并查看 eventlog 表，这时就可以看到又新增了一条记录，表示 JTA 事务提交成功。

在工程根目录中有个 transaction-logs 目录，存放了 JTA 事务相关信息，里面包含了 tmlog.lck 文件（方法执行完毕该 lck 文件会被自动删除）及 tmlog0.log 文件。tmlog0.log 文件内容实例为：

```
{"id":"127.0.0.1.tm153257563865200001","wasCommitted":true,"participants":[{"uri":"127.0.0.1.tm1","state":"COMMITTING","expires":1532575648744,"resourceName":"order"},{"uri":"127.0.0.1.tm2","state":"COMMITTING","expires":1532575648744,"resourceName":"log"}]}
{"id":"127.0.0.1.tm153257563865200001","wasCommitted":true,"participants":[{"uri":"127.0.0.1.tm1","state":"TERMINATED","expires":1532575648750,"resourceName":"order"},{"uri":"127.0.0.1.tm2","state":"TERMINATED","expires":1532575648750,"resourceName":"log"}]}
```

从上述内容中可以看到，其中记录了 JTA 相关子事务的信息，有兴趣的读者可以深入研究 atomikos 源码进行了解。

接下来执行 testJtaRollback 方法，来演示 JTA 事务的回滚，执行之后抛出 java.lang.RuntimeException:test jta rollback 异常。同理也可以打开 localhost:8080/h2-console，看到两个数据库都没有新插入数据，表示 JTA 已经回滚两边的数据。在 transaction-logs 目录下，可以看到 tmlog0.log 日志如下：

```
{"id":"127.0.0.1.tm153258201350400001","wasCommitted":false,"participants":[{"u
ri":"127.0.0.1.tm1","state":"TERMINATED","expires":1532582023656,"resourceN
ame":"order"},{"uri":"127.0.0.1.tm2","state":"TERMINATED","expires":1532582
023658,"resourceName":"log"}]}
```

由上述日志可以看到，其中记录了两边事务的状态为 TERMINATED。

24.3.2　TCC for REST

atomikos 开源的 transactions-essentials 提供了 TCC for REST 的 API 规范（https://github.com/atomikos/transactions-essentials/tree/master/public/transactions-tcc-rest-api），该 API 定义了 Coordinator 接口、ParticipantLink 类及 Transaction 类，分别对应 TCC 事务的协调者、参与者信息及分布式事务描述。本节我们以订单及库存为案，来演示如何在基于 REST 的微服务中进行 TCC 事务控制。案例的背景为订单及库存两个微服务，然后用户下单要在两个微服务之间进行分布式事务控制，即订单状态为确认时，需同时扣减库存。关于本案例的更多内容见 ch24-3-2 工程，该工程主要分为几个模块：

- tcc-coordinator-atomikos：使用 atomikos 的 transactions-tcc-rest 作为 coordinator 服务的实现。
- tcc-rest-participant-api：规范了事务参与者所需提供的 REST API。
- order-service：订单微服务工程模块。
- inventory-service：库存微服务工程模块。
- tcc-coordinator-example：演示了如何使用 REST TCC 进行分布式事务管理。

下面我们详细介绍这些工程模块。

1. tcc-coordinator-atomikos

这个微服务是 TCC REST 的 coordinator，用来协调参与者，进行事务的 confirm 与 cancel。其 pom 文件中的依赖部分如代码清单 24-10 所示。

代码清单24-10　ch24/ch24-tcc-rest/tcc-coordinator-atomikos/pom.xml

```xml
<dependencies>
    <dependency>
        <groupId>org.springframework.boot</groupId>
        <artifactId>spring-boot-starter-web</artifactId>
    </dependency>
    <dependency>
        <groupId>org.apache.cxf</groupId>
        <artifactId>cxf-rt-rs-client</artifactId>
        <version>${cxf.version}</version>
    </dependency>
    <dependency>
        <groupId>org.apache.cxf</groupId>
        <artifactId>cxf-rt-transports-http-netty-server</artifactId>
        <version>${cxf.version}</version>
    </dependency>
    <dependency>
        <groupId>io.netty</groupId>
```

```xml
        <artifactId>netty-handler</artifactId>
        <version>4.1.27.Final</version>
    </dependency>
    <dependency>
        <groupId>org.apache.cxf</groupId>
        <artifactId>cxf-rt-frontend-jaxrs</artifactId>
        <version>${cxf.version}</version>
    </dependency>
    <dependency>
        <groupId>org.apache.cxf</groupId>
        <artifactId>cxf-rt-rs-extension-providers</artifactId>
        <version>${cxf.version}</version>
    </dependency>
    <dependency>
        <groupId>com.fasterxml.jackson.jaxrs</groupId>
        <artifactId>jackson-jaxrs-json-provider</artifactId>
        <version>2.9.6</version>
    </dependency>
    <dependency>
        <groupId>org.codehaus.jettison</groupId>
        <artifactId>jettison</artifactId>
        <version>1.4.0</version>
    </dependency>
    <dependency>
        <groupId>com.atomikos</groupId>
        <artifactId>transactions-tcc-rest-api</artifactId>
        <version>${atomikos.version}</version>
    </dependency>
    <dependency>
        <groupId>com.atomikos</groupId>
        <artifactId>transactions-tcc-rest</artifactId>
        <version>${atomikos.version}</version>
    </dependency>
    <dependency>
        <groupId>com.atomikos</groupId>
        <artifactId>transactions-api</artifactId>
        <version>${atomikos.version}</version>
    </dependency>
    <dependency>
        <groupId>com.atomikos</groupId>
        <artifactId>transactions</artifactId>
        <version>${atomikos.version}</version>
    </dependency>
    <dependency>
        <groupId>com.atomikos</groupId>
        <artifactId>atomikos-util</artifactId>
        <version>${atomikos.version}</version>
    </dependency>
    <dependency>
        <groupId>org.springframework.boot</groupId>
        <artifactId>spring-boot-starter</artifactId>
    </dependency>

    <dependency>
        <groupId>org.springframework.boot</groupId>
```

```xml
        <artifactId>spring-boot-starter-test</artifactId>
        <scope>test</scope>
    </dependency>
</dependencies>
```

这里依赖了与 com.atomikos 相关的事务组件，比如 transactions、transactions-api、transactions-tcc-rest、transactions-tcc-rest-api。另外由于 atomikos 底层使用 cxf 来发布服务，然后用 cxf-rt-rs-client 进行 REST 服务调用，因此依赖了 org.apache.cxf 相关依赖。

由于 transactions-tcc-rest 没有提供对 springboot 的适配，这里我们适配一下，如代码清单 24-11 所示。

代码清单 24-11　ch24/ch24-tcc-rest/tcc-coordinator-atomikos/src/main/java/cn/springcloud/book/config/AtomikosTccConfig.java

```java
@Configuration
public class AtomikosTccConfig {

    @Bean
    public AtomikosTccSpringAdapter atomikosTccSpringAdpater(){
        return new AtomikosTccSpringAdapter();
    }

    public static class AtomikosTccSpringAdapter {
        @PostConstruct
        public void start(){
            com.atomikos.icatch.config.Configuration.init();
        }

        @PreDestroy
        public void shutdown(){
            com.atomikos.icatch.config.Configuration.shutdown(false);
        }
    }
}
```

这里通过 @PostConstruct 及 @PreDestroy 将 atomikos 的 Configuration 纳入 spring 容器的生命周期管理中，进行初始化及销毁相关动作。

另外我们重新使用 spring mvc 对外暴露 REST API，而不是 atomikos 原来使用的 apahce cxf 来暴露，具体如代码清单 24-12 所示。

代码清单 24-12　ch24/ch24-tcc-rest/tcc-coordinator-atomikos/src/main/java/cn/springcloud/book/controller/TccCoordinatorController.java

```java
@RestController
@RequestMapping(value = "/coordinator", consumes = "application/tcc+json")
public class TccCoordinatorController {

    private static final Logger LOGGER = LoggerFactory.getLogger(TccCoordinator
        Controller.class);
```

```java
        CoordinatorImp coordinatorImp = new CoordinatorImp();

    @PutMapping(path = "/confirm")
    public ResponseEntity confirm(@RequestBody Transaction transaction) {
        try {
            coordinatorImp.confirm(transaction);
            return ResponseEntity.noContent().build();
        } catch (Exception e) {
            LOGGER.error(e.getMessage(),e);
            return ResponseEntity
                .notFound()
                .build();
        }
    }

    @PutMapping(path = "/cancel")
    public ResponseEntity cancel(@RequestBody Transaction transaction) {
        try {
            coordinatorImp.cancel(transaction);
            return ResponseEntity.noContent().build();
        } catch (Exception e) {
            LOGGER.error(e.getMessage(),e);
            return ResponseEntity
                .notFound()
                .build();
        }
    }
}
```

这里有一点值得注意：coordinator 提供两个接口，分别是 PUT /coordinator/cancel 和 PUT /coordinator/confirm，接收的 media types 为 application/tcc+json 类型，请求参数为 Transaction 的 json。

2. tcc-rest-participant-api

tcc-rest-participant-api 主要是规范参与者的 try、cancel、confirm 的 REST API，具体如代码清单 24-13 所示。

代码清单24-13　ch24/ch24-tcc-rest/tcc-rest-participant-api/src/main/java/cn/springcloud/book/controller/TccParticipantController.java

```java
public abstract class TccParticipantController<T> {

    public static final String TCC_MEDIA_TYPE = "application/tcc";
    public static final String TRANSACTION_ID = "txId";
    protected static final Logger LOGGER = LoggerFactory.getLogger(TccParticipantController.class);

    @PostMapping(value = "/tcc/{txId}", consumes = MediaType.APPLICATION_JSON_UTF8_VALUE)
    public ResponseEntity tryOperation(@PathVariable String txId, @RequestBody T body) {
        LOGGER.info("{} begin to try transaction {}", getParticipantName(), txId);
        ResponseEntity result;
```

```java
        try {
            result = executeTry(txId, body);
        } catch (Exception e) {
            LOGGER.error(e.getMessage(), e);
            result = ResponseEntity.notFound().build();
        }
        LOGGER.info("{} finish try transaction {} ,result {}", getParticipantName(),
            txId, result.getStatusCode());
        return result;
    }

    @DeleteMapping(value = "/tcc/{txId}", consumes = TCC_MEDIA_TYPE, produces =
        TCC_MEDIA_TYPE)
    public ResponseEntity cancel(@PathVariable(TRANSACTION_ID) String txId) {
        LOGGER.info("{} begin to cancel transaction {}", getParticipantName(), txId);
        ResponseEntity result;
        try {
            result = executeCancel(txId);
        } catch (Exception e) {
            LOGGER.error(e.getMessage(), e);
            result = ResponseEntity.notFound().build();
        }
        LOGGER.info("{} finish cancel transaction {} ,result {}", getParticipantName(),
            txId, result.getStatusCode());
        return result;
    }

    @PutMapping(value = "/tcc/{txId}", consumes = TCC_MEDIA_TYPE, produces =
        TCC_MEDIA_TYPE)
    public ResponseEntity confirm(@PathVariable(TRANSACTION_ID) String txId) {
        LOGGER.info("{} begin to confirm transaction {}", getParticipantName(), txId);
        ResponseEntity result;
        try {
            result = executeConfirm(txId);
        } catch (Exception e) {
            LOGGER.error(e.getMessage(), e);
            result = ResponseEntity.notFound().build();
        }
        LOGGER.info("{} finish confirm transaction {} ,result {}", getParticipantName(),
            txId, result.getStatusCode());
        return result;
    }
    public abstract String getParticipantName();
    public abstract ResponseEntity executeTry(String txId, T body);
    public abstract ResponseEntity executeCancel(String txId);

    public abstract ResponseEntity executeConfirm(String txId);
}
```

这里设计了几个抽象接口，将具体的 executeTry、executeCancel、executeConfirm 的实现交给业务子类，另外参数头统一增加了一个 txId，即分布式事务的全局 id，用于方便后续事务管理协调及日志追踪。该类定义了三个 REST API 接口，分别是 POST /tcc/{txId}、DELETE /

tcc/{txId}、PUT /tcc/{txId}，它们一个用于 try 操作，一个用于 cancel 操作，一个用于 confirm 操作。try 操作使用 json 格式，cancel 及 confirm 操作使用的是 application/tcc，同时也规范了 cancel 及 confirm 操作失败时返回 404 状态码。

3. order-service

订单服务设计了一个 UserOrder 的实体类，如代码清单 24-14 所示。

代码清单24-14 ch24/ch24-tcc-rest/order-service/src/main/java/cn/springcloud/book/domain/UserOrder.java

```java
@Entity
@Data
@NoArgsConstructor
@Table(name = "user_order", uniqueConstraints = {@UniqueConstraint(name = "t_
    order_tx_idx", columnNames = {"txId"})})
@EntityListeners(AuditingEntityListener.class)
public class UserOrder {

    @Id
    @GeneratedValue(strategy = GenerationType.AUTO)
    private Integer id;

    private String txId;

    private String userId;

    private String productCode;

    private Integer quantity;

    @Enumerated(EnumType.STRING)
    private OrderState state;

    private LocalDateTime expireTime;

    @Version
    private Long version;

    @CreatedDate
    @Column(updatable = false)
    @JsonFormat(shape = JsonFormat.Shape.STRING, pattern = "yyyy-MM-dd HH:mm:ss")
    private LocalDateTime createTime;
}
```

这里设计了一个状态字段 OrderState，用于标识订单的状态，其枚举为 ORDERED、CONFIRMED、CANCELED，分别表示下单、确认（有库存）、取消。另外对 txId 进行唯一索引，方便进行幂等操作处理。

其对外暴露的 REST API 继承自 TccParticipantController，实现了 try、cancel、confirm 三个操作，具体如代码清单 24-15 所示。

代码清单24-15 ch24/ch24-tcc-rest/order-service/src/main/java/cn/springcloud/book/controller/OrderController.java

```java
@RestController
@RequestMapping("/order")
public class OrderController extends TccParticipantController<UserOrder> {

    @Autowired
    OrderDao orderDao;

    @Override
    public String getParticipantName() {
        return "order-service";
    }

    @Override
    public ResponseEntity executeTry(String txId, UserOrder body) {
        body.setTxId(txId);
        body.setState(OrderState.ORDERED);
        body.setExpireTime(LocalDateTime.now().plusMinutes(30));
        try{
            orderDao.save(body);
            return ResponseEntity.status(HttpStatus.CREATED).build();
        }catch (DataIntegrityViolationException e){
            return ResponseEntity.status(HttpStatus.CREATED).build();
        }
    }

    @Override
    public ResponseEntity executeCancel(String txId) {
        UserOrder userOrder = orderDao.findByTxId(txId);
        if (userOrder == null) {
//            return ResponseEntity.notFound().build();
            return ResponseEntity.status(HttpStatus.ACCEPTED).build();
        }
        userOrder.setState(OrderState.CANCELED);
        orderDao.save(userOrder);
        return ResponseEntity.status(HttpStatus.ACCEPTED).build();
    }

    @Override
    public ResponseEntity executeConfirm(String txId) {
        UserOrder userOrder = orderDao.findByTxId(txId);
        if (userOrder == null) {
//            return ResponseEntity.notFound().build();
            return ResponseEntity.status(HttpStatus.NO_CONTENT).build();
        }
        userOrder.setState(OrderState.CONFIRMED);
        orderDao.save(userOrder);
        return ResponseEntity.status(HttpStatus.NO_CONTENT).build();
    }
}
```

可以看到对于成功时的状态码，cancel 返回 202，confirm 返回 204。

4. inventory-service

库存服务设计了两个实体类，分别是 Inventory 和 FrozeRequest，Inventory 用于表示商品库存，具体如代码清单 24-16 所示。

代码清单24-16 ch24/ch24-tcc-rest/inventory-service/src/main/java/cn/springcloud/book/domain/Inventory.java

```java
@Entity
@Data
@Builder
@EntityListeners(AuditingEntityListener.class)
public class Inventory {

    @Tolerate
    public Inventory() {

    }

    @Id
    @GeneratedValue(strategy = GenerationType.AUTO)
    private Integer id;
    private String productCode;
    private Integer leftNum;
    @Version
    private Long version;
    @CreatedDate
    @Column(updatable = false)
    @JsonFormat(shape = JsonFormat.Shape.STRING, pattern = "yyyy-MM-dd HH:mm:ss")
    private LocalDateTime createTime;

}
```

Inventory 类设计了 leftNum 字段表示库存数量。

FrozeRequest 用于表示一个冻结商品库存的请求，具体如代码清单 24-17 所示。

代码清单24-17 ch24/ch24-tcc-rest/inventory-service/src/main/java/cn/springcloud/book/domain/FrozeRequest.java

```java
@Entity
@Data
@NoArgsConstructor
@SQLDelete(sql = "update froze_request set deleted = 1 where tx_id = ? and version = ?")
@Where(clause = "deleted = 0")
@EntityListeners(AuditingEntityListener.class)
public class FrozeRequest {

    @Id
    private String txId;

    @Column(name = "deleted")
    private Integer deleted = 0;

    private String productCode;
```

```
    private Integer frozenNum;

    @Version
    private Long version;

    @CreatedDate
    @Column(updatable = false)
    @JsonFormat(shape = JsonFormat.Shape.STRING, pattern = "yyyy-MM-dd HH:mm:ss")
    private LocalDateTime createTime;
}
```

这里设计了 deleted 字段，用于 cancel 操作时逻辑删除这个请求。

其对外暴露的 REST API 继承自 TccParticipantController，实现了 try、cancel、confirm 三个操作，具体如代码清单 24-18 所示。

代码清单24-18 ch24/ch24-tcc-rest/inventory-service/src/main/java/cn/springcloud/book/controller/InventoryController.java

```
@RestController
@RequestMapping("/inventory")
public class InventoryController extends TccParticipantController<FrozeRequest> {

    @Autowired
    InventoryDao inventoryDao;

    @Autowired
    FrozeRequestDao frozeRequestDao;

    @Autowired
    FrozeService frozeService;

    @Override
    public String getParticipantName() {
        return "inventory-service";
    }

    @Override
    public ResponseEntity executeTry(String txId, FrozeRequest body) {
        Inventory inventory = inventoryDao.findByProductCode(body.getProductCode());
        if (inventory == null) {
            return ResponseEntity.notFound().build();
        }
        if (inventory.getLeftNum() < body.getFrozenNum()) {
            return ResponseEntity.notFound().build();
        }
        body.setTxId(txId);
        try{
            frozeRequestDao.save(body);
            return ResponseEntity.status(HttpStatus.CREATED).build();
        }catch (DataIntegrityViolationException e){
            return ResponseEntity.status(HttpStatus.CREATED).build();
        }
    }
```

```java
@Override
public ResponseEntity executeCancel(String txId) {
    Optional<FrozeRequest> optional = frozeRequestDao.findById(txId);
    if (!optional.isPresent()) {
        return ResponseEntity.status(HttpStatus.ACCEPTED).build();
    }
    FrozeRequest frozeRequest = optional.get();
    Inventory inventory = inventoryDao.findByProductCode(frozeRequest.
        getProductCode());
    if (inventory == null) {
        return ResponseEntity.notFound().build();
    }
    frozeService.cancel(frozeRequest);
    return ResponseEntity.status(HttpStatus.ACCEPTED).build();
}

@Override
public ResponseEntity executeConfirm(String txId) {
    Optional<FrozeRequest> optional = frozeRequestDao.findById(txId);
    if (!optional.isPresent()) {
        return ResponseEntity.status(HttpStatus.NO_CONTENT).build();
    }
    FrozeRequest frozeRequest = optional.get();
    Inventory inventory = inventoryDao.findByProductCode(frozeRequest.
        getProductCode());
    if (inventory == null) {
        return ResponseEntity.notFound().build();
    }
    frozeService.confirm(frozeRequest, inventory);
    return ResponseEntity.status(HttpStatus.NO_CONTENT).build();
}
}
```

对于操作成功时的状态码，cancel 返回 202，confirm 返回 204，另外这里使用本地事务在 frozeService 的 confirm 方法中封装了删除冻结请求和减库存的操作，具体如代码清单 24-19 所示。

代码清单24-19 ch24/ch24-tcc-rest/inventory-service/src/main/java/cn/springcloud/book/service/FrozeService.java

```java
@Component
public class FrozeService {

    @Autowired
    InventoryDao inventoryDao;

    @Autowired
    FrozeRequestDao frozeRequestDao;

    @Transactional
    public void confirm(FrozeRequest request, Inventory inventory) {
        frozeRequestDao.delete(request);
        int left = inventory.getLeftNum() - request.getFrozenNum();
        if (left < 0) {
```

```
            throw new IllegalStateException("inventory left < 0");
        }
        inventory.setLeftNum(left);
        inventoryDao.save(inventory);
    }

    @Transactional
    public void cancel(FrozeRequest request) {
        frozeRequestDao.delete(request);
    }
}
```

5. tcc-coordinator-example

tcc-coordinator-example 展示了如何调用 coordinator 来完成整个 TCC 事务，这里我们把对 coordinator 方法的调用封装在 TccCoordinatorClient 类中，具体如代码清单 24-20 所示。

代码清单24-20　ch24/ch24-tcc-rest/tcc-coordinator-example/src/main/java/cn/springcloud/book/service/TccCoordinatorClient.java

```java
@Component
public class TccCoordinatorClient {

    private static final MediaType APPLICATION_TCC_JSON = new MediaType("application",
        "tcc+json");

    @Value("${tcc.coordinator.url}")
    String tccCoordinatorUrl;

    @Autowired
    RestTemplate restTemplate;

    public ResponseEntity<String> confirm(Transaction transaction) {
        RequestEntity<Transaction> requestEntity = RequestEntity.put(URI.
            create(tccCoordinatorUrl + "/confirm"))
            .contentType(APPLICATION_TCC_JSON)
            .body(transaction);
        return restTemplate.exchange(requestEntity, String.class);
    }

    public ResponseEntity<String> cancel(Transaction transaction) {
        RequestEntity<Transaction> requestEntity = RequestEntity.put(URI.
            create(tccCoordinatorUrl + "/cancel"))
            .contentType(APPLICATION_TCC_JSON)
            .body(transaction);
        return restTemplate.exchange(requestEntity, String.class);
    }
}
```

这里使用 RestTemplate 进行了客户端调用的封装。

使用 atomikos 的 transactions-tcc-rest-api 进行整个 TCC 事务的控制，实现代码如代码清单 24-21 所示。

代码清单24-21　ch24/ch24-tcc-rest/tcc-coordinator-example/src/main/java/cn/springcloud/book/service/TccOrderService.java

```java
@Component
public class TccOrderService {

    @Value("${tcc.participant.orderService}")
    String orderServiceUrlTemplate;
    @Value("${tcc.participant.inventoryService}")
    String inventoryServiceUrlTemplate;
    @Autowired
    TccCoordinatorClient tccCoordinatorClient;
    @Value("${tcc.transaction.timeoutInMs}")
    private long transactionTtimeoutInMs;
    @Autowired
    private RestTemplate restTemplate;

    public void newOrderWithTcc(OrderRequest orderRequest,String txId) {
        long expireTime = System.currentTimeMillis() + transactionTtimeoutInMs;

        List<ParticipantLink> participantLinks = new ArrayList<>(2);
        String orderServiceUrl = String.format(orderServiceUrlTemplate, txId);
        participantLinks.add(new ParticipantLink(orderServiceUrl, expireTime));

        String inventoryServiceUrl = String.format(inventoryServiceUrlTemplate, txId);
        participantLinks.add(new ParticipantLink(inventoryServiceUrl, expireTime));

        Transaction transaction = new Transaction(participantLinks);
        try {

            //1. try participant order-service
            restTemplate.postForEntity(orderServiceUrl, orderRequest, String.class);

            //2. try participant inventory-service
            FrozeRequest frozeRequest = FrozeRequest.builder()
                .productCode(orderRequest.getProductCode())
                .frozenNum(orderRequest.getQuantity())
                .build();
            restTemplate.postForEntity(inventoryServiceUrl, frozeRequest, String.class);

            //3. call coordinator to confirm
            tccCoordinatorClient.confirm(transaction);
        } catch (Exception e) {
            //4. call coordinator to cancel
            tccCoordinatorClient.cancel(transaction);
            String msg = e instanceof HttpStatusCodeException ? ((HttpStatusCodeException)
                e).getResponseBodyAsString() : e.getMessage();
            throw new RuntimeException(msg, e);
        }
    }
}
```

这里主要分为 3 步进行处理：

1）首先构建参与事务的 ParticipantLink，这里主要是告诉 coordinator，对于参与者的 cancel

及 confirm 操作需要调用哪个 URL。

2）按顺序挨个对参与者进行 try 操作的调用。

3）如果调用都成功，则最后一步告知 coordinator 进行 confirm 操作；如果中途有一步出现异常，则告知 coordinator 进行 cancel 操作。

6. 验证

场景 1：TCC 事务成功

该场景下，分别启动 tcc-coordinator-atomikos、order-service、inventroy-service，然后执行 tcc-coordinator-example 的 TccCoordinatorExampleApplicationTests.testTccOrder 方法，可以看到执行成功。

tcc-coordinator-atomikos 的 debug 日志部分如下：

```
2018-07-28 21:37:37.579 DEBUG 4372 --- [nio-9090-exec-1] c.a.i.i.CompositeTrans
    actionManagerImp         : createCompositeTransaction ( 27438 ): created new ROOT
    transaction with id 192.168.2.114.tm153278505756400001
2018-07-28 21:37:37.581 DEBUG 4372 --- [nio-9090-exec-1] c.a.icatch.imp.
    CompositeTransactionImp        : addParticipant ( http://localhost:8080/order/
    tcc/a9e42a69-f4b5-47f0-a934-7463096b6f7a ) for transaction 192.168.2.114.
    tm153278505756400001
2018-07-28 21:37:37.581 DEBUG 4372 --- [nio-9090-exec-1] c.a.icatch.imp.
    CompositeTransactionImp        : addParticipant ( http://localhost:8081/
    inventory/tcc/a9e42a69-f4b5-47f0-a934-7463096b6f7a ) for transaction
    192.168.2.114.tm153278505756400001
2018-07-28 21:37:37.584 DEBUG 4372 --- [nio-9090-exec-1] c.a.icatch.imp.
    CompositeTransactionImp        : commit() done (by application) of transaction
    192.168.2.114.tm153278505756400001
```

order-service 的部分日志为：

```
2018-07-28 21:37:36.222  INFO 4376 --- [nio-8080-exec-1] c.s.b.c.TccParticipa
    ntController        : order-service begin to try transaction a9e42a69-f4b5-
    47f0-a934-7463096b6f7a
Hibernate: call next value for hibernate_sequence
Hibernate: insert into user_order (create_time, expire_time, product_code, quantity,
    state, tx_id, user_id, version, id) values (?, ?, ?, ?, ?, ?, ?, ?, ?)
2018-07-28 21:37:36.356  INFO 4376 --- [nio-8080-exec-1] c.s.b.c.TccParticipan
    tController        : order-service finish try transaction a9e42a69-f4b5-
    47f0-a934-7463096b6f7a ,result 201
2018-07-28 21:37:38.075  INFO 4376 --- [nio-8080-exec-3] c.s.b.c.TccParticipan
    tController        : order-service begin to confirm transaction a9e42a69-
    f4b5-47f0-a934-7463096b6f7a
2018-07-28 21:37:38.092  INFO 4376 --- [nio-8080-exec-3] o.h.h.i.QueryTranslato
    rFactoryInitiator        : HHH000397: Using ASTQueryTranslatorFactory
Hibernate: select userorder0_.id as id1_0_, userorder0_.create_time as create_
    t2_0_, userorder0_.expire_time as expire_t3_0_, userorder0_.product_code
    as product_4_0_, userorder0_.quantity as quantity5_0_, userorder0_.state
    as state6_0_, userorder0_.tx_id as tx_id7_0_, userorder0_.user_id as user_
    id8_0_, userorder0_.version as version9_0_ from user_order userorder0_
    where userorder0_.tx_id=?
Hibernate: update user_order set expire_time=?, product_code=?, quantity=?,
    state=?, tx_id=?, user_id=?, version=? where id=? and version=?
```

```
2018-07-28 21:37:38.493  INFO 4376 --- [nio-8080-exec-3] c.s.b.c.TccParticipa
ntController                            : order-service finish confirm transaction a9e42a69-
f4b5-47f0-a934-7463096b6f7a ,result 204
```

inventory-service 的部分日志为：

```
2018-07-28 21:37:36.589  INFO 4378 --- [nio-8081-exec-1] c.s.b.c.TccParticipan
tController                             : inventory-service begin to try transaction a9e42a69-
f4b5-47f0-a934-7463096b6f7a
2018-07-28 21:37:36.615  INFO 4378 --- [nio-8081-exec-1] o.h.h.i.QueryTranslato
rFactoryInitiator  : HHH000397: Using ASTQueryTranslatorFactory
Hibernate: select inventory0_.id as id1_1_, inventory0_.create_time as create_
    t2_1_, inventory0_.left_num as left_num3_1_, inventory0_.product_code as
    product_4_1_, inventory0_.version as version5_1_ from inventory inventory0_
    where inventory0_.product_code=?
Hibernate: insert into froze_request (create_time, deleted, frozen_num, product_
    code, version, tx_id) values (?, ?, ?, ?, ?, ?)
2018-07-28 21:37:37.045  INFO 4378 --- [nio-8081-exec-1] c.s.b.c.TccParticipa
ntController                            : inventory-service finish try transaction a9e42a69-
f4b5-47f0-a934-7463096b6f7a ,result 201
2018-07-28 21:37:38.501  INFO 4378 --- [nio-8081-exec-2] c.s.b.c.TccPartici
pantController                          : inventory-service begin to confirm transaction
a9e42a69-f4b5-47f0-a934-7463096b6f7a
Hibernate: select frozereque0_.tx_id as tx_id1_0_0_, frozereque0_.create_time
    as create_t2_0_0_, frozereque0_.deleted as deleted3_0_0_, frozereque0_.
    frozen_num as frozen_n4_0_0_, frozereque0_.product_code as product_5_0_0_,
    frozereque0_.version as version6_0_0_ from froze_request frozereque0_ where
    frozereque0_.tx_id=? and ( frozereque0_.deleted = 0)
Hibernate: select inventory0_.id as id1_1_, inventory0_.create_time as create_
    t2_1_, inventory0_.left_num as left_num3_1_, inventory0_.product_code as
    product_4_1_, inventory0_.version as version5_1_ from inventory inventory0_
    where inventory0_.product_code=?
Hibernate: update inventory set left_num=?, product_code=?, version=? where
    id=? and version=?
Hibernate: update froze_request set deleted = 1 where tx_id = ? and version = ?
2018-07-28 21:37:38.536  INFO 4378 --- [nio-8081-exec-2] c.s.b.c.TccParticipant
Controller                              : inventory-service finish confirm transaction a9e42a69-
f4b5-47f0-a934-7463096b6f7a ,result 204
```

场景 2：参与者 try 操作超时

如果出现参与者 try 操作超时的情况，那么会调用 coordinator 进行 cancel 操作；如果 cancel 操作失败，则不做任何操作，已经预留的资源依赖过期时间来释放。

场景 3：coordinator confirm 超时

如果 try 操作成功，在 coordinator 进行 confirm 的时候，部分参与者出现超时（coordinator 重试之后仍然超时），这种异常称为 heuristic exception，出现的不一致性需要人工介入处理。

场景 4：coordinator 挂掉

在 try 操作成功，调用 coordinator 进行 confirm 时，coordinator 挂掉，这个时候，若流程转去调用 coordinator 的 cancel 依旧没能成功，则已经预留的资源依赖过期时间来释放，或者调用方进行重试。若此时 coordinator 已经恢复，则可以重试成功。

24.3.3 Servicecomb SAGA

本节我们主要使用 incubator-servicecomb-saga 组件来演示使用 SAGA 模式进行分布式事务管理的方法。incubator-servicecomb-saga 是华为开源的实现 SAGA 模式的分布式事务管理组件，目前已经在 Apache 中孵化。

incubator-servicecomb-saga 设计了一个分布式的 SAGA 协调器，参与者使用 Omega 与 Alpha 进行通信，而 Alpha 则负责协调管理事务的执行，它们之间的关系如图 24-6 所示。

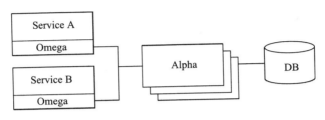

图 24-6　Omega 与 Alpha 关系图

这里我们以订单及库存为案，案例的背景为订单及库存两个微服务，然后用户下单要在两个微服务之间进行分布式事务控制，即订单状态为确认时，需同时扣减库存。关于本案例的更多内容见 ch24-3-3 工程，该工程主要分为几个模块：

- alpha-server：SAGA 协调器，类似 TCC 的 coordinator。
- omega-order-service：订单微服务工程。
- omega-inventory-service：库存微服务工程。
- saga-servicecomb-example：演示了如何使用 servicecombe saga 进行分布式事务管理。

下面我们详细介绍这些工程模块。

1. alpha-server

目前 alpha-server 的最新版本为 0.2.0，发布的 jar 文件名为 alpha-server-0.2.0-exec.jar，该服务使用 springboot 及 grpc 开发，依赖 MySQL 或者 Postgres 数据库，这里我们使用 Docker 启动 Postgres，命令如下：

```
docker run --rm --name saga-postgres -p 5432:5432 -e POSTGRES_DB=saga -e POSTGRES_USER=saga -e POSTGRES_PASSWORD=password postgres:9.5
```

之后还需初始化 SAGA 数据库，创建相关表。需要用到的 schema 如代码清单 24-22 所示。

代码清单24-22　ch24/ch24-saga-servicecomb/alpha-server/schema-postgresql.sql

```
CREATE TABLE IF NOT EXISTS TxEvent (
    surrogateId BIGSERIAL PRIMARY KEY,
    serviceName varchar(36) NOT NULL,
    instanceId varchar(256) NOT NULL,
    creationTime timestamp(6) NOT NULL DEFAULT CURRENT_DATE,
    globalTxId varchar(36) NOT NULL,
    localTxId varchar(36) NOT NULL,
```

```
    parentTxId varchar(36) DEFAULT NULL,
    type varchar(50) NOT NULL,
    compensationMethod varchar(256) NOT NULL,
    expiryTime timestamp(6) NOT NULL,
    retryMethod varchar(256) NOT NULL,
    retries int NOT NULL DEFAULT 0,
    payloads bytea
);
CREATE INDEX IF NOT EXISTS saga_events_index ON TxEvent (surrogateId,
    globalTxId, localTxId, type, expiryTime);
CREATE TABLE IF NOT EXISTS Command (
    surrogateId BIGSERIAL PRIMARY KEY,
    eventId bigint NOT NULL UNIQUE,
    serviceName varchar(36) NOT NULL,
    instanceId varchar(256) NOT NULL,
    globalTxId varchar(36) NOT NULL,
    localTxId varchar(36) NOT NULL,
    parentTxId varchar(36) DEFAULT NULL,
    compensationMethod varchar(256) NOT NULL,
    payloads bytea,
    status varchar(12),
    lastModified timestamp(6) NOT NULL DEFAULT CURRENT_DATE,
    version bigint NOT NULL
);
CREATE INDEX IF NOT EXISTS saga_commands_index ON Command (surrogateId, eventId,
    globalTxId, localTxId, status);
CREATE TABLE IF NOT EXISTS TxTimeout (
    surrogateId BIGSERIAL PRIMARY KEY,
    eventId bigint NOT NULL UNIQUE,
    serviceName varchar(36) NOT NULL,
    instanceId varchar(256) NOT NULL,
    globalTxId varchar(36) NOT NULL,
    localTxId varchar(36) NOT NULL,
    parentTxId varchar(36) DEFAULT NULL,
    type varchar(50) NOT NULL,
    expiryTime TIMESTAMP NOT NULL,
    status varchar(12),
    version bigint NOT NULL
);
CREATE INDEX IF NOT EXISTS saga_timeouts_index ON TxTimeout (surrogateId,
    expiryTime, globalTxId, localTxId, status);
```

之后使用如下命令启动 alpha server：

```
java -jar -Dspring.profiles.active=prd \
    alpha-server-0.2.0-exec.jar \
    --spring.datasource.url=jdbc:postgresql://192.168.99.100:5432/saga \
    --spring.datasource.username=saga \
    --spring.datasource.password=password \
    --alpha.server.port=9090 \
    --server.port=9091
```

为了方便演示，也可以使用笔者提供的 h2 数据库版本的 alpha-server，可以省去手工新建数据库及初始化表结构等操作，直接启动工程 ch24-3-3-saga-servicecomb/alpha-server 即可。

2. omega-order-service

参与分布式事务的微服务需要引入 omega 组件，进行本地注解拦截处理，以及向 alpha server 汇报及执行相关操作。部分依赖如代码清单 24-23 所示。

代码清单24-23　ch24/ch24-saga-servicecomb/omega-order-service/pom.xml

```xml
<dependency>
    <groupId>org.apache.servicecomb.saga</groupId>
    <artifactId>omega-spring-starter</artifactId>
</dependency>
<dependency>
    <groupId>org.apache.servicecomb.saga</groupId>
    <artifactId>omega-transport-resttemplate</artifactId>
</dependency>
```

这些组件主要用于 omega 与 alpha sever 的通信以及本地事务的切面拦截、处理等。同时在配置文件中通过 alpha.cluster.address 配置项指定 alpha-server 的地址，这里为 localhost:9090。

引入依赖及指定完 alpha server 地址之后，需要在启动类上标注开启 omega，如代码清单 24-24 所示。

代码清单24-24　ch24/ch24-saga-servicecomb/omega-order-service/src/main/java/cn/springcloud/book/OmegaOrderServiceApplication.java

```java
@SpringBootApplication
@EnableJpaRepositories(basePackages = "cn.springcloud.book.dao")
@EnableTransactionManagement
@EnableJpaAuditing
@EnableOmega
public class OmegaOrderServiceApplication {

    public static void main(String[] args) {
        SpringApplication.run(OmegaOrderServiceApplication.class, args);
    }
}
```

开启 EnableOmega 之后，就可以用 service 方法标记参与分布式事务管理的执行方法以及对应的补偿方法，具体如代码清单 24-25 所示。

代码清单24-25　ch24/ch24-saga-servicecomb/omega-order-service/src/main/java/cn/springcloud/book/service/OrderService.java

```java
@Service
public class OrderService {

    private Set<String> executedSet = new ConcurrentHashMap<>().newKeySet();

    private Set<String> canceledSet = new ConcurrentHashMap<>().newKeySet();

    @Autowired
    OrderDao orderDao;
```

```java
@Compensable(timeout = 1, compensationMethod = "cancel")
@Transactional
public UserOrder order(UserOrder userOrder){
    if(executedSet.contains(userOrder.getTxId()) || canceledSet.
        contains(userOrder.getTxId())){
        return orderDao.findByTxId(userOrder.getTxId());
    }
    userOrder.setState(OrderState.CONFIRMED);
    userOrder.setExpireTime(LocalDateTime.now().plusMinutes(30));
    UserOrder result = orderDao.save(userOrder);
    executedSet.add(userOrder.getTxId());
    return result;
}
@Transactional
public UserOrder cancel(UserOrder userOrder){
    if(canceledSet.contains(userOrder.getTxId()) || !executedSet.
        contains(userOrder.getTxId())){
        return orderDao.findByTxId(userOrder.getTxId());
    }
    UserOrder result = orderDao.findByTxId(userOrder.getTxId());
    if (result == null) {
        throw new IllegalStateException();
    }
    result.setState(OrderState.CANCELED);
    UserOrder saved = orderDao.save(result);
    canceledSet.add(userOrder.getTxId());
    return saved;
}
```

这里使用 Compensable 标注执行方法，同时通过注解的 compensationMethod 指定对应的补偿方法。由于 servicecomb 使用 grpc 来进行 omega 与 alpha server 之间的通信，因此 controller 只需要创建订单的接口，具体如代码清单 24-26 所示。

代码清单24-26 ch24/ch24-saga-servicecomb/omega-order-service/src/main/java/cn/springcloud/book/controller/OrderController.java

```java
@RestController
@RequestMapping("/order")
public class OrderController {

    @Autowired
    OrderService orderService;

    @Autowired
    OmegaContext omegaContext;

    @PostMapping
    public UserOrder order(@RequestBody UserOrder userOrder){
        userOrder.setTxId(omegaContext.globalTxId());
        return orderService.order(userOrder);
    }
}
```

3. omega-inventory-service

omega-inventory-service 的依赖及配置与 omega-order-service 类似，这里就不再赘述了。与 TCC 的 inventory-service 相比，这里就没有设计 FrozeRequest 实体类，请求操作直接扣减库存，具体如代码清单 24-27 所示。

代码清单24-27　ch24/ch24-saga-servicecomb/omega-inventory-service/src/main/java/cn/springcloud/book/service/InventoryService.java

```java
@Service
public class InventoryService {
    private Set<String> executedSet = new ConcurrentHashMap<>().newKeySet();
    private Set<String> canceledSet = new ConcurrentHashMap<>().newKeySet();
    @Autowired
    InventoryDao inventoryDao;
    @Compensable(timeout = 1, compensationMethod = "cancel")
    @Transactional
    public Inventory order(OrderRequest orderRequest){
        if(executedSet.contains(orderRequest.getTxId()) || canceledSet.
            contains(orderRequest.getTxId())){
            return inventoryDao.findByProductCode(orderRequest.getProductCode());
        }
        Inventory inventory = inventoryDao.findByProductCode(orderRequest.
            getProductCode());
        if(inventory == null){
            throw new IllegalStateException("product not found");
        }
        if(inventory.getLeftNum() < orderRequest.getQuantity()){
            throw new IllegalStateException("not enough product left");
        }
        inventory.setLeftNum(inventory.getLeftNum() - orderRequest.
            getQuantity());
        Inventory result = inventoryDao.save(inventory);
        executedSet.add(orderRequest.getTxId());
        return result;
    }
    @Transactional
    public Inventory cancel(OrderRequest orderRequest){
        if(canceledSet.contains(orderRequest.getTxId()) || !executedSet.
            contains(orderRequest.getTxId())){
            return inventoryDao.findByProductCode(orderRequest.getProductCode());
        }
        Inventory inventory = inventoryDao.findByProductCode(orderRequest.
            getProductCode());
        if(inventory == null){
            throw new IllegalStateException("product not found");
        }
        inventory.setLeftNum(inventory.getLeftNum() + orderRequest.
            getQuantity());
        Inventory result = inventoryDao.save(inventory);
        canceledSet.add(orderRequest.getTxId());
        return result;
    }
}
```

同样，由于 omega 与 alpha server 使用 grpc 通信，跟 TCC 的工程相比，这里只需要暴露一个扣减库存的 API 即可，具体如代码清单 24-28 所示。

代码清单24-28 ch24/ch24-saga-servicecomb/omega-inventory-service/src/main/java/cn/springcloud/book/controller/InventoryController.java

```java
@RestController
@RequestMapping("/inventory")
public class InventoryController {
    @Autowired
    OmegaContext omegaContext;
    @Autowired
    InventoryService inventoryService;
    @PostMapping
    public Inventory submitOrderRequest(@RequestBody OrderRequest orderRequest){
        orderRequest.setTxId(omegaContext.globalTxId());
        return inventoryService.order(orderRequest);
    }
}
```

4. saga-servicecomb-example

saga-servicecomb-example 展示了如何通过 omega 及 alpha 来完成整个 SAGA 事务，其调用方法如代码清单 24-29 所示。

代码清单24-29 ch24/ch24-saga-servicecomb/saga-servicecomb-example/src/main/java/cn/springcloud/book/controller/SagaOrderController.java

```java
@RestController
@RequestMapping("/saga")
public class SagaOrderController {
    @Value("${omega.orderService}")
    String orderServiceUrl;
    @Value("${omega.inventoryService}")
    String inventoryServiceUrl;
    @Autowired
    private RestTemplate restTemplate;
    @SagaStart
    @PostMapping("")
    public void sagaOrder(@RequestBody OrderRequest orderRequest){
        //1. order-service
        restTemplate.postForEntity(orderServiceUrl, orderRequest, String.class);
        //2. inventory-service
        restTemplate.postForEntity(inventoryServiceUrl, orderRequest, String.class);
    }
}
```

这里主要是通过 SagaStart 注解来标注一个 SAGA 的分布式事务的开启，之后依赖相关切面进行拦截处理并与 alpha server 进行通信。

5. 验证

分别启动 alpha-server、omega-order-service、omega-inventory-service、saga-servicecomb-example 工程，之后执行如下命令来验证：

```
curl -i -H "Content-Type: application/json" -X POST localhost:8082/saga -d
    '{"userId":"caibosi","productCode":"spring-cloud-in-action","quantity":10}'
```

场景 1：SAGA 事务成功

事务执行成功，可以看到 TXEVENT 表有 6 条记录，具体如图 24-7 所示。

SURROGATEID	SERVICENAME	INSTANCEID	CREATIONTIME	GLOBALTXID	LOCALTXID	PARENTTXID	TYPE
1	saga-servicecomb-example	saga-servicecomb-example-192.168.2.114	2018-07-29 23:23:58.882	0fdc5c71-fdd9-40b3-8ab2-6d61cff711f6	0fdc5c71-fdd9-40b3-8ab2-6d61cff711f6	null	SagaStartedEvent
2	omega-order-service	omega-order-service-192.168.2.114	2018-07-29 23:23:59.367	0fdc5c71-fdd9-40b3-8ab2-6d61cff711f6	be1cb638-a24a-4819-b455-52ea15693d99	0fdc5c71-fdd9-40b3-8ab2-6d61cff711f6	TxStartedEvent
3	omega-order-service	omega-order-service-192.168.2.114	2018-07-29 23:23:59.512	0fdc5c71-fdd9-40b3-8ab2-6d61cff711f6	be1cb638-a24a-4819-b455-52ea15693d99	0fdc5c71-fdd9-40b3-8ab2-6d61cff711f6	TxEndedEvent
4	omega-inventory-service	omega-inventory-service-192.168.2.114	2018-07-29 23:23:59.626	0fdc5c71-fdd9-40b3-8ab2-6d61cff711f6	8cd5bebf-a97c-4a6e-8acc-7f8fb34b5d3a	0fdc5c71-fdd9-40b3-8ab2-6d61cff711f6	TxStartedEvent
5	omega-inventory-service	omega-inventory-service-192.168.2.114	2018-07-29 23:24:00.12	0fdc5c71-fdd9-40b3-8ab2-6d61cff711f6	8cd5bebf-a97c-4a6e-8acc-7f8fb34b5d3a	0fdc5c71-fdd9-40b3-8ab2-6d61cff711f6	TxEndedEvent
6	saga-servicecomb-example	saga-servicecomb-example-192.168.2.114	2018-07-29 23:24:00.156	0fdc5c71-fdd9-40b3-8ab2-6d61cff711f6	0fdc5c71-fdd9-40b3-8ab2-6d61cff711f6	null	SagaEndedEvent

(6 rows, 5 ms)

图 24-7　TXEVENT 表记录

图 24-7 中所示的 6 个事件分别为 SagaStartedEvent、order-service 的 TxStartedEvent 及 TxEndedEvent、inventory-service 的 TxStartedEvent 及 TxEndedEvent、事务结束的 SagaEndedEvent。

场景 2：参与者业务方法超时或异常

默认情况下，使用 SagaStart 及 Compensable 注解，如果没有指定 timeout 参数，则默认为 0，表示永不超时，因此请求可能会一直阻塞。这里将 order-service 及 inventory-service 的 Compensable 的 timeout 设置为 1 秒，将 SagaStart 的 timeout 设置为 2 秒，然后对 inventory-service 的 API 模拟延时 5 分钟。此时，请求接口报错返回如下：

```
{"timestamp":"2018-07-30T02:07:26.045+0000","status":500,"error":"Internal
    Server Error","message":"transaction b0c6fd07-532b-4538-a8bc-95c2540bcbb6
    is aborted","path":"/saga"}
```

在 alpha-server 中可以看到如下日志：

```
2018-07-30 10:02:26.876  INFO 2418 --- [pool-2-thread-1] o.a.s.saga.alpha.
    core.EventScanner        : Found timeout event TxEvent{surrogateId=1,
    serviceName='saga-servicecomb-example', instanceId='saga-servicecomb-
```

```
          example-127.0.0.1', creationTime=Mon Jul 30 10:02:24 CST 2018,
          globalTxId='b0c6fd07-532b-4538-a8bc-95c2540bcbb6', localTxId='b0c6fd07-
          532b-4538-a8bc-95c2540bcbb6', parentTxId='null', type='SagaStartedEvent',
          compensationMethod='', expiryTime=Mon Jul 30 10:02:26 CST 2018,
          retryMethod='', retries=0}
2018-07-30 10:02:26.905  INFO 2418 --- [pool-2-thread-1] o.a.s.saga.alpha.
          core.EventScanner        : Found timeout event TxTimeout{eventId=1,
          serviceName='saga-servicecomb-example', instanceId='saga-servicecomb-
          example-127.0.0.1', globalTxId='b0c6fd07-532b-4538-a8bc-95c2540bcbb6',
          localTxId='b0c6fd07-532b-4538-a8bc-95c2540bcbb6', parentTxId='null',
          type='SagaStartedEvent', expiryTime=Mon Jul 30 10:02:26 CST 2018,
          status=NEW} to abort
2018-07-30 10:02:26.912  INFO 2418 --- [pool-2-thread-1] o.a.s.saga.alpha.
          core.EventScanner        : Found uncompensated event TxEvent{surrogateId=3,
          serviceName='omega-order-service', instanceId='omega-order-
          service-127.0.0.1', creationTime=Mon Jul 30 10:02:25 CST 2018,
          globalTxId='b0c6fd07-532b-4538-a8bc-95c2540bcbb6', localTxId='be8538df-
          b5ce-4d8d-a5b7-9180313114b0', parentTxId='b0c6fd07-532b-4538-a8bc-
          95c2540bcbb6', type='TxEndedEvent', compensationMethod='public
          cn.springcloud.book.domain.UserOrder cn.springcloud.book.service.
          OrderService.cancel(cn.springcloud.book.domain.UserOrder)', expiryTime=Fri
          Dec 31 08:00:00 CST 9999, retryMethod='', retries=0}
2018-07-30 10:02:26.916  INFO 2418 --- [pool-2-thread-1] o.a.s.s.a.s.Spring
          CommandRepository        : Saving compensation command Command{eventId=2,
          serviceName='omega-order-service', instanceId='omega-order-
          service-127.0.0.1', globalTxId='b0c6fd07-532b-4538-a8bc-95c2540bcbb6',
          localTxId='be8538df-b5ce-4d8d-a5b7-9180313114b0', parentTxId='b0c6fd07-
          532b-4538-a8bc-95c2540bcbb6', compensationMethod='public cn.springcloud.
          book.domain.UserOrder cn.springcloud.book.service.OrderService.cancel(cn.
          springcloud.book.domain.UserOrder)'}
2018-07-30 10:02:26.947  INFO 2418 --- [pool-2-thread-1] o.a.s.s.a.s.Spring
          CommandRepository        : Saved compensation command Command{eventId=2,
          serviceName='omega-order-service', instanceId='omega-order-
          service-127.0.0.1', globalTxId='b0c6fd07-532b-4538-a8bc-95c2540bcbb6',
          localTxId='be8538df-b5ce-4d8d-a5b7-9180313114b0', parentTxId='b0c6fd07-
          532b-4538-a8bc-95c2540bcbb6', compensationMethod='public cn.springcloud.
          book.domain.UserOrder cn.springcloud.book.service.OrderService.cancel(cn.
          springcloud.book.domain.UserOrder)'}
2018-07-30 10:02:26.954  INFO 2418 --- [pool-2-thread-1] o.a.s.saga.alpha.core.
          EventScanner        : Compensating transaction with globalTxId b0c6fd07-532b-
          4538-a8bc-95c2540bcbb6 and localTxId be8538df-b5ce-4d8d-a5b7-9180313114b0
2018-07-30 10:02:27.468  INFO 2418 --- [pool-2-thread-1] o.a.s.saga.alpha.
          core.EventScanner        : Found timeout event TxEvent{surrogateId=4,
          serviceName='omega-inventory-service', instanceId='omega-inventory-
          service-127.0.0.1', creationTime=Mon Jul 30 10:02:25 CST 2018,
          globalTxId='b0c6fd07-532b-4538-a8bc-95c2540bcbb6', localTxId='1123d098-
          66c3-4a68-8618-e2133a5b4806', parentTxId='b0c6fd07-532b-4538-a8bc-
          95c2540bcbb6', type='TxStartedEvent', compensationMethod='public
          cn.springcloud.book.domain.Inventory cn.springcloud.book.service.
          InventoryService.cancel(cn.springcloud.book.dto.OrderRequest)',
          expiryTime=Mon Jul 30 10:02:26 CST 2018, retryMethod='', retries=0}
2018-07-30 10:02:27.473  INFO 2418 --- [pool-2-thread-1] o.a.s.saga.alpha.
          core.EventScanner        : Found timeout event TxTimeout{eventId=4,
          serviceName='omega-inventory-service', instanceId='omega-inventory-
          service-127.0.0.1', globalTxId='b0c6fd07-532b-4538-a8bc-95c2540bcbb6',
```

```
localTxId='1123d098-66c3-4a68-8618-e2133a5b4806', parentTxId='b0c6fd07-
532b-4538-a8bc-95c2540bcbb6', type='TxStartedEvent', expiryTime=Mon Jul 30
10:02:26 CST 2018, status=NEW} to abort
2018-07-30 10:02:28.044  INFO 2418 --- [pool-2-thread-1] o.a.s.saga.alpha.
   core.EventScanner        : Found compensated event TxEvent{surrogateId=7,
   serviceName='omega-order-service', instanceId='omega-order-
   service-127.0.0.1', creationTime=Mon Jul 30 10:02:27 CST 2018,
   globalTxId='b0c6fd07-532b-4538-a8bc-95c2540bcbb6', localTxId='be8538df-
   b5ce-4d8d-a5b7-9180313114b0', parentTxId='b0c6fd07-532b-4538-a8bc-
   95c2540bcbb6', type='TxCompensatedEvent', compensationMethod='public
   cn.springcloud.book.domain.UserOrder cn.springcloud.book.service.
   OrderService.cancel(cn.springcloud.book.domain.UserOrder)', expiryTime=Fri
   Dec 31 08:00:00 CST 9999, retryMethod='', retries=0}
2018-07-30 10:02:28.047  INFO 2418 --- [pool-2-thread-1] o.a.s.saga.alpha.
   core.EventScanner        : Transaction with globalTxId b0c6fd07-532b-4538-
   a8bc-95c2540bcbb6 and localTxId be8538df-b5ce-4d8d-a5b7-9180313114b0 was
   compensated
2018-07-30 10:02:28.055  INFO 2418 --- [pool-2-thread-1] o.a.s.saga.alpha.core.
   EventScanner             : Marked end of transaction with globalTxId b0c6fd07-
   532b-4538-a8bc-95c2540bcbb6
2018-07-30 10:02:28.589  INFO 2418 --- [pool-2-thread-1] o.a.s.saga.alpha.
   core.EventScanner        : Found compensated event TxEvent{surrogateId=8,
   serviceName='omega-inventory-service', instanceId='omega-inventory-
   service-127.0.0.1', creationTime=Mon Jul 30 10:02:27 CST 2018,
   globalTxId='b0c6fd07-532b-4538-a8bc-95c2540bcbb6', localTxId='1123d098-
   66c3-4a68-8618-e2133a5b4806', parentTxId='b0c6fd07-532b-4538-a8bc-
   95c2540bcbb6', type='TxCompensatedEvent', compensationMethod='public
   cn.springcloud.book.domain.Inventory cn.springcloud.book.service.
   InventoryService.cancel(cn.springcloud.book.dto.OrderRequest)',
   expiryTime=Fri Dec 31 08:00:00 CST 9999, retryMethod='', retries=0}
2018-07-30 10:02:28.591  INFO 2418 --- [pool-2-thread-1] o.a.s.saga.alpha.
   core.EventScanner        : Transaction with globalTxId b0c6fd07-532b-4538-
   a8bc-95c2540bcbb6 and localTxId 1123d098-66c3-4a68-8618-e2133a5b4806 was
   compensated
2018-07-30 10:02:28.595  INFO 2418 --- [pool-2-thread-1] o.a.s.saga.alpha.core.
   EventScanner             : Marked end of transaction with globalTxId b0c6fd07-
   532b-4538-a8bc-95c2540bcbb6
2018-07-30 10:07:25.951  INFO 2418 --- [ault-executor-4] o.a.s.s.alpha.core.
   TxConsistentService      : Transaction event SagaEndedEvent rejected, because its
   parent with globalTxId b0c6fd07-532b-4538-a8bc-95c2540bcbb6 was already aborted
```

从日志中可以看出，触发事件的先后顺序是：SagaStartedEvent、order-service 的 TxStartedEvent、order-service 的 TxEndedEvent、inventory-service 的 TxStartedEvent、saga-servicecomb-example 的 TxAbortedEvent、inventory-service 的 TxAbortedEvent、order-service 的 TxCompensatedEvent、inventory-service 的 TxCompensatedEvent、最后是 SagaEndedEvent。

查看 SAGA 数据库的 TXTIMEOUT 表，可以看到有 2 条记录，如图 24-8 所示。

可以看到启动 omega-inventory-service 的事务超时，会导致 SAGA 事务也超时。对于这种情况，只要标注 Compensable 的方法以及补偿方法能合理实现并能够成功执行，通过 SAGA 就可以保持数据一致性。

如果是 order-service 的 order 方法出现异常，则 inventory-service 方法不会执行，产生的事件为 SagaStartedEvent、order-service 的 TxStartedEvent、order-service 的 TxAbortedEvent、最

后是 SagaEndedEvent。如果补偿方法出现异常，则需要人工介入处理数据一致性问题。

SURROGATEID	EVENTID	SERVICENAME	INSTANCEID	GLOBALTXID	LOCALTXID	PARENTTXID	TYPE	EXPIRYTIME	STATUS	VERSION
1	1	saga-servicecomb-example	saga-servicecomb-example-127.0.0.1	b0c6fd07-532b-4538-a8bc-95c2540bcbb6	b0c6fd07-532b-4538-a8bc-95c2540bcbb6	null	SagaStartedEvent	2018-07-30 10:02:26.416	DONE	3
2	4	omega-inventory-service	omega-inventory-service-127.0.0.1	b0c6fd07-532b-4538-a8bc-95c2540bcbb6	1123d098-66c3-4a68-8618-e2133a5b4806	b0c6fd07-532b-4538-a8bc-95c2540bcbb6	TxStartedEvent	2018-07-30 10:02:26.68	DONE	3

图 24-8 TXTIMEOUT 表记录

场景 3：alpha server 挂掉

如果 alpha server 挂掉，则请求直接返回 500，SAGA 事务无法进行下去，因此生产上需要保证 alpha server 的高可用。另外 omega 会对 alpha server 自动进行断开重连，只要 alpha server 出故障之后可以迅速恢复，则事务参与者无须再次重启进行连接。

24.4 本章小结

本章首先简述了单体架构数据库 ACID 的特性、XA 事务模型以及 BASE 定理，然后梳理了 JTA、TCC 及 SAGA 三种分布式事务的模式，最后通过三个案例分别演示了如何使用 JTA、TCC 以及 SAGA 模式来解决分布式事务的一致性问题。通过本章的学习，相信读者对分布式事物处理就能有一个基本的认知了。由于时间及精力限制，本章并没有对相关分布式事务管理组件进行源码解析，有兴趣的读者可以在本章的基础上继续深入研究。

第 25 章

Spring Cloud 与领域驱动实践

本书前面的章节主要从实战的角度介绍了 Spring Cloud 中常用的核心组件，以及一些扩展和坑点。但是 Spring Cloud 组件很多，更像是一套中间件体系，是实现微服务架构的基础实施。不同行业和不同业务，对应着不同复杂度的业务架构。不管是由于技术债堆积的臃肿系统、直接做到 Dao 层的三层架构，还是创业初期为了快速试错而经历过野蛮生长的业务系统，都有一些共性：复杂、bug 居多、新人难以上手、线上故障频发、恶性循环。针对这种业务架构需要进行架构治理。从本质来看，复杂度的治理其实就是想办法控制程序员的随心所欲。本章基于微服务架构设计的思想，结合领域驱动模型及 Spring Cloud 推出方法论，用应用框架去指导实际架构治理，为企业 IT 架构变革稳健发展保驾护航。

25.1 领域驱动概述

微服务系统的设计自然离不开 DDD（Domain-Driven Design，领域驱动设计），它由 Eric Evans 提出，是一种全新的系统设计和建模方法。DDD 事实上是针对面向对象分析和设计的一个扩展和延伸，对技术架构进行了分层规划，同时对每个类进行了策略和类型的划分。领域模型是领域驱动的核心。领域模型通过聚合（Aggregate）组织在一起，聚合间有明显的业务边界，这些边界将领域划分为一个个限界上下文（Bounded Context）。采用 DDD 的设计思想，业务逻辑不再集中在几个大型的类上，而是由大量相对小的领域对象（类）组成，这些类具备自己的状态和行为，每个类是相对完整的独立体，并与现实领域的业务对象映射。领域模型就是由许多这样的细粒度的类组成的。基于领域驱动的设计，保证了系统的可维护性、可扩展性和可复用性，在处理复杂业务逻辑方面有着先天的优势。

25.1.1 Spring Cloud 与领域驱动

在微服务（MicroServices）架构实践中，大量借用了 DDD 中的概念和技术，比如一个微服务应该对应 DDD 中的一个限界上下文（Bounded Context）；在微服务设计中应该首先识别出 DDD 中的聚合根（Aggregate Root）；还有在微服务之间集成时应该采用 DDD 中的防腐层（Anti-Corruption Layer, ACL）。我们甚至可以说 DDD 和微服务有着天生的默契。

Spring Cloud 基于 Spring Boot 提供了一套完善的微服务解决方案，作为微服务架构的基础设施，快速帮助企业开发者搭建微服务架构。更准确地来说，Spring Cloud 更像是一个中间件，解决了框架层面的问题。但是业务怎么开发？业务架构怎么治理？架构怎么防腐？怎么解决应用架构的复杂性？这些问题都需要方法论去指导实践。

在微服务架构落地的过程中，Spring Cloud 和领域驱动相辅相成，Spring Cloud 解决架构分布式等问题，领域驱动作为业务治理和架构防腐的方法论，两者并驾齐驱，为企业 IT 架构变革微服务改造保驾护航。

25.1.2 为什么需要领域建模

领域模型有助于团队创建一个业务部门与 IT 部门都能理解的通用模型，并用该模型来沟通业务需求、数据实体、过程模型。模型是模块化、可扩展、易于维护的，同时设计还反映了业务模型，提高了业务领域对象的可重用性和可测性。反过来，如果 IT 团队在开发大中型企业软件应用时不遵循领域模型方法，不投放资源去建立和开发领域模型，会导致应用架构出现"肥服务层"和"贫血的领域模型"，在这样的架构中，会积聚越来越多的业务逻辑。我们希望领域对象能够准确地表达出业务意图，但是多数时候，我们看到的却是充满 getter 和 setter 的领域对象。此时的领域对象已经不是领域对象了，它们只是个数据载体，也就是 Martin Fowler 所说的贫血对象。这种做法会导致领域特定业务逻辑分散在一堆 service 层中，软件架构随业务开发常年累积野蛮生长，从而腐败，无法维护。

领域驱动设计告诉我们，在通过软件实现一个业务系统时，建立一个领域模型是非常重要和必要的，因为领域模型具有以下特点：

- ❑ 领域模型是对具有某个边界的领域的一个抽象，反映了领域内用户业务需求的本质；领域模型是有边界的，只反映了我们在领域内所关注的部分。
- ❑ 领域模型只反映业务，和任何技术实现无关；领域模型不仅能反映领域中的一些实体概念，如货物、书本、应聘记录、地址等；还能反映领域中的一些过程概念，如资金转账等。
- ❑ 领域模型确保了我们的软件业务逻辑都在一个模型中，这样对提高软件的可维护性、业务可理解性以及可重用性都有帮助。
- ❑ 领域模型能够帮助开发人员相对平滑地将领域知识转化为软件构造。
- ❑ 领域模型贯穿软件分析、设计及开发的整个过程；领域专家、设计人员、开发人员通过领域模型进行交流，彼此共享知识与信息；因为大家面向的都是同一个模型，所以可以防止需求走样，可以让软件设计开发人员做出来的软件真正满足需求。
- ❑ 要建立正确的领域模型并不简单，需要领域专家、设计人员、开发人员积极沟通共同努力，然后才能使大家对领域的认识不断深入，从而不断细化和完善领域模型。

- 为了让领域模型看得见，我们需要用一些方法来表示它；图是表达领域模型最常用的方式，但不是唯一的方式，代码或文字描述也能表达领域模型。
- 领域模型是整个软件的核心，是软件中最有价值和最具竞争力的部分；设计足够精良且符合业务需求的领域模型能够更快速地响应需求的变化。

25.2 领域驱动核心概念

25.2.1 实体概述

实体（Entity）是领域中需要唯一标识的领域概念，因为我们有时需要区分是哪个实体。如果有两个实体，且唯一标识不一样，那么即便实体的其他所有属性都一样，我们也认为它们是不同的实体；因为实体有生命周期，实体被创建后可能会被持久化到数据库，然后某个时候又会被取出来。所以，如果我们不为实体定义一种可以唯一区分的标识，那我们就无法区分到底是这个实体还是那个实体。

另外，不应该给实体定义太多的属性或行为，而应该寻找关联，发现其他一些实体或值对象，将属性或行为转移到其他关联的实体或值对象上。比如 Customer 实体，它有一些地址信息，由于地址信息是一个完整的有业务含义的概念，所以我们可以定义一个 Address 对象，然后把 Customer 中与地址相关的信息转移到 Address 对象上。如果没有 Address 对象，而把这些地址信息直接放在 Customer 对象上，并且把其他类似 Address 的信息也都直接放在 Customer 上，会导致 Customer 对象很混乱，结构不清晰，最终导致它难以维护和理解。

25.2.2 值对象概述

在领域中，并不是每一个事物都必须有一个唯一标识，也就是说我们不关心对象是哪个，只关心对象是什么。就以上面的地址对象 Address 为例，如果有两个 Customer 的地址信息是一样的，我们就会认为这两个 Customer 的地址是同一个。也就是说只要地址信息一样，我们就认为是同一个地址。

用程序的方式来表达就是，如果两个对象的所有属性的值都相同，我们会认为它们是同一个对象，那么我们就可以把这种对象设计为值对象（Value Object）。因此，值对象没有唯一标识，这是它和实体的最大不同。另外值对象在判断是否是同一个对象时是通过它们的所有属性是否相同实现的，如果相同则认为是同一个值对象；而我们在区分是否是同一个实体时，只看实体的唯一标识是否相同，不管实体的属性是否相同。值对象另外一个明显的特征是不可变，即所有属性都是只读的。因为属性是只读的，所以可以被安全共享。当共享值对象时，一般有复制和共享两种做法，具体采用哪种做法还要根据实际情况而定。另外，我们应该将值对象设计得尽量简单，不要让它引用很多其他对象，因为它只是一个值。实体和值对象的对比，如表25-1所示。

表 25-1　实体与值对象对比

	概　　念	区　　别	举例说明
实体	实体表示那些具有生命周期并且会在其生命周期中发生改变的东西	实体是有唯一标识的，只要唯一标识不同就是两个不同的实体	央行发行了一些 100 元的钞票，每张钞票都有唯一识别，此时两张 100 元的钞票就是不同的实体
值对象	值对象则表示起描述性作用的并且可以相互替换的概念	值对象是没有唯一标识的，只是数据传输的载体。用描述性属性字段来实现	我们花了 100 元买了一本书，我们只是关心货币的数量，而不会关心具体使用了哪一张 100 元的钞票，即两张 100 元的钞票是可以互换的，此时的钞票就是值对象

25.2.3　领域服务

领域中的一些概念不适合建模为对象，即不适合归类到实体对象或值对象，因为它们本质上就是一些操作或动作，而不是实物。这些操作或动作往往会涉及多个领域的对象，并且需要协调这些领域对象共同完成这个操作或动作。如果强行将这些操作职责分配给任何一个对象，则被分配的对象就会承担一些不该承担的职责，从而会导致对象的职责不明确。但是基于类的面向对象的语言规定，任何属性或行为都必须放在对象里面。所以我们需要寻找一种新的模式来表示这种跨多个对象的操作，DDD 认为服务是一个很自然的范式，可用来对应这种跨多个对象的操作，所以就有了领域服务（Domain Service）这个模式。领域服务本来就是来处理这种场景的。比如要对密码进行解密，可以创建一个 PasswordService 来专门处理加解密的问题。

领域服务还有一个很重要的功能，就是可以避免领域逻辑泄露到应用层。因为如果没有领域服务，那么应用层会直接调用领域对象完成本该属于领域服务做的操作，这样一来，领域层可能会把一部分领域泄露到应用层。因此，引入领域服务可以有效防止领域层的逻辑泄露到应用层。对于应用层，从可理解的角度来讲，通过调用领域服务提供的简单、易懂、明确的接口肯定要比直接操纵领域对象容易得多。

那如何去识别领域服务呢？主要看它是否满足以下三个特征：
- ❑ 服务执行的操作代表了一个领域概念，这个领域概念无法自然隶属于一个实体或者值对象。
- ❑ 被执行的操作涉及领域中的其他的对象。
- ❑ 操作是无状态的。

25.2.4　聚合及聚合根

聚合通过定义对象之间清晰的所属关系和边界来实现领域模型的内聚，并避免了错综复杂的、难以维护的对象关系网的形成。聚合定义了一组具有内聚关系的相关对象的集合，我们把聚合看作一个修改数据的单元。聚合中所包含的对象之间具有密不可分的联系，它们是内聚在一起的。比如一辆汽车（Car）包含了引擎（Engine）、车轮（Wheel）和油箱（Tank）等组件，缺一不可。一个聚合中可以包含多个实体和值对象，因此聚合也被称为根实体。图 25-1 所示就是一个聚合，Customer 是聚合根也是实体，address 是值对象，ContactInfo 也是值对象。

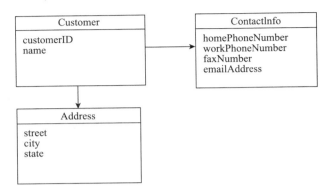

图 25-1　示例聚合

聚合根（Aggregate Root）是 DDD 中的一个概念，是一种更大范围的封装，其把一组有相同生命周期、在业务上不可分隔的实体和值对象放在一起考虑，只有根实体可以对外暴露引用，也是一种内聚性的表现。但是要确定聚合边界要满足固定规则（Invariant），也就是在数据变化时必须保持一致性规则，具体规则如下：

- 根实体具有全局标识，最终负责检查规定规则。
- 聚合内的实体具有本地标识，这些标识在 Aggregate 内部才是唯一的。
- 外部对象不能引用除根 Entity 之外的任何内部对象。
- 只有 Aggregate 的根 Entity 才能直接通过数据库查询获取，其他对象必须通过遍历关联来发现。
- Aggegate 内部的对象可以保持对其他 Aggregate 根的引用。
- 对 Aggregate 边界内的任何对象进行修改时，整个 Aggregate 的所有固定规则都必须满足。

25.2.5　边界上下文

领域实体是有边界上下文的，系统获取的数据是有界上下文（Bounded Context）下的数据。边界上下文（Bounded Context）在 DDD 里面是一个非常重要的概念，Bounded Context 明确限定了模型的应用范围。在 Context 中，要保证模型在逻辑上统一，而不用考虑它是不是适用于边界之外的情况。在其他 Context 中，会使用其他模型，这些模型具有不同的术语、概念、规则和 Ubiquitous Language。那么不同 Context 下的业务要互相通信怎么办？这就涉及跨边界的集成了，集成不能是简单的 RPC 服务调用，而需要一个专门的防腐层（Anti-Corruption）做转化。防腐层主要是对外部依赖解耦，以及避免外部领域概念污染 Context 内部实体语义。以我们真实的业务场景举个例子，比如会员这个概念在 ICBU 网站上指网站上的买主，但是在 CRM 领域中指客户，虽然很多的属性都是一样的，但是二者在不同的 Context 下其语义和概念是有差别的，我们需要用防腐层做一下转换，如图 25-2 所示。

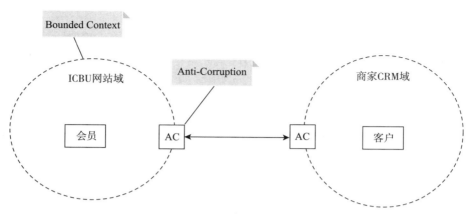

图 25-2 防腐层转换

25.2.6 工厂

DDD 中的工厂（Factory）也是一种体现封装思想的模式。DDD 中引入工厂模式的原因是：有时创建一个领域对象是一件比较复杂的事情，而不是仅仅进行简单的 new 操作就可以。正如对象封装了内部实现一样（我们无须知道对象的内部实现就可以使用对象的行为），工厂则用来封装创建一个复杂对象的操作。工厂的作用是将创建对象的细节隐藏起来。

工厂在创建一个复杂的领域对象时，通常会知道该满足什么业务规则（它知道先怎样实例化一个对象，然后对这个对象做哪些初始化操作，这些规则就是创建对象的细节），如果传递进来的参数符合创建对象的业务规则，则可以顺利创建相应的对象；但是如果由于参数无效等不能创建出期望的对象，则应该抛出一个异常，以确保不会创建出一个错误的对象。

当然也不是所有都需要通过工厂来创建对象，当构造器很简单或者构造对象不依赖于其他对象来创建时，我们只需要简单地使用构造函数创建对象就可以。隐藏创建对象的好处是显而易见的，这样可以不让领域层的业务逻辑泄露到应用层，同时也减轻了应用层的负担，它只需要简单地调用领域工厂创建符合期望的对象即可。

25.2.7 仓储 / 资源库

领域模型中的对象自从被创建出来后不会一直在内存中活动，当它不活动时会被持久化到数据库中，然后当需要的时候我们会重建该对象。重建对象就是根据数据库中已存储的对象的状态重新创建对象。所以重建对象是一个和数据库打交道的过程。从更广义的角度来理解，我们经常会像集合一样从某个类似集合的地方根据某个条件获取一个或一些对象，往集合中添加对象或移除对象。也就是说，我们需要提供一种机制，可以提供类似集合的接口来帮助我们管理对象。仓储（Repository）就是基于这样的思想被设计出来的。

仓储里面存放的对象一定是聚合，原因是领域模型中是以聚合的概念去划分边界的。聚合是我们更新对象的一个边界，事实上我们把整个聚合看成一个整体概念，要么一起被取出来，要么一起被删除。我们永远不会单独对某个聚合内的子对象进行单独查询或做更新操作。因此,我们只为聚合设计仓储。

仓储还有一个重要的特征就是分为仓储定义部分和仓储实现部分,在领域模型中我们定义仓储的接口,而在基础设施层实现具体的仓储。这样设计的原因是:仓储背后的实现都是在和数据库打交道,但是我们又不希望调用方(如应用层)把重点放在如何从数据库获取数据的问题上,因为这样做会导致调用方(应用层)代码混乱,很可能会因此而忽略了领域模型的存在。所以我们需要提供一个简单明了的接口供调用方使用,确保客户能以最简单的方式获取领域对象,从而可以让它在不被数据访问代码打扰的情况下协调领域对象以完成业务逻辑。这种通过接口来隔离封装变化的做法其实很常见。由于对外暴露的是抽象的接口并不是具体的实现,所以可以随时替换仓储的真实实现。

25.2.8 CQRS 架构

CQRS 的核心思想是将应用程序的查询部分和命令部分完全分离,这两部分可以用完全不同的模型和技术去实现。比如命令部分可以通过领域驱动设计来实现;查询部分可以直接用最快的非面向对象的方式来实现,比如用 SQL。这样的思想有很多好处:

- 实现命令部分的领域模型,不用经常为了考虑领域对象可能会被如何查询而做一些折中处理。
- 由于命令和查询是完全分离的,所以这两部分可以用不同的技术架构实现,包括数据库设计理论上都可以分开设计,每一部分可以充分发挥其长处。
- 因为命令端没有返回值,所以可以像消息队列一样接受命令,放在队列中,慢慢处理;处理完后,可以通过异步的方式通知查询端,这样查询端可以做数据同步的处理。

CQRS 架构的优缺点如表 25-2 所示。

表 25-2 CQRS 架构优缺点

架构	优点	缺点
CQRS	Command 和 Query 两端架构分离、相互不受束缚,各自独立设计、扩展; Command 端通常结合 DDD,解决复杂的业务逻辑; Query 端轻量级查询,多种不同的查询视图通过订阅事件来更新; Command 端通过分布式消息队列水平扩展,天然支持削峰技术架构,业务代码完全分离,程序员不用关心技术问题,更方便的分工合作	不是强一致性,而是面向最终一致性;强依赖高性能可靠的分布式消息队列;必须有强大可靠的 CQRS 框架,从头做起成本高、风险大;必须结合 Event Sourcing 模式,否则 CQ 分离意义不大,CQRS 的最佳原则提高了开发人员的门槛

25.2.9 领域事件

领域事件(Domain Event)是最近几年才加入 DDD 生态系统的,通过领域事件的方式达到各个组件之间的数据一致性。领域事件的额外好处在于它可以记录发生在软件系统中的所有重要修改,这样可以很好地支持程序调试和商业智能化。在 CQRS 架构的软件系统中,领域事件还用于写模型和读模型之间的数据同步。再进一步发展,事件驱动架构可以演变成事件源

（Event Sourcing），即对聚合的获取并不是通过加载数据库中的瞬时状态实现的，而是通过重放发生在聚合生命周期中的所有领域事件完成的。

事件溯源（Event Sourcing）是基于 DDD 设计的，对于聚合，不保存聚合的当前状态，而是保存对象上所发生的每个事件。当要重建一个聚合对象时，可以通过回溯这些事件（即让这些事件重新发生）来让对象恢复到某个特定的状态；因为有时一个聚合可能会发生很多事件，所以如果每次要在重建对象时都从头回溯事件，会导致性能低下，所以我们会在一定时候为聚合创建一个快照。这样，我们就可以基于某个快照开始创建聚合对象了。

25.2.10 领域驱动模型的设计步骤

领域驱动模型的设计步骤如下：

（1）根据需求建立一个初步的领域模型，识别出一些明显的领域概念及它们之间的关联，关联可以暂时没有方向但需要有一对一、一对多、多对多这些关系。可以用文字精确且没有歧义地描述出每个领域概念的涵义及包含的主要信息。

（2）分析主要的软件应用程序功能，识别出主要的应用层的类，这样有助于及早发现哪些是应用层的职责，哪些是领域层的职责。

（3）进一步分析领域模型，识别出哪些是实体，哪些是值对象，哪些是领域服务。

（4）分析关联，通过对业务进行更深入分析及各种软件设计原则、性能方面的权衡，明确关联的方向或者去掉一些不需要的关联。

（5）找出聚合边界及聚合根，这是一件很有难度的事情，因为在分析的过程中往往会碰到很多难以清晰判断的问题，此时需要我们凭借经验找出正确的聚合根。

（6）为聚合根配备仓储，一般情况下是为一个聚合分配一个仓储，此时只要设计好仓储的接口即可。

（7）捋顺实际业务应用场景，确定我们设计的领域模型能够有效解决业务需求。

（8）考虑如何创建领域实体或值对象，明确是通过工厂还是直接通过构造函数实现。

虽然上面介绍了设计领域模型的步骤，但是领域建模是一个不断重构、持续完善模型的过程。大家会在讨论中将变化的部分反映到模型中，从而使模型不断细化并朝正确的方向走。领域建模是领域专家、设计人员、开发人员之间沟通交流的过程，是大家工作和思考问题的基础。

25.3 Halo 框架的设计

25.3.1 DDD 应用框架的意义

工欲善其事，必先利其器。DDD 已经抽象出了维护领域模型的方法论、分层方式及各个组件的职责。理论是有了，但是在落地到实践的代码编写中，特别是多人协作开发时，会出现风格各异的情况。于是我们需要维护模型的干净，约束开发的行为，使模型对应到 DDD 中的职责显性化（具体的来讲你需要清楚这个类是聚合根还是仓储）。

拥有一个看上去正确的模型不代表模型能被直接转换成代码，或者它的实现可能会违背某些软件设计原则。那么我们该如何实现从模型到代码的转换，并让代码具有可扩展性、可维护性和高性能等指标呢？另外，如实反映领域的模型可能会导致对象持久化问题，或者引起不可接受的性能问题，那么我们应该怎么做呢？我们应该紧密关联领域建模和设计，将领域模型和软件编码实现捆绑在一起，在模型构建时就考虑到软件和设计，并通过开发满足需要的 DDD 应用框架去解决开发效率低下、开发规范不统一的问题。

综合上所述，我们需要设计一个 DDD 框架，做到架构即约定和防腐层架构防腐。下面将详细介绍如何设计开发一个 DDD 应用框架。

25.3.2　领域驱动框架现状

自从 Eric Evan 提出 DDD 领域驱动设计以来已经过了很多年了，现在已经有很多人在学习或实践 DDD。但是目前来看能够支持 DDD 开发的框架并不多，至少在国内比较罕见。在 Java 平台上，国外比较受欢迎的领域驱动框架是 Axon Framework（https://github.com/AxonFramework/AxonFramework），该框架发展至今相对来说比较活跃，目前 Github 上星标已经超过 1000。还有就是 banq 的 Jdon framework（https://github.com/banq/jdonframework），这是基于 DDD+CQRS+EventSourcing 的开发，也是基于 Java 平台的。对比分析归纳如表 25-3 所示。

表 25-3　领域驱动框架对比

框　　架	平台	星标	文档	活跃程度
Jdon framework	Java	559	比较少	一般
Axon Framework	Java	1194	中等	活跃
enode	.Net	1163	比较少	一般

上表中列举的领域驱动框架各有优点和缺点。如果就 Java 平台来讲，可以尝试使用 Axon Framework，目前已经支持 Spring Cloud。但是它不是目前最好的领域驱动框架，下面将介绍的 Halo 框架会更有优势。

25.3.3　Halo 框架概述

Halo 框架是基于领域驱动 +CQRS+ 扩展点 + 流程编排的应用框架，致力于采用领域驱动的设计思想，规范控制程序员的随心所欲，从而解决软件的复杂性问题。架构设计原则非常简单，即在高内聚、低耦合、可扩展、易理解的大的指导思想下，尽可能贯彻面向对象的设计思想和原则。架构如图 25-3 所示。

Halo 框架目前已经对外开源，目前托管于 github 上，github 地址为 https://github.com/softwareking/halo。

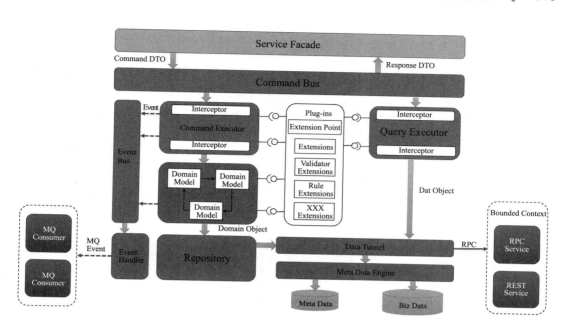

图 25-3　Halo 架构图

25.3.4　Halo 框架分层设计

一个经典的基于领域驱动设计的应用，分层结构如图 25-4 所示。

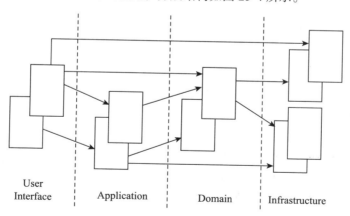

图 25-4　领域驱动分层结构图

由图 25-4 可知：

- 用户界面/展现层：负责向用户展现信息及解释用户命令。更细的方面来讲就是请求应用层以获取用户需要展现的数据；发送命令给应用层要求其执行某个用户命令。
- 应用层（Application Layer）：该层的职责是为展现层提供各种应用功能（包括查询或命令），对内调用领域层（领域对象或领域服务）完成各种业务逻辑，应用层不包含业务逻辑。应用层采用命令查询职责分离（Command Query Responsibility Separation，CQRS）的设计思想，将查询和命令的实现分离，这样可以对这两部分进行单独设计。对于查询 Request，会委托给一个专门负责查询的服务去完成；对于命令的 Request，会委托给一个专门负责处理命令的领域服务去完成。
- 领域层：负责表达业务概念、业务状态信息及业务规则，领域模型处于这一层，是业务软件的核心。领域层会包含领域服务、实体、值对象、聚合（根）、工厂、仓储。
- 基础设施层：本层为其他层提供通用的技术能力，提供层间的通信，为领域层实现持久化机制。总之，基础设施层可以通过架构和框架来支持其他层的技术需求。

Halo 框架在实际应用中也划分为三个大的层次，分别是 App 层、Domain 层和 Infrastructure 层，如图 25-5 所示。

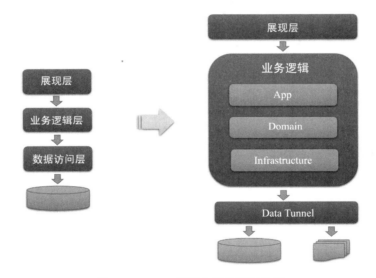

图 25-5　Halo 框架应用分层结构图

由上图可知：
- App 层主要负责获取输入、组装 context、做输入校验、发送消息给领域层做业务处理、监听确认消息，以及如果需要的话使用 MQ 进行消息通知。
- Domain 层主要是通过领域服务（Domain Service）及领域对象（Domain Object）的交互，对上层提供业务逻辑处理，然后调用下层 Repository 做持久化处理。
- Infrastructure 层主要负责数据的 CRUD 操作，在这一层抽取了数据通道（Tunnel）的概念，通过 Tunnel 的抽象概念来屏蔽具体的数据来源。数据来源可以是 MySQL、NoSQL、Search、RPC 服务或者 Spring Cloud 提供的 Rest 服务等。

在判断所属分层的时候，需要注意的是，决定一个服务应该归属哪一层是很困难的。总地来说，涉及重要领域概念的行为应该放在 Domain 层，而其他非领域逻辑的技术代码放在 App 层，例如参数的解析、上下文的组装调用领域服务等。

由于分层设计及防腐层的需要，会在各种类型的 DTO 之间相互转换。在 Halo 框架中涉及三种 DTO 之间的转换，分别是 Client Object、Data Object 和 Domain Object 之间的转换，如图 25-6 所示。

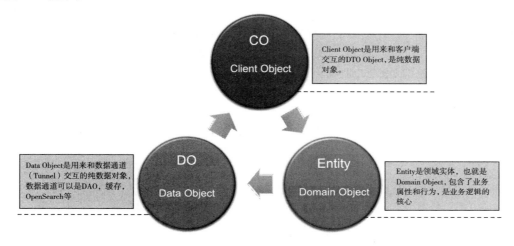

图 25-6　数据转换

25.3.5　Halo 框架中的 CQRS 设计

命令查询职责分离（Command Query Responsibility Segregation，CQRS）是一种应用架构模式，它会将应用分为两部分：查询部分（查看模型）和命令部分（写入模型）。每一部分都负责处理特定的操作集。CQRS 概念最初是由 Greg Young 提出和积极推动的。它是 CQS（命令 - 查询分离）理念的自然延伸、CQS 理念由 Bertrand Meyers 提出，主张将方法分为命令和查询。CQRS 使用了相同的原则，不过其扩大到了整个系统中。按照这种架构，业务逻辑层的两个组件将会相互独立运行。因此，读取模型（Read Model）将会处理用户的查询——在处理能力方面的要求会更少一些；而写入模型（Write Model）将会经历一个很长的处理路径，包括校验、队列、消息以及用户命令要执行的业务规则处理。

在常用的三层架构中，通常都是通过数据访问层来修改或者查询数据的，一般修改和查询使用的是相同的实体。这在一些业务逻辑简单的系统中可能没有什么问题，但是随着系统逻辑变得复杂，用户增多，这种设计就会出现一些性能问题。虽然在 DB 上可以做一些读写分离的设计，但在业务上如果把读写混合在一起的话，仍然会出现一些问题。

命令查询职责分离模式（Command Query Responsibility Segregation，CQRS）是一种架构体系模式，该模式从业务上分离修改（Command，增、删、改，会对系统状态进行修改）和查询（Query，查，不会对系统状态进行修改）的行为，从而使得逻辑更加清晰，便于对不同部分进行针对性优化。

这样的设计思想有很多好处，比如，实现命令部分的领域建模不用为了顾及领域对象可能会被如何查询而做一些折中处理，由于命令和查询是完全分离的，所以这两部分可以用不同的技术和架构实现，包括数据库设计都可以分开设计，每一部分可以充分发挥其长处。命令端因为没有返回值，可以像消息队列一样接受命令，放在队列中，慢慢处理；处理完后，可以通过异步的方式通知查询端，这样查询端可以做到数据同步处理。

从纵向来看，如图 25-7 所示，可以看出一切的业务动作皆为命令，一切界面的查询皆为 Query。UI 层发起一次操作命令，命令被发送到 Command Bus（命令总线），总线找到对应的 command handler，handler 封装了 UI 请求所需要处理的业务。接下来 command handler 会协调各个粒度的聚合来完成聚合的加载、聚合状态的改变及聚合的持久化（落库）。同时 Query 组件直接面对 DB，做最优化的查询方式，不再受模型所限。

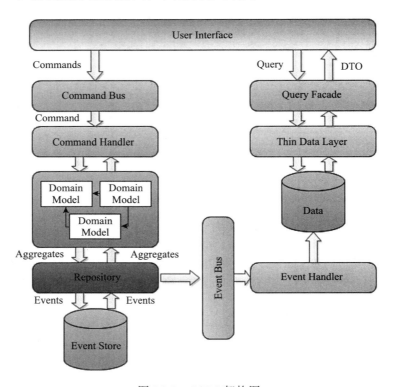

图 25-7　CQRS 架构图

到 domain 层面上，值得关注的是 Event Bus，从实现层面看，与上面的 Command Bus 是一样的；从职责方面来看，Event Bus 接收的就是事件，同样最终会将事件转化为具体的 Event 的 handler 来做具体的事情。

25.3.6　Command 与 Command Bus 设计

1. Command 命令

什么是命令（Command）？用户界面 / 展现层的读取或者写入操作都可被封装为一个命令，Command 中不会带有具体的业务逻辑。Command 的实现涉及三个东西，分别是命令对象

(Command Object)、命令执行器（CommandExecutor）和命令总线（Command Bus）。后面会做详细介绍。

在给 Command 类命名的时候，由于 Command 表示的是想要执行的命令，所以 Command 类的名称应当是动词的形式。例如 RegisterCommand、ChangePasswordCommand 等。其中的 Command 后缀是可选的，只要在系统中统一规范命名即可。

在实际项目中，我们需要注意 Command 的类名的重要作用，每个 Command 类的名称都清晰地表达了一个意图，例如 ChangePasswordCommand 清晰地表达了这个命令是要修改密码，所以千万不要随意复用 Command，这里的"复用"指的是看到某两个 Command 中有完全一样的属性，为了减少几行代码就觉得没有必要使用两个 Command，而把它们合并成一个 Command。这样的"复用"会让系统变得越来越难以理解。

2. Command 对象

Command 对象的作用是用来封装命令数据，所以这类对象以属性为主，包括少量简单的方法。但注意，这些方法中不能包含业务逻辑。在 Halo 框架中我们高度抽象了一个 Command 对象，其继承自 DTO，org.xujin.halo.dto.DTO 主要实现了序列化接口，相关代码如下所示：

```
public class DTO implements Serializable{
    private static final long serialVersionUID = 1L;
}
```

org.xujin.halo.dto.Command 继承自 DTO 并增加了 Command 对象需要的基本属性 set 和 get 方法，相关代码如下所示：

```
public abstract class Command extends DTO{
    private static final long serialVersionUID = 1L;
    /**
     * 不需要操作人设置，此值为true
     */
    private boolean operaterIsNotNeeded = false;
    /**
     * command的操作人
     */
    private String operater;
    public String getOperater() {
        return operater;
    }
    public void setOperater(String operater) {
        this.operater = operater;
    }
    public boolean isOperaterIsNotNeeded() {
        return operaterIsNotNeeded;
    }
    public void setOperaterIsNotNeeded(boolean operaterIsNotNeeded) {
        this.operaterIsNotNeeded = operaterIsNotNeeded;
    }
}
```

3. Command Bus

Command Bus 就是一个命令的执行总线。执行 Command 的是命令执行器，但 Command-

Executor 的调用不是通过在用户界面层直接依赖注入调用，用户界面层会把请求转变为 Command 对象，然后放入 CommandBus 中即可实现调用。

　　Command Bus 的作用是将一个 Command 分发给对应的 CommandExecutor 去执行。CommandBus 的出现使开发用户界面层时不再需要关心 Command 会被哪个 Executor 执行。Commandbus 屏蔽了底层的细节，让分工协作开发更加独立明确。下面我看一下 Halo 框架中的命令总线声明接口的例子，如下代码所示：

```java
public interface CommandBusI {
    /**
     * Send command to CommandBus, then the command will be executed by CommandExecutor
     *
     * @param Command or Query
     * @return Response
     */
    public Response send(Command cmd);
}
```

命令总线接口对应的主要实现代码如下所示：

```java
@Component
public class CommandBus implements CommandBusI{
    Logger logger = LoggerFactory.getLogger(CommandBus.class);
    @Autowired
    private CommandHub commandHub;
    @SuppressWarnings("unchecked")
    @Override
    public Response send(Command cmd) {
        Response response = null;
        try {
            //从commandHub中获取对应的命令去调用命令的execute方法
            response = commandHub.getCommandInvocation(cmd.getClass()).invoke(cmd);
        }
        catch (Exception exception) {
            //统一的Command异常处理器
            response = handleException(cmd, response, exception);
        }
        finally {
            //Clean up context
            HaloContext.remove();
        }
        return response;
    }
}
```

4. Command 执行器

　　CommandExecutor 就是一个命令执行器，它的作用是执行一个命令。CommandExecutor 主要完成两部分工作：一是验证传入的 Command 对象是否合法；二是调用领域服务完成相关操作。Command 和 CommandExecutor 是一一对应的，也就是说，一个 Command 只会对应一个 CommandExecutor。

　　命令执行器抽象接口的声明，如下代码所示：

```
public interface CommandExecutorI<R extends Response, C extends Command> {
    public R execute(C cmd);
}
```

下面我们以增加客户的命令执行器为例，看一下命令执行器接口对应的实现，如下代码所示：

```
@Command
public class AddCustomerCmdExe implements CommandExecutorI<Response, AddCustomerCmd> {
    private Logger logger = LoggerFactory.getLogger(AddCustomerCmd.class);
    @Autowired
    private ValidatorExecutor validatorExecutor;
    @Autowired
    private ExtensionExecutor extensionExecutor;
    @Override
    public Response execute(AddCustomerCmd cmd) {
        logger.info("Start processing command:" + cmd);
        validatorExecutor.validate(AddCustomerValidatorExtPt.class, cmd);
        //Convert CO to Entity
        CustomerEntity customerEntity = extensionExecutor.execute(CustomerConvertorExtPt.
            class, extension -> extension.clientToEntity(cmd.getCustomerCO()));
        //Call Domain Entity for business logic processing
        logger.info("Call Domain Entity for business logic processing..."+customerEntity);
        customerEntity.addNewCustomer();
        logger.info("End processing command:" + cmd);
        return Response.buildSuccess();
    }
}
```

5. Command 执行结果

命令执行器会执行命令，经过一系列的命令处理之后，会返回命令处理结果。Halo 框架为了统一规范，对返回命令处理结果进行封装，目前支持不带返回值的 Response、单值的 Response、多值的 Response 和带分页结果的 Response。可以查看源码文件，如图 25-8 所示。

25.3.7　Event 与 Event Bus 设计

针对领域驱动中的事件来说，Halo 框架对事件的设计分别是 EventHub（事件中枢）、EventBus（事件总线）、EventHandlerI（事件处理器抽象接口）。

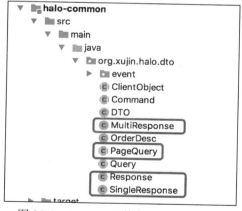

图 25-8　Command 执行封装 Response

EventHub 是事件控制中枢，主要用于注册事件，从事件仓库中获取对应的事件处理器，代码如下所示：

```
@Component
public class EventHub {
    @Getter
```

```java
@Setter
private ListMultimap<Class, EventHandlerI> eventRepository = ArrayListMultimap.
    create();

@Getter
private Map<Class, Class> responseRepository = new HashMap<>();

public List<EventHandlerI> getEventHandler(Class eventClass) {
    List<EventHandlerI> eventHandlerIList = findHandler(eventClass);
    if (eventHandlerIList == null || eventHandlerIList.size() == 0) {
        throw new InfraException(eventClass + "is not registered in
            eventHub, please register first");
    }
    return eventHandlerIList;
}

/**
 * 注册事件
 * @param eventClz
 * @param executor
 */
public void register(Class<? extends Event> eventClz, EventHandlerI
    executor){
    eventRepository.put(eventClz, executor);
}

private List<EventHandlerI> findHandler(Class<? extends Event> eventClass){
    List<EventHandlerI> eventHandlerIList = null;
    Class cls = eventClass;
    eventHandlerIList = eventRepository.get(cls);
    return eventHandlerIList;
}

}
```

Event Bus 事件总线主要是从 Event Hub 中获取对应的事件处理器，主要代码如下所示：

```java
@Component
public class EventBus implements EventBusI{
    Logger logger = LoggerFactory.getLogger(EventBus.class);
    /**
     * 默认线程池
     *     如果处理器无定制线程池，则使用此默认设置
     */
    ExecutorService defaultExecutor = new ThreadPoolExecutor(Runtime.
        getRuntime().availableProcessors() + 1,
Runtime.getRuntime().availableProcessors() + 1,
0L,TimeUnit.MILLISECONDS,
        new LinkedBlockingQueue<Runnable>(1000),new ThreadFactoryBuilder().
            setNameFormat("event-bus-pool-%d").build());

    @Autowired
    private EventHub eventHub;
    @SuppressWarnings("unchecked")
    @Override
    public Response fire(Event event) {
        Response response = null;
```

```
            EventHandlerI eventHandlerI = null;
            try {
                eventHandlerI = eventHub.getEventHandler(event.getClass()).get(0);
                response = eventHandlerI.execute(event);
            }catch (Exception exception) {
                response = handleException(eventHandlerI, response, exception);
            }
            return response;
        }

        @Override
        public void fireAll(Event event){
    eventHub.getEventHandler(event.getClass()).parallelStream().map(p->{
            Response response = null;
            try {
                response = p.execute(event);
            }catch (Exception exception) {
                response = handleException(p, response, exception);
            }
            return response;
        }).collect(Collectors.toList());
    }
}
```

事件处理器主要是执行与事件对应的处理动作，事件处理器的抽象接口代码如下所示：

```
public interface EventHandlerI<R extends Response, E extends Event> {
    default public ExecutorService getExecutor(){
        return null;
    }
    public R execute(E e);
}
```

25.3.8　Extend 扩展点设计

1. 扩展点概述

在电商系统或者大型业务系统中都是按业务域的维度对系统进行划分，比如交易下单系统有十几个域：交付域、支付域、库存域、仓储域、价格域、配送域等。在应用框架层面，域的信息主要用于分类聚合，在应用系统开发中通过手动加注解的方式统一收集信息，在运营平台可以统一管控信息，从业务域的角度去看哪些功能可以扩展定制，或者对其进行增强。域描述定义的注解可以查看源码 org.xuJin.halo.annotation.domain.Domain.Java。

每个功能域都会对外提供一系列的能力，我们一般称之为对外提供的服务，比如库存域提供库存查询、扣减库存等服务。每次交易过程可能跨多个域完成，如果我们想知道一次交易参与其中的每个域的调用顺序，或者域和域之间的依赖关系，我们可以通过注解的方式去定义域提供的服务。域服务定义的注解可以查看源码 org.xuJin.halo.annotation.domain.DomainService.Java。

每个功能域都会对外提供一系列的能力，比如订单域对外提供了订单拆单的能力。而每一个能力往往又是由多个扩展点组成的。第一次接触扩展点，可能大家理解起来比较抽象。这里举个例子帮助大家理解：支付域提供了对外支付的能力，而支付方式有银行卡支付、银联支付、苹果支付、支付宝支付和微信支付等多种第三方支付，而与支付能力对应的多种支付方式就是扩展点。

很多应用架构分层不合理，怎么解决呢？用扩展点的设计思想去解决。什么是扩展点的设计思想？其实设计思想一直在使用，比如经常使用的策略模式，还有就是简单系统和简单场景编写的 if…else 代码。因此基于扩展点的设计思想是应用框架必备的设计思想。在 Halo 框架中扩展点的设计是通过自定义注解的方式实现的，扩展点注解的代码如下所示：

```
@Inherited
@Retention(RetentionPolicy.RUNTIME)
@Documented
@Target({ElementType.TYPE})
@Component
public @interface Extension {

    // 扩展的Id
    String id() default "";

    // 对应的扩展点Id
    String extensionPointId() default "";

    // 扩展对应的描述
    String desc() default "";

    boolean defaultExtension() default false;

    // 一级业务code
    String bizCode()   default CoreConstant.DEFAULT_BIZ_CODE;

    // 可扩展业务code，属于预留，并不是每个系统都有
    String extBizCode() default CoreConstant.DEFAULT_EXT_BIZ_CODE;
}
```

扩展点的问题解决了，那当一个业务流程请求处理进来了，怎么判断它要走哪个扩展？这个时候就需要确定扩展点的注册与发现机制，一般来说我们都是通过业务身份识别去发现并路由到对应扩展的。框架层面主要是解决扩展点的注册和发现机制，以及扩展点的采集。作为业务应用开发方，是怎么从业务角度去制定扩展点的注册与发现机制并使之落地的？下面会介绍扩展点的设计原理和发现机制。

2. 扩展点设计原理

扩展点的设计原理是这样的：所有的扩展点（ExtensionPoint）必须通过接口声明。而扩展点的实现是通过 @Extension 注解的方式标注的，Extension 里面使用 BizCode 和 ExtBizCode 两个属性来标识业务身份，关于什么是业务身份，后面会介绍。其中 BizCode 和 ExtBizCode 为非必填参数。@Extension 注解下面还有其他可选项，主要用于 Halo 管控平台的可视化展示。Halo 框架的 Bootstrap 类会在 Spring 启动的时候做类扫描，进行 Extension 注册，在 Runtime 的时候，通过 HaloContext 来路由发现要使用的 Extension。而 HaloContext 是通过 Interceptor 在调用业务逻辑之前进行初始化的。根据业务身份发现对应扩展点的流程，如图 25-9 所示。

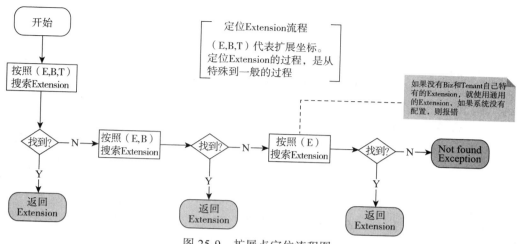

图 25-9　扩展点定位流程图

25.3.9　业务身份设计

业务身份识别在应用中是非常重要的。Halo 框架在进行此设计时是通过自定义 @ Extension 注解的方式，定义了扩展点对应的实现。三维的扩展坐标由 BizCode、ExtBizCod、ExtensionPiont 组成，它们分别是业务身份 Code、可扩展预留的业务身份 Code、扩展点。其实在一般的系统中二维扩展坐标（BizCode,ExtensionPiont）就可以解决。但是在基于云计算 SaaS 模式的多租户系统下，需要三位的扩展坐标解决，在这种场景下与租户 Id 的值对应 Halo 框架中预留了 ExtBizCode。比如 CRM 系统要服务不同的业务方，而且每个业务方又有多个租户，而传统的基于多租户（TenantId）的业务身份识别还不能满足需求，于是在此基础上我们又引入了可扩展的业务码（ExtBizCode）来标识。此时的业务身份实际上是（BizCode，ExtBizCode）二元组。在每一个业务身份下，又可以有多个扩展点（ExtensionPoint），所以一个扩展点对应的实现，实际上是一个三维空间中的向量。借鉴 Maven Coordinate 的概念我给它起了个名字叫扩展坐标（Extension Coordinate），这个坐标可以用（BizCode，ExtBizCode，ExtensionPoint）来唯一标识，如图 25-10 所示。

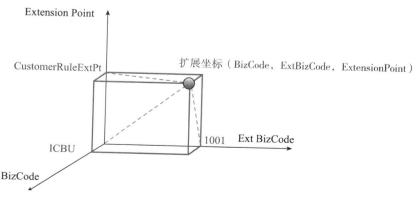

图 25-10　扩展坐标

25.3.10 规范设计

我们进行规范设计主要是为了满足收纳原则的两个约束：

1）对应的 Java 类要放对位置，也就是类不要随便乱放，我们的每一个组件（Module）、每一个包（Package）都有明确的职责定义和范围，不可以放错。例如，extension 包只是放扩展实现的，不允许放其他东西；而 Interceptor 包只是放拦截器的；validator 包只是放校验器的。

2）类名需要按照规范合理命名，例如和数据有关的 Object，主要有 Client Object、Domain Object 和 Data Object，其中 Client Object 是放在二方库中，用来与外部交互使用的 DTO，其命名必须以 CO 结尾；相应的 Data Object 主要是持久层使用的，命名必须以 DO 结尾。这个类名应该是自明的（self-evident），也就是看到类名就知道里面是干了什么事情，这也就反向要求我们的类必须是单一职责（Single Responsibility）的。如果我们的 Class Name 是自明的，Package Name 是自明的，Module Name 也是自明的，那么我们整个应用系统就会很容易被理解，看起来就会很舒服，维护效率会提高很多。推荐的命名规则如表 25-4 所示（仅供参考）。

表 25-4 规范设计

	描述	示例分层	命名规范
Client Request	客户的请求	crm-sales-client	xxxCmd.java
Client Object	客户端对象	crm-sales-client	xxxCo.java
API Service	二方包接口	crm-sales-client	xxxServiceI.java
Command Executor	命令执行器	crm-sales-app	xxxCmdExe.java
Query Executor	查询执行器	crm-sales-app	xxxQueryExe.java
Command Interceptor	命令拦截器	crm-sales-app	xxxInterceptor
Extension Point	扩展点	任意层	xxxExtPt.java
Extension	扩展点的实现	任意层	xxxExt.java
Parameter Validator	参数校验	crm-sales-app	xxxValidator.java
Object Convertor	对象转换器		xxxConvertor.java
Parameter Assembler	参数装配	crm-sales-app	xxxAssembler.java
Domain Object	域对象	crm-sales-domain	xxxE.java
Value Object	值对象	crm-sales-domain	xxxV.java
Domain Factory	域工厂	crm-sales-domain	xxxFactory.java
Repository	仓储	crm-sales-infrastructure	xxxRepository.java
Business Rule	业务规则	crm-sales-domain	xxxRule.java
Domain Service	域服务	crm-sales-domain	xxxDomainService.java
Data Object	数据对象	crm-sales-infrastructure	xxxDO.java
DB Dao	数据操作对象	crm-sales-infrastructure	xxxTunnel.java

25.4 Spring Cloud 与 Halo 实战

25.4.1 事件风暴寻找模型和聚合

领域模型（Domain Model）通过聚合（Aggregate）组织在一起，聚合间有明显的业务边界，这些边界将领域划分为一个个限界上下文（Bounded Context）。理论清楚了，问题来了，怎么来找模型和聚合呢？有一个非常流行的方法就是 Event Storming（事件风暴），它是由 Alberto Brandolini 发明的，经历了 DDD 社区和很多团队的实践，也是一种非常有参与感的团队活动。下面简单介绍一下事件风暴。

Event Storming 是一项团队活动，旨在通过领域事件识别出聚合根，进而划分微服务的限界上下文。在活动中，团队先通过头脑风暴的形式罗列出领域中所有的领域事件，整合之后形成最终的领域事件集合，然后对于每一个事件标注出导致该事件的命令（Command），然后为每个事件标注出命令发起方的角色，命令可以是用户发起，也可以是第三方系统调用或者是定时器触发等。最后对事件进行分类，整理出聚合根以及边界上下文。事件风暴还有一个额外的好处，可以加深参与人员对领域的认识。但是需要注意的一点是：在事件风暴活动中，需要有领域专家在现场进行评估，事件风暴活动的形式如图 25-11 所示。

图 25-11　事件风暴活动

> 提示　Event Storming 是一项非常有创造性的活动，也是一个持续讨论和反复改进的过程，不同的团队关注的核心域（Core Domain）不同，得到的最终结果也会有差异。

图 25-12 所示是对地产 CRM 这个场景使用 Event Storming 探索的示例结果，将边界上下文清晰梳理出来的示例如图 25-13 所示。

图 25-12　事件风暴结构

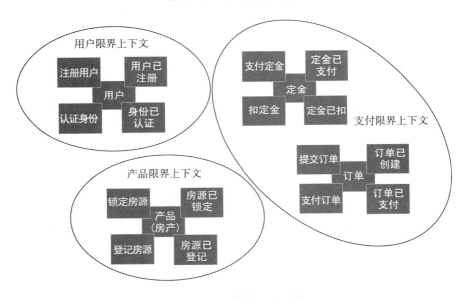

图 25-13　示例边界上下文

25.4.2　Spring Cloud 与 Halo 实战案例

安排 Spring Cloud 与 Halo 框架的综合实战案例的目的主要是对 CRM 中客户这个聚合根进行说明。新建表 25-5 所示的 Maven 工程和模块。详细代码可直接导入 IDE 中。

表 25-5　案例工程模块

模块	说明	备注
crm-sales(ch25)	父模块	管理子模块，依赖版本
crm-sales-start	用户，接口层	注册到 Eureka Server，用于模拟服务提供者

(续)

模　块	说　明	备　注
crm-sales-app	应用层	进行简单的业务检查，基于领域驱动的 CQRS 把 query 和 command 发送到 CommandBus 中
crm-sales-domain	领域层	领域聚合核心业务逻辑处理
crm-sales-infrastructure	基础实施层	包含数据通道层等
crm-sales-client	提供的二方包	通过对 Feign 的封装提供方包
crm-sales-consumer	服务消费者	用于模拟服务消费者
eureka-server	注册中心，读者自己提供	使用在线地址 http://eureka.springcloud.cn/

本案例使用 MySQL 数据库，需要创建数据库和表，表的 SQL 文件如下路径所示：ch25/crm-sales-infrastructure/src/main/resources/customer.sql，创建完数据库和表之后，需要到 ch25/crm-sales-start/src/main/resources/application.yml 中修改数据库的连接信息。

25.4.3　新建二方包工程模块

在微服务开发过程中，服务之间的调用需要提供接口或 SDK，我们统一称为二方包。一方包指的是本项目中的依赖，二方包指的是公司内部其他项目提供的依赖，三方包指的是其他组织、公司等来自第三方的依赖，比如 Spring 提供的 jar。二方包的包结构规划如图 25-14 所示（仅供参考）。

```
├── crm-sales-client
│   └── src
│       └── main
│           └── java
│               └── cn
│                   └── springcloud
│                       └── book
│                           └── crm
│                               └── sales
│                                   ├── api         //二方包接口
│                                   ├── dto         //对应领域驱动框架中的Command对象
│                                   └── clientobject //客户端对象CO
```

图 25-14　二方包结构图

1）新建 Maven 工程 crm-sales-client，核心 Maven 依赖如代码清单 25-1 所示。

代码清单25-1　ch25/crm-sales-client/pom.xml

```xml
<!-- halo框架对Spring Cloud Feign增强的jar-->
<dependency>
    <groupId>org.xujin.halo</groupId>
    <artifactId>halo-feign</artifactId>
    <version>${halo.framework.version}</version>
</dependency>
<!-- spring cloud openfeign -->
<dependency>
    <groupId>org.springframework.cloud</groupId>
    <artifactId>spring-cloud-starter-openfeign</artifactId>
</dependency>
```

2）编写二方包接口 CustomerServiceI，代码如代码清单 25-2 所示。

代码清单25-2　ch25/crm-sales-client/src/main/java/cn/springcloud/book/crm/sales/api/CustomerServiceI.java

```java
@FeignClient("crm-sales-provider")
public interface CustomerServiceI {
    @GetMapping("/add")
    public Response addCustomer(CustomerAddCmd customerAddCmd);
    @GetMapping("/checkConflict")
    public MultiResponse<CustomerCO> checkConflict(CustomerCheckConflictCmd
        customerCheckConflictCmd);

    @GetMapping("/list")
    public MultiResponse<CustomerCO> findByCriteria(CustomerFindByCriteriaQry
        CustomerFindByCriteriaQry);
}
```

3）编写 Command 和 Query 类

CustomerAddCmd 用于说明 CQRS 中的写命令，CustomerFindByCriteriaQry 用于说明读 Query。

CustomerAddCmd 用于增加 Customer，代码如代码清单 25-3 所示。

代码清单25-3　ch25/crm-sales-client/src/main/java/cn/springcloud/book/crm/sales/dto/CustomerAddCmd.java

```java
@Data
public class CustomerAddCmd extends Command{
    private CustomerCO customer;
}
```

说明　CustomerAddCmd 继承了 Halo 框架中的 Command 对象，增加、修改、删除命令都需要继承 Halo 框架中的 Command 对象。

CustomerFindByCriteriaQry 主要用于根据条件查询 Customer，代码如代码清单 25-4 所示。

代码清单25-4　ch25/crm-sales-client/src/main/java/cn/springcloud/book/crm/sales/dto/CustomerFindByCriteriaQry.java

```java
package cn.springcloud.book.crm.sales.dto;
import org.xujin.halo.dto.Query;
public class CustomerFindByCriteriaQry extends Query{
    //目前为空，读者根据自己需要补充

}
```

25.4.4　新建 DDD 基础设施层

在前面的小节中介绍过，领域驱动的基础设施层主要负责数据的 CRUD 操作，在这一层抽取了数据通道（Tunnel）的概念，通过 Tunnel 的抽象概念来屏蔽具体的数据来源，来源可以是 MySQL、NoSQL、Search、RPC 服务或 Spring Cloud 提供的 Rest 服务等，由于数据访问层和通道层相对简单，在这里就不做详细说明，请读者自行阅读源码工程进行学习。

crm-sales-infrastructure 工程和包结构规划如图 25-15 所示。

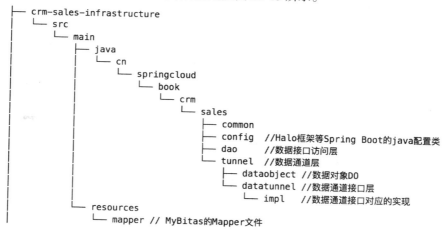

图 25-15　基础设施层

1）新建 Maven 工程模块 crm-sales-infrastructure，主要依赖如代码清单 25-5 所示。

代码清单25-5　ch25/crm-sales-infrastructure/pom.xml

```xml
<!-- Halo Framework -->
<dependency>
    <groupId>org.xujin.halo</groupId>
    <artifactId>halo-core</artifactId>
<!-- Halo Framework End-->
<dependency>
    <groupId>org.mybatis</groupId>
    <artifactId>mybatis</artifactId>
</dependency>

<dependency>
    <groupId>org.mybatis.spring.boot</groupId>
    <artifactId>mybatis-spring-boot-starter</artifactId>
```

```
            <version>1.3.0</version>
    </dependency>
```

2）配置 Halo 框架启动的扫描包的 Java 配置类，如代码清单 25-6 所示。

代码清单25-6　　ch25/crm-sales-infrastructure/src/main/java/cn/springcloud/book/crm/sales/config/HaloConfig.java

```
@Configuration
public class HaloConfig {
    @Bean(initMethod = "init")
    public Bootstrap bootstrap() {
        Bootstrap bootstrap = new Bootstrap();
        List<String> packages = new ArrayList<>();
        packages.add("cn.springcloud.book.crm.sales.command");
        packages.add("cn.springcloud.book.crm.sales.interceptor");
        packages.add("cn.springcloud.book.crm.sales.validator");
        packages.add("cn.springcloud.book.crm.sales.event.handler");
        packages.add("cn.springcloud.book.crm.marketing.domain.customer.rule");
        packages.add("cn.springcloud.book.crm.sales.domain.customer.convertor");
        packages.add("cn.springcloud.book.crm.sales.domain.customer.rule");
        bootstrap.setPackages(packages);
        return bootstrap;
    }
}
```

25.4.5　新建 DDD 领域层

领域层包含领域的核心业务规则，各个领域按聚合分包，对应用层只暴露领域聚合（entity），创建领域实体的工厂（factory）、仓储接口（repository），持久化领域对象和 Tunnel 层交互。由于在领域层的值对象、聚合实体，以及创建领域聚合的工厂相对来说比较简单，在这里就不做过多说明，请读者自行查看源码工程进行学习。

crm-sales-infrastructure 工程和包结构规划如图 25-16 所示。

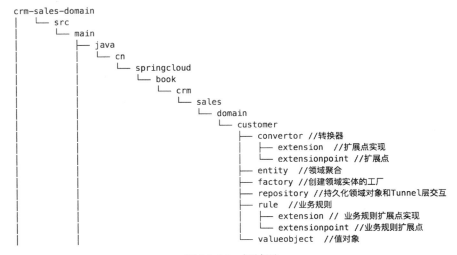

图 25-16　领域层

1）新建 Maven 模块工程 crm-sales-infrastructure，主要 Maven 依赖如代码清单 25-7 所示。

代码清单25-7 ch25/crm-sales-domain/pom.xml

```xml
<dependency>
    <groupId>cn.springcloud.book</groupId>
    <artifactId>crm-sales-infrastructure</artifactId>
</dependency>
<dependency>
    <groupId>cn.springcloud.book</groupId>
    <artifactId>crm-sales-client</artifactId>
</dependency>
```

2）创建扩展点和与扩展点对应的实现。定义扩展点 CustomerConvertorExtPt，需要继承 ExtensionPointI 和 ConvertorI 接口，该扩展点用于提供一个 Customer 的转换扩展，如代码清单 25-8 所示。

代码清单25-8 ch25/crm-sales-domain/src/main/java/cn/springcloud/book/crm/sales/domain/customer/convertor/extensionpoint/CustomerConvertorExtPt.java

```java
public interface CustomerConvertorExtPt extends ConvertorI, ExtensionPointI {
    public CustomerE clientToEntity(CustomerCO customerCO);
}
```

其中一个扩展点的实现，用 CustomerCGSConvertorExt 来说明。其需要加上注解 @Extension，用于标识这是一个扩展点实现，并要加上业务身份 BizCode 或者 ExtBizCode 等。注解中的部分信息可选填，主要用于收集可视化展示。CustomerCGSConvertorExt 的代码如代码清单 25-9 所示。

代码清单25-9 ch25/crm-sales-domain/src/main/java/cn/springcloud/book/crm/sales/domain/customer/convertor/extensionpoint/CustomerConvertorExtPt.java

```java
@Extension(bizCode = BizCode.CGS)
public class CustomerCGSConvertorExt implements CustomerConvertorExtPt {
    @Autowired
    private CustomerConvertor customerConvertor;//Composite basic convertor to do basic conversion
    @Override
    public CustomerE clientToEntity(CustomerCO customerCO){
        CustomerE customerEntity = customerConvertor.clientToEntity(customerCO);
        //In this business, if customers from RFQ and Advertisement are both regarded as Advertisement
        if(AppConstants.SOURCE_AD.equals(customerCO.getSource()) || AppConstants.SOURCE_RFQ.equals(customerCO.getSource()))
        {
            customerEntity.setSourceType(SourceType.AD);
        }
        return customerEntity;
    }
    public CustomerCO dataToClient(CustomerDO customerDO){
        return customerConvertor.dataToClient(customerDO);
    }
}
```

25.4.6 新建 DDD 应用层

App 层主要采用 CQRS 架构模型，主要分为 Command 和 Query。Command 负责获取输入、组装 context、做输入校验、调用领域层做业务处理。Query 在应用层直接通过 Tunnel 的数据访问接口，通过查询条件，获取 Data Object，然后转换为 Client Object 后返回给客户端。crm-sales-app 工程和包结构规划如图 25-17 所示。

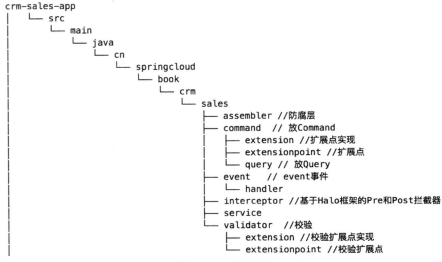

图 25-17 应用层

1）新建 Maven 工程模块 crm-sales-app，添加主要依赖，如代码清单 25-10 所示。

代码清单25-10　ch25/crm-sales-app/pom.xml

```xml
<!-- 服务提供者基于Feign提供的二方包-->
<dependency>
    <groupId>cn.springcloud.book</groupId>
    <artifactId>crm-sales-client</artifactId>
</dependency>
<!-- 领域层模块-->
<dependency>
    <groupId>cn.springcloud.book</groupId>
    <artifactId>crm-sales-domain</artifactId>
</dependency>
```

2）创建与 Command 对象和 Query 对象对应的命令执行器。

Command 对象的命令执行器需要实现 CommandExecutorI 高度抽象的命令执行器接口，在这里以 CustomerAddCmdExe 为例说明，如代码清单 25-11 所示。

代码清单25-11　ch25/crm-sales-app/src/main/java/cn/springcloud/book/crm/sales/command/CustomerAddCmdExe.java

```
@Command
```

```java
public class CustomerAddCmdExe implements CommandExecutorI<Response, CustomerAddCmd>{
    /**
     * 依赖注入校验执行器
     */
    @Autowired
    private ValidatorExecutor  validatorExecutor;
    /**
     * 依赖注入扩展点执行器
     */
    @Autowired
    private ExtensionExecutor extensionExecutor;

    @Override
    public Response execute(CustomerAddCmd cmd) {
        //1, validation
        validatorExecutor.validate(CustomerAddValidatorExtPt.class, cmd);

        //2, invoke domain service or directly operate domain to do business logic process
        CustomerE customerEntity = extensionExecutor.execute(CustomerConvertorExtPt.
            class, extension -> extension.clientToEntity(cmd.getCustomer()));
        customerEntity.addNewCustomer();

        //3, notify by sending message out
        return Response.buildSuccess();
    }
}
```

Query 需要对应实现 QueryExecutorI 执行接口，与 CustomerFindByCriteriaQueryExe 对应的代码如代码清单 25-12 所示。

代码清单25-12　ch25/crm-sales-app/src/main/java/cn/springcloud/book/crm/sales/command/query/CustomerFindByCriteriaQueryExe.java

```java
@Command
public class CustomerFindByCriteriaQueryExe implements QueryExecutorI<MultiResponse<CustomerCO>, CustomerFindByCriteriaQry> {
    @Autowired
    CustomerTunnelI customerDBTunnel;
    @Autowired
    CustomerRepository customerRepository;
    @Autowired
    CustomerConvertor customerConvertor;
    @Override
    public MultiResponse<CustomerCO> execute(CustomerFindByCriteriaQry cmd) {
        List<CustomerE> customerEList = customerRepository.findByCriteria("");
        List<CustomerCO> customerCos = new ArrayList<>();
        for(CustomerE entity:customerEList ) {
            customerCos.add(customerConvertor.entityToClient(entity));
        }
        return MultiResponse.of(customerCos, customerCos.size());
    }

}
```

3）应用层的核心拦截器，相对比较简单，汇总后如表 25-6 所示。

表 25-6　核心拦截器

拦截器名称	使用注解	执行顺序	说　　明
HaloContextPostInterceptor	@PostInterceptor	默认	命令执行完之后清理上下文
HaloContextPreInterceptor	@PreInterceptor	默认	命令执行之前根据业务身份设置上下文
LoggerPostInterceptor	@PostInterceptor	@Order(100)	命令之后进行日志处理
LoggerPreInterceptor	@PreInterceptor	@Order(1)	命令执行之前进行日志预处理
ValidationInterceptor	@PreInterceptor	默认	统一的 Command 校验拦截器

25.4.7　启动测试

启动测试的步骤如下：

1）启动服务提供者 crm-sales-start，并将其注册到 http://eureka.springcloud.cn。
2）启动服务消费者 crm-sales-consumer。
3）访问 http://localhost:8010/customer/list，测试 CQRS 中的 Query 结果，如图 25-18 所示。

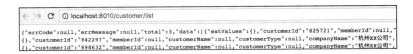

图 25-18　查询 Query 测试结果

4）访问 http://localhost:8010/customer/add，测试 CQRS 中的 Command，如图 25-19 所示。

图 25-19　增加 Command 测试结果

25.5　本章小结

本章首先介绍了什么是领域驱动，解释了微服务架构的落地需要 Spring Cloud 与领域驱动完美融合。本章采用通俗易懂的方式阐述了领域驱动中的核心概念，然后基于对领域驱动概念的理解，从 DDD 的角度剖析 Halo 框架的设计和核心原理，最后通过案例驱动，采用 Feign+Command 对象编写二方包，无缝将 Halo 框架和 Spring Cloud 融合，提供快速有效的 Spring Cloud 与 DDD 落地的方法论，以指导生产实践。

推荐阅读

中兴通讯技术丛书

Ceph设计原理与实现

本书是中兴Clove团队多年研究和实践经验的总结，Ceph创始人Sage Weil的高度评价并亲自作序。

Clove团队是Ceph项目的核心贡献者，从贡献的Commit数上看，连续多个版本贡献在中国排名第一，世界排名第二，对Ceph有非常深入的研究，在中兴通讯内部进行了大量的生产实践。

本书同时从设计者和使用者的角度系统剖析了Ceph的整体架构、核心设计理念，以及各个组件的功能与原理；同时，结合大量在生产环境中积累的真实案例，展示了大量实战技巧。每一章都从基本原理切入，采用循序渐进的方式自然过渡至Ceph，并结合Ceph的核心设计理念指出需要进行哪些必要的改进和裁剪，使得读者不但能够知其然，而且能够知其所以然，真正做到了"源于Ceph，高于Ceph"。此外，写作时尽量避免涉及到过多、非必要的专业术语，做到深入浅出并且每章相对独立，以最大程度的减少阅读障碍。

RRU设计原理与实现

这是一部以工程实践为导向，以信号流为方向，自顶向下详细讲解RRU的系统架构、功能组件、设计方法和实现原理的著作。作者团队来自中兴通讯，都是在无线通信领域有10余年工作经验的资深专家。

本书既适合无线通信领域新人入门，初步了解RRU的概貌、基础理论和系统架构，也适合有经验的通信工程师全面掌握无线射频子系统设计方法。此外，本书还适合无线通信专业的高年级本科生和研究生学习，架起学校和企业应用之间的一座桥梁，旨在系统学习与工程实现方面提高学生的实践能力。

即将出版：

《OpenStack CI/CD：原理与实践》

《Ceph之RADOS设计原理与实现》